D1326406

INTRODUCTION TO AIRCRAFT AEROELASTICITY AND LOADS

Aerospace Series List

Introduction to Aircraft Aeroelasticity and Loads, 2nd Edition	Wright and Cooper	December 2014
Aircraft Aerodynamic Design: Geometry and Optimization	Sóbester and Forrester	October 2014
Theoretical and Computational Aerodynamics	Sengupta	September 2014
Aerospace Propulsion	Lee	October 2013
Aircraft Flight Dynamics and Control	Durham	August 2013
Civil Avionics Systems, 2nd Edition	Moir, Seabridge and Jukes	August 2013
Modelling and Managing Airport Performance	Zografos, Andreatta and Odoni	July 2013
Advanced Aircraft Design: Conceptual Design, Analysis and Optimization of Subsonic Civil Airplanes	Torenbeek	June 2013
Design and Analysis of Composite Structures: With Applications to Aerospace Structures, 2nd Edition	Kassapoglou	April 2013
Aircraft Systems Integration of Air-Launched Weapons	Rigby	April 2013
Design and Development of Aircraft Systems, 2nd Edition	Moir and Seabridge	November 2012
Understanding Aerodynamics: Arguing from the Real Physics	McLean	November 2012
Aircraft Design: A Systems Engineering Approach	Sadraey	October 2012
Introduction to UAV Systems 4e	Fahlstrom and Gleason	August 2012
Theory of Lift: Introductory Computational Aerodynamics with MATLAB and Octave	McBain	August 2012
Sense and Avoid in UAS: Research and Applications	Angelov	April 2012
Morphing Aerospace Vehicles and Structures	Valasek	April 2012
Gas Turbine Propulsion Systems	MacIsaac and Langton	July 2011
Basic Helicopter Aerodynamics, 3rd Edition	Seddon and Newman	July 2011
Advanced Control of Aircraft, Spacecraft and Rockets	Tewari	July 2011
Cooperative Path Planning of Unmanned Aerial Vehicles	Tsourdos et al	November 2010
Principles of Flight for Pilots	Swatton	October 2010
Air Travel and Health: A Systems Perspective	Seabridge et al	September 2010
Design and Analysis of Composite Structures: With applications to aerospace Structures	Kassapoglou	September 2010
Unmanned Aircraft Systems: UAVS Design, Development and Deployment	Austin	April 2010
Introduction to Antenna Placement & Installations	Macnamara	April 2010
Principles of Flight Simulation	Allerton	October 2009
Aircraft Fuel Systems	Langton et al	May 2009
The Global Airline Industry	Belobaba	April 2009
Computational Modelling and Simulation of Aircraft and the Environment: Volume 1 – Platform Kinematics and Synthetic Environment	Diston	April 2009
Handbook of Space Technology	Ley, Wittmann Hallmann	April 2009
Aircraft Performance Theory and Practice for Pilots	Swatton	August 2008
Aircraft Systems, 3rd Edition	Moir & Seabridge	March 2008
Introduction to Aircraft Aeroelasticity and Loads	Wright & Cooper	December 2007
Stability and Control of Aircraft Systems	Langton	September 2006
Military Avionics Systems	Moir & Seabridge	February 2006
Design and Development of Aircraft Systems	Moir & Seabridge	June 2004
Aircraft Loading and Structural Layout	Howe	May 2004
Aircraft Display Systems	Jukes	December 2003
Civil Avionics Systems	Moir & Seabridge	December 2002

INTRODUCTION TO AIRCRAFT AEROELASTICITY AND LOADS

Second Edition

Jan R. Wright
University of Manchester and
J2W Consulting, UK

Jonathan E. Cooper
University of Bristol, UK

© 2015 John Wiley & Sons, Ltd

Registered Office
John Wiley & Sons Ltd, The Atrium, Southern Gate, Chichester, West Sussex, PO19 8SQ, United Kingdom

For details of our global editorial offices, for customer services and for information about how to apply for permission to reuse the copyright material in this book please see our website at www.wiley.com.

The right of the author to be identified as the author of this work has been asserted in accordance with the Copyright, Designs and Patents Act 1988.

All rights reserved. No part of this publication may be reproduced, stored in a retrieval system, or transmitted, in any form or by any means, electronic, mechanical, photocopying, recording or otherwise, except as permitted by the UK Copyright, Designs and Patents Act 1988, without the prior permission of the publisher.

Wiley also publishes its books in a variety of electronic formats. Some content that appears in print may not be available in electronic books.

Designations used by companies to distinguish their products are often claimed as trademarks. All brand names and product names used in this book are trade names, service marks, trademarks or registered trademarks of their respective owners. The publisher is not associated with any product or vendor mentioned in this book

Limit of Liability/Disclaimer of Warranty: While the publisher and author have used their best efforts in preparing this book, they make no representations or warranties with respect to the accuracy or completeness of the contents of this book and specifically disclaim any implied warranties of merchantability or fitness for a particular purpose. It is sold on the understanding that the publisher is not engaged in rendering professional services and neither the publisher nor the author shall be liable for damages arising herefrom. If professional advice or other expert assistance is required, the services of a competent professional should be sought.

MATLAB® is a trademark of The MathWorks, Inc. and is used with permission. The MathWorks does not warrant the accuracy of the text or exercises in this book. This book's use or discussion of MATLAB® software or related products does not constitute endorsement or sponsorship by The MathWorks of a particular pedagogical approach or particular use of the MATLAB® software.

Library of Congress Cataloging-in-Publication Data

Wright, Jan R.
Introduction to aircraft aeroelasticity and loads / Jan R. Wright, Jonathan E. Cooper. – Second edition.
   pages   cm
  Includes bibliographical references and index.
  ISBN 978-1-118-48801-0 (cloth)
1. Aeroelasticity. I. Cooper, Jonathan E.   II. Title.
  TL574.A37W75 2014
  629.132'362–dc23

                                   2014027710

A catalogue record for this book is available from the British Library

Set in 10/12pt Times by SPi Publisher Services, Pondicherry, India
Printed and bound in Malaysia by Vivar Printing Sdn Bhd

1   2015

UNIVERSITY
OF SHEFFIELD
LIBRARY

To our children and grandchildren
Jodi, Peter, Laura, Rhys & Erin
&
William, Rory, Tyler & Ella

Contents

Series Preface

The field of aerospace is multi-disciplinary and wide-ranging, covering a large variety of products, disciplines and domains, not merely in engineering but in many related supporting activities. These combine to enable the aerospace industry to produce exciting and technologically advanced vehicles. The wealth of knowledge and experience that has been gained by expert practitioners in the various aerospace fields needs to be passed on to others working in the industry, including those just entering from University.

The *Aerospace Series* aims to be a practical, topical and relevant series of books aimed at people working in the aerospace industry, including engineering professionals and operators, allied professions such as commercial and legal executives, and also engineers in academia. The range of topics is intended to be wide-ranging, covering the design and development, manufacture, operation and support of aircraft, as well as topics such as infrastructure operations and developments in research and technology.

Aeroelasticity and loads are important interdisciplinary topics, involving the interaction of aerodynamic, elastic and inertia forces, which have a significant effect on aircraft designs and flight performance. Important phenomena include those that are critical for structural stability e.g. flutter, divergence and shimmy, the shape of aircraft wings in-flight, and the critical design loads due to the aircraft response to turbulence and flight / ground manoeuvres that have a significant influence on aircraft structural designs.

This book, *Introduction to Aircraft Aeroelasticity and Loads*, provides a welcome update to the first edition. Containing a range of revised material, the same approach is used as before, employing simple mathematical models to guide the reader towards an understanding of key underlying concepts relating to both aeroelasticity and loads. Of particular note is the frequent reference to the airworthiness certification procedures for civil aircraft, with the final section providing an introduction to current industrial practice. The companion website provides a number of computer codes that can be used to gain further understanding of the mathematical models that are discussed within the book.

Peter Belobaba, Jonathan Cooper and Alan Seabridge

Preface to the Second Edition

Following publication of the first edition of *Introduction to Aircraft Aeroelasticity and Loads,* the authors have run a series of in-house short courses, based upon the book, for several companies within the world-wide aerospace industry. These lectures enabled the material to be tested on engineers from industry in a classroom situation; this, combined with both authors also using some of the content to teach undergraduate modules at the Universities of Sheffield and Bristol, resulted in a number of changes being made to the course content. These changes have resulted in this second edition and are aimed at improving the presentation and content throughout the book, while maintaining the philosophy of illustrating the underlying concepts of aeroelasticity and loads using as simple mathematical models as possible.

The most significant changes are:

- Use of the same binary wing model for the static and dynamic aeroelasticity chapters, based on an unswept, untapered cantilever with one bending and one torsion assumed shape, rather than the previous use of a pitching and flapping wing – this has meant that the models used for aeroelastics and loads are now more closely related.
- Revised treatment of swept wing divergence.
- Major revision of the flutter and aeroservoelasticity chapters.
- Changing the chapter order so that the chapters involving a discrete treatment of structures and aerodynamics all sit together.
- Inclusion of more elements relevant to modelling of aircraft structures.
- Inclusion of simpler examples relating to potential flow aerodynamics.
- Revision to multiple degree of freedom damped free vibration.
- Revision to the Ground Maneouvres chapter to include redundancy in gear attachments, Craig-Bampton modes and bookcase landing examples.
- Combining the two chapters in the first edition on the Flight Mechanics Model and Dynamic Manoeuvres, and revising the lateral bookcase modelling material.
- Introduce unsteady aerodynamics effects into the examples in the Gust and Turbulence Encounters chapter.
- Revision of the introduction to the Flight Control System (FCS).
- Update of the certification requirements.

The authors would like to thank a number of their colleagues for their valued comments on the content of the industrial short courses, and this second edition. Particular mention needs to be made of Mark Hockenhull, Tom Wilson, Gido Brendes, Tobias Mauermann and Hans-Gerd Giesseler (all Airbus), Raj Nangia (Nangia Associates), George Constandinidis (Messier-Bugatti-Dowty), Dorian Jones, Arthur Richards, Olivia Stodieck and Luke Lambert (all University of Bristol).

Preface to the First Edition

Aeroelasticity is the study of the interaction of aerodynamic, elastic and inertia forces. For fixed wing aircraft there are two key areas: (a) static aeroelasticity, where the deformation of the aircraft influences the lift distribution can lead to the statically unstable condition of divergence and will normally reduce the control surface effectiveness, and (b) dynamic aeroelasticity, which includes the critical area of flutter where the aircraft can become dynamically unstable in a condition where the structure extracts energy from the air stream.

Aircraft are also subject to a range of static and dynamic loads resulting from flight manoeuvres (equilibrium/steady and dynamic), ground manoeuvres and gust/turbulence encounters. These load cases are responsible for the critical design loads over the aircraft structure and hence influence the structural design. Determination of such loads involves consideration of aerodynamic, elastic and inertia effects and requires the solution of the dynamic responses; consequently there is a strong link between aeroelasticity and loads.

The aircraft vibration characteristics and response are a result of the flexible modes combining with the rigid body dynamics, with the inclusion of the Flight Control System (FCS) if it is present. In this latter case, the aircraft will be a closed loop system and the FCS affects both the aeroelasticity and loads behaviour. The interaction between the FCS and the aeroelastic system is often called aeroservoelasticity.

This book aims to embrace the range of basic aeroelastic and loads topics that might be encountered in an aircraft design office and to provide an understanding of the main principles involved. Colleagues in industry have often remarked that it is not appropriate to give some of the classical books on aeroelasticity to new graduate engineers as many of the books are too theoretical for a novice aeroelastician. Indeed, the authors have found much of the material in them to be too advanced to be used in the undergraduate level courses that they have taught. Also, the topics of aeroelasticity and loads have tended to be treated separately in textbooks, whereas in industry the fields have become much more integrated. This book is seen as providing some grounding in the basic analysis techniques required which, having been mastered, can then be supplemented via more advanced texts, technical papers and industry reports.

Some of the material covered in this book developed from undergraduate courses given at Queen Mary College, University of London and at the University of Manchester. In the UK, many entrants into the aerospace industry do not have an aerospace background, and almost certainly will have little knowledge of aeroelasticity or loads. To begin to meet this need, during the early 1990s the authors presented several short courses on Aeroelasticity and Structural Dynamics to young engineers in the British aerospace industry, and this has influenced the content and approach of this book. A further major influence was the work by Hancock, Simpson and Wright (1985) on the teaching of flutter, making use of a simplified flapping and pitching wing model with strip theory aerodynamics (including a simplified unsteady aerodynamics model) to illustrate the fundamental principles of flutter. This philosophy has been employed here for the treatment of static aeroelasticity and flutter, and has been extended into the area of loads by focusing on a simplified flexible whole aircraft model in order to highlight key features of modelling and analysis.

The intention of the book is to provide the reader with the technical background to understand the underlying concepts and application of aircraft aeroelasticity and loads. As far as possible, simplified mathematical models for the flexible aircraft are used to illustrate the phenomena and also to demonstrate the link between these models, industrial practice and the certification process. Thus, fairly simple continuum models based upon a small number of assumed modes (so avoiding partial differential equations) have been used. Consequently, much of the book is based upon strip theory aerodynamics and the Rayleigh–Ritz assumed modes method. By using this approach, it has been possible to illustrate most concepts using a maximum of three degrees of freedom. Following on from these continuum models, basic discretized structural and aerodynamic models are introduced in order to demonstrate some underlying approaches in current industrial practice. The book aims to be suitable for final year undergraduate or Masters level students, or engineers in industry who are new to the subject. For example, it could provide the basis of two taught modules in aeroelasticity and loads. It is hoped that the book will fill a gap in providing a broad and relatively basic introductory treatment of aeroelastics and loads.

A significant number of different topics are covered in order to achieve the goals of this book, namely structural dynamics, steady and unsteady aerodynamics, loads, control, static aeroelastic effects, flutter, flight manoeuvres (both steady/equilibrium and dynamic), ground manoeuvres (e.g. landing, taxiing), gust and turbulence encounters, calculation of loads and, finally, Finite Element and three-dimensional panel methods. In addition, a relatively brief explanation is given as to how these topics might typically be approached in industry when seeking to meet the certification requirements. Most of the focus is on commercial and not military aircraft, though of course all of the underlying principles, and much of the implementation, are common between the two.

The notation employed has not been straightforward to define, as many of these disciplines have tended to use the same symbols for different variables and so inevitably this exercise has been a compromise. A further complication is the tendency for aeroelasticity textbooks from the USA to use the reduced frequency k for unsteady aerodynamics, as opposed to the frequency parameter ν that is often used elsewhere. The reduced frequency has been used throughout this textbook to correspond with the classical textbooks of aeroelasticity.

The book is split into three parts. After a brief introduction to aeroelasticity and loads, Part A provides some essential background material on the fundamentals of single and multiple degree of freedom (DoF) vibrations for discrete parameter systems and continuous systems (Rayleigh–Ritz and Finite Element), steady aerodynamics, loads and control. The presentation is not very detailed, assuming that a reader having a degree in engineering will have some background in most of these topics and can reference more comprehensive material if desired.

Part B is the main part of the book, covering the basic principles and concepts required to provide a bridge to begin to understand current industry practice. The chapters on aeroelasticity include static aeroelasticity (lift distribution, divergence and control effectiveness), unsteady aerodynamics, dynamic aeroelasticity (i.e. flutter) and aeroservoelasticity; the treatment is based mostly on a simple two-DoF flapping/pitching wing model, sometimes attached to a rigid fuselage free to heave and pitch. The chapters on loads include equilibrium and dynamic flight manoeuvres, gusts and turbulence encounters, ground manoeuvres and internal loads. The loads analyses are largely based on a three-DoF whole aircraft model with heave and pitch rigid body motions and a free–free flexible mode whose characteristics may be varied, so allowing fuselage bending, wing bending or wing torsional deformation to be dominant. Part B concludes with an introduction to three-dimensional aerodynamic panel methods and simple coupled discrete aerodynamic and structural models in order to move on from the Rayleigh–Ritz assumed modes and strip theory approaches to more advanced methods, which provide the basis for much of the current industrial practice.

The basic theory introduced in Parts I and II provides a suitable background to begin to understand Part III, which provides an outline of industrial practice that might typically be involved in aircraft design and certification, including aeroelastic modelling, static aeroelasticity and flutter, flight manoeuvre and gust/turbulence loads, ground manoeuvre loads and finally testing relevant to aeroelastics and loads. A number of MATLAB/SIMULINK programs are available on a companion website for this book at http://www.wiley/go/wright&cooper.

The authors are grateful to the input from a number of colleagues in the UK university sector: John Ackroyd, Philip Bonello, Grigorios Dimitriadis, Zhengtao Ding, Dominic Diston, Barry Lennox and Gareth Vio. The authors greatly valued the input on industrial practice from Mark Hockenhull, Tom Siddall, Peter Denner, Paul Bruss, Duncan Pattrick, Mark Holden and Norman Wood. The authors also appreciated useful discussions with visiting industrial lecturers to the University of Manchester (namely Rob Chapman, Brian Caldwell, Saman Samarasekera, Chris Fielding and Brian Oldfield). Some of the figures and calculations were provided by Colin Leung, Graham Kell and Gareth Vio. Illustrations were provided with kind agreement of Airbus, Messier-Dowty, DLR, DGA/CEAT, ONERA, Royal Aeronautical Society and ESDU. Use of software was provided by MATLAB.

The authors would also like to acknowledge the encouragement they have received over the years in relation to research and teaching activities in the areas of structural dynamics, aircraft structures, loads and aeroelasticity, namely Alan Simpson (University of Bristol), Mike Turner (deceased) (British Aircraft Corporation), Geoff Hancock and David Sharpe (Queen Mary College), Colin Skingle, Ian Kaynes and Malcolm Nash (RAE Farnborough, now QinetiQ), Jer-Nan Juang (NASA Langley), Peter Laws (University of Manchester), Geof Tomlinson (University of Sheffield, formerly at Manchester), Otto Sensburg (MBB) and Hans Schwieger (EADS).

Jan R. Wright and Jonathan E. Cooper
Manchester, UK

Abbreviations

AC	aerodynamic centre
AC	Advisory Circular
AIC	Aerodynamic Influence Coefficient
AMC	Additional Means of Compliance
AR	aspect ratio
CFD	Computational Fluid Dynamics
CoM	centre of mass
COTS	commercial off the shelf
CRI	Certification Review Item
CS	Certification Specifications
DL	Doublet Lattice
DoF	degree of freedom
EAS	Equivalent Air Speed
EASA	European Aviation Safety Agency
FAA	Federal Aviation Administration
FAR	Federal Aviation Regulation
FBD	Free Body Diagram
FCS	Flight Control System
FD	frequency domain
FE	Finite Element
FFT	Flight Flutter Test
FRF	Frequency Response Function
FT	Fourier Transform
GVT	Ground Vibration Test
GLA	Gust Load Alleviation
IRF	Impulse Response Function
ISA	International Standard Atmosphere
JAA	Joint Airworthiness Authorities
LCO	limit cycle oscillation
LDHWG	Loads and Dynamics Harmonization Working Group
LE	leading edge
MDoF	multiple degree of freedom
MLA	Manoeuvre Load Alleviation
NPA	Notice of Proposed Amendment

PSD	Power Spectral Density
RMS	root mean square
SDoF	single degree of freedom
SRF	step response function
TAS	True Air Speed
TD	time domain
TE	trailing edge
TF	Transfer Function
WA	wing aerodynamic (axis)
WE	wing elastic (axis)
WM	wing mass (axis)

Introduction

Aeroelasticity is the subject that describes the interaction of aerodynamic, inertia and elastic forces for a flexible structure and the phenomena that can result. This field of study is summarized most clearly by the classical Collar aeroelastic triangle (Collar, 1978), seen in Figure I.1, which shows how the major disciplines of stability and control, structural dynamics and static aeroelasticity each result from the interaction of two of the three types of force. However, all three forces are required to interact in order for dynamic aeroelastic effects to occur.

Aeroelastic effects have had a major influence upon the design and flight performance of aircraft, even before the first controlled powered flight of the Wright Brothers. Since some aeroelastic phenomena (e.g. flutter and divergence) can lead potentially to structural failure, aircraft structural designs have had to be made heavier (the so-called aeroelastic penalty) in order to ensure that structural integrity has been maintained through suitable changes in the structural stiffness and mass distributions. The first recorded flutter problem to be modelled and solved (Bairstow and Fage, 1916; Lanchester, 1916) was the Handley–Page O/400 bomber in 1916, shown on the front cover of the first edition of this book. Excellent histories about the development of aeroelasticity and its influence on aircraft design can be found in Collar (1978), Garrick and Reed (1981), Flomenhoft (1997) and Weisshaar (2010), with surveys of more recent applications given in Friedmann (1999) and Livne (2003).

Of course, aeroelasticity is not solely concerned with aircraft, and the topic is extremely relevant for the design of structures such as bridges, Formula 1 racing cars, wind turbines, turbo-machinery blades, helicopters, etc. However, in this book only fixed wing aircraft will be considered, with the emphasis being on large commercial aircraft, but the underlying principles have relevance to other applications.

It is usual to classify aeroelastic phenomena as being either static or dynamic. *Static aeroelasticity* considers the *non-oscillatory* effects of aerodynamic forces acting on the flexible aircraft structure. The flexible nature of the wing will influence the in-flight wing shape and hence the lift distribution in a steady (or so-called equilibrium) manoeuvre (see below) or in the special case of cruise. Thus, however accurate and sophisticated any aerodynamic calculations carried out might be, the final in-flight shape could be in error if the structural stiffness were to be modelled inaccurately; drag penalties could result and the aircraft range could reduce. Static aeroelastic effects can also often lead to a reduction in the effectiveness of the control surfaces and eventually to the phenomenon of control reversal; for example, an aileron has the opposite effect to that intended because the rolling moment it generates is

Introduction to Aircraft Aeroelasticity and Loads, Second Edition. Jan R. Wright and Jonathan E. Cooper.
© 2015 John Wiley & Sons, Ltd. Published 2015 by John Wiley & Sons, Ltd.

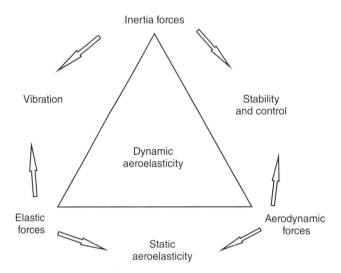

Figure I.1   Collar's aeroelastic triangle.

negated by the wing twist that accompanies the control rotation. There is also the potentially disastrous phenomenon of divergence to consider, where the wing twist can increase without limit when the aerodynamic pitching moment on the wing due to twist exceeds the structural restoring moment. It is important to recognize that the lift distribution and divergence are influenced by the trim of the aircraft, so strictly speaking the wing cannot be treated on its own.

Dynamic aeroelasticity is concerned with the *oscillatory* effects of the aeroelastic interactions, and the main area of interest is the potentially catastrophic phenomenon of flutter. This instability involves two or more modes of vibration and arises from the unfavourable coupling of aerodynamic, inertial and elastic forces; it means that the aircraft structure can effectively extract energy from the air stream. The most difficult issue, when seeking to predict the flutter phenomenon, is that of the unsteady nature of the aerodynamic forces and moments generated when the aircraft oscillates, and the effect the motion has on the resulting forces, particularly in the transonic regime. The presence of flexible modes influences the dynamic stability modes of the rigid aircraft (e.g. short period) and so affects the flight dynamics. Also of serious concern is the potential unfavourable interaction of the Flight Control System (FCS; Pratt, 2000) with the flexible aircraft, considered in the topic of *aeroservoelasticity* (also known as structural coupling).

There are a number of textbooks on aeroelasticity, e.g. Broadbent (1954), Scanlan and Rosenbaum (1960), Fung (1969), Bisplinghoff *et al.* (1996), Hodges and Pierce (2011), Dowell *et al.* (2004) and Rodden (2011). These offer a comprehensive and insightful mathematical treatment of more fundamental aspects of the subject. However, the approach in most of these books is on the whole somewhat theoretical and often tends to restrict coverage to static aeroelasticity and flutter, considering cantilever wings with fairly sophisticated analytical treatments of unsteady aerodynamics. All, except Hodges and Pierce (2011), Dowell *et al.* (2004) and Rodden (2011), were written in the 1950s and 1960s. The textbook by Forsching (1974) must also be mentioned as a valuable reference but there is no English translation from the German original. There is some material relevant to static aeroelasticity in the ESDU Data Sheets. A further useful source of reference is the AGARD *Manual on Aeroelasticity* 1950–1970, but again this was written nearly 50 years ago. Further back in history are the key references on aeroelasticity by Frazer and Duncan (1928) and Theodorsen (1935).

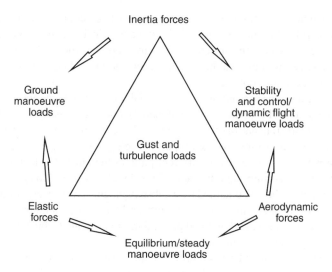

Figure I.2   Loads triangle.

Aeroelastic considerations influence the aircraft design process in a number of ways. Within the design flight envelope, it must be ensured that flutter and divergence cannot occur and that the aircraft is sufficiently controllable. The in-flight wing shape influences drag and performance and so must be accurately determined; this requires careful consideration of the jig shape used in manufacture (i.e. the shape in the absence of gravitational and aerodynamic forces). The aircraft handling is affected by the aeroelastic deformations, especially where the flexible modes are close in frequency to the rigid body modes.

Apart from the consideration of aeroelasticity, the other significant related topic covered in this book is that of loads. Collar's aeroelastic triangle may be modified to cater for loads (mainly dynamic) to generate a loads triangle, as shown in Figure I.2. Equilibrium (or steady, trimmed or balanced) manoeuvres involve the interaction of elastic and aerodynamic effects (cf. static aeroelasticity), dynamic manoeuvres involve the interaction of aerodynamic and inertia effects (cf. stability and control, but for a flexible aircraft, elastic effects may also be important), ground manoeuvres primarily involve the interaction of inertia and elastic effects (cf. structural dynamics) and gust/turbulence encounters involve the interaction of inertia, aerodynamic and elastic effects (cf. flutter).

Equilibrium manoeuvres concern the aircraft undergoing steady longitudinal or lateral manoeuvres, e.g. a steady pull-out from a dive involving acceleration normal to the flight path and a steady pitch rate. *Dynamic manoeuvres* involve the aircraft responding dynamically to transient control inputs from the pilot or else to failure conditions. *Ground manoeuvres* cover a significant number of different steady and dynamic load conditions (e.g. landing, taxiing, braking, turning) where the aircraft is in contact with the ground via the landing gear. Finally, *gust/turbulence encounters* involve the aircraft responding to discrete gusts (represented in the time domain) or continuous turbulence (usually represented in the frequency domain).

The different load cases required for certification under these four main headings may be considered in one of two categories: the term '*bookcase*' refers to a load case where in essence a relatively artificial state of the aircraft, in which applied and inertia loads are in equilibrium, is considered whereas the term '*rational*' refers to a condition in which the aircraft dynamic behaviour is modelled and simulated as realistically as possible. Bookcase load cases apply primarily to equilibrium manoeuvres and to some ground manoeuvres, whereas rational load cases apply to most dynamic and ground manoeuvres as well as to gust/turbulence encounters. These load cases in the certification process provide *limit* loads, which are the maximum loads to be expected in service and which the structure needs to support without 'detrimental permanent deformation'; the structure must also be able to support *ultimate* loads (limit loads × factor of safety of 1.5) without failure/rupture.

The resulting distributions of bending moment, axial force, shear force and torque along each component (referred to in this book as 'internal loads'), due to the distribution of aerodynamic and inertial forces (and, for that matter, propulsive forces) acting on the aircraft, need to be determined as a function of time for each type of loading across the entire design envelope. The critical internal loads for design of different parts of the aircraft structure are then found via a careful process of sorting the multitude of results obtained; the load paths and stresses within the structure may then be obtained by a subsequent analysis process for the critical load cases in order to allow an assessment of the strength and fatigue life or damage tolerance of the aircraft. The aircraft response in taxiing and particularly in gust and turbulence encounters will influence crew and passenger comfort.

The Flight Control System (FCS) is a critical component for aircraft control that has to be designed so as to provide the required stability and carefree handling qualities, and to avoid unfavourable couplings with the structure; it will in turn influence the loads generated and must be represented in the loads calculations. Manoeuvre Load Alleviation (MLA) and Gust Load Alleviation (GLA) systems are often fitted to the aircraft to reduce loads and improve ride comfort.

There are a number of textbooks on classical aircraft structural analysis, but in the area of loads there are far fewer relevant textbooks. The AIAA Education Series book on structural loads analysis (Lomax, 1996) is extremely useful and deals with aircraft loads in an applied manner, using relatively simplified aircraft models that may be used to check results from more sophisticated approaches employed when seeking to meet certification requirements; it also aims to present somewhat of an historical perspective. The book on aircraft loading and structural layout (Howe, 2004) covers approximate loading action analysis for the rigid aircraft, together with use of results provided in initial load estimates and hence layout and sizing of major structural members in aircraft conceptual design. The AIAA Education Series book on gusts and turbulence (Hoblit, 1988) provides a comprehensive introductory treatment of loads due to gusts and particularly due to continuous turbulence. Some of the classical books on aeroelasticity also include an introductory treatment of gust response (Scanlan and Rosenbaum, 1960; Fung, 1969; Bisplinghoff *et al.*, 1996) and so partially bridge the aeroelasticity and loads fields. The ESDU Data Sheets provide some coverage of loads in steady manoeuvres, linked to static aeroelastic effects, and also an introductory item on gusts and turbulence. The AIAA Education Series book on landing gear design (Currey, 1988) provides a very practical treatment of design issues but is not aimed at addressing the associated mathematical modelling required for estimation of loads in ground manoeuvres. Niu (1988) provides a useful chapter on aircraft loads but the main focus is on the practicalities of airframe structural design. Donaldson (1993) and Megson (1999) are primarily aimed at covering a wide range of aircraft structural analysis methods, but also provide introductory chapters on loads and aeroelasticity.

Historically, the topics of loads and aeroelasticity have often been treated separately in industry, whereas in recent years they have been considered in a much more integrated manner; indeed now they are often covered by a single department. This is because the mathematical model for a flexible aircraft has traditionally been developed for flutter calculations and the aircraft static and dynamic aeroelastic effects have gradually become more important to include in the flight/ground manoeuvre and gust/turbulence load calculations. Also, as the rigid body and flexible mode frequencies have grown closer together, the rigid body and FCS effects have had to be included in the flutter solution. The flight mechanics model used for dynamic manoeuvres would be developed in conjunction with the departments that consider stability and control, handling and FCS issues since the presence of flexible modes would affect the aircraft dynamic stability and handling. There also needs to be close liaison with the aerodynamics and structures departments when formulating mathematical models. The models used in loads and aeroelastic calculations are becoming ever more advanced. The model of the structure has progressed from a 'beam-like' model based on the Finite Element (FE) method to a much more representative 'box-like' FE model. The aerodynamic model has progressed from one based on two-dimensional strip theory to three-dimensional panel methods and, in an increasing number of cases, Computational Fluid Dynamics (CFD).

The airworthiness certification process requires that all possible aeroelastic phenomena and a carefully defined range of load cases should be considered in order to ensure that any potentially disastrous scenario cannot occur or that no critical load value is exceeded. The analysis process must be supported by a ground and flight test programme to validate the aerodynamic, structural, aeroelastic and aeroservoelastic models. The certification requirements for large aircraft in Europe and the United States are described in CS-25 and FAR-25 respectively. The requirements from Europe and the United States are very similar and use essentially the same numbering system; here, for convenience, reference is made mostly to the European version of the requirements.

In recent years there has also been an increasing interest in the effect of aerodynamic and structural non-linearities and the effect they have on the aeroelastic behaviour. Of particular interest are phenomena such as limit cycle oscillations (LCO) and also the transonic aeroelastic stability boundaries. In addition the FCS has non-linear components. More advanced mathematical models are required to predict and characterize the non-linear phenomena, which cannot be predicted using linear representations. However, in this book non-linear effects will only be mentioned briefly.

This book is organized into three parts. Part I provides some essential background material on the fundamentals of single and multiple degree of freedom vibrations for discrete parameter systems, continuous systems using the Rayleigh–Ritz method, steady aerodynamics, loads and control. The presentation is relatively brief, on the assumption that a reader can reference more comprehensive material if desired.

Chapter 1 introduces the vibration of single degree of freedom discrete parameter systems, including setting up equations of motion using Lagrange's equations, and in particular determining the response to various types of forced vibrations. Chapter 2 presents the equivalent theory for multiple degree of freedom systems with reference to modes of vibration and modelling in modal space, as well as free and forced vibration. Chapter 3 employs the Rayleigh–Ritz assumed shapes approach for continuous systems, primarily slender structures in bending and torsion, but also considering use of branch modes and whole aircraft 'free–free' modes.

Chapter 4 introduces a number of basic steady aerodynamics concepts that will be used to determine the flows, lift forces and moments acting on simple two-dimensional aerofoils and three-dimensional wings, including two-dimensional strip theory which is used for convenience, albeit recognizing its limitations. Chapter 5 describes simple dynamic solutions for a particle or body using Newton's laws of motion and d'Alembert's principle, and then introduces the use of inertia loads to generate an equivalent static representation of a dynamic problem, leading to internal loads for slender members experiencing non-uniform acceleration. Chapter 6 introduces some basic concepts of control for open and closed loop feedback systems.

Part II is the main part of the book, covering the basic principles and concepts required to provide a link to begin to understand current industry practice. A Rayleigh–Ritz approach for flexible modes is used to simplify the analysis and to allow the equations to be almost entirely limited to three degrees of freedom to aid understanding. A strip theory representation of the aerodynamics is employed to simplify the mathematics, but it is recognized that strip theory is not particularly accurate and that three-dimensional panel methods are more commonly used in practice. The static and dynamic aeroelastic content makes use of wing models, sometimes attached to a rigid fuselage. The loads chapters combine a rigid body heave/pitch model with a whole aircraft free–free flexible mode (designed to permit fuselage bending, wing bending or wing torsional motions as dominant), and consider a range of flight/ground manoeuvre and gust/turbulence cases.

Chapter 7 considers the effect of static aeroelasticity on the aerodynamic load distribution, resulting deflections and potential divergence for a flexible wing, together with the influence of wing sweep and aircraft trim. Chapter 8 examines the impact of wing flexibility on aileron effectiveness. Chapter 9 introduces the concept of quasi-steady and unsteady aerodynamics and the effect that the relative motion between an aerofoil and the flow has on the lift and moment produced. Chapter 10 explores the critical area of flutter and also how aeroelastic calculations are performed where frequency-dependent aerodynamics is involved. Chapter 11 introduces aeroservoelasticity and illustrates the implementation of a simple feedback control on an aeroelastic system for flutter suppression and gust loads alleviation.

Chapter 12 considers the behaviour of rigid and flexible aircraft undergoing symmetric equilibrium manoeuvres. Chapter 13 introduces the two-dimensional flight mechanics model with body fixed axes and extends it to include a flexible mode. It goes on to show how the flight mechanics model may be used to examine dynamic manoeuvres in heave/pitch, and how the flexibility of the aircraft can affect the response, dynamic stability modes and control effectiveness. Simple lateral bookcase manoeuvres are considered for a rigid aircraft. Chapter 14 considers discrete gust and continuous turbulence analysis approaches in the time and frequency domains respectively. Chapter 15 presents a simple model for the non-linear landing gear and considers taxiing, landing, braking, wheel 'spin-up' and 'spring-back', turning and shimmy. Chapter 16 introduces the evaluation of internal loads from the aircraft dynamic response and any control/gust input, applied to continuous and discretized components, and also loads sorting. Chapter 17 describes the Finite Element method, the most common discretization approach for modelling the vibration of continuous structures. Chapter 18 describes potential flow aerodynamic approaches and how they lead to determination of Aerodynamic Influence Coefficients (AICs) for three-dimensional panel methods in both steady and unsteady flows. Chapter 19 considers the development of simple coupled two- and three-dimensional structural and aerodynamic models in steady and unsteady flows.

Finally, Part III provides an outline of industrial practice that might typically be involved in aircraft design and certification. It references the earlier two parts of the book and indicates how the processes illustrated on simple mathematical models might be applied in practice to 'real' aircraft.

Chapter 20 introduces the design and certification process as far as aeroelasticity and loads are concerned. Chapter 21 explains how the mathematical models used for aeroelasticity and loads analyses can typically be constructed. Chapter 22 considers the calculations undertaken to meet the requirements for static aeroelasticity and flutter. Chapter 23 presents the calculation process involved for determination of loads in meeting the requirements for equilibrium and dynamic flight manoeuvres and gust/continuous turbulence encounters. Chapter 24 introduces the analyses required for determining the ground manoeuvre loads and loads post-processing. It should be noted that the book stops at the determination of internal loads, and so does not extend to the evaluation of component loads and stresses. Finally, Chapter 25 describes briefly the range of ground and flight tests performed to validate mathematical aeroelastic and loads models and demonstrate aeroelastic stability.

Part I

Background Material

1

Vibration of Single Degree of Freedom Systems

In this chapter, some of the basic concepts of vibration analysis for single degree of freedom (SDoF) discrete parameter systems will be introduced. The term 'discrete (or sometimes lumped) parameter' implies that the system is a combination of discrete rigid masses (or components) interconnected by flexible/elastic stiffness elements. Later it will be seen that a single DoF representation may be employed to describe the behaviour of a particular characteristic (or mode) shape of the system via what are known as modal coordinates. Multiple degree of freedom (MDoF) discrete parameter systems will be considered in Chapter 2. The alternative approach to modelling multiple DoF systems, as so-called 'continuous' systems where components of the system are flexible and deform in some manner, is considered later in Chapters 3 and 17.

Much of the material in this introductory part of the book on vibrations is covered in detail in many other texts, such as Tse *et al.* (1978), Newland (1987), Rao (1995), Thomson (1997) and Inman (2006); it is assumed that the reader has some engineering background and so should have met many of the ideas before. Therefore, the treatment here will be as brief as is consistent with the reader being reminded, if necessary, of various concepts used later in the book. Such introductory texts on mechanical vibration should be referenced if more detail is required or if the reader's background understanding is limited.

There are a number of ways of setting up the equations of motion for an SDoF system, e.g. Newton's laws or d'Alembert's principle. However, consistently throughout the book, Lagrange's energy equations will be employed, although in one or two cases other methods are adopted as they offer certain advantages. In this chapter, the determination of the free and forced vibration response of an SDoF system to various forms of excitation relevant to aircraft loads will be considered. The idea is to introduce some of the core dynamic analysis methods (or tools) to be used later in aircraft aeroelasticity and loads calculations.

1.1 Setting up Equations of Motion for SDoF Systems

A single DoF system is one whose motion may be described by a single coordinate, that is, a displacement or rotation. All systems that may be described by a single DoF may be shown to have an identical form of governing equation, albeit with different symbols employed in each

Introduction to Aircraft Aeroelasticity and Loads, Second Edition. Jan R. Wright and Jonathan E. Cooper.
© 2015 John Wiley & Sons, Ltd. Published 2015 by John Wiley & Sons, Ltd.

case. Two examples will be considered here, a classical mass/spring/damper system and an aircraft control surface able to rotate about its hinge line but restrained by an actuator. These examples will illustrate translational and rotational motions.

1.1.1 Example: Classical SDoF System

The classical form of an SDoF system is shown in Figure 1.1, and comprises a mass m, a spring of stiffness k and a viscous damper whose damping constant is c; a viscous damper is an idealized energy dissipation device where the force developed is linearly proportional to the relative velocity between its ends (note that the alternative approach of using hysteretic/structural damping will be considered later). The motion of the system is a function of time t and is defined by the displacement $x(t)$. A time-varying force $f(t)$ is applied to the mass.

Lagrange's energy equations are differential equations of the system expressed in what are sometimes termed 'generalized coordinates' but written in terms of energy and work quantities (Wells, 1967; Tse et al., 1978; Rao, 1995). These equations will be suitable for a specific physical coordinate or a coordinate associated with a shape (see Chapter 3). Now, Lagrange's equation for an SDoF system with a displacement coordinate x may be written as

$$\frac{\mathrm{d}}{\mathrm{d}t}\left(\frac{\partial T}{\partial \dot{x}}\right) - \frac{\partial T}{\partial x} + \frac{\partial \mathfrak{I}}{\partial \dot{x}} + \frac{\partial U}{\partial x} = Q_x = \frac{\partial(\delta W)}{\partial(\delta x)} \tag{1.1}$$

where T is the kinetic energy, U is the elastic potential (or strain) energy, $\mathfrak{I}$ is the dissipative function, Q_x is the so-called generalized force and W is a work quantity. The overdot indicates the derivative with respect to time, namely $\mathrm{d}/\mathrm{d}t$.

For the SDoF example, the kinetic energy of the mass is given by

$$T = \frac{1}{2} m \dot{x}^2 \tag{1.2}$$

and the elastic potential (or strain) energy in the spring is

$$U = \frac{1}{2} k x^2 \tag{1.3}$$

The damper contribution may be treated as an external force, or else may be defined by the dissipative function

$$\mathfrak{I} = \frac{1}{2} c \dot{x}^2 \tag{1.4}$$

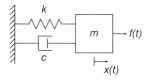

Figure 1.1 SDoF mass/spring/damper system.

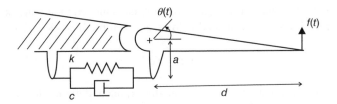

Figure 1.2   SDoF control surface/actuator system.

Finally, the effect of the force is included in Lagrange's equation by considering the incremental work done δW obtained when the force moves through the incremental displacement δx, namely

$$\delta W = f\,\delta x \tag{1.5}$$

Substituting Equations (1.2) to (1.5) into Equation (1.1) yields the classical ordinary second-order differential equation

$$m\ddot{x} + c\dot{x} + kx = f(t) \tag{1.6}$$

1.1.2   Example: Aircraft Control Surface

As an example of a completely different SDoF system that involves a rotational coordinate, consider the control surface/actuator model shown in Figure 1.2. The control surface has a moment of inertia J about the hinge, the values of the effective actuator stiffness and damping constant are k and c respectively, and the rotation of the control surface is θ rad. The actuator lever arm has length a. A force $f(t)$ is applied to the control surface at a distance d from the hinge. The main surface of the wing is assumed to be fixed rigidly as shown.

The energy, dissipation and work done functions corresponding to Equations (1.2) to (1.5) may be shown to be

$$T = \frac{1}{2}J\dot{\theta}^2 \quad U = \frac{1}{2}k(a\theta)^2 \quad \Im = \frac{1}{2}c(a\dot{\theta})^2 \quad \delta W = (f(t)d)\delta\theta \tag{1.7}$$

where the angle of rotation is assumed to be small, so that, for example, $\sin\theta = \theta$. The work done term is the torque multiplied by an incremental rotation. Then, applying the Lagrange equation with coordinate θ, it may be shown that

$$J\ddot{\theta} + ca^2\dot{\theta} + ka^2\theta = df(t) \tag{1.8}$$

Clearly, this equation is of the same general form as that in Equation (1.6). Indeed, all SDoF systems have equations of a similar form, albeit with different symbols and units.

1.2   Free Vibration of SDoF Systems

In free vibration, an initial condition is imposed and motion then occurs in the absence of any external force. The motion takes the form of a non-oscillatory or oscillatory decay; the latter corresponds to the low values of damping normally encountered in aircraft, so only this case will be considered. The solution method is to assume a form of motion given by

$$x(t) = X\mathrm{e}^{\lambda t} \tag{1.9}$$

where X is the amplitude and λ is the characteristic exponent defining the decay. Substituting Equation (1.9) into Equation (1.6), setting the applied force to zero and simplifying, yields the quadratic equation

$$\lambda^2 m + \lambda c + k = 0 \tag{1.10}$$

The solution of this so-called 'characteristic equation' for the case of oscillatory motion produces a complex conjugate pair of roots, namely

$$\lambda_{1,2} = -\frac{c}{2m} \pm i\sqrt{\left(\frac{k}{m}\right) - \left(\frac{c}{2m}\right)^2} \tag{1.11}$$

where the complex number $i = \sqrt{-1}$. Equation (1.11) may be rewritten in the non-dimensional form

$$\lambda_{1,2} = -\zeta\omega_n \pm i\omega_n\sqrt{1-\zeta^2} = -\zeta\omega_n \pm i\omega_d \tag{1.12}$$

where

$$\omega_n = \sqrt{\frac{k}{m}} \qquad \omega_d = \omega_n\sqrt{1-\zeta^2} \qquad \zeta = \frac{c}{2m\omega_n} \tag{1.13}$$

Here ω_n is the undamped natural frequency (frequency in rad/s of free vibration in the absence of damping), ω_d is the damped natural frequency (frequency of free vibration in the presence of damping) and ζ is the *damping ratio* (i.e. c expressed as a proportion of the critical damping constant c_{crit}, the value at which motion becomes non-oscillatory); these parameters are fundamental and unique properties of the system.

Because there are two roots to Equation (1.10), the solution for the free vibration motion is given by the sum

$$x(t) = X_1\mathrm{e}^{\lambda_1 t} + X_2\mathrm{e}^{\lambda_2 t} \tag{1.14}$$

After substitution of Equation (1.12) into Equation (1.14), the motion may be expressed in the form

$$x(t) = \mathrm{e}^{-\zeta\omega_n t}[(X_1 + X_2)\cos\omega_d t + i(X_1 - X_2)\sin\omega_d t] \tag{1.15}$$

Since the displacement $x(t)$ must be a real quantity, then X_1 and X_2 must be a complex conjugate pair and Equation (1.15) simplifies to one of the classical forms

$$x(t) = \mathrm{e}^{-\zeta\omega_n t}[A_1\sin\omega_d t + A_2\cos\omega_d t] \quad \text{or} \quad x(t) = A\mathrm{e}^{-\zeta\omega_n t}\sin(\omega_d t + \psi) \tag{1.16}$$

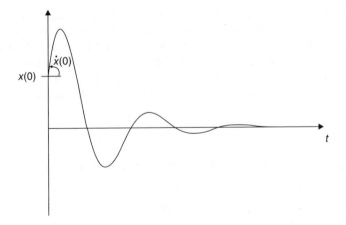

Figure 1.3 Free vibration response for an underdamped SDoF system.

where the amplitude A and phase ψ (or amplitudes A_1, A_2) are unknown values to be determined from the initial conditions for displacement and velocity. Thus this 'underdamped' motion is sinusoidal with an exponentially decaying envelope, as shown in Figure 1.3 for a case with general initial conditions.

1.2.1 Example: Aircraft Control Surface

Using Equation (1.8) for the control surface actuator system and comparing the expressions with those for the simple system, the undamped natural frequency and damping ratio may be shown by inspection to be

$$\omega_n = \sqrt{\frac{ka^2}{J}} \quad \text{and} \quad \zeta = \frac{ca}{2\sqrt{kJ}} \tag{1.17}$$

1.3 Forced Vibration of SDoF Systems

In determining aircraft loads for gusts and manoeuvres (see Chapters 12 to 15), the aircraft response to a number of different types of forcing functions needs to be considered. These may be divided into three categories.

1. *Harmonic excitation* is primarily concerned with excitation at a single frequency (for engine or rotor out-of-balance, control failure or as a constituent part of continuous turbulence analysis).
2. *Non-harmonic deterministic excitation* includes the '1-cosine' input (for discrete gusts or runway bumps) and various shaped inputs (for flight manoeuvres); such forcing functions often have clearly defined analytical forms and tend to be of short duration, often called transient.
3. *Random excitation* includes continuous turbulence and general runway profiles, the latter required for taxiing. Random excitation can be specified using a time or frequency domain description (see later).

The aircraft dynamics are sometimes non-linear (e.g. doubling the input does not double the response), which complicates the solution process; however, in this chapter only the linear case will be considered. The treatment of non-linearity will be covered in later chapters, albeit only fairly briefly. In the following sections, the determination of the responses to harmonic, transient and random excitation will be considered, using both time and frequency domain approaches. The extension to multiple degree of freedom (MDoF) systems will be covered later in Chapter 2.

1.4   Harmonic Forced Vibration – Frequency Response Functions

The most important building block for forced vibration requires determination of the response to a harmonic (i.e. sinusoidal) force with excitation frequency ω rad/s (or $\omega/(2\pi)$ Hz). The relevance to aircraft loads is primarily in helping to lay important foundations for the behaviour of dynamic systems, such as continuous turbulence analysis. However engine, rotor or propeller out-of-balance or control system failure can introduce harmonic excitation to the aircraft.

1.4.1   Response to Harmonic Excitation

When a harmonic excitation force is applied, there is an initial transient response, followed by a steady-state where the response will be sinusoidal at the same frequency as the excitation but lagging it in phase; only the steady-state response will be considered here, though the transient response may often be important.

The excitation input force is defined by

$$f(t) = F \sin \omega t \tag{1.18}$$

and the steady-state response is given by

$$x(t) = X \sin (\omega t - \phi), \tag{1.19}$$

where F, X are the amplitudes and ϕ is the amount by which the response 'lags' the excitation in phase (so-called 'phase lag'). In one approach, the steady-state response may be determined by substituting these expressions into the equation of motion and then equating sine and cosine terms using trigonometric expansion.

However, an alternative approach uses complex algebra and will be adopted since it is more powerful and commonly used. In this approach, the force and response are rewritten, somewhat artificially, in a complex notation such that

$$f(t) = Fe^{i\omega t} = F \cos \omega t + iF \sin \omega t$$
$$x(t) = Xe^{i(\omega t - \phi)} = \left(Xe^{-i\phi}\right)e^{i\omega t} = \tilde{X}e^{i\omega t} = \tilde{X} \cos \omega t + i\tilde{X} \sin \omega t \tag{1.20}$$

Here, the excitation is rewritten in terms of sine and cosine components and the phase lag is embedded in a new complex amplitude quantity $\tilde{X}$. Only the imaginary part of the excitation and response will be used for sine excitation; an alternative way of viewing this approach is that the solutions for both the sine and cosine excitation will be found simultaneously. The solution process is straightforward once the concepts have been grasped. The complex expressions in Equations (1.20) are now substituted into Equation (1.6).

Noting that $\dot{x} = i\omega \tilde{X} e^{i\omega t}$ and $\ddot{x} = -\omega^2 \tilde{X} e^{i\omega t}$ and cancelling the exponential term, then

$$\left(-\omega^2 m + i\omega c + k\right)\tilde{X} = F \tag{1.21}$$

Thus the complex response amplitude may be solved from this algebraic equation so that

$$\tilde{X} = X e^{-i\phi} = \frac{F}{k - \omega^2 m + i\omega c} \tag{1.22}$$

and equating real and imaginary parts from the two sides of this equation yields the amplitude and phase as

$$X = \frac{F}{\sqrt{\left(k - \omega^2 m\right)^2 + (\omega c)^2}} \quad \text{and} \quad \phi = \tan^{-1}\left(\frac{\omega c}{k - \omega^2 m}\right) \tag{1.23}$$

Hence, the steady-state time response may be calculated using X, ϕ from this equation.

1.4.2 Frequency Response Functions

An alternative way of writing Equation (1.22) is

$$H_D(\omega) = \frac{\tilde{X}}{F} = \frac{1}{k - \omega^2 m + i\omega c} \tag{1.24}$$

or in non-dimensional form

$$H_D(\omega) = \frac{1/k}{1 - (\omega/\omega_n)^2 + i2\zeta(\omega/\omega_n)} = \frac{1/k}{1 - r^2 + i2\zeta r} \quad \text{where} \quad r = \frac{\omega}{\omega_n} \tag{1.25}$$

Here $H_D(\omega)$ is known as the displacement Frequency Response Function (FRF) of the system and is a system property; it dictates how the system behaves under harmonic excitation at any frequency. The equivalent velocity and acceleration FRFs are given by

$$H_V = i\omega H_D \qquad H_A = -\omega^2 H_D \tag{1.26}$$

since multiplication by $i\omega$ in the frequency domain is equivalent to differentiation in the time domain, noting again that $i = \sqrt{-1}$.

The quantity $kH_D(\omega)$ from Equation (1.25) is a non-dimensional expression, or dynamic magnification, relating the dynamic amplitude to the static deformation for several damping ratios. The well-known 'resonance' phenomenon is shown in Figure 1.4 by the amplitude peak that occurs when the excitation frequency ω is at the 'resonance' frequency, close in value to the undamped natural frequency ω_n; the phase changes rapidly in this region, passing through 90° at resonance. Note that the resonance peak increases in amplitude as the damping ratio reduces and that the dynamic magnification (approximately $1/2\zeta$) can be extremely large.

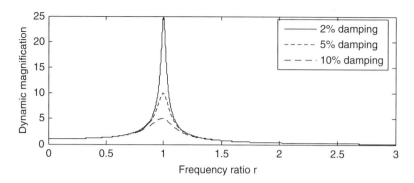

Figure 1.4   Displacement Frequency Response Function for an SDoF system.

1.4.3   Hysteretic (or Structural) Damping

So far, a viscous damping representation has been employed, based on the assumption that the damping force is proportional to velocity (and therefore to frequency). However, in practice, damping measurements in structures and materials have sometimes shown that damping is independent of frequency but acts in quadrature (i.e. is at 90° phase) to the displacement of the system. Such an internal damping mechanism is known as hysteretic (or sometimes structural) damping (Rao, 1995). It is common practice to combine the damping and stiffness properties of a system having hysteretic damping into a so-called complex stiffness, namely

$$k^* = k(1+ig) \tag{1.27}$$

where g is the structural damping coefficient or loss factor (not to be confused with the same symbol used for acceleration due to gravity). The complex number indicates that the damping force is in quadrature with the stiffness force. The SDoF equation of motion, amended to employ hysteretic damping, may then be written as

$$m\ddot{x} + k(1+ig)x = f(t) \tag{1.28}$$

This equation is strictly invalid, being expressed in the time domain but including the complex number; it is not possible to solve the equation in this form. However, it is feasible to write the equation in the time domain as

$$m\ddot{x} + c_{eq}\dot{x} + kx = f(t) \tag{1.29}$$

where $c_{eq} = gk/\omega$ is the equivalent viscous damping. This equation of motion may be used if the motion is dominantly at a single frequency. The equivalent viscous damping ratio expression may be shown to be

$$\zeta_{eq} = \frac{g}{2}\left(\frac{\omega_n}{\omega}\right) \tag{1.30}$$

or, if the system is actually vibrating at the natural frequency, then

$$\zeta_{eq} = \frac{g}{2} \tag{1.31}$$

Thus the equivalent viscous damping ratio is half of the loss factor, and this factor of 2 is often used when plotting flutter damping plots (see Chapter 10).

An alternative way of considering hysteretic damping is to convert Equation (1.28) into the frequency domain, using the methodology employed earlier in Section 1.4.1, so yielding the FRF in the form

$$H_D(\omega) = \frac{\tilde{X}}{F} = \frac{1}{k(1+ig) - \omega^2 m} \tag{1.32}$$

and now the complex stiffness takes a more suitable form. Thus, a frequency domain solution of a system with hysteretic damping is acceptable, but a time domain solution needs to assume motion at essentially a single frequency. The viscous damping model, despite its drawbacks, does lend itself to more simple analysis, though both viscous and hysteretic damping models are widely used. It will be seen in Chapter 9 that the damping inherent in aerodynamic motion behaves as viscous damping.

1.5   Transient/Random Forced Vibration – Time Domain Solution

When a transient or random excitation is present, the time response may be calculated in one of three ways.

1.5.1   Analytical Approach

If the excitation is deterministic, having a relatively simple mathematical form (e.g. step, ramp), then an analytical method suitable for ordinary differential equations may be used, i.e. expressing the solution as a combination of complementary function and particular integral. Such an approach is impractical for more general forms of excitation.

For example, a *unit step* force applied to the system initially at rest may be shown to give rise to the response (i.e. so-called 'Step Response Function')

$$s(t) = x_{SRF}(t) = \frac{1}{k}\left[1 - \frac{e^{-\zeta\omega_n t}}{\sqrt{1-\zeta^2}}\sin(\omega_d t + \psi)\right] \quad \text{with} \quad \tan\psi = \frac{\sqrt{1-\zeta^2}}{\zeta} \tag{1.33}$$

Note that the term in square brackets is the ratio of the dynamic-to-static response and this ratio is shown in Figure 1.5 for different damping ratios. Note that there is a tendency for the transient response to 'overshoot' the steady-state value, but this initial peak response is not greatly affected by damping; this behaviour will be referred to later as 'dynamic overswing' when considering dynamic manoeuvres in Chapters 13 and 23.

Another important excitation quantity is the *unit impulse* of force. This may be idealized crudely as a very narrow rectangular force–time pulse of unit area (i.e. strength) of 1 N s (the ideal impulse is the so-called Dirac-δ function, having zero width and infinite height). Because this impulse imparts an instantaneous change in momentum, the velocity changes by an amount equal to the impulse strength/mass). Thus the unit impulse case is equivalent to free vibration with a finite initial velocity and zero initial displacement. It may be shown that the response to a unit impulse (i.e. so-called 'Impulse Response Function') is

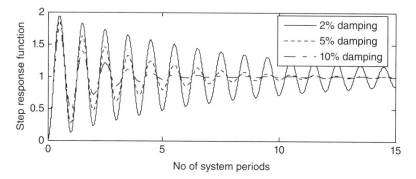

Figure 1.5   Dynamic-to-static ratio of step response for an SDoF system.

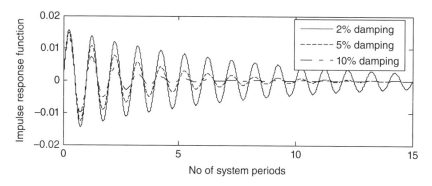

Figure 1.6   Impulse Response Function for an SDoF system.

$$h(t) = x_{\mathrm{IRF}}(t) = \frac{1}{m\omega_{\mathrm{d}}} e^{-\zeta\omega_{\mathrm{n}}t} \sin\omega_{\mathrm{d}}t \qquad (1.34)$$

The Impulse Response Function (IRF) is shown plotted against non-dimensional time for several damping ratios in Figure 1.6; the response starts and ends at zero. The y axis values depend upon the mass and natural frequency. The IRF may be used in the convolution approach described later in Section 1.5.3.

1.5.2   Principle of Superposition

The principle of superposition, only valid for linear systems, states that if the responses to separate forces $f_1(t)$ and $f_2(t)$ are $x_1(t)$ and $x_2(t)$ respectively, then the response $x(t)$ to the sum of the forces $f(t) = f_1(t) + f_2(t)$ will be the sum of their individual responses, namely $x(t) = x_1(t) + x_2(t)$.

1.5.3   Example: Single Cycle of Square Wave Excitation – Response Determined by Superposition

Consider an SDoF system with an effective mass of 1000 kg, natural frequency 2 Hz and damping ratio 5% excited by a transient excitation force consisting of a single cycle of a square wave

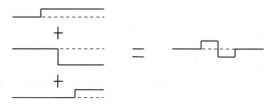

Figure 1.7   Single cycle of a square wave described by the principle of superposition.

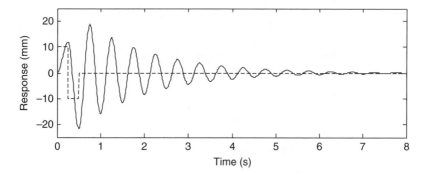

Figure 1.8   Response to a single cycle of a square wave, using superposition.

with amplitude 1000 N and period τ_{square}. The response may be found by superposition of a step input of amplitude 1000 N at $t = 0$, a negative step input of amplitude 2000 N at $t = \tau_{\text{square}}/2$ and a single positive step input of amplitude 1000 N at $t = \tau_{\text{square}}$, as illustrated in Figure 1.7. The response may be calculated using the MATLAB program given in the companion website.

Figure 1.8 shows the response when $\tau_{\text{square}} = 0.5$ s, the period of the system; the dashed line shows the corresponding input. In this case, the square wave pulse is nearly 'tuned' to the system (i.e. near to the resonance frequency) and so the response is significantly larger (by almost a factor of 2) than for a single on/off pulse. This magnification of response is the reason why the number of allowable pilot control input reversals in a manoeuvre is strictly limited.

1.5.4   Convolution Approach

The principle of superposition illustrated above may be employed in the solution of the response to a general transient or random excitation. The idea here is that a general excitation input may be represented by a sequence of very narrow (ideal) impulses of different heights (and therefore strengths), as shown in Figure 1.9. A typical impulse occurring at time $t = \tau$ is of height $f(\tau)$ and width $d\tau$. Thus the corresponding impulse strength is $f(\tau)\,d\tau$ and the response to this impulse, using the unit Impulse Response Function $h(t)$ in Equation (1.34), is

$$x_\tau(t) = \{f(\tau)\,d\tau\}\,h(t-\tau) = \frac{\{f(\tau)\,d\tau\}}{m\omega_n}e^{-\zeta\omega_n(t-\tau)}\sin\omega_d(t-\tau) \quad \text{for } t \geq \tau$$

$$x_\tau(t) = 0 \quad \text{for } t < \tau$$

(1.35)

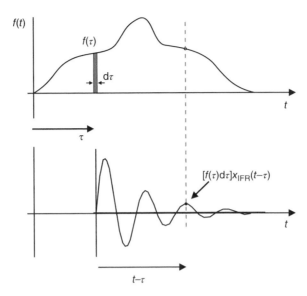

Figure 1.9 Convolution process.

Note that the response is only non-zero *after* the impulse at $t = \tau$. The response to the entire excitation time history is equal to the summation of the responses to all the constituent impulses. Given that each impulse is $d\tau$ wide, and allowing $d\tau \rightarrow 0$, then the summation effectively becomes an integral, given by

$$x(t) = \int_{\tau=0}^{t} f(\tau) h(t-\tau) \, d\tau. \qquad (1.36)$$

This is known as the convolution integral (Newland, 1989; Rao, 1995) or, alternatively, the Duhamel integral (Fung, 1969). A shorthand way of writing this integral, where * denotes convolution, is

$$x(t) = h(t)^* f(t). \qquad (1.37)$$

An alternative form of the convolution process may be written by treating the excitation as a combination of on/off steps and using the Step Response Function $s(t)$, thus yielding a similar convolution expression (Fung, 1969)

$$x(t) = f(t)s(0) + \int_{\tau=0}^{t} f(\tau) \frac{ds}{dt}(t-\tau) \, d\tau. \qquad (1.38)$$

This form of convolution will be encountered in Chapters 9 and 14 for unsteady aerodynamics and gusts.

In practice, the convolution integrations would be performed numerically and not analytically. Thus the force input and Impulse (or Step) Response Function would need to be obtained in discrete, and not continuous, time form. The Impulse Response Function may in fact be obtained numerically via the inverse Fourier transform of the Frequency Response Function.

1.5.5   Direct Solution of Ordinary Differential Equations

An alternative approach to obtaining the response to general transient or random excitation is to solve the relevant ordinary differential equation by employing numerical integration such as the Runge–Kutta or Newmark-β algorithms (Rao, 1995). To present one or both of these algorithms in detail is beyond the scope of this book. Suffice it to say that, knowing the response at the jth time value, the differential equation expressed at the $(j+1)$th time value is used, together with some assumption for the variation of the response within the length of the step, to predict the response at the $(j+1)$th time value.

   In this book, time responses are sometimes calculated using numerical integration in the SIMULINK package called from a MATLAB program. The idea is illustrated in the next section using the earlier superposition example.

1.5.6   Example: Single Cycle of Square Wave Excitation – Response Determined by Numerical Integration

Consider again the SDoF system excited by the single square wave cycle as used in Section 1.5.3. The response may be found using numerical integration and may be seen to overlay the exact result in Figure 1.8 provided an adequately small time step size is used (typically corresponding to at least 30 points per cycle). The response is calculated using a Runge–Kutta algorithm in a MATLAB/SIMULINK program (see companion website).

1.6   Transient Forced Vibration – Frequency Domain Solution

The analysis leading up to the definition of the Frequency Response Function in Section 1.4 considered only the response to an excitation input comprising a single sine wave at frequency ω rad/s. However, if the excitation was made up of several sine waves with different amplitudes and frequencies, the total steady-state response could be found by superposition of the responses to each individual sine wave, using the appropriate value of the FRF at each frequency. Again, because superposition is implied, the approach only applies for linear systems.

1.6.1   Analytical Fourier Transform

In practice, the definition of the FRF may be extended to cover a more general excitation by employing the Fourier Transform (FT), so that

$$H(\omega) = \frac{X(\omega)}{F(\omega)} = \frac{\text{Fourier Transform of } x(t)}{\text{Fourier Transform of } f(t)} \tag{1.39}$$

where, for example, $X(\omega)$, the Fourier Transform of $x(t)$, is given for a continuous infinite length signal by

$$X(\omega) = \int_{-\infty}^{+\infty} x(t) e^{-i\omega t}\, dt \tag{1.40}$$

The Fourier Transform $X(\omega)$ is a complex function of frequency (i.e. a spectrum) whose real and imaginary parts define the magnitude of the components of $\cos \omega t$ and $-\sin \omega t$ in the signal

$x(t)$. The units of $X(\omega)$, $F(\omega)$ in this definition are typically ms and Ns and the units of $H(\omega)$ are m/N. The Inverse Fourier Transform (IFT), not defined here, allows the frequency function to be transformed back into the time domain.

Although the Fourier Transform is initially defined for an infinite continuous signal, and this would appear to challenge its usefulness, in practice signals of finite length T may be considered using the different definition

$$X(\omega) = \frac{1}{T} \int_0^T x(t) \mathrm{e}^{-i\omega t} \, \mathrm{d}t \tag{1.41}$$

In this case, the units of $X(\omega)$, $F(\omega)$ become m and N respectively, while units of $H(\omega)$ remain m/N.

What is being assumed by using this expression is that $x(t)$ is periodic with period T; i.e. the signal keeps repeating itself in a cyclic manner. Provided there is no discontinuity between the start and end of $x(t)$, then the analysis may be applied for a finite length excitation such as a pulse. If a discontinuity does exist, then a phenomenon known as 'leakage' occurs and additional incorrect Fourier amplitude components are introduced to represent the discontinuity; in practice, window functions (e.g. Hanning, Hamming, etc.) are often applied to minimize this effect (Newland, 1987). The choice of the parameters defining the excitation time history in the analysis must be made carefully to minimize this error; to avoid leakage, the aim would be for the excitation and response signals to start and stop at zero.

1.6.2 Frequency Domain Response – Excitation Relationship

It may be seen that rearranging Equation (1.39) leads to

$$X(\omega) = H(\omega) F(\omega) \tag{1.42}$$

and it is interesting to relate this to the time domain convolution in Equation (1.37). The FRF and IRF are in fact Fourier Transform pairs, e.g. the FRF is the Fourier Transform of the IRF. Further, it may also be shown that by taking the Fourier Transform of both sides of Equation (1.37), then Equation (1.42) results, i.e. convolution in the 'time domain' is equivalent to multiplication in the 'frequency domain'. The extension of this approach for an MDoF system will be considered in Chapter 2.

A useful feature of Equation (1.42) is that it may be used to determine the response of a system, given the excitation time history, by taking a frequency domain route. Thus the response $x(t)$ of a linear system to a finite length transient excitation input $f(t)$ may be found by the following procedure.

1. Fourier Transform $f(t)$ to find $F(\omega)$.
2. Determine the FRF $H(\omega)$ for the system.
3. Multiply the FRF by $F(\omega)$ using Equation (1.42) to determine $X(\omega)$.
4. Inverse Fourier Transform $X(\omega)$ to find $x(t)$.

1.6.3 Example: Single Cycle of Square Wave Excitation – Response Determined via Fourier Transform

Consider again the SDoF system excited by a single square wave cycle as used in Section 1.5.3. The response is calculated using a MATLAB program employing the Discrete Fourier

Transform (see companion website). Note that only a limited number of data points are used in order to allow the discrete values in the frequency and time domains to be seen; only discrete data points are plotted in the frequency domain functions. The results agree well with those in Figure 1.8 but the accuracy improves as more data points are used to represent the signals.

1.7   Random Forced Vibration – Frequency Domain Solution

There are two cases in aircraft loads where response to a random-type excitation is required: (a) flying through continuous turbulence and (b) taxiing on a runway with a non-smooth profile. Continuous turbulence is considered as a stationary random variable with a Power Spectral Density (PSD) description available from extensive flight testing so it is normal practice to use a frequency domain spectral approach based on a linearized model of the aircraft (see Chapter 14); however, when the effect of significant non-linearity is to be explored, a time domain computation would need to be used (see Chapters 19 and 23). For taxiing (see Chapter 15), no statistical description is usually available for the runway input so the solution would be carried out in the time domain using numerical integration of the equations of motion, as they are highly non-linear due to the presence of the landing gear.

When random excitation is considered in the frequency domain, then a statistical approach is normally employed by defining the PSD of the excitation and response (Newland, 1987; Rao, 1995). For example, the PSD of response $x(t)$ is defined by

$$S_{xx}(\omega) = \frac{T}{2\pi}X(\omega)^*X(\omega) = \frac{T}{2\pi}\left|X(\omega)^2\right| \qquad (1.43)$$

where * denotes the complex conjugate (not to be confused with the convolution symbol). Thus the PSD is essentially proportional to the modulus squared of the Fourier amplitude at each frequency and would have units of density (m^2 per rad/s if $x(t)$ were a displacement). It is a measure of how the 'power' in $x(t)$ is distributed over the frequency range of interest. In practice, the PSD of a time signal could be computed from a long data record by employing some form of averaging of finite length segments of the data.

If Equation (1.42) is multiplied on both sides by its complex conjugate then

$$X(\omega)X^*(\omega) = H(\omega)F(\omega)\,H^*(\omega)F^*(\omega) = |H(\omega)|^2\,F(\omega)F^*(\omega) \qquad (1.44)$$

and if the relevant scalar factors present in Equation (1.43) are accounted for, then Equation (1.44) becomes

$$S_{xx}(\omega) = |H(\omega)|^2\,S_{ff}(\omega) \qquad (1.45)$$

Thus, knowing the definition of the excitation PSD $S_{ff}(\omega)$ (units N^2 per rad/s for force), the response PSD may be determined given the FRF for the system (m/N for displacement per force). It may be seen from Equation (1.45) that the spectral shape of the excitation is carried through to the response, but is filtered by the system dynamic characteristics. The extension of this approach for an MDoF system will be considered in Chapter 2. This relationship between the response and excitation PSDs is very useful, but phase information is lost.

In the analysis shown so far, the PSD $S_{xx}(\omega)$, for example, has been 'two-sided' in that values exist at both positive and negative frequencies; negative frequencies are somewhat artificial but derive from the resulting mathematics in that a positive frequency corresponds to a vector rotating anticlockwise at ω, whereas a negative frequency corresponds to rotation in the opposite direction. However, in practice the 'two-sided' (or double-sided) PSD is often converted into a 'one-sided' (or single-sided) function $\Phi_{xx}(\omega)$, existing only at non-negative frequencies and calculated using

$$\Phi_{xx}(\omega) = 2S_{xx}(\omega) \quad 0 \le \omega < \infty \tag{1.46}$$

Single-sided spectra are in fact used in determining the response to continuous turbulence (considered in Chapter 14), since the continuous turbulence PSD is defined in this way.

The mean square value is the corresponding area under the one-sided or two-sided PSD, so

$$\overline{x^2} = \int_0^{+\infty} \Phi_{xx}(\omega)\, d\omega \quad \text{or} \quad \overline{x^2} = \int_{-\infty}^{+\infty} S_{xx}(\omega)\, d\omega \tag{1.47}$$

where clearly only a finite, not infinite, frequency range is used in practice. The root mean square value is the square root of the mean square value.

1.8 Examples

Note that these examples may be useful preparation when carrying out the examples in later chapters.

1. An avionics box may be idealized as an SDoF system comprising a mass m supported on a mounting base via a spring k and damper c. The system displacement is $y(t)$ and the base displacement is $x(t)$. The base is subject to acceleration $\ddot{x}(t)$ from motion of the aircraft. Show that the equation of motion for the system may be written in the form $m\ddot{z} + c\dot{z} + kz = -m\ddot{x}(t)$ where $z = y - x$ is the relative displacement between the mass and the base (i.e. spring extension).

2. In a flutter test (see Chapter 25), the acceleration of an aircraft control surface following an explosive impact decays to a quarter of its amplitude after 5 cycles, which corresponds to an elapsed time of 0.5 s. Estimate the undamped natural frequency and the percentage of critical damping.
   [10.01 Hz, 4.4 %]

3. Determine an expression for the response of a single degree of freedom undamped system undergoing free vibration following an initial condition of zero velocity and displacement x_0.
   $$\left[x(t) = Ae^{-\zeta\omega_n t} \sin(\omega_d t + \psi) \quad A = x_0 / \sqrt{1-\zeta^2} \quad \psi = \tan^{-1}\left(\sqrt{1-\zeta^2}/\zeta \right) \right]$$

4. Determine an expression for the time t_p at which the response of a damped SDoF system to excitation by a step force F_0 reaches a maximum $\left(\omega_n t_p = \pi / \sqrt{1-\zeta^2} \right)$. Show that the maximum response is given by the expression $xk/F_0 = 1 + \exp\left(-\zeta\pi / \sqrt{1-\zeta^2} \right)$, noting the insensitivity to damping at low values.

5. Using the complex algebra approach for harmonic excitation and response, determine an expression for the transmissibility (i.e. system acceleration per base acceleration) for the base excited system in Example 1.

$$[([k+i\omega c]/[k-\omega^2 m + i\omega c])]$$

6. A motor supported in an aircraft on four anti-vibration mounts may be idealized as an SDoF system of effective mass 20 kg. Each mount has a stiffness of 5000 N/m and a damping coefficient of 200 Ns/m. Determine the natural frequency and damping ratio of the system. Also, estimate the displacement and acceleration response of the motor when it runs with a degree of imbalance equivalent to a sinusoidal force of ±40 N at 1200 rpm (20 Hz). Compare this displacement value to the static deflection of the motor on its mounts.

   [5.03 Hz, 63.2%, 0.128 mm, 2.02 m/s^2, 9.8 mm]

7. A machine of mass 1000 kg is supported on a spring/damper arrangement. In operation, the machine is subjected to a force of 750 cosωt, where ω (rad/s) is the operating frequency. In an experiment, the operating frequency is varied and it is noted that resonance occurs at 75 Hz and that the magnitude of the FRF is 2.5. However, at its normal operating frequency this value is found to be 0.7. Find the normal operating frequency and the support stiffness and damping coefficient.

   [118.3 Hz, 2.43 × 10^8 N/m, 1.97 × 10^5 N s/m]

8. An aircraft fin may be idealized in bending as an SDoF system with an effective mass of 200 kg, undamped natural frequency of 5 Hz and damping 3% critical. The fin is excited via the control surface by an 'on/off' force pulse of magnitude 500 N. Using MATLAB and one or more of the (a) superposition, (b) simulation and (c) Fourier Transform approaches, determine the pulse duration that will maximize the resulting response and the value of the response itself.

9. Using MATLAB, generate a time history of 16 data points with a time interval Δt of 0.05 s and composed of a DC value of 1, a sine wave of amplitude 3 at 4 Hz and cosine waves of amplitude −2 at 2 Hz and 1 at 6 Hz. Perform the Fourier Transform and examine the form of the complex output sequence as a function of frequency to understand how the data are stored and how the frequency components are represented. Then perform the Inverse FT and examine the resulting sequence, comparing it to the original signal.

10. Generate other time histories with a larger number of data values, such as (a) single (1-cosine) pulse, (b) multiple cycles of a sawtooth waveform and (c) multiple cycles of a square wave. Calculate the FT of each and examine the amplitude of the frequency components to see how the power is distributed.

2

Vibration of Multiple Degree of Freedom Systems

In this chapter, some of the basic concepts of vibration analysis for multiple degree of freedom (MDoF) discrete parameter systems will be introduced, since there are many significant differences to single degree of freedom (SDoF) systems. The term 'discrete (or sometimes lumped) parameter' implies that the system in question is a combination of discrete rigid masses (or components) interconnected by stiffness and damping elements. Note that the equations may be expressed in a modal coordinate system (see later). On the other hand, 'continuous' systems, considered later in Chapters 3 and 17, are those where parts of the system have distributed mass and stiffness.

The focus of this chapter will be in setting up the equations of motion, finding natural frequencies and mode shapes for free vibration, considering damping and determining the forced vibration response with various forms of excitation relevant to aircraft loads. Some of the core solution methods introduced in Chapter 1 will be considered for MDoF systems. For simplicity, the ideas will be illustrated for only two degrees of freedom. The general form of equations will be shown in matrix form to cover any number of degrees of freedom, since matrix algebra unifies all MDoF systems. Treatment may also be found in Tse *et al.* (1978), Newland (1989), Rao (1995), Thomson (1997), Meirovitch (1986) and Inman (2006).

2.1 Setting up Equations of Motion

There are a number of ways of setting up the equations of motion for an MDoF system. As in Chapter 1, Lagrange's energy equations will be employed. Two examples will be considered: a classical 'chain-like' discrete parameter system and later a rigid aircraft capable of undergoing heave and pitch motion while supported on its landing gears. The latter example will also be used in Chapter 15 when considering the taxiing case.

A classical form of a 2DoF chain-like system is shown in Figure 2.1. All other systems that may be described by multiple degrees of freedom may be shown to have an identical form of governing equation, albeit with different parameters. This basic system comprises masses m_1, m_2, springs of stiffness k_1, k_2 and viscous dampers of damping constants c_1, c_2. The motion of

---

Introduction to Aircraft Aeroelasticity and Loads, Second Edition. Jan R. Wright and Jonathan E. Cooper.
© 2015 John Wiley & Sons, Ltd. Published 2015 by John Wiley & Sons, Ltd.

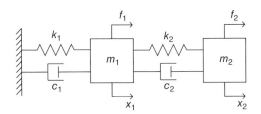

Figure 2.1   Two-DoF 'chain-like' mass/spring/damper system.

the system is a function of time t and is defined by the displacements $x_1(t)$, $x_2(t)$. Time varying forces $f_1(t)$, $f_2(t)$ are also applied to the masses as shown.

Although there are now two DoF, and therefore two equations of motion, the energy and work terms required are obviously still scalar and therefore additive quantities. Firstly, the kinetic energy is given by

$$T = \frac{1}{2} m_1 \dot{x}_1{}^2 + \frac{1}{2} m_2 \dot{x}_2{}^2 \tag{2.1}$$

The elastic potential (or strain) energy in the springs depends upon the relative extension/compression of each and is given by

$$U = \frac{1}{2} k_1 x_1{}^2 + \frac{1}{2} k_2 (x_2 - x_1)^2 \tag{2.2}$$

and the dissipative term for the dampers depends upon the relative velocities and is written as

$$\Im = \frac{1}{2} c_1 \dot{x}_1{}^2 + \frac{1}{2} c_2 (\dot{x}_2 - \dot{x}_1)^2 \tag{2.3}$$

Finally, the effect of the forces is included in Lagrange's equation by considering the incremental work done δW obtained when the two forces move through incremental displacements δx_1, δx_2, namely

$$\delta W = f_1 \, \delta x_1 + f_2 \, \delta x_2 \tag{2.4}$$

Now, Lagrange's equation for a system with multiple degrees of freedom N may be written as

$$\frac{d}{dt}\left(\frac{\partial T}{\partial \dot{x}_j}\right) - \frac{\partial T}{\partial x_j} + \frac{\partial \Im}{\partial \dot{x}_j} + \frac{\partial U}{\partial x_j} = Q_j = \frac{\partial(\delta W)}{\partial(\delta x_j)} \text{ for } j = 1, 2, \ \dots \ , N \tag{2.5}$$

Substituting Equations (2.1) to (2.4) into Equation (2.5) and performing the differentiations for $N = 2$ yields the ordinary second-order differential equations

$$m_1 \ddot{x}_1 + (c_1 + c_2)\dot{x}_1 - c_2 \dot{x}_2 + (k_1 + k_2)x_1 - k_2 x_2 = f_1(t)$$
$$m_2 \ddot{x}_2 - c_2 \dot{x}_1 + c_2 \dot{x}_2 - k_2 x_1 + k_2 x_2 = f_2(t) \tag{2.6}$$

These equations of motion are usually expressed in matrix form as

$$\begin{bmatrix} m_1 & 0 \\ 0 & m_2 \end{bmatrix} \begin{Bmatrix} \ddot{x}_1 \\ \ddot{x}_2 \end{Bmatrix} + \begin{bmatrix} c_1 + c_2 & -c_2 \\ -c_2 & c_2 \end{bmatrix} \begin{Bmatrix} \dot{x}_1 \\ \dot{x}_2 \end{Bmatrix} + \begin{bmatrix} k_1 + k_2 & -k_2 \\ -k_2 & k_2 \end{bmatrix} \begin{Bmatrix} x_1 \\ x_2 \end{Bmatrix} = \begin{Bmatrix} f_1 \\ f_2 \end{Bmatrix} \qquad (2.7)$$

where the mass matrix is diagonal (so the system is uncoupled inertially in its physical degrees of freedom) whereas the damping and stiffness matrices are coupled. In general matrix notation this becomes

$$\mathbf{M}\ddot{x} + \mathbf{C}\dot{x} + \mathbf{K}x = f(t) \qquad (2.8)$$

where $\mathbf{M}$, $\mathbf{C}$, $\mathbf{K}$ are the mass, damping and stiffness matrices respectively, and x, f are the column vectors of displacements and forces. Note that the matrices are symmetric. All MDoF systems may be expressed in this matrix form. In this book, the bold symbol will be used to indicate a matrix quantity and bold italics for a vector, as seen in the above equation. It is assumed that the reader is familiar with basic matrix concepts, and if not, another reference should be consulted (e.g. Stroud and Booth, 2007).

2.2   Undamped Free Vibration

Initially, free vibration for the *undamped* MDoF system will be considered; the damped case will be introduced later. Because of the compact nature of matrix algebra in illustrating the theory, where possible any analysis will be carried out in matrix form for N degrees of freedom, and an example shown for $N = 2$. The free undamped vibration case yields the natural frequencies and so-called mode shapes.

2.2.1   Direct Approach

As for the SDoF system, the solution method is to seek a form of free vibration motion given by

$$x(t) = X \sin \omega t \qquad (2.9)$$

where X is the amplitude vector and ω is the frequency of free vibration. All coordinates are assumed to move in or out of phase at the same frequency during free vibration. Substituting Equation (2.9) into Equation (2.8), setting the damping and forcing values to zero and simplifying, yields the expression

$$\left[\mathbf{K} - \omega^2 \mathbf{M} \right] X = 0 \qquad (2.10)$$

The solution to this equation recognizes that X must be non-trivial and therefore that the matrix in brackets must be singular (i.e. have a zero determinant). By setting the determinant $|\mathbf{K} - \omega^2 \mathbf{M}|$ to zero, an Nth-order polynomial in ω^2 is obtained. The solution of this polynomial yields roots ω_j, $j = 1, 2, \ldots, N$. These are the so-called *undamped natural frequencies* of the system and are the frequencies at which motion of the type described by Equation (2.9) may be found. They are a property of the system and there are as many natural frequencies as there are DoF.

For each natural frequency ω_j, the response may be characterized by the vector X_j, given by the solution of

$$\left[\mathbf{K} - \omega_j^2 \mathbf{M}\right] X_j = 0 \quad \text{for } j = 1, 2, \ldots, N \tag{2.11}$$

These characteristic vectors may be found by solving Equation (2.11) directly, though only ratios of the vector elements and not absolute values are obtained. This process will be illustrated later by an example.

2.2.2 Eigenvalue Approach

As an alternative approach, Equation (2.10) may be rewritten as

$$\mathbf{K}X = \omega^2 \mathbf{M}X \quad \text{or} \quad \mathbf{M}^{-1}\mathbf{K}X = \omega^2 X \tag{2.12}$$

which is equivalent to the classical eigenvalue expressions

$$\mathbf{A}X = \lambda \mathbf{B}X \quad \text{or} \quad \mathbf{A}X = \lambda X \tag{2.13}$$

where $\mathbf{A}$, $\mathbf{B}$ are symmetric matrices and $\lambda \ (=\omega^2)$ is referred to as an eigenvalue. The eigenvalues are $\lambda_j, j = 1, 2, \ldots, N$ (leading to the corresponding natural frequency values $\omega_j, j = 1, 2, \ldots, N$), and these are readily obtainable using matrix methods (Golub and van Loan, 1989). Also, it may be recognized that the corresponding vectors $X_j, j = 1, 2, \ldots, N$, are in fact the eigenvectors; these are also known as the *undamped normal modes*, a further property of the system. Each mode shape yields the relative displacements of each of the physical coordinates when the system vibrates at the corresponding natural frequency.

The so-called *modal matrix* is defined as the matrix having the mode shapes (i.e. eigenvectors) as columns, so

$$\mathbf{\Phi} = [X_1 \, X_2, \cdots, X_N] \tag{2.14}$$

The undamped free vibration motion can be shown to be comprised of the sum of all the mode shapes, each vibrating at its corresponding natural frequency, with amplitudes and phases determined by the initial conditions. Because there is no damping present, the motion will not decay.

2.2.3 Example: 'Chain-like' 2DoF System

Consider the example of the 2DoF 'chain-like' system shown in Figure 2.1 and introduced earlier in Section 2.1. However, now assign numerical values as follows: $m_1 = 2$ kg, $m_2 = 1$ kg, $k_1 = 2000$ N/m and $k_2 = 1000$ N/m. Damping and force values are set to zero. Substituting these values into Equation (2.7) to determine the equation of motion, and then using the mass and stiffness matrices in Equation (2.10) yields

$$\begin{bmatrix} 3000 - 2\omega^2 & -1000 \\ -1000 & 1000 - \omega^2 \end{bmatrix} \begin{Bmatrix} X_1 \\ X_2 \end{Bmatrix} = 0 \tag{2.15}$$

Setting the determinant of the matrix equal to zero,

$$(3000-2\omega^2)(1000-\omega^2)-(-1000)^2=0 \tag{2.16}$$

leads to a quadratic equation in ω^2, namely

$$2(\omega^2)^2-5000(\omega^2)+2000000=0 \tag{2.17}$$

The two equation roots are $\omega_1^2 = 500$, $\omega_2^2 = 2000$, so $\omega_1 = 22.36$ rad/s, $\omega_2 = 44.72$ rad/s and so 3.559 Hz and 7.118 Hz are the undamped natural frequencies at which the system will respond in free vibration.

To determine the mode shapes, the characteristic vector must be solved for each natural frequency. For this 2DoF system, Equation (2.15) may be used to find the ratio of X_1/X_2 from either of the two equations, so

$$\frac{X_1}{X_2}=\frac{1000}{3000-2\omega^2} \tag{2.18}$$

Now, substituting the values for each natural frequency into this equation yields the ratios

$$\left(\frac{X_1}{X_2}\right)_{\text{Mode 1}} = 0.5 \quad \text{and} \quad \left(\frac{X_1}{X_2}\right)_{\text{Mode 2}} = -1 \tag{2.19}$$

These ratios imply that in mode 1 the second mass moves twice as much as the first mass but in phase with it; however, in the second mode both masses move an equal amount but are out-of-phase. It is not possible to assign absolute values to X_1, X_2 for each mode. The ratios may be written as mode shape vectors by choosing some suitable form of normalization. For example, the vector may be written with a maximum element of unity, so

$$X_1 = \begin{Bmatrix} 0.5 \\ 1 \end{Bmatrix} \quad \text{and} \quad X_2 = \begin{Bmatrix} 1 \\ -1 \end{Bmatrix} \tag{2.20}$$

and thus the modal matrix is

$$\Phi = \begin{bmatrix} 0.5 & 1 \\ 1 & -1 \end{bmatrix} \tag{2.21}$$

The same result would be obtained using an eigenvalue solver (see the MATLAB program in the companion website).

2.3 Damped Free Vibration

Now consider the free vibration of a *damped* MDoF system. The behaviour of an MDoF viscously damped system is dependent upon the relationship between the damping matrix and the mass and stiffness matrices. If the physical damping matrix **C** can be written as a linear combination of the physical mass and stiffness matrices (**M**, **K**), e.g.

$$C = \alpha M + \beta K \tag{2.22}$$

where α, β are scalar coefficients, then such a damping model is known as *proportional* (or Rayleigh) damping (Rao, 1995; Thomson, 1997). If this relationship is not satisfied, then the damping is said to be *non-proportional*. The 2DoF 'chain-like' example will be used to illustrate the effect of these two damping models on the free damped vibration behaviour.

The mathematical form of the motion of the damped system following release from an initial condition is rather complicated so will not be covered in detail here. Rather as for the SDoF system, the response is assumed to be

$$x(t) = X e^{\lambda t} \tag{2.23}$$

If this solution is substituted into Equation (2.8) with no excitation and the exponent term cancelled, then

$$[\lambda^2 \mathbf{M} + \lambda \mathbf{C} + \mathbf{K}] X = 0 \tag{2.24}$$

By setting the matrix determinant to zero for a nontrivial solution as before, a $2N$th-order characteristic polynomial in λ may be obtained and the $2N$ roots determined. Alternatively, the roots may be obtained by employing a second order eigenvalue approach to solve Equation (2.24) or by transforming to a first-order (state space) eigenvalue form (illustrated later for the flutter solution in Chapter 10).

For systems where the damping is small enough for oscillatory motion to occur in each mode (i.e. each mode is underdamped), then the roots will occur in N complex conjugate pairs. These roots will be of the form $\lambda_j = -a_j + ib_j$, $\lambda_j^* = -a_j - ib_j$ for $j = 1, 2, ..., N$. For each pair of roots, there will be corresponding complex conjugate pairs of eigenvectors $\tilde{X}_j, \tilde{X}_j^*$ for $j = 1, 2, ..., N$. The solutions for the free vibration in Equation (2.23) will be governed by the decay terms $a_j \pm ib_j = -\zeta_j \omega_j \pm i\omega_{dj}$ where $\zeta_j = -a_j/\omega_j$ is the damping ratio, $\omega_{dj} = \omega_j \sqrt{1 - \zeta_j^2}$ is the damped natural frequency and $\omega_j = \sqrt{a_j^2 + b_j^2}$. This is somewhat akin to the SDoF approach used in Chapter 1.

If the damping is *proportional*, these modes that are found will be the same as the real normal modes and ω_j will be equal to the relevant undamped natural frequency. Then, following any initial condition, the free vibration response will consist of the summation of decaying oscillations in each of the normal modes.

However, if the damping is *non-proportional*, the modes will be complex and so will differ from the normal modes. Also, ω_j will not be equal to the undamped natural frequency. The response will now be a summation of the complex modes, each oscillating at the relevant damped natural frequency ω_{dj}. In a complex mode, each coordinate will have a fixed relative amplitude and phase with respect to the other coordinates (usually different to $0°$ or $180°$); this means that points will reach their maximum excursion at different instants of time and that nodal points will move about during each cycle.

Complex modes become important when aerodynamic effects are involved (e.g. flutter – see Chapter 10), when discrete dampers are present (e.g. viscoelastic components) and in particular where natural frequencies are close.

2.3.1 Example: 2DoF 'Chain-Like' System with Proportional Damping

Consider the same 2DoF 'chain-like' example as earlier, but now define the damping matrix as $\mathbf{C} = 0.001\mathbf{K}$ (implying $c_1 = 2$ Ns/m, $c_2 = 1$ Ns/m); this describes a *proportional damping* model with Rayleigh coefficients $\alpha = 0$ and $\beta = 0.001$. Thus the mass, proportional damping and stiffness matrices are

$$\mathbf{M} = \begin{bmatrix} 2 & 0 \\ 0 & 1 \end{bmatrix} kg, \quad \mathbf{C} = \begin{bmatrix} 3 & -1 \\ -1 & 1 \end{bmatrix} Ns/m \quad \mathbf{K} = \begin{bmatrix} 3000 & -1000 \\ -1000 & 1000 \end{bmatrix} N/m \quad (2.25)$$

Solution of the relevant first or second order eigenvalue problem for the damped 2DoF system gives $\lambda_1 = -0.25 \pm 22.36i$ and $\lambda_2 = -1 \pm 44.71i$, leading to damped natural frequencies of 3.559 and 7.118 Hz, with damping ratio values of 0.011 (i.e. 1.1% critical) and 0.022 for the two modes. The corresponding damped eigenvectors are found to be

$$\tilde{X}_1 = \begin{Bmatrix} 0.5 \\ 1 \end{Bmatrix} = X_1 \quad \text{and} \quad \tilde{X}_2 = \begin{Bmatrix} 1 \\ -1 \end{Bmatrix} = X_2 \quad (2.26)$$

These vectors $\tilde{X}_j$ may be termed the damped modes; for the proportional damping case they are real and are exactly the same as the undamped normal modes X_j. Thus motion in any such mode will involve each coordinate being in-phase or out-of-phase, with an invariant nodal point location and simultaneous maximum and minimum excursion of all the coordinates.

The implication of these results is that the free vibration of a proportionally damped MDoF system will consist of the summation of decaying responses in each of N normal modes with shape X_j, damped natural frequency ω_{dj}(rad/s) and damping ratio ζ_j; thus in effect the general free vibration response is expressed as the combination of modes, each behaving like a SDoF system. Each component in the summation will have an amplitude and phase that depend upon the initial conditions.

2.3.2   Example: 2DoF 'Chain-Like' System with Non-proportional Damping

Now consider the 2DoF chain-like system with the first damping value c_1 increased to 12 Ns/m; this means that the damping will be *non-proportional*, with the damping matrix

$$C = \begin{bmatrix} 13 & -1 \\ -1 & 1 \end{bmatrix} Ns/m \quad (2.27)$$

The solution of the relevant eigenvalue problem for this damped MDoF system gives complex conjugate roots $\lambda_1 = -1.08 \pm 22.38i$ and $\lambda_2 = -2.66 \pm 44.56i$, leading to damped natural frequencies of 3.562 Hz and 7.092 Hz, with damping ratio values of 0.048 and 0.060 for the two modes. Note that the damped natural frequencies still govern the free vibration decay. The damping ratio values have increased as expected. The corresponding complex conjugate eigenvectors are found to be

$$\tilde{X}_1 = \begin{Bmatrix} 0.499 \pm 0.037i \\ 1 \end{Bmatrix} \quad \text{and} \quad \tilde{X}_2 = \begin{Bmatrix} 1 \\ -0.987 \pm 0.149i \end{Bmatrix} \quad (2.28)$$

and these are clearly seen to be complex and different to the values in Equation (2.26). The vectors are termed complex (or damped) mode shapes and differ to the undamped normal modes for this non-proportionally damped case. A complex mode involves each coordinate having a fixed relative amplitude and phase with respect to the other coordinates (usually different to $0°$ or $180°$), points reaching their maximum excursion at different instants of time and nodal point locations (in continuous systems) that vary with time during a cycle of vibration. Note that the presence of complex modes will also be seen in the flutter solution (see Chapter 10).

The free vibration response will now be a summation of the decaying complex mode responses as opposed to normal mode responses for the proportionally damped system.

2.4 Transformation to Modal Coordinates

A particularly powerful feature of undamped normal mode shapes of vibration is that they may be used to transform the coupled equations of motion in physical coordinates into a different (principal/modal) coordinate form where coupling is absent. The analysis will be presented in matrix form for a general MDoF system and illustrated for the 2DoF 'chain-like' system. Damping and excitation terms are now included since the approach is applicable generally.

2.4.1 Modal Coordinates

Firstly, define a coordinate transformation based on the modal matrix and 'modal' (or 'principal') coordinates q, namely

$$x = \Phi q \tag{2.29}$$

Now substitute for x using Equation (2.29) in Equation (2.8) and pre-multiply by the transpose of the modal matrix, therefore

$$\Phi^{T}M\Phi\ddot{q} + \Phi^{T}C\Phi\dot{q} + \Phi^{T}K\Phi q = \Phi^{T}f$$
$$M_q\ddot{q} + C_q\dot{q} + K_q q = \Phi^{T}f = f_q \tag{2.30}$$

where

$$M_q = \Phi^{T}M\Phi \quad C_q = \Phi^{T}C\Phi \quad K_q = \Phi^{T}K\Phi \quad f_q = \Phi^{T}f \tag{2.31}$$

The matrices M_q, C_q, K_q are known as the modal mass, damping and stiffness matrices, and f_q is the modal force vector. It may be shown that the modal mass and modal stiffness matrices are in fact diagonal (i.e. uncoupled), with diagonal elements equal to the modal mass m_j and modal stiffness k_j for the jth mode. This diagonalization occurs because the modes of vibration are 'orthogonal' with respect to the mass and stiffness matrices (Rao, 1995); this is an extremely useful feature, as will be illustrated later. The statement of orthogonality with respect to the mass matrix, for example, may be expressed as

$$X_i^{T}MX_j = \begin{cases} 0, & i \neq j \\ m_j, & i = j \end{cases} \tag{2.32}$$

The properties of the modal damping matrix are less clear-cut. Provided that the physical damping matrix $\mathbf{C}$ can be written as a linear combination of the physical mass and stiffness matrices ($\mathbf{M}$, $\mathbf{K}$), as described above in Section 2.3, then the damping is *proportional* and the modal damping matrix $\mathbf{C}_q$ will also be diagonal. However, if the damping is *non-proportional*, then the modal damping matrix will include modal cross-coupling terms. At the initial analysis stage, it is normal to assume proportional damping so that the equations of motion expressed in 'modal space' in Equation (2.30) are in fact fully uncoupled.

The power of the modal transformation defined by Equation (2.29) may now be seen by writing out the modal equation of motion for the jth mode in Equation (2.30), with the assumption of proportional damping, so that

$$m_j \ddot{q}_j + c_j \dot{q}_j + k_j q_j = f_{qj}(t) \quad \text{for} \quad j = 1, 2, \ldots, N \tag{2.33}$$

where m_j, c_j, k_j and f_{qj} are the modal mass, damping, stiffness and force for the jth mode. Using the SDoF concepts introduced in Chapter 1, the damping ratio for each mode is then given by $\zeta_j = c_j/(2m_j\omega_j)$, where $\omega_j = \sqrt{k_j/m_j}$ is the jth mode natural frequency. The modal equation, expressed in non-dimensional form, is

$$\ddot{q}_j + 2\zeta_j\omega_j\dot{q}_j + \omega_j^2 q_j = \frac{f_{qj}(t)}{m_j} \quad \text{for} \quad j = 1, 2, \ldots, N \tag{2.34}$$

The *coupled* MDoF equations of motion originally derived in physical coordinates have now been expressed as a set of *uncoupled* SDoF equations in modal coordinates. An MDoF system may therefore now be treated as a summation of SDoF systems. All the SDoF concepts (e.g. forced response) introduced in Chapter 1 may then be applied to each modal equation and Equation (2.29) used to combine the modal results back into physical coordinates. Such a transformation will later be seen to be a fundamental part of the analysis approach used for aircraft aeroelasticity and loads calculations.

2.4.2   Example: 2DoF 'Chain-like' System with Proportional Damping

To illustrate the transformation to modal coordinates, the 2DoF 'chain-like' system introduced earlier in Section 2.1 will be used, together with proportional damping defined by $\alpha = 0$ and $\beta = 0.001$, so $c_1 = 2$ and $c_2 = 1$ Ns/m. The general physical forces f_1, f_2 will be included, without numerical values defined as yet. Firstly, the modal mass matrix defined in Equation (2.31) may be calculated as

$$\mathbf{M}_q = \mathbf{\Phi}^{\mathrm{T}} \mathbf{M} \mathbf{\Phi} = \begin{bmatrix} 0.5 & 1 \\ 1 & -1 \end{bmatrix} \begin{bmatrix} 2 & 0 \\ 0 & 1 \end{bmatrix} \begin{bmatrix} 0.5 & 1 \\ 1 & -1 \end{bmatrix} = \begin{bmatrix} 1.5 & 0 \\ 0 & 3 \end{bmatrix} \tag{2.35}$$

and it is clearly diagonal. Repeating this procedure for the modal damping and stiffness matrices yields

$$\mathbf{C}_q = \begin{bmatrix} 0.75 & 0 \\ 0 & 6 \end{bmatrix} \quad \text{and} \quad \mathbf{K}_q = \begin{bmatrix} 750 & 0 \\ 0 & 6000 \end{bmatrix} \tag{2.36}$$

Also, the modal force vector is given by

$$f_q = \mathbf{\Phi}^{\mathrm{T}} f = \begin{bmatrix} 0.5 & 1 \\ 1 & -1 \end{bmatrix} \begin{Bmatrix} f_1 \\ f_2 \end{Bmatrix} = \begin{Bmatrix} 0.5f_1 + f_2 \\ f_1 - f_2 \end{Bmatrix} \tag{2.37}$$

Thus, writing out the two modal equations separately as illustrated in Equation (2.33) yields

$$1.5\ddot{q}_1 + 0.75\dot{q}_1 + 750q_1 = 0.5f_1(t) + f_2(t)$$
$$3\ddot{q}_2 + 6\dot{q}_2 + 6000q_2 = f_1(t) - f_2(t) \tag{2.38}$$

The coupled 2DoF equations in physical coordinates have therefore been transformed into uncoupled SDoF equations in modal coordinates. Each SDoF equation has the natural frequency for the corresponding undamped normal mode, with each mode having an effective damping ratio of $\zeta_j = c_j/(2m_j\omega_j)$, so $\zeta_1 = 0.011$ and $\zeta_2 = 0.022$. These results may be seen when running the MATLAB program in the companion website.

It is clear from Equation (2.38) that values of force may be chosen to excite one or both modes; e.g. if the forces f_1, f_2 are equal then mode 1 can be excited but mode 2 will not be excited. This is the principle of multiple exciter testing sometimes used to isolate and measure modes in the aircraft Ground Vibration Test (GVT – see Chapter 25).

2.4.3   Example: 2DoF 'Chain-like' System with Non-proportional Damping

Applying the modal transformation to the non-proportional damping matrix in Equation (2.27) yields

$$C_q = \begin{bmatrix} 3.25 & 5 \\ 5 & 16 \end{bmatrix} \tag{2.39}$$

so clearly the resulting modal damping matrix is not diagonal. Thus the equations of motion in modal space become

$$1.5\ddot{q}_1 + 3.25\dot{q}_1 + 5\dot{q}_2 + 750q_1 = 0.5f_1(t) + f_2(t)$$
$$3\ddot{q}_2 + 5\dot{q}_1 + 16\dot{q}_2 + 6000q_2 = f_1(t) - f_2(t) \tag{2.40}$$

Modal damping cross-coupling terms are now present, thus implying that excitation of one mode causes a response in the other. Clearly, the equations have not been uncoupled using the classical normal mode transformation; it is only possible to generate uncoupled equations by converting to first order form (Tse *et al.*, 1978).

2.4.4   Mode Shape Normalization

Note that it is important to recognize that the values of modal mass, and therefore modal damping, stiffness and force, depend upon the normalization used in defining the modal matrix. Thus, for example, modal mass does not have a unique value and the statement 'this mode has a high modal mass' needs qualifying; it is quite meaningless unless the mode shape normalization employed when generating the modal mass is also defined. By choosing different mode shape normalizations it is possible to generate massive or minute values of modal mass. However, provided the definitions used are consistent throughout the analysis, it does not matter what normalization is used and the same final result will emerge for, say, the response to an excitation.

Common normalization approaches for mode shapes are:

• mode shape normalized so as to generate a unit modal mass (i.e. 'mass normalized' mode shape);
• mode shape normalized to a maximum value of unity; or
• mode shape normalized such that the vector norm is unity.

In the above example, the modal quantities corresponded to the mode shapes normalized to a maximum value of unity. However, if, for example, the first mode for the 'chain-like' system was to be defined by a unit modal mass, then the mode shape vector $\{0.5 \quad 1\}^T$ would need to be multiplied by $\sqrt{1/1.5}$.

2.4.5   Meaning of Modal Coordinates

At this point, it is helpful to consider the physical meaning of modal (or principal) coordinates. The coordinate q_j indicates the amount of the jth mode present in the motion. In the chain-like example, q_1 describes the in-phase motion of the two masses while q_2 describes the out-of-phase motion. Thus it is not possible to place a transducer on the system and measure a modal coordinate – it defines a characteristic 'shape' and the absolute value of the modal coordinate in any given response depends upon the mode shape normalization employed.

2.4.6   Dimensions of Modal Coordinates

2.4.6.1   Consistent Coordinates

The units of mode shapes, modal coordinates and other modal quantities are interesting to consider as they can cause confusion. Consider the 2DoF 'chain-like' system, where all the physical coordinates have the same dimensions. It is sensible to think of the mode shape vectors as being dimensionless since they have no absolute values; the consequence of this choice is that the modal mass has dimensions of mass (kg), the modal coordinates have dimensions of displacement (m) and the modal equation is then a force equation (N).

If, instead, the mode shapes were taken as having displacement units, then the modal coordinates would be dimensionless, modal mass would be in kg m^2 units and the equation would be expressed in terms of moments; this is not consistent with a classical description of the terms, so the former approach is preferable.

2.4.6.2   Mixed Coordinates

However, the position is less clear when considering an example where the mode shapes have mixed (i.e. both translational and rotational) coordinates e.g. an aircraft undergoing heave and pitch motions (see later example). In order for the modal transformation and the modal equation to be dimensionally consistent and produce a modal mass in kg and modal force in N, the mode shapes will need to be treated rather differently to the earlier consistent coordinate case. In fact, the mode shape vector needs to be considered as dimensionless in displacement but having units of rad/m in rotation (Meirovitch, 1986). This approach leads to standard modal quantities as before.

2.4.7 Model Order Reduction

A further benefit in working with modal coordinates for systems with a large number of DoF (and therefore many modes of vibration) is that it allows the number of modes included in a solution to be considerably reduced. Since the frequency range of interest is limited, it may be advantageous to reduce the scale of the analysis by only considering a subset of the modes in the modal transformation. By reducing the number of modes, the residual effect of higher frequency modes would be omitted, so it is normal to include modes with natural frequencies somewhat higher than the maximum frequency of interest to make some allowance for this effect. Consider only including n ($<N$) of the modes; therefore $\mathbf{\Phi}_n = [X_1 X_2 \; \ldots \; X_n]$ would be the reduced modal matrix and the transformation to the reduced set of principal coordinates would be

$$x = \mathbf{\Phi}_n q_n \qquad (2.41)$$

Thus the final set of transformed equations would simply be reduced to n instead of N SDoF equations. For example, a system with 200 000 (N) physical degrees of freedom from a Finite Element analysis (see Chapter 17) could be analyzed using only 20 (n) modal equations.

2.5 Two-DoF Rigid Aircraft in Heave and Pitch

As an example of a completely different 2DoF system that involves both translational and rotational coordinates, consider the rigid aircraft supported on linear landing gears as shown in Figure 2.2. The aircraft has mass m, pitch moment of inertia about the centre of mass I_y, nose and main landing gear stiffnesses K_N, K_M respectively, and viscous damping constants C_N, C_M. In order to demonstrate how excitation forces are treated, consider an arbitrary input excitation force $f(t)$ to be applied vertically downwards at the tail. The coordinates chosen to describe the aircraft motion are the centre of mass heave displacement z_C (downwards positive to be

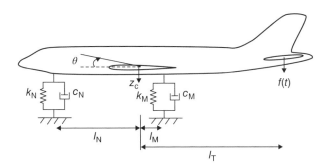

Figure 2.2   Two-DoF rigid aircraft in heave/pitch supported on landing gear.

consistent with axes systems used often later on) and pitch angle θ (nose up positive). The geometry is shown in the figure.

The energy, dissipation and work functions corresponding to Equations (2.1) to (2.4) depend upon the expressions for the extension/compression of the springs and dampers, and for small angles are given by

$$T = \frac{1}{2} m \dot{z}_C^2 + \frac{1}{2} I_y \dot{\theta}^2 \quad U = \frac{1}{2} K_N (z_C - l_N \theta)^2 + \frac{1}{2} K_M (z_C + l_M \theta)^2$$

$$\Im = \frac{1}{2} C_N (\dot{z}_C - l_N \dot{\theta})^2 + \frac{1}{2} C_M (\dot{z}_C + l_M \dot{\theta})^2 \quad \delta W = f (\delta z_C + l_T \delta \theta) \tag{2.42}$$

Then, applying Lagrange's equations with physical coordinates z_C and θ, the aircraft equations of motion are

$$\begin{bmatrix} m & 0 \\ 0 & I_y \end{bmatrix} \begin{Bmatrix} \ddot{z}_C \\ \ddot{\theta} \end{Bmatrix} + \begin{bmatrix} C_N + C_M & -l_N C_N + l_M C_M \\ -l_N C_N + l_M C_M & l_N^2 C_N + l_M^2 C_M \end{bmatrix} \begin{Bmatrix} \dot{z}_C \\ \dot{\theta} \end{Bmatrix}$$

$$+ \begin{bmatrix} K_N + K_M & -l_N K_N + l_M K_M \\ -l_N K_N + l_M K_M & l_N^2 K_N + l_M^2 K_M \end{bmatrix} \begin{Bmatrix} z_C \\ \theta \end{Bmatrix} = \begin{Bmatrix} f(t) \\ l_T f(t) \end{Bmatrix} \tag{2.43}$$

It can be seen clearly that this equation is of the same general form as that in Equation (2.8). Depending upon the parameter values, the damping and stiffness matrices will in general be coupled whereas there is no inertia coupling for this choice of coordinate system. Further analysis of this problem will be shown later in this chapter and also in Chapter 15, where the taxiing problem will be examined.

Now, in order to solve for the natural frequencies and mode shapes, assign numerical values as follows: $m = 4000$ kg, $I_y = 12\,000$ kg m^2, $l_N = 4$ m, $l_M = 1$ m, $K_N = 40\,000$ N/m and $K_M = 120\,000$ N/m. Damping and force values are set to zero for the determination of natural frequencies and mode shapes. Substituting these values into Equation (2.43) yields the mass and stiffness matrices

$$\mathbf{M} = \begin{bmatrix} 4000 & 0 \\ 0 & 12000 \end{bmatrix} \quad \text{and} \quad \mathbf{K} = \begin{bmatrix} 160000 & -40000 \\ -40000 & 760000 \end{bmatrix} \tag{2.44}$$

For free undamped vibration, the determinant $|\mathbf{K} - \omega^2 \mathbf{M}|$ must be set to zero, so

$$\begin{vmatrix} 160000 - 4000\omega^2 & -40000 \\ -40000 & 760000 - 12000\omega^2 \end{vmatrix} = 0 \tag{2.45}$$

Expanding the determinant and simplifying by dividing through by 10^6 yields a quartic polynomial which is actually a quadratic in ω^2:

$$48\omega^4 - 4960\omega^2 + 120000 = 0 \tag{2.46}$$

This equation has roots $\omega_1^2 = 38.65$ and $\omega_2^2 = 64.68$ rad/s, so the undamped natural frequencies are 0.989 and 1.280 Hz. Solving Equation (2.13) for the mode shape vectors yields

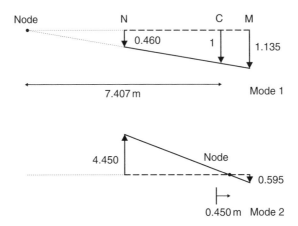

Figure 2.3 Mode shapes for the rigid aircraft example.

$$X_1 = \begin{Bmatrix} 1 \\ 0.135 \end{Bmatrix} \quad \text{and} \quad X_2 = \begin{Bmatrix} -0.405 \\ 1 \end{Bmatrix} \tag{2.47}$$

and thus the modal matrix is

$$\Phi = \begin{bmatrix} 1 & -0.405 \\ 0.135 & 1 \end{bmatrix} \tag{2.48}$$

The mode shape vectors need to be interpreted physically since the two values in the vector refer to the motion of the centre of mass (downwards positive) and the pitch angle (nose up positive). The downwards motion at the nose and main landing gear positions, for example, may be found using $z_C - l_N \theta$ and $z_C + l_M \theta$ respectively (i.e. $z_C - 4\theta$ and $z_C + \theta$); thus the values 1 and 0.135 for mode 1 imply that the corresponding nose and main gear displacements in the mode shape are 0.460 and 1.135, whereas the values –0.405 and 1 for mode 2 imply nose and main gear modal displacements of –4.405 and 0.595. These values may be shown graphically, as in Figure 2.3, though it should be noted that these shapes have unknown absolute values and only show the ratio between the deflections.

Each displaced shape is essentially a snapshot in time of the motion in the mode. Mode 1 is a motion of heave down/up and pitch nose up/down with a stationary point (or 'node') at a position 7.407 m in front of the centre of mass, whereas mode 2 is primarily a pitching motion with a node point 0.405 m behind the centre of mass. Altering the value of the nose gear stiffness to 30 000 N/m would eliminate coupling in the stiffness matrix and mean that the two modes would be pure heave and pure pitch respectively.

From this point onwards the damped vibration and transformation to modal coordinates may be considered using the same process as for the chain-like system.

2.6 'Free–Free' Systems

A 'free–free' (or semi-definite) system is one that is not connected to 'earth' via any support stiffness, i.e. it is effectively freely floating in space. An aircraft in flight is a typical example of

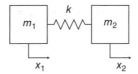

Figure 2.4   Two-DoF free–free system.

a free–free system and it is therefore important to recognize the particular features of such systems. Consider the unsupported chain-like 2DoF system shown in Figure 2.4.

The equations of motion may be shown to be given by

$$\begin{bmatrix} m_1 & 0 \\ 0 & m_2 \end{bmatrix} \begin{Bmatrix} \ddot{x}_1 \\ \ddot{x}_2 \end{Bmatrix} + \begin{bmatrix} k & -k \\ -k & k \end{bmatrix} \begin{Bmatrix} x_1 \\ x_2 \end{Bmatrix} = 0 \tag{2.49}$$

where the stiffness matrix is singular. Applying the usual method for calculating natural frequencies yields the quadratic equation in ω^2, namely

$$\omega^2 \left[m_1 m_2 \omega^2 - k(m_1 + m_2) \right] = 0 \tag{2.50}$$

The natural frequencies of the two modes are $\omega_1 = 0$ and $\omega_2 = \sqrt{k(m_1 + m_2)/(m_1 m_2)}$. The first mode shape is $\{1 \ 1\}^{\mathrm{T}}$, which is known as a 'rigid body' mode, with both masses moving together and having zero natural frequency. The second mode shape is $\{1 \ -\mu\}^{\mathrm{T}}$, where $\mu = m_1/m_2$ is the mass ratio; this is a flexible mode with the two masses moving in opposite directions in such a way that there is no net inertia force acting on the system. For these mode shape normalizations, the two modal masses are $m_1 + m_2$, the total mass, and $m_1 (1 + \mu)$ respectively. Similar features will be seen later in Chapter 3 when considering a free–free continuous system.

2.7   Harmonic Forced Vibration

The response to harmonic excitation may be determined via equations expressed in physical or modal coordinates using a similar approach to that in Chapter 1, except that matrix algebra is employed.

2.7.1   Equations in Physical Coordinates

In this section, the solution will be based upon the equations of motion expressed in physical coordinates. The similarity of the SDoF and MDoF expressions will be seen. The excitation and response are now column vectors and in complex algebra form are assumed to be

$$f(t) = F e^{i\omega t} \quad \text{and} \quad x(t) = \tilde{X} e^{i\omega t} \tag{2.51}$$

where again ~ indicates a complex quantity. The complex expressions in Equations (2.51) are now substituted into the equation of motion (2.8) and after cancelling the exponential term, the result is

UNIVERSITY
OF SHEFFIELD
LIBRARY

$$\left[-\omega^2 \mathbf{M} + i\omega \mathbf{C} + \mathbf{K}\right]\tilde{X} = F \qquad (2.52)$$

Thus the response may be solved using a matrix inverse operation so that

$$\tilde{X} = \left[-\omega^2 \mathbf{M} + i\omega \mathbf{C} + \mathbf{K}\right]^{-1} F \quad \text{or} \quad \tilde{X} = \mathbf{H}(\omega)F \qquad (2.53)$$

Here $\mathbf{H}(\omega)$ is the Frequency Response Function (FRF) matrix, given by

$$\mathbf{H}(\omega) = \left[\mathbf{K} - \omega^2 \mathbf{M} + i\omega \mathbf{C}\right]^{-1} \qquad (2.54)$$

where the matrix inversion must be carried out at every frequency considered. A typical term $H_{rs}(\omega)$ in the FRF matrix is a complex quantity representing the modulus and phase of coordinate r when a unit harmonic force is applied at coordinate s at a frequency ω. The diagonal terms $H_{rr}(\omega)$ are known as direct or driving point FRFs whereas the off-diagonal terms $H_{rs}(\omega)$, $r \neq s$, are transfer FRFs. Sample driving point and transfer FRFs are shown in Figure 2.5 for the 2DoF chain-like system considered earlier.

The direct FRF shows anti-resonance behaviour (i.e. a trough) between each pair of modal peaks, which is characteristic for all vibrating systems; the transfer FRF behaviour depends upon the number of nodal points between the excitation and response positions. The phase behaviour could also be examined using the MATLAB program in the companion website.

2.7.2   Equations in Modal Coordinates

In Section 2.4, it was shown how a transformation to modal coordinates, based on the modal matrix, could yield uncoupled SDoF equations of motion provided the damping was proportional. With such uncoupled equations, a different approach to forced vibration is possible. Basically, the response of each mode may be determined from Equation (2.33) or (2.34) using the SDoF methods introduced in Chapter 1; the results can then be transformed back into physical coordinates using Equation (2.29). However, it is also possible to write the response vector and the FRF directly from the full set of modal equations defined in

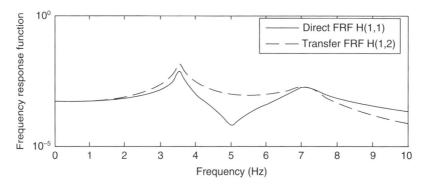

Figure 2.5   Sample driving point and transfer FRFs for the chain-like 2DoF system.

Equation (2.30) using a similar approach to that taken above in Section 2.7.1. Now, in addition to the physical force and response vectors defined in Equation (2.51) above, the modal response may be written as

$$q(t) = \tilde{Q} e^{i\omega t} \tag{2.55}$$

Applying the methodology used in Section 2.7.1 to Equation (2.30) then an equation of similar form to Equation (2.53) will result and it may be seen that

$$\tilde{Q} = \left[-\omega^2 \mathbf{M}_q + i\omega \mathbf{C}_q + \mathbf{K}_q\right]^{-1} \mathbf{\Phi}^{\mathrm{T}} F \tag{2.56}$$

Transforming back to physical coordinates using Equation (2.29) then yields

$$\tilde{X} = \mathbf{\Phi} \, \tilde{Q} = \mathbf{\Phi} \left[-\omega^2 \mathbf{M}_q + i\omega \mathbf{C}_q + \mathbf{K}_q\right]^{-1} \mathbf{\Phi}^{\mathrm{T}} F \tag{2.57}$$

so by inspection the FRF matrix in physical space is

$$\mathbf{H}(\omega) = \mathbf{\Phi} \left[\mathbf{K}_q - \omega^2 \mathbf{M}_q + i\omega \mathbf{C}_q\right]^{-1} \mathbf{\Phi}^{\mathrm{T}} \tag{2.58}$$

Note that if the damping is proportional, then the FRF matrix inverse seen in Equations (2.58) is straightforward to compute since the matrix is diagonal. The FRF matrix may then be calculated by a summation of the modal contributions and it may be shown that a typical [r, s] element of the matrix for an N DoF system is

$$H_{rs}(\omega) = \sum_{j=1}^{N} \frac{\Phi_{rj}\Phi_{sj}}{k_j - \omega^2 m_j + i\omega c_j} \tag{2.59}$$

where Φ_{rj} is the jth mode shape value at coordinate r. This expression is often used in curve fitting experimentally derived FRF data in Ground Vibration Testing (GVT – see Chapter 25). The FRF numerator shows the importance of the mode shape at the excitation and response points in determining the contribution of a particular mode to the FRF; the denominator shows how each mode contributes to the resonant peaks.

2.8   Transient/Random Forced Vibration – Time Domain Solution

In Chapter 1, the methods available for solution of the response to transient excitation for an SDoF system were discussed. In this section, the suitability of these methods for an MDoF system is considered briefly.

2.8.1   Analytical Approach

An analytical approach may still be used for the solution of the response of linear MDoF systems to transient excitation, provided the excitation has a relatively simple mathematical form such that a closed form solution is possible. In particular, when the damping is proportional, then the uncoupled SDoF modal equations may be used to solve for the response to transient excitation. For example, since the response to a step excitation is known for a linear SDoF system, then the modal response for each mode of an MDoF system to a step modal force may be determined and the results combined using the modal transformation in Equation (2.29).

2.8.2 Convolution Approach

Convolution for a linear SDoF system (see Chapter 1) may be extended to a linear MDoF system to relate the response vector $x(t)$ to the excitation vector $f(t)$ using the matrix form of the convolution equation, namely

$$x(t) = \int_{\tau=0}^{t} \mathbf{h}(t-\tau)f(\tau)\,d\tau \qquad (2.60)$$

where $\mathbf{h}(t)$ is the IRF (Impulse Response Function) matrix, the Inverse Fourier Transform of the FRF matrix. A typical term in the IRF matrix is $h_{rs}(t)$, the response of the rth coordinate due to a unit impulse at the sth coordinate. The FRF and IRF matrices can also be calculated where aerodynamic terms are present.

2.8.3 Solution of Ordinary Differential Equations

It is possible to solve the equations of motion for an MDoF system directly using a numerical integration approach as explained briefly in Chapter 1. The algorithms are adapted to handle response and excitation vectors instead of scalars. The approach is powerful and suitable for highly non-linear systems, such as would be encountered when landing gear dynamics or Flight Control Systems are present.

2.9 Transient Forced Vibration – Frequency Domain Solution

In Chapter 1, it was shown for an SDoF system that the response to a finite length general excitation input could be determined by a process based upon the Fourier Transform (FT) and multiplication in the frequency domain. A similar approach is possible for an MDoF system, and potentially for a multiple input–multiple output (MIMO) system, except that the analysis needs to be expressed in matrix form and the FT of the excitation and response vectors is involved.

2.10 Random Forced Vibration – Frequency Domain Solution

In Chapter 1, it was shown that the Power Spectral Density (PSD) of the response of an SDoF system to a random excitation input could be determined using a spectral approach, with the response and excitation PSDs being related via the modulus squared value of the FRF. In the case of an MDoF system, multiple independent random sources may be applied simultaneously and a matrix spectral relationship would then be developed; however, in this book, such a case is not required. Since turbulence acts as a single excitation source, each response can be treated separately using the relevant MDoF FRF. Thus the PSD relationship is

$$S_{x_r x_r}(\omega) = |H_{ra}(\omega)|^2 S_{aa}(\omega), \quad r = 1, 2, \ldots, N \tag{2.61}$$

where the rth response term is being considered, $H_{ra}(\omega)$ is the FRF relating the rth response to the source and $S_{aa}(\omega)$ is the source PSD. It may be seen that all the modes will be included in the response via the FRF.

2.11 Examples

Note that these examples may be useful preparation when carrying out the examples in later chapters.

1. For the 2DoF lumped parameter system shown in Figure 2.6, determine (a) the equations of motion in matrix form, (b) the undamped natural frequencies and mode shapes, (c) the modal masses, (d) the modal dampings and (e) the modal stiffnesses. Write the uncoupled equations in modal space. Repeat the calculations using MATLAB (see companion website).

   [(a) $\mathbf{M} = \mathrm{diag}\,[m\ m]$, $\mathbf{K} = [2\,k\,-k;\,-k\,2\,k]$, (b) $\omega_1 = \sqrt{k/m}$, $\omega_2 = \sqrt{3k/m}$ and $\{1\ 1\}$, $\{1\ -1\}$, (c) $2\,m$, $2\,m$, (d) $2c$, $6c$, (e) $2\,k$, $6\,k$]

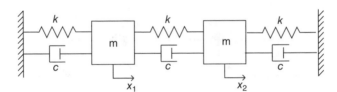

Figure 2.6

2. For the aircraft plus landing gear example investigated in this chapter, obtain the equations of motion written in terms of the vertical displacements z_N, z_M at the nose and main gears instead of using z_C, θ. (Note that the pitch rotation angle will need to be written in terms of the displacements in order to obtain the kinetic energy of rotation.) For the parameter values used in Section 2.5, obtain the undamped natural frequencies and mode shapes and show that they are the same as those determined earlier. Determine the modal masses. Repeat the calculations using MATLAB.

   $\left[\mathbf{M} = \left[\left(I_y + ml_M^2 \right),\ \left(ml_N l_M - I_y \right);\ \left(ml_N l_M - I_y \right),\ \left(I_y + ml_N^2 \right) \right] / l^2,\ l = l_N + l_M,\right.$

   $\mathbf{K} = [K_N, 0; 0, K_M], 0.989$ and $1.280\,\mathrm{Hz}$, $\{0.405\ 1\}$, $\{1 - 0.135\}$, $3275\,\mathrm{kg}, 652\,\mathrm{kg}]$

3. An aerofoil section has a mass m and moment of inertia I_O about the point O, where it is supported in heave and pitch by a linear spring k and a rotational spring K respectively,

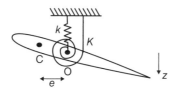

Figure 2.7

as shown in Figure 2.7. The centre of mass C is a distance e ahead of O. Determine the equations of motion for two coordinate sets (a) z_C, θ and (b) z_O, θ, where z_C, z_O are measured downwards from points C and O respectively and θ is the nose up pitch angle. Note the different types of coupling term in the equations of motion.

[(a) stiffness coupled $\mathbf{M} = [m, 0; 0, (I_O - me^2)]$, $\mathbf{K} = [k, ke; ke, (K + ke^2)]$ and

(b) inertia coupled $\mathbf{M} = [m, -me; -me, I_O + me^2]$, $\mathbf{K} = [k, 0; 0, K]$]

4. The system shown below in Figure 2.8 consists of two masses ($m = 1$ kg) mounted on a rigid member of length $3a$ ($a = 1$ m), supported by springs ($k = 1000$ N/m) and dampers ($c = 2$ N s/m). Determine (a) the equations of motion in matrix form, (b) the undamped natural frequencies and mode shapes, (c) the modal masses and (d) the modal damping ratios. Note that expressions need to be determined for the compressions of the springs in terms of $z_{1, 2}$.

[$\mathbf{M} = \mathrm{diag}[2, 0; 0, 1]$ and $\mathbf{K} = [5000, -4000; -4000, 5000]$, mode 1 : 4.08 Hz, $\{1, 0.921\}$ 2.848 kg, 0.0256 and mode 2 : 13.17 Hz, $\{1 - 2.171\}$, 6.713 kg, 0.0826]

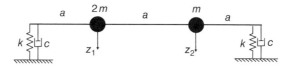

Figure 2.8

5. For the system in Example 4, write down the modal equations. Then, ignoring damping and using a superposition of the modal responses, obtain an expression for the response of the system at point 1 due to (a) a 100 N step input and (b) a sine input of 100 N at a frequency of 5 Hz (close to the resonance of mode 1), both applied at point 2. Single DoF results from Chapter 1 may be helpful.

[(a) $z_1(t) = 0.0443 - 0.0492 \cos 25.65t + 0.0047 \cos 82.72t$ and (b) $z_1(t) = -0.092 \sin(31.4t)$]

6. For the system in Example 4 without damping, determine expressions for the FRFs H_{12}, H_{22} using both the (a) physical and (b) modal coordinate models. Sketch the amplitudes of these functions against frequency.

7. Describe how the analyses in Examples 5 and 6 would change for the case where modal damping is present in each mode. Sketch the changes in the responses and FRFs.

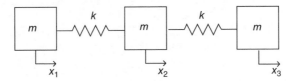

Figure 2.9

8. For the three-DoF free–free system shown in Figure 2.9, with $m = 100$ kg and $k = 10\,000$ N/m, determine expressions for the undamped natural frequencies, mode shapes and modal masses.

$[0, 1.591 \text{ and } 2.757\,\text{Hz}, \{1\ 1\ 1\}, \{-1\ 0\ 1\}, \{-0.5\ 1 - 0.5\} \text{ and } 300, 200 \text{ and } 150\,\text{kg}]$

3

Vibration of Continuous Systems – Assumed Shapes Approach

3.1 Continuous Systems

In Chapters 1 and 2, some basic concepts were introduced for single and multiple DoF 'discrete parameter' systems, where motion was defined via discrete displacement or rotation coordinates. However, for most problems encountered in aircraft aeroelasticity and loads, the systems are 'continuous', involving mass and stiffness properties distributed spatially over all or part of the system. An aircraft wing, tailplane or fuselage may be considered as elastic continuum members able to bend and twist, but these require a different analysis approach to that employed in Chapter 2.

The mode shapes for discrete parameter systems are defined by a finite length vector. However, mode shapes for continuous systems are continuously varying functions and strictly require an infinite number of values to define them spatially. As an example of mode shapes for a simple continuous system in 2D (Blevins, 2001), consider a slender member (i) clamped at one end and free at the other (i.e. clamped-free) and (ii) entirely unsupported (i.e. free-free). Figure 3.1(i) shows the first 3 mode shapes (all flexible) for the clamped-free member; they may be seen to satisfy the boundary conditions at the clamped end. Figure 3.1(ii) shows 2 rigid body modes and the first 3 flexible modes for the free-free member. The form of any higher frequency modes of interest may be deduced since the number of nodal points (see Chapter 2) will increase with the mode number.

Note that the description 'slender member' is used here in place of the many terms often used to describe what are essentially similar members that experience different types of loading, namely beams, shafts, bars and rods; it makes sense to unify them. The term 'slender' implies that their length is significantly greater than their cross-section dimensions.

3.2 Modelling Continuous Systems

There are several ways of modelling 'continuous' systems, namely:

a. an exact approach using the partial differential equations of the system to determine exact natural frequencies and mode shapes,

Introduction to Aircraft Aeroelasticity and Loads, Second Edition. Jan R. Wright and Jonathan E. Cooper.
© 2015 John Wiley & Sons, Ltd. Published 2015 by John Wiley & Sons, Ltd.

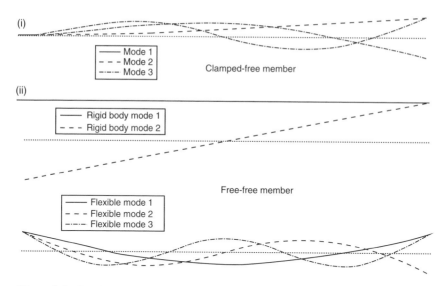

Figure 3.1   Mode shapes for (i) clamped-free and (ii) free-free slender members in 2D.

b. an approximate approach using some form of spatial 'discretization' (Finite Elements) or
c. an approximate approach using a series of assumed shapes to represent the deformation
   (Rayleigh–Ritz).

The exact approach (Tse *et al.*, 1978; Rao, 1995) is satisfactory for simple systems such as uniform slender members under bending, torsional or axial deformation, but is impractical for 'real' systems with complex stiffness and mass distributions. Mode shapes and natural frequencies for a variety of relatively simple continuous systems may be found in Blevins (2001).

The approximate method involving spatial discretization is the well-known Finite Element (FE) method (Cook *et al.*, 1989) in which the system is sub-divided into a number of elements joined at discrete points known as nodes. The element stiffness and mass matrices, expressed in terms of nodal coordinates, are obtained and assembled for the entire system. This method, which is that used in practice for aircraft aeroelasticity and loads calculations, will be considered later in Chapter 17 in Part III of the book.

However, in this chapter and in Part II of the book, the Rayleigh–Ritz approach (Tse *et al.*, 1978; Rao, 1995) for modelling a system using a series of assumed shapes will be introduced as a way of representing continuous systems with relatively simple geometries (e.g. uniform cantilever wings). Also, special cases where the assumed shapes are in fact normal modes of the whole aircraft, or 'branch modes' of parts of the aircraft, will be considered.

All these approaches can yield models of the system expressed in terms of so-called 'generalized', and not physical, coordinates; the generalized coordinates define the amount of a particular shape (e.g. assumed or normal mode shape) present in the motion. Such models will be used to demonstrate the basic concepts of aircraft aeroelasticity and loads in Part II of the book, since by doing so the number and complexity of equations will be minimized to avoid obscuring the underlying principles. It will then be seen that the MDoF methods introduced in Chapter 2 may be employed to determine the vibration characteristics of a set of simultaneous ordinary differential equations in the unknown generalized coordinates. It should be pointed out that a particular advantage of the Rayleigh–Ritz approach is that the resulting model can be of small order and can show explicitly the influence of the system physical parameters (e.g. dimensions, moduli, aerodynamic coefficients).

Before progressing with the Rayleigh–Ritz analysis, the difficult issue of which appropriate symbols to use in expressing system deformation will be discussed. It is normal in stress/ structural analysis textbooks to consider the geometric position coordinate using the symbol z and the bending deformation using the symbol w. However, later in this book it will be seen that the symbol w is used in the flight mechanics model to denote downwards velocity and the symbol w_g denotes gust velocity. Also, in aeroelastic calculations, the symbol z is often used to describe the downwards deformation of the aircraft. It has therefore been decided to use a notation throughout that is more consistent with the aircraft usage. Therefore in this chapter the bending deformation will be denoted using the symbol z; the context of a particular analysis should clarify what is being considered in any given case. For convenience, when comparing the Rayleigh–Ritz treatments given here with that in other books, the bending displacement will be considered as positive upwards. Later in Part II, when aeroelastic and loads models are considered, a positive downwards displacement will be used.

3.3   Elastic and Flexural Axes

In this chapter and in Part II of the book, the wing, tailplane and fuselage are mainly treated as uniform, straight and thin-walled slender members. In order to carry out a dynamic analysis involving such members, it is helpful to understand the so-called *shear* and *flexural centres* and the corresponding *axes*. These terms have often led to confusion and will be explained here with reference to a wing.

The *shear centre* is defined as the point on a particular 2D wing cross-section at which a transverse shear load must be applied in order to produce *zero rate of twist* (i.e. zero twist per unit span) *at that section*; it is important to note that it is a section property. On the other hand, the *flexural centre* is the point on a particular wing cross-section at which a transverse shear load must be applied in order to produce *zero twist* of that section *relative to the wing root*.

The *elastic axis* is then defined as the locus of shear centres along the wing, whilst the *flexural axis* (or line) is the locus of flexural centres, and so is a wing property. Classically, the elastic axis has been used in textbooks (Fung, 1969; Scanlan and Rosenbaum, 1960; Megson, 1999; Bisplinghoff *et al.*, 1996; Hodges and Pierce, 2011).

For the simple case of a uniform unswept homogeneous wing, these axes will be straight and the elastic and flexural axes will coincide, whereas for a more realistic wing where sweep, taper, warping and cut-outs etc. are present, then the two axes will be different and sometimes difficult to define. Analyses of such wings will nowadays be carried out using the Finite Element method (see Part III). Arguably, the flexural axis is more relevant than the elastic axis for the more complex cases and can help in understanding aeroelastic phenomena.

However, for the simple Rayleigh–Ritz analyses carried out in this book, it is reasonable to consider the wing as being uniform, homogeneous and unswept; in such a case, in order to be consistent with the literature, the elastic axis will be used. The bending and torsion deformation of the wing can then be defined about the elastic axis and the elastic potential (or strain) energies evaluated in terms of the flexural rigidity EI and torsional rigidity GJ, both defined about the elastic axis. Thus the use of the elastic axis permits the static decoupling of bending and torsion. However, if the mass axis does not coincide with the elastic axis then dynamic coupling will occur through inertia terms.

3.4 Rayleigh–Ritz 'Assumed Shapes' Method

The Rayleigh–Ritz approach is used to represent the deformation of the system by a finite series of known assumed deformation shapes, each multiplied by an unknown coefficient. The method was introduced when a practical approximate methodology was required in the absence of computers. It is an extension of Rayleigh's method (where only a single shape is employed).

3.4.1 One-dimensional Systems

For a system where the deformation varies in only one dimension, the bending deformation $z(y, t)$ with reference to the elastic axis (see earlier remarks on notation) can be represented by the series

$$z(y,t) = \sum_{j=1}^{N} \psi_j(y)\, q_j(t) \qquad (3.1)$$

where $\psi_j(y)$ is the jth assumed deformation shape (a function of y), $q_j(t)$ is the jth unknown coefficient (the 'generalized coordinate'), which is a function of time, and N is the number of terms in the series. The idea is that this combination of shapes represents the true deformation of the system as closely as possible, as shown in Figure 3.2 for $N = 2$. The more shapes used, the more accurate will be the approximation. Also, the degree to which the shapes satisfy the boundary conditions is important (see later). If the assumed shapes are identical to the undamped normal mode shapes, then the generalized coordinate $q_j(t)$ is equivalent to the modal coordinate introduced in Chapter 2. The principle of assumed shapes is somewhat akin to using a Fourier series to represent a time signal by the summation of a series of sinusoids of different amplitude and phase.

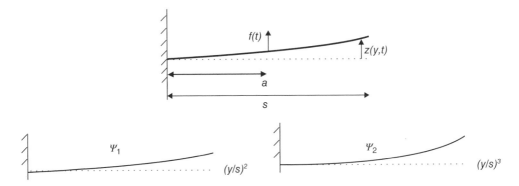

Figure 3.2 Deformation of a slender clamped-free member in bending (N = 2).

$$z(y,t) = \psi_1(y)\, q_1(t) + \psi_2(y)\, q_2(t)$$

3.4.2   Two-dimensional Systems

Where the deformation varies in two dimensions (e.g. a thin flat plate that can both bend and twist), the assumed shapes may be a product of functions that vary along each dimension, for example $\phi(x)\psi(y)$, or else may be a combined shape accounting for deformation in both coordinates simultaneously, for example $\chi(x, y)$.

3.4.3   Choice of Assumed Shapes

The assumed shapes are classically of polynomial, trigonometric or hyperbolic form. It is essential that each shape satisfies the geometric (or 'kinematic') boundary conditions of the system. For a member under bending there is no transverse deformation at a simple or built-in support and also no rotation at a built-in support. For a member under torsion, there is zero twist at a built-in end.

However, the accuracy of the representation will be improved if the shapes also satisfy the load (or 'natural') boundary conditions. For a member under bending, the load boundary conditions at a free end are zero bending moment and shear force (i.e. zero curvature and zero rate of change of curvature). For a member under torsion, the load boundary condition is a zero torque (i.e. zero rate of twist) at a free end.

A better choice of shapes means that fewer terms will be required for the same accuracy. However, whereas the kinematic boundary conditions are relatively simple to satisfy, it is much more difficult to satisfy the load (or natural) boundary conditions.

The choice of shapes is better served by example than by attempting to generalize further. However, because the aim in this book is simply to generate systems with a small number of equations that may be used to demonstrate aircraft aeroelasticity and load concepts, only simple polynomial assumed shapes will be employed. It is recognized that the results will be less accurate than if more terms in the series, and better shapes, were to be used but the principles are unaffected and the results are of a simpler form.

Having defined the assumed shapes, an energy principle is employed to minimize the error in the approximation and so generate equations in the unknown generalized coordinates; Lagrange's equations will be used again.

3.4.4   Normal Modes for a Continuous System

When a continuous system is considered, then theoretically there are a near-infinite number of normal modes, with each mode defined by a continuous mode shape and having its own natural frequency, damping ratio and modal mass. When using a finite number N of assumed shapes, then the analysis will yield estimates of N normal modes, with the accuracy being superior for the lower frequency modes.

3.5   Generalized Equations of Motion – Basic Approach

In this section, an analysis will be performed for a uniform built-in member (or 'wing') under bending or torsional vibration, with one or two simple polynomial terms used in the series. Results will be compared to those from an exact analysis. Later, the use of matrix algebra to set up the equations will be shown.

3.5.1   Clamped–Free Member in Bending – Single Assumed Shape

The member shown in Figure 3.2 is clamped at the root and has length s, mass per length μ, material Young's modulus E and relevant section second moment of area for vertical bending I (sometimes incorrectly termed the second moment of inertia); the product EI is known as the flexural rigidity. A force $f(t)$ is applied at position $y = a$ as shown. No damping is included though an estimate could be added in subsequently. Firstly, only one term in the series will be used and the polynomial will be a simple quadratic function, namely

$$z(y,t) = \psi(y)\, q(t) = \left(\frac{y}{s}\right)^2 q(t) \tag{3.2}$$

This function is shown in Figure 3.2 and satisfies the requirement for zero displacement and zero slope/rotation at the built-in end (i.e. the kinematic boundary condition) but not the load boundary condition at the free end (because the curvature there is finite, not zero). In this case, it may be seen that the assumed shape has been chosen to be dimensionless (and normalized to a maximum value of unity), so the generalized coordinate has dimensions of displacement. The shape is an approximation to the exact first mode in Figure 3.1(i).

The use of Lagrange's equations, as described in Chapters 1 and 2, requires various energy and work terms to be determined for discrete systems, but in this continuous case the quantities need to be found by integration over the member. The kinetic energy dT for an element of length dy and mass $\mu\, dy$ is

$$dT = \frac{1}{2}(\mu\, dy)\dot{z}^2 \tag{3.3}$$

and the total kinetic energy T is calculated by summing up (i.e. integrating) the elemental energies over the length of the member; thus

$$T = \frac{1}{2}\int_0^s \mu \dot{z}^2 \, dy \tag{3.4}$$

Substituting the expression for $z(y, t)$ from Equation (3.2) into Equation (3.4) and performing the integration yields

$$T = \frac{1}{2}\int_0^s \mu \left[\left(\frac{y}{s}\right)^2 \dot{q}\right]^2 dy = \frac{\mu s}{10}\dot{q}^2 \tag{3.5}$$

The elastic potential (or strain) energy in bending depends upon the curvature and flexural rigidity (Benham et al., 1996) and is

$$U = \frac{1}{2}\int_0^s EI \left(\frac{\partial^2 z}{\partial y^2}\right)^2 dy \tag{3.6}$$

Thus, substituting the expression for $z(y, t)$ from Equation (3.2) into Equation (3.6) and integrating yields the elastic potential energy

$$U = \frac{1}{2}\int_0^s EI \left(\frac{2}{s^2}q\right)^2 dy = \frac{2EI}{s^3}q^2 \tag{3.7}$$

Finally, the work done by the applied force moving through an incremental displacement δz at $y = a$ will be

$$\delta W = f(t)\,\delta z(a,t) = f(t)\left(\frac{a}{s}\right)^2 \delta q \qquad (3.8)$$

where it should be noted that the incremental physical displacement $\delta z(a, t)$ may be expressed in terms of an increment δq in the generalized coordinate. The effectiveness of the force depends upon the value of the assumed shape at the point of application; e.g. applying a force at a nodal point will have no effect.

Taking Lagrange's equation (see Chapters 1 and 2), rewritten in terms of generalized coordinates q_j, gives

$$\frac{d}{dt}\left(\frac{\partial T}{\partial \dot{q}_j}\right) - \frac{\partial T}{\partial q_j} + \frac{\partial \mathfrak{I}}{\partial \dot{q}_j} + \frac{\partial U}{\partial q_j} = Q_j = \frac{\partial(\delta W)}{\partial(\delta q_j)} \quad \text{for } j = 1, 2, \dots, N \qquad (3.9)$$

and substituting the energy and work expressions for the single generalized coordinate $q(t)$ ($N = 1$) yields

$$\frac{\mu s}{5}\ddot{q} + \frac{4EI}{s^3}q = \left(\frac{a}{s}\right)^2 f(t) \qquad (3.10)$$

By inspection of this SDoF differential equation (see Chapter 1), the Rayleigh–Ritz estimate of the undamped natural frequency is given by considering just the left hand side of Equation (3.10) and thus

$$\omega_{1_{RR}} = 4.47\sqrt{\frac{EI}{\mu s^4}} \qquad (3.11)$$

This is an overestimate by 27% on the exact value of $\omega_{1_{Exact}} = 3.516\sqrt{EI/(\mu s^4)}$ (Blevins, 2001), the difference occurring because the member is effectively constrained (or forced) into the assumed shape and so the mathematical model is over-stiff. Note that this assumed shape has led to an estimate of the lowest, or fundamental, natural frequency. One reason that this estimate is so much in error is that relatively small errors in the assumed shape can make a significant difference when differentiated twice within the elastic potential energy expression. Note also that the resulting model is an analytical expression that shows clearly the influence of the system parameters.

For a slender member problem, it can be shown (Thompson, 1997) that, by evaluating the elastic potential energy via the bending moment that corresponds to the distributed inertia loading, the errors are much smaller. However, the standard approach adopted here is simpler to apply and using sufficient shapes will yield adequate results.

3.5.2   Clamped–Free Member in Bending – Two Assumed Shapes

To show how the analysis changes and the accuracy improves when more than a single shape is used, consider the expression for displacement of the clamped-free member in bending with two assumed shapes, given by

$$z(y,t) = \psi_1(y)\,q_1(t) + \psi_2(y)\,q_2(t) = \left(\frac{y}{s}\right)^2 q_1(t) + \left(\frac{y}{s}\right)^3 q_2(t) \qquad (3.12)$$

The second shape is now a cubic polynomial which also satisfies the kinematic boundary condition, as shown in Figure 3.2, but not the load condition. The energy and work done terms may be determined in a similar way to that in Section 3.5.1 above, but now there are two terms in the series. Thus the kinetic and elastic potential energy expressions are

$$ T = \frac{1}{2} \int_0^s \mu \left[\left(\frac{y}{s} \right)^2 \dot{q}_1 + \left(\frac{y}{s} \right)^3 \dot{q}_2 \right]^2 dy = \frac{\mu s}{10} \dot{q}_1^2 + \frac{\mu s}{14} \dot{q}_2^2 + \frac{\mu s}{6} \dot{q}_1 \dot{q}_2 $$

and

$$ U = \frac{1}{2} \int_0^s EI \left(\frac{2}{s^2} q_1 + \frac{6y}{s^3} q_2 \right)^2 dy = \frac{2EI}{s^3} q_1^2 + \frac{6EI}{s^3} q_2^2 + \frac{6EI}{s^3} q_1 q_2 \tag{3.13} $$

The work done term is

$$ \delta W = f(t)\, \delta z(a,t) = f(t) \left[\left(\frac{a}{s} \right)^2 \delta q_1 + \left(\frac{a}{s} \right)^3 \delta q_2 \right] \tag{3.14} $$

where now increments in both generalized coordinates are required.

Finally, applying Lagrange's equations for the generalized coordinates q_1 and q_2 ($N = 2$) yields the simultaneous differential equations of motion

$$ \frac{\mu s}{5} \ddot{q}_1 + \frac{\mu s}{6} \ddot{q}_2 + \frac{4EI}{s^3} q_1 + \frac{6EI}{s^3} q_2 = \left(\frac{a}{s} \right)^2 f(t) $$

$$ \frac{\mu s}{6} \ddot{q}_1 + \frac{\mu s}{7} \ddot{q}_2 + \frac{6EI}{s^3} q_1 + \frac{12EI}{s^3} q_2 = \left(\frac{a}{s} \right)^3 f(t) \tag{3.15} $$

Note that it is possible to perform the differentiations required by Lagrange's equations prior to carrying out the integrals for the kinetic and elastic potential energies; this would reduce the amount of integration involved. This idea will be used later when the matrix approach is introduced. Equations (3.15) may be rewritten in matrix form as

$$ \begin{bmatrix} \dfrac{\mu s}{5} & \dfrac{\mu s}{6} \\ \dfrac{\mu s}{6} & \dfrac{\mu s}{7} \end{bmatrix} \begin{Bmatrix} \ddot{q}_1 \\ \ddot{q}_2 \end{Bmatrix} + \begin{bmatrix} \dfrac{4EI}{s^3} & \dfrac{6EI}{s^3} \\ \dfrac{6EI}{s^3} & \dfrac{12EI}{s^3} \end{bmatrix} \begin{Bmatrix} q_1 \\ q_2 \end{Bmatrix} = \begin{Bmatrix} \left(\dfrac{a}{s} \right)^2 \\ \left(\dfrac{a}{s} \right)^3 \end{Bmatrix} f(t) \tag{3.16} $$

The equations are in the classical MDoF form shown in Chapter 2, with 'mass' and 'stiffness' matrices and a 'force' vector. However, in this case both the matrices are coupled; this would not be the case if the assumed shapes were identical to the normal mode shapes because of their orthogonality. Also, the 'mass' matrix, for example, is not a classical mass matrix since the generalized coordinates are multipliers of assumed shapes and not physical coordinates.

Using the approach introduced in Chapter 2 for determining the natural frequencies and mode shapes of an MDoF system, the natural frequencies predicted using this approximate method based on two simple shapes are

$$\omega_{1_{RR}} = 3.533 \sqrt{\frac{EI}{\mu s^4}} \quad \text{and} \quad \omega_{2_{RR}} = 34.81 \sqrt{\frac{EI}{\mu s^4}} \tag{3.17}$$

The frequency values in Equation (3.17) may be compared to the exact values of

$$\omega_{1_{Exact}} = 3.516 \sqrt{\frac{EI}{\mu s^4}} \quad \text{and} \quad \omega_{2_{Exact}} = 22.03 \sqrt{\frac{EI}{\mu s^4}} \tag{3.18}$$

The first natural frequency is now predicted much more accurately (only 0.5% overestimated) because the combination of shapes approximate the true first mode shape more accurately; however, the second natural frequency is overestimated by 58%. To improve this latter estimate would require further or better shapes; e.g. including a quartic shape yields the second natural frequency to within 1% error.

The eigenvalue approach introduced in Chapter 2, when applied to Equation (3.16), yields mode shape vectors in terms of the generalized coordinates $\{q_1 \; q_2\}^{\mathrm{T}}$. These may be converted into continuous mode shapes by using them to factor the assumed shapes. Note that the mode shapes for the first two bending modes of a clamped-free member estimated using the Rayleigh–Ritz approach are almost indistinguishable by eye from the exact modes in Figure 3.1(i).

It can be seen that to obtain an estimate for the natural frequencies of a tapered built-in member would be fairly straightforward since all that would be required would be to include the mass per length μ and the flexural rigidity EI as functions of y/s, with the same assumed shapes being used. If the mass and/or stiffness properties varied along the member in a piecewise manner, this could be handled by piecewise integration.

To obtain equivalent exact results using the partial differential equation approach for either of these non-uniform scenarios would be much more difficult.

3.5.3   Clamped–Free Member in Torsion – One Assumed Shape

A related problem to bending is that of torsion, since commercial aircraft have high aspect ratio wings that have often been treated as slender members (or 'sticks') under combined bending and torsion. Later in the book, the importance of wing twist will become apparent, so an introduction to torsional vibration analysis using the Rayleigh–Ritz approach is appropriate.

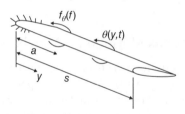

Figure 3.3   Slender clamped-free member in torsion.

The uniform member in Figure 3.3, clamped at one end and free at the other, now has a moment of inertia in twist per unit length of χ and a torsional rigidity GJ where G is the material shear modulus. J is the section torsion constant, which is not equal to the polar second moment of area, as is sometimes incorrectly stated, except for the special case of a circular section. Here, the assumed shapes describe the twist θ (y, t) about the elastic axis, so for a single assumed shape then typically

$$\theta(y,t) = \gamma(y)q(t) = \left(\frac{y}{s}\right)q \tag{3.19}$$

where a linear twist shape is assumed and q is effectively the tip twist. This shape satisfies the kinematic condition of zero twist at the root but not the load condition at the tip (zero torque and rate of twist $d\theta/dy$).

Given that the moment of inertia in torsion of an element dy is $\chi\,dy$, then the torsional kinetic energy is given by

$$T = \frac{1}{2}\int_0^s \chi\dot{\theta}^2\,dy = \frac{1}{2}\int_0^s \chi\left(\frac{y}{s}\dot{q}\right)^2 dy = \frac{\chi s}{6}\dot{q}^2 \tag{3.20}$$

and the elastic potential energy is (Benham et al., 1996)

$$U = \frac{1}{2}\int_0^s GJ\left(\frac{\partial\theta}{\partial y}\right)^2 dy = \frac{1}{2}\int_0^s GJ\left(\frac{1}{s}q\right)^2 dy = \frac{GJ}{2s}q^2 \tag{3.21}$$

Finally, a torque of value $f_\theta(t)$ is to be applied at position $y = a$, so the incremental work done is

$$\delta W = f_\theta(t)\,\delta\theta(a) = f_\theta(t)\frac{a}{s}\delta q \tag{3.22}$$

When Lagrange's equations are used, the differential equation of motion becomes

$$\frac{\chi s}{3}\ddot{q} + \frac{GJ}{s}q = \frac{a}{s}f_\theta(t) \tag{3.23}$$

The estimated natural frequency is therefore $\omega_{1_{RR}} = 1.732\sqrt{GJ/(\chi s^2)}$ and this value is a 10% overestimate compared to the exact value of $\omega_{1_{Exact}} = 1.571\sqrt{GJ/(\chi s^2)}$. Clearly, the application to more than one assumed shape is a straightforward extension and will improve the accuracy. Also, the analysis of a combined bending/torsion problem is possible, using both bending and torsion shapes, since the energy terms are scalar and therefore additive.

3.6 Generalized Equations of Motion – Matrix Approach

Having seen how the process works for more than one assumed shape, it is possible to approach the problem in a general form using matrix algebra. This approach is particularly useful when using a large number of shapes or considering exact modal representations (see later). The idea will be illustrated for the slender member in bending with two assumed shapes, considered earlier in Section 3.5.2.

3.6.1   Representation of Deformation

Firstly, the assumed series expression must be written in matrix form, so the bending displacement is

$$z(y,t) = \sum_{j=1}^{N} \psi_j(y)\, q_j(t) = \mathbf{\psi}^T \mathbf{q} \ \text{ or } \ \mathbf{q}^T \mathbf{\psi}$$

where   (3.24)

$$\mathbf{\psi}(y) = \{ \psi_1(y)\psi_2(y) \cdots \psi_N(y) \}^T \qquad \mathbf{q}(t) = \{ q_1(t)q_2(t) \cdots q_N(t) \}^T$$

Note that because $\mathbf{\psi}$ and $\mathbf{q}$ are column vectors and $z(y, t)$ is a scalar, the inner product of these vectors may be written in either order, as shown in Equation (3.22). In the above $N = 2$ example, these vectors will be given by

$$\mathbf{\psi} = \left\{ \begin{array}{c} (y/s)^2 \\ (y/s)^3 \end{array} \right\} \qquad \mathbf{q} = \left\{ \begin{array}{c} q_1 \\ q_2 \end{array} \right\}$$ (3.25)

3.6.2   Kinetic Energy

The kinetic energy may now be written in matrix form as

$$T = \frac{1}{2} \int_0^s \mu \dot{z}^2 \mathrm{d}y = \frac{1}{2} \int_0^s \mu \left(\dot{\mathbf{q}}^T \mathbf{\psi} \right) \left(\mathbf{\psi}^T \dot{\mathbf{q}} \right) \mathrm{d}y$$

$$= \frac{1}{2} \dot{\mathbf{q}}^T \left[\int_0^s \mu \left(\mathbf{\psi}\mathbf{\psi}^T \right) \mathrm{d}y \right] \dot{\mathbf{q}} = \frac{1}{2} \dot{\mathbf{q}}^T \mathbf{M}_q \dot{\mathbf{q}}$$ (3.26)

where the order of the vector products is chosen so that the vectors of generalized coordinates may be taken outside the integral since they are not functions of y. Note that the expression for T is a quadratic form involving the generalized mass matrix $\mathbf{M}_q$. In the two-shape notation, the kinetic energy would be given by

$$T = \frac{1}{2} \{ \dot{q}_1 \ \dot{q}_2 \} \begin{bmatrix} \dfrac{\mu s}{5} & \dfrac{\mu s}{6} \\[2mm] \dfrac{\mu s}{6} & \dfrac{\mu s}{7} \end{bmatrix} \left\{ \begin{array}{c} \dot{q}_1 \\ \dot{q}_2 \end{array} \right\}$$ (3.27)

3.6.3   Elastic Potential Energy

The elastic potential energy may be written in a similar way, involving the generalized stiffness matrix $\mathbf{K}_q$, as

$$U = \frac{1}{2}\int_0^s EI\left(\frac{\partial^2 z}{\partial y^2}\right)^2 dy = \frac{1}{2}\int_0^s EI\left(q^\mathrm{T}\psi''\right)\left(\psi''^\mathrm{T}q\right) dy$$

$$= \frac{1}{2}q^\mathrm{T}\left[\int_0^s EI\left(\psi''\psi''^\mathrm{T}\right) dy\right]q = \frac{1}{2}q^\mathrm{T}\mathbf{K}_q q \tag{3.28}$$

where the dash notation $'$ indicates partial differentiation with respect to y.

3.6.4 Incremental Work Done

The incremental work done for force $f(t)$ applied at position $y = a$ is expressed as a vector inner product

$$\delta W = f(t)\,\delta z(a,t) = \left[\delta q^\mathrm{T}\psi(a)\right]f(t) \tag{3.29}$$

where $\psi(a)$ is the assumed shape vector at $y = a$. For the two assumed shape example, $\psi(a) = \{(a/s)^2\ (a/s)^3\}^\mathrm{T}$.

3.6.5 Differentiation of Lagrange's Equations in Matrix Form

Lagrange's equations may also be expressed in matrix form, namely

$$\frac{\mathrm{d}}{\mathrm{d}t}\left(\frac{\partial T}{\partial \dot{q}}\right) - \frac{\partial T}{\partial q} + \frac{\partial \Im}{\partial \dot{q}} + \frac{\partial U}{\partial q} = Q = \frac{\partial(\delta W)}{\partial(\delta \dot{q})} \tag{3.30}$$

and since energy and work terms have been expressed in matrix form, an efficient approach is to use matrix differentiation rules (Graupe, 1972). Energy terms E (scalar) may be expressed in the quadratic form, namely $E = x^\mathrm{T}\mathbf{A}x$, and if $\mathbf{A}$ is symmetric then $dE/dx = 2\mathbf{A}x$. Work terms W (scalar) are in the inner product form $W = a^\mathrm{T}x$ or $x^\mathrm{T}a$ and then $dW/dx = a$. These expressions may be proven by expanding into scalar form and carrying out the differentiations.

Following the rules of matrix differentiation for the kinetic and elastic potential energies (quadratic form) and work terms (inner product), the N equations in matrix form for this slender member in bending are given by

$$\left[\int_0^s \mu\left(\psi\psi^\mathrm{T}\right) dy\right]\ddot{q} + \left[\int_0^s EI\left(\psi''\psi''^\mathrm{T}\right) dy\right]q = \psi(a)f(t) \tag{3.31}$$

It should be noted that the integrals for the torsion case or for any other type of structure would differ but the principles would be the same. Equation (3.31) may then be written in the generalized matrix form

$$\mathbf{M}_q\ddot{q} + \mathbf{K}_q q = \psi(a)f(t) \tag{3.32}$$

When the series with two assumed shapes is used, the result using this equation may be shown to be identical to that in Equation (3.16), the equations obtained by the longhand method. However, such a matrix approach is more compact, and because the $N \times N$ matrices are symmetric some integrals need not be calculated. If the chosen assumed shapes $\boldsymbol{\psi}$ had corresponded to the *exact* normal mode shapes, which are orthogonal, then both the 'mass' and 'stiffness' matrices in Equation (3.32) would be diagonal since the generalized coordinates would in fact be the same as modal coordinates (see Chapter 2).

The remainder of this chapter addresses two approaches used for determining the free-free modes for a whole aircraft made up of slender members in bending and torsion.

3.7 Generating Whole Aircraft 'Free–Free' Modes from 'Branch' Modes

In the first approach, the aircraft is divided into a number of separate 'branches', each constrained at one end, as illustrated in Figure 3.4; the normal modes of each branch are determined. The modes are then treated as assumed shapes for a Rayleigh–Ritz analysis, together with additional rigid assumed shapes (i.e. effectively rigid body modes). This combination of assumed shapes allows the branch root constraints to be 'freed' in order to generate a model for a complete free–free (unconstrained) aircraft. The process is also known as Component Mode Synthesis (Cook *et al.*, 1989), where the structure is divided into components or substructures.

Consider the simple example of an aircraft consisting of two uniform flexible wings of mass per length μ_W and flexural rigidity EI, plus a rigid fuselage of mass m_F as shown in Figure 3.5; only symmetric modes are sought. The aircraft of total mass m is assumed not to undergo any

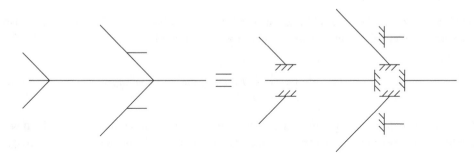

Figure 3.4   Aircraft 'branches'.

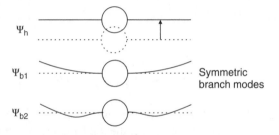

Figure 3.5   Aircraft with 'branch mode' representation for the wing.

pitch motion; thus the wings only bend and the fuselage only heaves. It will be assumed that the first two *exact* normal 'branch' mode shapes (subscript b), for each wing constrained/built-in at its root, are known and given by the functions ψ_{b1} and ψ_{b2}.

In order to 'free' the aircraft so that it behaves as a free–free structure, and so be able to determine the equivalent whole aircraft free–free flexible modes, the constraints must be 'released'. This can be achieved by assuming that the displacement of the aircraft is a combination of the exact flexible branch modes $\psi_{b1,2}$ and a rigid body heave assumed shape (or heave mode) ψ_h. Thus the assumed total displacement along the wing ($y \geq 0$) is given by

$$z(y,t) = \psi_h(y)q_h(t) + \psi_{b1}(y)q_{b1}(t) + \psi_{b2}(y)q_{b2}(t) \quad \text{where} \quad \psi_h(y) = 1 \qquad (3.33)$$

The constituent shapes are shown in Figure 3.5. Now, recognizing that the two wings move in-phase (if only symmetric modes are required) and that the fuselage width is ignored in the integrals, the total kinetic energy is

$$T_{\text{Aircraft}} = T_{\text{Wings}} + T_{\text{Fuselage}} \qquad (3.34)$$

where

$$T_{\text{Wings}} = 2 \left(\frac{1}{2} \int_0^s \mu_W \dot{z}^2 \, dy \right) = 2 \left[\frac{1}{2} \int_0^s \mu_W (\psi_h \dot{q}_h + \psi_{b1} \dot{q}_{b1} + \psi_{b2} \dot{q}_{b2})^2 \, dy \right] \qquad (3.35)$$

$$T_{\text{Fuselage}} = \frac{1}{2} m_F \dot{z}(0)^2 = \frac{1}{2} m_F (\psi_h \dot{q}_h + \psi_{b1}(0)\dot{q}_{b1} + \psi_{b2}(0)\dot{q}_{b2})^2 = \frac{1}{2} m_F (\psi_h \dot{q}_h)^2 \qquad (3.36)$$

Note that the value of the branch mode shapes is zero at $y = 0$ (i.e. built-in at the root). Also, since elastic potential energy is only present in wing bending then

$$U = 2 \left(\frac{1}{2} \int_0^s EI z''^2 \, dy \right) = 2 \left[\frac{1}{2} \int_0^s EI (\psi_h'' q_h + \psi_{b1}'' q_{b1} + \psi_{b2}'' q_{b2})^2 \, dy \right] \qquad (3.37)$$

However, since the additional rigid shape has no elastic deformation, then $\psi_h'' = 0$ which will simplify the final expression. When Lagrange's equations are used with the coordinates q_h, q_{b1}, q_{b2}, it may be shown that the generalized equations of motion for the free-free aircraft are

$$\begin{bmatrix} m_h & 2m_{hb1} & 2m_{hb2} \\ 2m_{hb1} & 2m_{b1} & 0 \\ 2m_{hb2} & 0 & 2m_{b2} \end{bmatrix} \begin{Bmatrix} \ddot{q}_h \\ \ddot{q}_{b1} \\ \ddot{q}_{b2} \end{Bmatrix} + \begin{bmatrix} 0 & 0 & 0 \\ 0 & 2k_{b1} & 0 \\ 0 & 0 & 2k_{b2} \end{bmatrix} \begin{Bmatrix} q_h \\ q_{b1} \\ q_{b2} \end{Bmatrix} = 0 \qquad (3.38)$$

where

$$m_h = m = m_F + 2\mu_W s \quad m_{hbj} = \int_0^s \mu_W \psi_{bj} dy, \quad m_{bj} = \int_0^s \mu_W \psi_{bj}^2 dy \quad \text{and} \quad k_{bj} = \int_0^s EI \psi_{bj}''^2 dy \quad j = 1, 2$$

Firstly, it may be seen that there are no stiffness terms associated with the rigid shape as there is no corresponding elastic potential energy (i.e. the top left-hand corner term in the stiffness matrix is zero). Then, because the branch mode shape ψ_{bj} is the exact *j*th normal mode shape for the built-in wing, it should be noted that the terms m_{bj} and k_{bj} are in fact the modal mass and stiffness for the jth branch mode and so $k_{bj} = \omega_{bj}^2 m_{bj}$, where ω_{bj} is the natural frequency of the jth branch mode. The orthogonality of these modes means that there are no mass or stiffness cross-coupling terms between the branch modes as seen in Equation (3.38). However, there is an inertia coupling term m_{hbj} between the rigid body and the jth branch mode; it is this coupling that enables the constrained branches to be 'released' and whole aircraft free–free modes produced.

If the eigenvalue solution of Equation (3.38) is carried out, then mode shapes will result for the whole free-free aircraft, expressed as the proportion of each of the three constituent shapes. In this present example, there will be one rigid body mode (with no contribution from the flexible branch modes) and two free–free flexible modes (involving flexible branch and rigid heave components). Note that the analysis approach may be extended to more modes, to include overall aircraft motions other than heave (e.g. pitch, roll), and also antisymmetric modes.

Now consider the above example with values of $m_F = 1200$ kg, $\mu_W = 50$ kg/m, $s = 6$ m and $EI = 500\,000$ N m^2. However, in order to solve the eigenvalue problem in Equation (3.38), the exact mode shapes must be known for a continuous member in bending built-in at one end. From the exact partial differential equation analysis approach (Bishop and Johnson, 1979; Rao, 1995; Thomson, 1997; Blevins, 2001), not covered in this book, it has been shown that for the member of length s and mass per length μ_W, built-in at one end and free at the other, the jth mode natural frequency is given by

$$\omega_{bj} = \left(\beta_j s\right)^2 \sqrt{\frac{EI}{\mu_W s^4}} \tag{3.39}$$

where for the first two modes $\beta_1 s = 1.875$, $\beta_2 s = 4.694$. The branch mode natural frequencies corresponding to the parameters chosen in this example are 1.55 and 9.74 Hz. The corresponding exact mode shapes (Bishop and Johnson, 1979) are given by

$$\psi_{bj}(y) = \left(\cosh \beta_j y - \cos \beta_j y\right) - \sigma_j \left(\sinh \beta_j y - \sin \beta_j y\right) \tag{3.40}$$

where

$$\sigma_j = \left(\cos \beta_j s + \cosh \beta_j s\right) / \left(\sin \beta_j s + \sinh \beta_j s\right)$$

The modal mass values for these mode shapes are given in Bishop and Johnson (1979) as $m_1 = m_2 = \mu_W s$. Thus, knowing the branch mode natural frequencies, the modal stiffness values may be calculated. By integrating the mode shapes, the mass coupling terms may be shown to equal $m_{hb1} = 0.734\ \mu_W s$, $m_{hb2} = 1.018\ \mu_W s$.

Then, solving the eigenvalue problem based on Equation (3.38), the rigid body heave mode has a frequency of 0 Hz and generalized mode shape {1, 0, 0}. The natural frequencies of the two free–free flexible modes of the whole aircraft are 1.74 and 10.15 Hz and the generalized mode shapes are {−0.261, 1, −0.004} and {−0.183, 0.147, 1}. These generalized mode shapes are the proportion of each of the three shapes in the series used; when the three shapes are weighted by these values and the results summed, then the resulting free–free flexible mode shapes are sketched in Figure 3.6. When additional branch modes are added in, then further free-free modes are obtained.

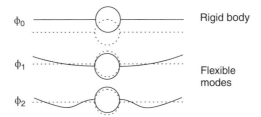

Figure 3.6   Whole aircraft 'free–free' mode shapes.

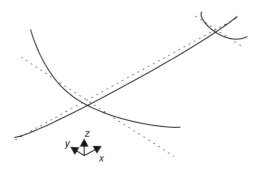

Figure 3.7   Flexible aircraft with free–free symmetric modes.

3.8   Whole Aircraft 'Free–Free' Modes

In Section 3.7, the first approach to determining whole aircraft free–free modes was explained. It was shown that the branch mode shapes for one or more built-in components, such as a wing, could be combined with rigid body shapes for the whole aircraft to generate free–free rigid body and flexible modes. The number of modes calculated for each branch could be reduced, before combining with modes from the other branches and with the rigid body displacements, so as to generate the final solution. This reduction of modes thus economized on computing requirements.

However, in practice nowadays, the aircraft is usually modelled as a whole using a second approach based on a discretization of the structure using the Finite Element method (see Chapter 17); here rigid body and free–free flexible modes are produced for the whole aircraft in a single calculation from a single FE model. In later parts of the book, simple whole aircraft flexible models composed of rigid body and free–free flexible modes will be used, particularly in manoeuvre and gust load calculations, to illustrate the effect of flexibility. In this section, the form of such models will be introduced.

Consider the 'stick' representation of a flexible aircraft shown in Figure 3.7, with only symmetric deformations shown for simplicity. However, this time the displacement will be expressed in matrix form as a summation of the whole free–free aircraft rigid body (subscript r) and flexible/elastic (subscript e) normal modes

$$z(x,y,t) = \boldsymbol{\phi}_{\mathrm{r}}^{\mathrm{T}}\boldsymbol{q}_{\mathrm{r}} + \boldsymbol{\phi}_{\mathrm{e}}^{\mathrm{T}}\boldsymbol{q}_{\mathrm{e}} = \boldsymbol{\phi}^{\mathrm{T}}\boldsymbol{q} \qquad (3.41)$$

where ϕ are free-free normal mode shapes and q are generalized/modal coordinates; simple examples of the aircraft free–free modes were shown in Figure 3.6. Following the usual approach of writing the kinetic and elastic potential energy terms and using Lagrange's equations, it may be shown that the equation of motion is

$$\begin{bmatrix} \mathbf{m_r} & \mathbf{0} \\ \mathbf{0} & \mathbf{m_e} \end{bmatrix} \begin{Bmatrix} \ddot{q}_r \\ \ddot{q}_e \end{Bmatrix} + \begin{bmatrix} \mathbf{0} & \mathbf{0} \\ \mathbf{0} & \mathbf{k_e} \end{bmatrix} \begin{Bmatrix} q_r \\ q_e \end{Bmatrix} = \mathbf{0} \tag{3.42}$$

where the modal mass matrices for the rigid body and flexible modes are given by

$$\mathbf{m_r} = \int \boldsymbol{\phi}_r^T \boldsymbol{\phi}_r dm = \mathrm{diag}[m_{r1}, m_{r2}...] \quad \mathbf{m_e} = \int \boldsymbol{\phi}_e^T \boldsymbol{\phi}_e dm = \mathrm{diag}[m_{e1}, m_{e2}...] \tag{3.43}$$

and where integrations are taken over the entire aircraft. It should be noted that these modal mass matrices are diagonal and that the mass coupling terms seen for the branch mode analysis in Equation (3.38) are now zero, because the rigid body and flexible modes are orthogonal. The modal stiffness matrix $\mathbf{k_e}$ is also diagonal, being found from the modal mass and the modal frequencies; there is no rigid body modal stiffness.

As an example of such equations, consider an aircraft undergoing heave and pitch motion in two dimensions. If the rigid body free–free mode shapes are given by pure heave and pitch motion about the centre of mass, namely $\phi_{r1} = 1$ and $\phi_{r2} = -x$, it may be shown that the modal mass terms are equal to the aircraft mass m and pitch moment inertia I_y respectively (see Appendix A). Clearly, the corresponding rigid body generalized coordinates are equal to the vertical motion of the centre of mass and the nose up pitch angle, namely $q_{r1} = z_C$ and $q_{r2} = \theta$. Adding a single flexible mode, governed by the generalized coordinate q_e with modal mass m_e and stiffness k_e, would then lead to the equations of motion (without aerodynamics or excitation terms present)

$$\begin{bmatrix} m & 0 & 0 \\ 0 & I_y & 0 \\ 0 & 0 & m_e \end{bmatrix} \begin{Bmatrix} \ddot{z}_C \\ \ddot{\theta} \\ \ddot{q}_e \end{Bmatrix} + \begin{bmatrix} 0 & 0 & 0 \\ 0 & 0 & 0 \\ 0 & 0 & k_e \end{bmatrix} \begin{Bmatrix} z_C \\ \theta \\ q_e \end{Bmatrix} = 0 \tag{3.44}$$

The effect of introducing further flexible modes would be to add more modal mass and stiffness terms along the diagonals. This result, and others similar to it, will be used later in Part II of the book to show how the flexible aircraft may be treated when experiencing ground or flight manoeuvres, or when encountering gusts or turbulence.

For loads calculations, then aerodynamic, control, gust and landing gear terms would need to be added. It is important to recognize that Equation (3.44) is in the same format whether Rayleigh–Ritz or Finite Element approaches are employed; all that is different is that the modal data would be more accurate for the Finite Element approach.

3.9 Examples

Note that some of these examples may be useful preparation when carrying out the examples in later chapters.

1. A wing is idealized as a uniform slender member of semi-span s, mass per length μ and flexural rigidity EI, built-in at one end and free at the other. Find an expression for the natural

frequency of the fundamental bending mode using each of the following assumed bending deformation shapes:

$$\text{(a)} \quad z(y,t) = (1 - \cos \pi y/s)q(t) \quad \text{and} \quad \text{(b)} \quad z(y,t) = (y/s)^2(3 - y/s)q(t)$$

Compare the results with those for the basic quadratic shape and the exact result, both quoted earlier in this chapter. Consider how well these three assumed shapes satisfy the kinematic and load boundary conditions.

$\left[\text{(a)} \, 5.70 \sqrt{EI/(\mu s^4)} \text{ and (b)} \, 3.57 \sqrt{EI/(\mu s^4)}; \text{ compare with } 4.47 \text{ for polynomial shape} \right.$
$\left. \text{and } 3.52 \text{ exact} \right]$

2. A wing is idealized as a uniform slender member of semi-span s, moment of inertia in twist per length χ and torsional rigidity GJ, built-in at one end and free at the other. Find an expression for the natural frequency of the fundamental torsional mode using each of the following assumed torsional deformation shapes:

$$\text{(a)} \quad \theta(y,t) = \sin\left(\frac{\pi y}{2s}\right)q(t) \quad \text{and} \quad \text{(b)} \quad \theta(y,t) = (y/s)\left[3 - (y/s)^2\right]q(t)$$

Compare the results with those for the basic linear shape and the exact result, both quoted earlier in this chapter. Consider how well these three assumed shapes satisfy the kinematic and load boundary conditions.

$\left[\text{(a)} \, 1.571 \sqrt{GJ/(\chi s^2)} \text{ and (b)} \, 1.572 \sqrt{GJ/(\chi s^2)}; \text{ compare with } 1.732 \text{ for linear shape and} \right.$
$\left. 1.571 \text{ exact} \right]$

3. For a uniform slender member of length s, moment of inertia in twist per length χ and torsional rigidity GJ, built-in at one end and free at the other, find an expression for the natural frequency of the fundamental torsional mode using a deformation with two assumed shapes, namely $\theta(y, t) = (y/s)\, q_1(t) + (y/s)^2 \, q_2(t)$. Repeat the analysis using matrix algebra. Note that the eigenvalue calculation could be performed using the 'eig' MATLAB function.

$\left[1.576 \sqrt{GJ/(\chi s^2)}; \text{ compare with } 1.571 \text{ exact} \right]$

4. An aircraft is idealized as a rigid fuselage of mass 1200 kg with two wings, each represented as a flexible member of mass per length 50 kg/m, length (or semi-span) 6 m and flexural rigidity 500 000 Nm^2, built-in at the fuselage. Assume that the series representation for the bending deformation of the wings is a combination of a rigid body heave and a flexible 'branch' shape, namely $z(y, t) = q_0(t) + (y/s)^2 q_1(t)$. Obtain an estimate for the frequency of the free–free bending mode of the aircraft. If desired, the matrix form of analysis may be used. Note that the mass matrix is not diagonal because the assumed shape for the wing deformation is not a free–free mode shape and therefore is not orthogonal to the rigid body shape. Note also that the result may be compared to the example in Section 3.5 where exact mode shapes for the built-in member were used.
   [2.19 Hz; cf. 1.74 Hz when the exact built-in mode shape was used]

5. An unswept, rectangular wing has a semi-span s, chord c, flexural rigidity EI and torsional rigidity GJ. The shear centre (centre of twist) and the mass centre lie at distances of $0.35c$ and $0.45c$ respectively aft of the leading edge. The mass per unit length of the wing is μ and the moment of inertia per unit length about a spanwise axis through the mass centre is $\chi = 0.1\mu c^2$. Assuming that the wing is built-in at one end and free at the other, and that the

bending deflection z, measured from the elastic axis, and angle of twist θ (positive nose up) are given by $z(y, t) = (y/s)^2 q_b(t)$ and $\theta(y, t) = (y/s)q_t(t)$, obtain the coupled equations of motion in generalized coordinates. For values of the parameters given by mass per length of 50 kg/m, semi-span 6 m, chord 1.2 m, torsional rigidity 240 000 N m^2 and flexural rigidity 500 000 N m^2, determine estimates for the first two natural frequencies. By noting the mode shape in generalized coordinates, indicate whether the modes are dominantly bending or torsion. Note that, as a check, the coupling term in the generalized mass matrix is $- 0.1 \, \mu c s/4$.

[2.16 and 8.39 Hz with generalized mode shapes {10.045} and {−0.161}]

6. Using Example 5, determine the expressions for the two generalized forces corresponding to a force F acting upwards on the leading edge at mid-span.

$[Q_b = F/4 \text{ and } Q_t = 0.35 \, Fc/2]$

7. A wing/tip store combination may be idealized as a uniform member, built-in at one end and free at the other, with an offset tip store (e.g. fuel tank). The wing has a mass per length of 75 kg/m, moment of inertia in twist per length 25 kg m^2/m, span 6 m, flexural rigidity 2×10^6 N m^2 and torsional rigidity 5×10^5 N m^2. The tip store has a mass of 100 kg and moment of inertia in pitch of 25 kg m^2 about its centre of mass, which itself is offset by 0.5 m forward of the wing centre line. Using simple quadratic bending and linear torsional assumed shapes, estimate the first two natural frequencies of the combination and sketch the expected mode shapes. Assume that the elastic and mass axes both lie along the mid-chord. Note that the inertia coupling term will be given by tip mass × distance forward of the mid-chord = +50 kg m, with the sign depending upon the twist sign convention; a sign error will be shown by incorrect mode shapes.

[2.18 and 5.48 Hz]

8. An idealized wing structure built-in at the root ($y = 0$) has a semi-span of $2\,L$ and the internal structure is such that the flexural rigidity is non-uniform, being $2EI_0$ for the inner half of the wing ($L > y \geq 0$) and EI_0 for the outer half ($2\,L \geq y > L$), whereas the mass per unit length μ is constant. Using two assumed shapes in the form of simple quadratic and cubic functions, determine the fundamental natural frequency and mode shape. Write down the additional terms that would need to be included in the analysis if a landing gear of stiffness K was positioned at mid-span.

$[0.2\sqrt{EI_0/(\mu L^3)}$; mode shape in generalized coordinates {1, −0.174} and in physical coordinates {0.33 at mid-span and 1 at the tip}]

9. For a non-uniform tapering member, built-in at one end and free at the other, with mass per length and flexural stiffness distributions $\mu(y) = \mu_0(1 - y/s)$, $EI(y) = EI_0(1 - y/s)\mu(y)$, use a quadratic assumed shape to estimate the fundamental natural frequency in bending.

$\left[3.87\sqrt{EI_0/(\mu_o s^3)} \right]$

10. A free-free aircraft in anti-symmetric motion is idealized as a rigid fuselage of moment of inertia in roll I_0 and two flexible wings cantilevered from the fuselage, as shown in Figure 3.8. Each wing has total mass M, span s and effective flexural rigidity EI where E is Young's modulus and I is the relevant second moment of area in bending. The two wings vibrate anti-symmetrically. The displacement of each wing is assumed to be given by the combination of rigid body rotation and elastic bending, namely

$\nu(x,t) = a_0(t)\left(^x/_s\right) + a_1(t)\left(^x/_s\right)^2 \quad x \geq 0$ where $a_0(t), a_1(t)$ are unknown coefficients that can vary with time. The physical size of the fuselage may be assumed to be sufficiently

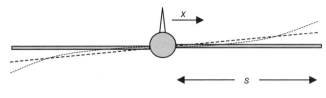

Figure 3.8

small compared to the span that the fuselage displacement is $v(0, t)$ and fuselage rotation (i.e. slope at the wing root) is $v'(0, t)$, where $' = \partial/\partial x$. Show that the equations of motion are given by

$$\begin{bmatrix} I_0/s^2 + 2M/3 & M/2 \\ M/2 & 2M/5 \end{bmatrix} \begin{Bmatrix} \ddot{a}_0 \\ \ddot{a}_1 \end{Bmatrix} + \begin{bmatrix} 0 & 0 \\ 0 & 8EI/s^3 \end{bmatrix} \begin{Bmatrix} a_0 \\ a_1 \end{Bmatrix} = \begin{Bmatrix} 0 \\ 0 \end{Bmatrix}$$

Assuming that $I_0 = Ms^2$, obtain expressions for the estimated rigid body and flexible anti-symmetric mode natural frequencies, Sketch the expected mode shapes.

$$\left[0; \sqrt{96EI/5Ms^3} \right]$$

4

Introduction to Steady Aerodynamics

Aircraft are able to fly because the lift generated by the airflow over the wings and horizontal tail surfaces supports their weight. For a flexible aircraft, these lift forces give rise to deflections in the aerodynamic shape, which in turn change the characteristics of the airflow, hence leading to aeroelastic phenomena and affecting the dynamic loads. An understanding of how the aerodynamic flow around a two-dimensional aerofoil (i.e. the section of a typical wing profile), or a three-dimensional aerodynamic surface, generates the forces and moments that occur during flight is very important in order to be able to develop mathematical models that describe the aeroelastic behaviour. In this book, the majority of the mathematical treatment concerns lifting surfaces that are described as continuous, but in some cases the surfaces are discretized into strips, sections or panels.

In this chapter, some of the fundamentals of fluid mechanics and aerodynamics are reviewed, with particular emphasis on the lift/drag forces and moments that occur when air flows around a two-dimensional aerofoil or three-dimensional aerodynamic surface (i.e. wing, tailplane or fin). Other parts of the chapter examine the atmosphere, as well as the effect of adding camber and control surfaces to the wing. The final section briefly discusses transonic flows. Both continuous and discretized wings are considered. More detail about the material in this chapter can be found in Anderson (2001) and Houghton and Carpenter (2001).

4.1 The Standard Atmosphere

Aircraft fly at a range of altitudes and air speeds. It will be shown that the aerodynamic forces and moments that act upon the lifting surfaces (e.g. wings, tail) depend in part upon the air density and pressure, and therefore these quantities must be determined at all altitudes. However, the characteristics of the atmosphere vary with altitude, position on the globe, time of day and time of year (Anderson, 2001). Consequently, the International Standard Atmosphere (ISA) has been defined, which enables aircraft performance to be related to a common reference. The ISA has been determined from experimental measurements and relates temperature, air density and pressure to the altitude above sea level. At extremely high altitudes, the varying values of the acceleration due to gravity also need to be considered, but they will be ignored

Introduction to Aircraft Aeroelasticity and Loads, Second Edition. Jan R. Wright and Jonathan E. Cooper.
© 2015 John Wiley & Sons, Ltd. Published 2015 by John Wiley & Sons, Ltd.

Table 4.1 International Standard Atmosphere properties

T_0 (sea level temperature)	288.15 K (°Kelvin)	518.69 R (°Rankine)
P_0 (sea level air pressure)	101325 N/m^2	21162 lbf/ft^2
ρ_0 (sea level air density)	1.225 kg/m^3	0.0023769 slug/ft^3
a_0 (speed of sound at sea level)	340.29 m/s	1116.43 ft/s
R (gas constant)	287.05 m^2/s^2 K	1716 ft^2/s^2 R
$\gamma = \dfrac{c_p}{c_v} = \dfrac{\text{specific heat at constant pressure}}{\text{specific heat at constant volume}}$	1.4	1.4

here and gravity will be assumed to remain constant at all altitudes. Table 4.1 shows the values of a number of important atmospheric parameters for the International Standard Atmosphere (ISA) in both SI and Imperial units.

Here, only the range of altitudes from sea level to 11 000 m (33 528 ft), known as the *troposphere*, will be considered. In this range, the temperature T (in degrees Kelvin) of the standard atmosphere decreases with altitude above sea level h (in m) in a linear manner such that

$$T = 288.15 - 0.0065\,h = T_0 - \chi h, \tag{4.1}$$

where constant $\chi = 0.0065$ (i.e. the temperature decreases by 6.5 °C for each 1000 m climbed). Equation (4.1) can be rewritten to relate the change in temperature dT to the height above sea level, so that

$$dT = T - T_0 = -\chi h. \tag{4.2}$$

Assuming that the atmosphere behaves as a perfect gas, then the state equation, and the hydrostatic equation that relates the change in pressure dP due to a change in height, can be written as

$$P = \rho R T \quad \text{and} \quad dP = P - P_0 = -\rho g h, \tag{4.3}$$

where P is pressure, ρ is air density, R is the gas constant and g is the acceleration due to gravity. Then, combining Equations (4.2) and (4.3) to eliminate altitude, and integrating, gives an expression that relates the pressure to the temperature, such that

$$\int_{P_0}^{P} \frac{dP}{P} = \frac{g}{\chi R} \int_{T_0}^{T} \frac{dT}{T} \quad \Rightarrow \quad \frac{P}{P_0} = \left(\frac{T}{T_0}\right)^{g/\chi R}. \tag{4.4}$$

A similar approach can also be used to determine how the density changes with temperature. Applying the state Equation (4.4) at some given altitude and at sea level gives

$$\frac{P}{P_0} = \frac{\rho T}{\rho_0 T_0} = \left(\frac{T}{T_0}\right)^{g/\chi R} \quad \Rightarrow \quad \frac{\rho}{\rho_0} = \left(\frac{T}{T_0}\right)^{g/(\chi R)-1} \approx \frac{(20-H)}{(20+H)}. \tag{4.5}$$

The final expression is a simple, but reasonably accurate, approximation for the density ratio where H is the altitude in km. Thus, using Equations (4.1), (4.4) and (4.5) it is possible to determine the temperature, pressure and air density for any altitude within the troposphere.

At 11000 m, the temperature, pressure and air density reduce to 75.2, 22.3 and 29.7% respectively of the sea level values.

A further property of the atmosphere that has an important effect on the aerodynamic properties is the speed of sound, defined by the symbol a. The speed of sound is a function of the ratio of specific heats of air γ, the gas constant R and the ambient absolute temperature T, and is defined as

$$a = \sqrt{\gamma RT} = \sqrt{\frac{\gamma P}{\rho}} = a_0 \sqrt{\frac{T}{T_0}} \tag{4.6}$$

where a_0 is the speed of sound at sea level. Hence, the speed of sound reduces with increasing altitude, for instance at 11 000 m it is 86.7% of the sea level value.

4.2 Effect of Air Speed on Aerodynamic Characteristics

The airflow and the resulting pressure distribution around a two-dimensional aerofoil changes depending upon the air speed and altitude. These characteristics can be defined in terms of several dimensionless quantities.

4.2.1 Mach Number

One particularly important influence upon the characteristics of all fluid flows is the compressibility of the air, which alters depending upon the ratio between the local flow velocity V at some point in the flow and the speed of sound a. This ratio is known as the Mach number (M) and is defined as

$$M = \frac{V}{a} \tag{4.7}$$

The value of M has a significant effect on the flow characteristics around aerofoils and aerodynamic surfaces, and specific flow regimes can be defined approximately as shown in Table 4.2 (note that the symbol M will be used in different ways in the book, particularly for pitching or bending moment, but the context will indicate the usage).

Shock waves (or 'shocks') appear in transonic and supersonic flows and effectively act as boundaries across which there are significant abrupt changes in Mach number and pressure.

Table 4.2 Flow regimes defined by Mach number

$M < 0.75$	Subsonic	No shocks present in the flow	Gliders / propeller aircraft / some jet transports
$0.75 < M < 1.2$	Transonic	Shocks are attached to the aerofoil	Civil transports (typically $M = 0.8$ to 0.9)
$M = 1$	Sonic	Flow at the speed of sound	Fighter aircraft
$1.2 < M < 5$	Supersonic	Shocks present but not attached to the aerofoil	Fighter aircraft
$M > 5$	Hypersonic	Viscous interaction, entropy layer, high temperature effects become important	Missiles

A commercial jet aircraft will typically cruise in the transonic regime at around $M = 0.85$, while fighter aircraft often fly at around $M = 2$.

4.2.2   Reynolds Number

The Reynolds Number (Re) is a further non-dimensional quantity that influences the flow around aerofoils and is defined as

$$Re = \frac{\rho V c}{\mu} \tag{4.8}$$

where c and μ are the aerofoil chord and air viscosity respectively. The Reynolds number defines whether a viscous flow, particularly in the boundary layer (the region close to the aerofoil surface where the flow velocity is slowed down due to surface friction) is laminar (i.e. flow velocity varies smoothly close to the surface of the aerofoil) or turbulent (i.e. flow velocity varies randomly and irregularly close to the surface). The Reynolds number represents the ratio of inertia to viscous forces in the flow.

4.2.3   Inviscid/Viscous and Incompressible/Compressible Flows

The simplest form of aerodynamic modelling, the so-called *inviscid* flow, assumes that there are no effects from the viscosity of the air. This assumption implies that the flow past an aerofoil, even at the surface, incurs no friction. In practice, viscosity does have an effect on the flow (*viscous* flow) and this is most notably demonstrated by the presence of the boundary layer, where the flow slows down from the velocity in the free stream to zero velocity on the surface.

A common simplification is to assume that the density of the air is constant (i.e. *incompressible*) throughout the flow, and this is valid for flows where $M < 0.3$. Above this Mach number, *compressibility* effects need to be taken into account and the density will vary through the flow field.

4.2.4   Dynamic Pressure

The dynamic pressure q is defined as $\frac{1}{2}\rho V^2$, where the density ρ and velocity V need to be defined consistently. It is common practice to define velocity in terms of the *equivalent air speed* V_{EAS}, which is the speed at sea level that gives the same dynamic pressure as at some altitude, i.e.

$$\frac{1}{2}\rho V^2 = \frac{1}{2}\rho_0 V_{EAS}^2 \quad \Rightarrow \quad V_{EAS} = \sqrt{\frac{\rho}{\rho_0}} V = \sqrt{\sigma} V \tag{4.9}$$

where σ is the ratio of the air density at some altitude to the sea level density ρ_0. Strictly, V should be referred to as V_{TAS}, the *true air speed*. These air speeds will be referred to later in the book when aeroelasticity and loads are considered. Note that speed is sometimes referred to in knots (nautical miles per hour); the conversion factors are 1 knot to 0.5144 m/s or 1.1508 mph.

4.3 Flows and Pressures Around a Symmetric Aerofoil

An aerofoil is a two-dimensional shape that is the cross-section of some three-dimensional aerodynamic surface; two-dimensional flows are fundamental for gaining understanding, whereas three-dimensional flows are more complex but are what occur in practice. The flow over an aerofoil moving in a fluid at rest is said to be *steady* when the velocity at any fixed point is constant with time. Figure 4.1 shows how 'streamlines' map the fluid motion around a symmetric aerofoil at zero angle of incidence. There is no flow across the streamlines; however, the velocity and pressure can change along them. Any element of fluid experiences a static pressure from adjacent elements as it is moving.

Figure 4.2 shows the same streamlines but this time with the aerofoil at some small positive angle of incidence α, defined as the angle between the chord line and the free-stream direction of the oncoming flow. It can be seen how the flow is altered by the change of incidence, and how the symmetry of Figure 4.1 is lost. Lift occurs because the flow is deflected downwards by the aerofoil; this leads to the flow over the upper surface being faster than that on the lower surface. Note that this difference in speeds is *not* because the upper surface flow has further to travel, but rather due to the aerofoil shape. Point S is the stagnation point where the flow is brought to rest.

Making use of Newton's laws of motion (see Chapter 5) and neglecting gravitational effects, then Bernoulli's equation (Anderson, 2001) for constant ρ (incompressible flow) can be derived to relate the pressure, density and velocity such that

$$P + \frac{1}{2}\rho V^2 = \text{constant} \tag{4.10}$$

For compressible flow it can be shown that Bernoulli's equation, again neglecting gravitational effects, becomes

$$\left(\frac{\gamma}{\gamma-1}\right)\frac{P}{\rho} + \frac{1}{2}V^2 = \text{constant} \tag{4.11}$$

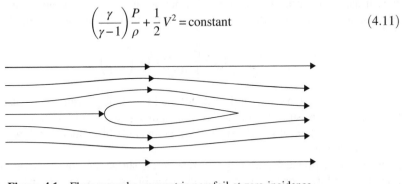

Figure 4.1   Flow around a symmetric aerofoil at zero incidence.

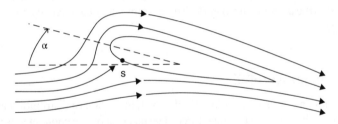

Figure 4.2   Flow around a symmetric aerofoil at a small angle of incidence to the flow.

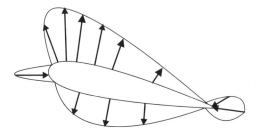

Figure 4.3 Typical pressure distribution for a symmetric aerofoil at a small angle of incidence.

Consider the flow along a typical streamline starting in the free stream at pressure P_∞ and velocity V_∞ and then changing to pressure P and velocity V at some point close to the aerofoil. Applying Bernoulli's equation gives

$$P_\infty + \frac{1}{2}\rho V_\infty^2 = P + \frac{1}{2}\rho V^2 \quad \Rightarrow \quad P = P_\infty + \frac{1}{2}\rho\left(V_\infty^2 - V^2\right) \qquad (4.12)$$

Hence for a velocity $V > V_\infty$, pressure $P < P_\infty$, an increase in velocity leads to suction (pressure reduction); for velocity $V = V_\infty$, pressure $P = P_\infty$; and for velocity $V < V_\infty$, pressure $P > P_\infty$, so a decrease in velocity leads to compression (pressure increase).

The maximum pressure occurs at the *stagnation point* S where the flow comes to rest on the aerofoil ($V = 0$), so the stagnation pressure is

$$P_s = P_\infty + \frac{1}{2}\rho V_\infty^2 \qquad (4.13)$$

It is usual to describe the pressure distribution in terms of the non-dimensional *pressure coefficient* C_p, which is defined for a point in the flow (or on the aerofoil) as

$$C_p = \frac{P - P_\infty}{\frac{1}{2}\rho V_\infty^2} = 1 - \left(\frac{V}{V_\infty}\right)^2 \qquad (4.14)$$

This coefficient is a measure of the ratio of the local static pressure on the aerofoil (relative to the free stream pressure P_∞) to the free stream dynamic pressure. In terms of C_p, the pressure distribution of a typical symmetric aerofoil at an angle of incidence below stall (see later) is plotted in Figure 4.3.

Note that the pressure always acts normal to the surface. A common way of presenting the pressure distribution on both surfaces is shown in Figure 4.4, where it can be seen that the lift is dominated by suction on the upper surface. The ratio of chordwise distance from the leading edge normalized to the aerofoil chord is x/c. There is a greater rate of change in pressure close to the leading edge.

4.4 Forces on an Aerofoil

For an aerofoil moving at velocity V in a fluid at rest, the pressure distribution acting over the surface of the aerofoil gives rise to a total force. The position on the chord at which the resultant force acts is called the *centre of pressure,* as shown in Figure 4.5. If the angle of incidence α

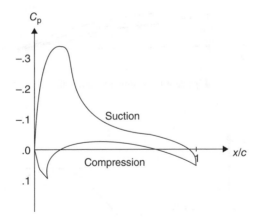

Figure 4.4 Pressure coefficient representation for a symmetric aerofoil at a small angle of incidence.

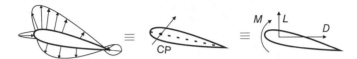

Figure 4.5 Resultant aerodynamic force acting at the centre of pressure.

(angle between the mean airflow and the chord line of the aerofoil, measured in radians) alters, then the pressure distribution over the aerofoil changes, which leads to a repositioning of the centre of pressure. The changing centre of pressure position with respect to different angles of incidence leads to difficulties in any simple aeroelastic analysis, since the forces and moments need to be recalculated continually. For convenience, the net force is usually replaced by two resultant orthogonal forces, acting at a chosen reference point on the aerofoil, and a moment as seen in Figure 4.5.

The lift (L) is the force normal to the relative velocity of the aerofoil and fluid, the drag (D) is the force in the direction of relative velocity of the aerofoil and fluid, and the pitching moment (M) is the moment due to offset between the centre of pressure and the reference point (positive when pushing the nose upwards as shown in Figure 4.5).

It is usual to use non-dimensional coefficients which relate the above quantities to the dynamic pressure and chord for a unit span of aerofoil (since it is two-dimensional), so that the lift, drag and moment coefficients are defined as

$$C_L = \frac{\text{Lift } L}{\frac{1}{2}\rho V^2 c} \quad C_D = \frac{\text{Drag } D}{\frac{1}{2}\rho V^2 c} \quad C_M = \frac{\text{Pitching moment } M}{\frac{1}{2}\rho V^2 c^2}$$

respectively, where c is the aerodynamic aerofoil chord and the lift, drag and pitching moments are defined per unit span of the aerofoil. It is often more useful to use the coefficients rather than the total lift, drag and pitching moment per unit span as they are normalized by dynamic pressure and the aerofoil chord. Note that the forces and pitching moment can be defined with reference to any point on the chord.

Aerofoil sections usually have unsymmetric cross-sections and incorporate camber, as shown in Figure 4.6, to improve the lift performance. Later on, equivalent coefficients will

Figure 4.6 Cambered aerofoil.

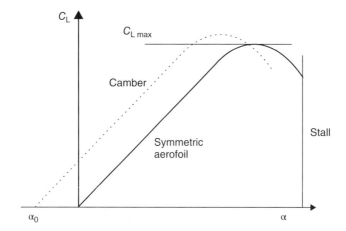

Figure 4.7   Variation of the lift coefficient with the angle of incidence.

be defined for the entire three-dimensional lifting surface (e.g. wing) and these will be based upon the total force (or moment) and normalized by the wing area instead of the chord for a unit span of the two-dimensional aerofoil.

4.5   Variation of Lift for an Aerofoil at an Angle of Incidence

Figure 4.7 shows the variation of lift coefficient with incidence. The lift coefficient C_L is seen to increase linearly with an increase in the angle of incidence α, from the zero lift angle until stall is reached, when the flow detaches from the upper surface and the lift drops off. The maximum lift coefficient obtained is $C_{L_{MAX}}$. Also, at some angle of incidence known as the *zero lift angle α_0*, all aerofoils have zero lift, with $\alpha_0 = 0$ for a symmetric aerofoil. Note how the use of a cambered aerofoil enables C_L to be increased, but at the expense of stall occurring at lower angles of incidence.

Hence, in the linear range

$$C_L = a_1(\alpha - \alpha_0) \quad \Rightarrow \quad C_L = a_0 + a_1\alpha \tag{4.15}$$

where $a_1 = dC_L/d\alpha$ is the two-dimensional *lift curve slope*, which has a theoretical value of 2π per radian. However, measurements show that in practice this has a value between 5.5 and 6 per radian, and a value of $a_1 = 5.73$ is often taken since this corresponds to 0.1 per degree (Houghton and Brock, 1960).

As the Mach number M increases, compressibility of the air has a greater influence on the aerodynamic forces and changes the two-dimensional lift curve slope to become

$$a_1' = \frac{1}{\sqrt{1-M^2}}a_1 \qquad (4.16)$$

It is usual to ignore compressibility effects for $M < 0.3$, and in this book, for simplicity, Mach number effects will be ignored in any modelling.

4.6 Pitching Moment Variation and the Aerodynamic Centre

If the pitching moment coefficient C_M is determined about the leading edge for a varying angle of incidence and hence lift coefficient, then the results shown in Figure 4.8 are found (note that stall is not shown). Note that C_{M_0} is the moment coefficient at the zero lift condition.

The relationship between the moment and lift coefficients below stall can be represented by a straight line, so

$$C_{M_{LE}} = C_{M_0} + bC_L \qquad (4.17)$$

where it is found that $b = dC_{M_{LE}}/dC_L \approx -0.25$ for all aerofoils and therefore

$$C_{M_{LE}} = C_{M_0} - 0.25C_L \qquad (4.18)$$

The pitching moment value and coefficient depend upon where the reference point for the lift and drag are chosen, with Figure 4.9 showing three possible arrangements that must be

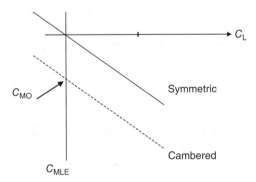

Figure 4.8   Variation of the moment coefficient about the leading edge with the lift coefficient.

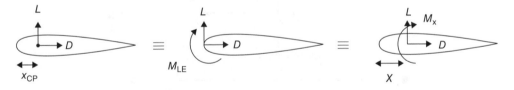

Figure 4.9   Forces and moments acting on an aerofoil for different reference points.

statically equivalent, i.e. at the centre of pressure, leading edge and a general point. Considering the forces and moments acting on the aerofoil shown in Figure 4.9 and taking moments about the leading edge, then

$$-Lx_{CP} = M_{LE} = M_x - Lx \qquad (4.19)$$

Dividing by $\frac{1}{2}\rho V^2 c_2$ gives the equivalent expression in coefficient form:

$$-C_L \frac{x_{CP}}{c} = C_{M_{LE}} = C_{M_x} - C_L \frac{x}{c} \qquad (4.20)$$

and comparison of Equations (4.18) and (4.20) shows that for $x = 0.25c$ (quarter chord) then $C_{M_x} = C_{M_0} =$ constant.

This quarter chord point is called the *aerodynamic centre* and is where the pitching moment coefficient C_{M_x} equals C_{M_0} and does not vary with C_L or incidence, unlike any other point on the chord. The aerodynamic centre position is independent of incidence and section shape, and is the point where any incremental lift due to any incremental change in incidence acts. Note that for a symmetric aerofoil section, C_{M_0} is zero and therefore the centre of pressure is at the quarter chord. These characteristics make it convenient to use the aerodynamic centre for the aeroelastic and load modelling covered in later chapters.

4.7 Lift on a Three-dimensional Wing

So far, most of the attention has been focused on the flow around two-dimensional aerofoil sections and the corresponding forces and moment; however, now the aerodynamic behaviour of a wing will be considered. In practice, the lifting surfaces on an aircraft (e.g. wing) are three-dimensional and there will be changes in the behaviour. Most of the focus in this book will be on unswept and untapered wings to keep the mathematics as simple as possible.

4.7.1 Wing Dimensions

Dimensions of a tapered but unswept wing are shown in Figure 4.10; here the semi-span is s and the root and tip chords are c_R, c_T. The mean chord is c and the wing planform area is given by

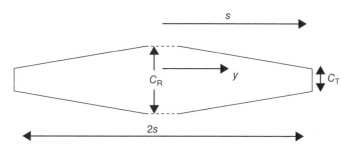

Figure 4.10 Dimensions of an unswept wing.

$S_W = 2sc$, thus including the section of wing passing though the fuselage. The aspect ratio of the wing is a measure of the slenderness of the wing planform and is given by

$$AR = \frac{2s}{c} = \frac{(2s)^2}{S_W} \tag{4.21}$$

Commercial aircraft tend to have wings of relatively high aspect ratio (~6 to 8) for reasons that will become apparent later when drag is considered.

4.7.2 Lift Curve Slope of a Three-dimensional Wing

There are a number of simple adjustments that can be made to the value of the lift curve slope for a two-dimensional aerofoil in order to account approximately for finite-span wings and also the effects of compressibility. For three-dimensional finite-span rigid wings, the value of the lift curve slope will be given the symbol a_W. (Note that this is a different quantity to the speed of sound defined earlier.)

Assuming that the lift distribution across a three-dimensional wing of aspect ratio AR is elliptically shaped, with lift falling off to zero at the wing tips (see later), then the effective wing lift curve slope can be shown to take the form (Fung, 1969)

$$a_W = \frac{a_1}{1 + a_1/(\pi AR)} \tag{4.22}$$

so the lift curve slope reduces for a finite span wing, with the largest reduction occurring for low aspect ratio wings.

4.7.3 Force and Moment Coefficients for a Three-dimensional Wing

Sectional lift, drag and pitching moment coefficients were defined earlier for the two-dimensional aerofoil in terms of force/moment per unit span. For a three-dimensional wing, the equivalent coefficients may be defined in terms of the total lift, moment, etc., over the wing area. The wing lift and pitching moment coefficients are defined by

$$C_L = \frac{L}{\frac{1}{2}\rho V^2 S_W} \qquad C_M = \frac{M}{\frac{1}{2}\rho V^2 S_W c} \tag{4.23}$$

where L, M are the total lift and pitching moment for both wings. The drag coefficient C_D may be defined similarly.

4.7.4 Strip Theory for a Continuous Wing

There are a number of different ways of modelling the spanwise lift distribution of a wing. In this first part of the book the simplest, known as *strip theory*, will be considered, initially for a continuous untapered wing and then for a discretized wing.

In strip theory, the wing is considered to be composed of a number of elemental chordwise 'strips'. It is assumed that the lift coefficient on each chordwise strip of the wing is proportional

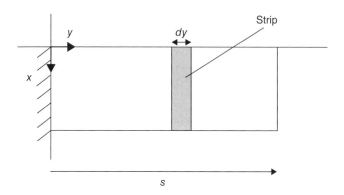

Figure 4.11 Aerodynamic 'strip' on a continuous rectangular wing.

to the local angle of incidence $\alpha(y)$ at distance y from the root and that the lift on one strip has no influence upon another. In its basic form, root and tip effects are ignored along with the effects of compressibility. In practice these assumptions imply that the air speed is low ($M < 0.3$) and that the wing has a high aspect ratio (AR ≥ 6). Note that strip theory cannot be used for drag calculations.

Consider an elemental strip of the wing at distance y from the root, having width dy and chord c as shown in Figure 4.11. Then the lift dL on the strip is taken to act at its aerodynamic centre (i.e. quarter chord) and is defined as

$$dL = \frac{1}{2}\rho V^2 c \, dy \, a_1 \alpha(y) \tag{4.24}$$

making use of the two-dimensional lift curve slope. The total lift acting on a single wing of semi-span s is found by integrating the effect of all the strips, so

$$L_{TOTAL} = \int_0^s dL = \frac{1}{2}\rho V^2 c a_1 \int_0^s \alpha(y)\,dy \tag{4.25}$$

If the wing had been tapered, then the chord would be a function of the spanwise coordinate $c(y)$ and would be included under the integral. Note that an expression for the wing pitching moment about, for example, the leading edge or axis of aerodynamic centres (i.e. quarter chord) can be obtained using a similar approach, and also the rolling moment about the root.

Strip theory in its basic form assumes that the lift on each strip of the wing is the same as if the strip were part of an infinite span two-dimensional wing, i.e. that aerofoil sectional properties would be used. However, when applied to a finite wing, the presence of lift in the tip region implies a pressure discontinuity at the tip that cannot occur in practice. The suction on the upper surface and compression on the lower surface must be equal at the wing tip. In practice, as shown in Figure 4.12, the spanwise lift distribution falls off to zero at the tip. A consequence of the difference in pressures between the upper and lower surfaces in the tip region is the 'trailing tip vortex' that occurs on all wings due to the flow around the tip from lower to upper surfaces.

For elliptically shaped or tapered wings, where the lift predicted by strip theory also drops off towards the wing tip due to the reduced chord, the finite wing effect can be accounted for through the use of Equation (4.22) to adjust the value of the lift curve slope and then using

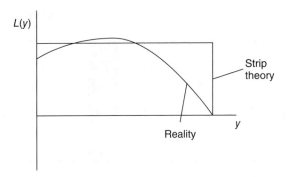

Figure 4.12   Spanwise lift distribution for a realistic wing and strip theory model.

a_W in place of a_1 in Equations (4.24) and (4.25). For wings without a significant taper, then strip theory may be modified (Yates, 1966) to account for the reduction towards the tip. In this case, the lift curve slope may be varied across the span, either by assuming functions of the form

$$a_W(y) = a_1 \left[1 - \left(\frac{y}{s}\right)^2 \right] \quad \text{or} \quad a_W(y) = a_1 \left[1 - \left(\frac{y}{s}\right) \right]^2 \quad \text{or} \quad a_W(y) = a_1 \cos\left(\frac{\pi y}{2s}\right) \qquad (4.26)$$

to replace a_1, or by using a more sophisticated aerodynamic theory. Alternatively, experimental pressure measurements can be used to determine the variation of lift curve slope and aerodynamic centre position along the span.

4.7.5   Strip Theory for a Discretized Wing

If a method such as strip theory (or modified strip theory) is to be employed in conjunction with a Finite Element 'beam-like' model (see later in the book, Chapters 17 and 19), then the continuous expressions used above will need to revised and the wing treated as if it is divided into N sections (i.e. finite width 'strips') of width Δy.

Consider the kth 'strip' (or section) located at a distance of y_k from the root. Following the previous approach for the elemental strip on the continuous wing, the lift force acting upon the kth strip will be given by the expression (Figure 4.13)

$$L_k = \frac{1}{2} \rho V^2 c \Delta y \, a_W \alpha(y_k) \qquad (4.27)$$

where again the local chord could have been used for a tapered wing and the lift curve slope a_W may be corrected in some way. Thus the total lift on the single starboard wing is given by

$$L_{\text{TOTAL}} = \sum_{k=1}^{N} L_k \qquad (4.28)$$

A pitching moment per section could also be calculated; again, the lift curve slope could be adjusted over the sections and the section width need not be constant.

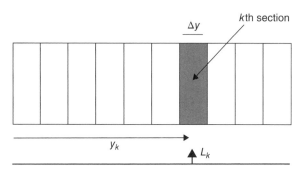

Figure 4.13   Aerodynamic 'strips' on a discretized rectangular wing.

4.7.6   Panel Methods

Even with the use of modified strip theory, there are still inaccuracies primarily due to the lack of interaction effects between different parts of the lifting surfaces (for instance, swept wings and T-tails cannot be modelled accurately with a strip theory approach) and more sophisticated approaches must be used instead. Current industry practice is to make use of the so-called *panel methods* whereby the lifting surface is divided up into panels and the lift distribution is modelled using potential flow elements such as vortices or doublets acting over each panel. The key calculation in such an approach is to determine the Aerodynamic Influence Coefficients (AICs) that determine the aerodynamic effect that each panel has on the others. These methods are described in more detail for both steady and unsteady flows in Chapter 18 and applied to simple aeroelastic models in Chapter 19. Further comments are made on their practical implementation in Chapter 21.

4.8   Drag on a Three-dimensional Wing

As well as generating lift, wings also produce drag. Commercial aircraft designers aim to achieve a maximum lift/drag ratio as this gives the maximum flight range. There are two main contributions to drag: profile drag and induced drag. Noting that the total drag D can be defined as

$$D = \frac{1}{2}\rho V^2 S C_D \tag{4.29}$$

then the drag coefficient can be defined as

$$C_D = C_{D_0} + C_{D_i} = C_{D_0} + \frac{C_L^2}{\pi e' \text{AR}} \tag{4.30}$$

Now C_{D_0} is the *profile drag* (the drag that is inherent from the aerofoil shape, i.e. when there is zero lift, $C_L = 0$). However, C_{D_i} is the *induced* or lift-dependent drag that is mainly due to the presence of the wing tip trailing vortex, but also includes the effects of the fuselage and engine nacelles and is proportional to the square of the lift (L^2). The value e' is the span efficiency factor, which is unity if the wing planform is elliptical. For typical commercial aircraft, the value is within the range $0.85 < e' < 0.95$. Note that the induced drag can be decreased by

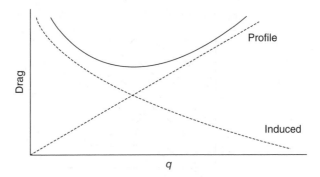

Figure 4.14 Variation of drag with dynamic pressure q.

Figure 4.15 Two-dimensional aerofoil with a control surface.

increasing the aspect ratio and by using as close to an elliptical wing planform as possible (tapered wings can give a good approximation of an elliptical planform). The addition of winglets effectively increases the aspect ratio and so also reduces the drag.

Figure 4.14 shows how both types of drag vary with dynamic pressure. At low dynamic pressures, the induced drag dominates, whereas at high dynamic pressures the profile drag dominates. Although the drag does not directly affect most aeroelastic calculations, the static aeroelastic wing bending deflection and twist in flight do have a significant effect on the drag and are one of the key components in efficient aerodynamic wing design (see Chapters 7 and 22). Drag will also have an impact upon aircraft handling via the flight mechanics equations (see Chapter 13) and therefore indirectly upon loads. Once the transonic flight regime is reached, the presence of shock waves creates the onset of *wave drag*, which produces a significant increase in the overall drag.

4.9 Control Surfaces

Control surfaces are used primarily to manoeuvre the aircraft by changing the pressure distribution over the aerofoil. Consider the two-dimensional aerofoil and control surface, shown in Figure 4.15, where the control angle β (not to be confused with later usage for a sideslip angle in Chapter 13) is taken as positive downwards.

Figure 4.16 shows how the C_L versus α curves vary for changing β. It can be seen that the lift coefficient is increased by increasing β while the slope of each curve remains the same. Consequently, the application of the control surface *increases the effective camber* of the aerofoil. Note also that the stall speed decreases and $C_{M_{LE}}$ becomes more negative.

Figure 4.17 shows how the pressure distribution changes with the applied control surface, with the centre of pressure moving aft. Application of the control surface therefore increases the

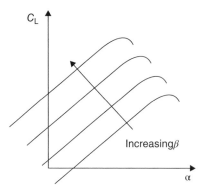

Figure 4.16   Variation of the lift coefficient for different incidence and control surface angles.

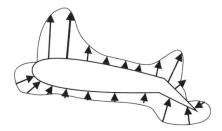

Figure 4.17   Pressure distribution for an aerofoil with an applied control surface angle.

lift but imparts an additional nose down pitching moment. The influence of the control surface on the lift coefficient and pitching moment about any point in the aerofoil section can be modelled as

$$C_L = a_0 + a_1\alpha + a_2\beta \quad \text{and} \quad C_M = b_0 + b_1\alpha + b_2\beta \tag{4.31}$$

where a_2 is the control surface lift slope and b_2 is the control surface pitching moment slope. The coefficients a_2 and b_2 for a two-dimensional aerofoil are defined (Fung, 1969) as

$$a_2 = \frac{a_1}{\pi}\left[\cos^{-1}(1-2E) + 2\sqrt{E(1-E)}\right] \quad \text{and} \quad b_2 = -\frac{a_1}{\pi}(1-E)\sqrt{E(1-E)} \tag{4.32}$$

where E is the ratio of the control surface chord to the total chord.

4.10   Transonic Flows

The transonic flight regime is characterized by the presence of shock waves on the wing surface. The shocks represent a sudden change in the pressure and their position is dependent upon the flight condition and also the chordwise wing geometry. It is consequently not possible to model accurately the transonic pressure distributions using either strip theory or the potential flow panel methods that will be covered in Chapter 18, since neither approach enables shock waves to be modelled. No further mention will be made in this book about transonic flows, except to consider their effect on the flutter speed and possible nonlinear aeroelastic effects in Chapter 10 and to mention some relevant applications of Computational Fluid Dynamics (CFD) in Part III.

4.11   Examples

1. Write a MATLAB program to determine:
   a. the temperature, air density and pressure at any given altitude;
   b. the Mach number for a given air speed and altitude;
   c. the air speed for a given altitude and Mach number.
2. Calculate P/P_0, ρ/ρ_0 and T/T_0 for the standard atmosphere at $h = 3$, 7 and 11 km.
3. The aerodynamic centre for an aerofoil is at 27% chord, and the pitching moment coefficient at zero lift C_{M_0} is –0.05. What is the pitching moment coefficient about the mid-chord point when $C_L = 1.5$, assuming that C_L and C_M vary linearly with incidence?
   [0.295]
4. Find the values of C_L for which the centre of pressure of an aerofoil is (a) at 30% chord behind the leading edge and (b) at the trailing edge, if the aerodynamic centre is at 25% chord and $C_{M_0} = -0.03$.
   [0.6, 0.04]
5. An aerofoil was suspended from a balance in a wind tunnel and its nominal lift and drag were measured normal to and along the wind tunnel axis. The balance readings indicated that at a value of C_L of 0.6 the lift/drag ratio for the aerofoil was 20. However, it was found subsequently that in the working section in the region where the model was suspended the air stream was inclined downwards by 0.5° relative to the tunnel axis. Find the correct value of the lift/drag ratio. (Note that since the model is suspended in the tunnel, positive lift is downwards.)
   [17]
6. An aerofoil of 2 m chord has $C_{M_0} = -0.02$, $\alpha_0 = -1°$, $a_1 = 5.7/\text{rad}$. The aerodynamic centre is at $0.25c$ behind the leading edge. It is at an incidence of 5° in a wind speed of 50 m/s ($\rho = 1.225 \text{ kg/m}^3$). Find the lift and pitching moment about the leading edge per unit span when a trailing edge flap angle is set at 10°. Take $b_1 = 2.0/\text{rad}$ and assume that the lift increment due to the flap acts through the mid-chord point.
   [2897 N, 2107 Nm (nose down)]
7. The aerofoil in Example 6 with the flap at 10° is found to stall at $\alpha = 12°$ and C_L varies linearly with α up to the stall. The aerofoil is then fitted with a leading edge slat of chord $0.15c$ and the stall is consequently delayed to $\alpha = 17°$ (a_1 and b_1 being unchanged by the presence of the slat). Find $C_{L_{max}}$ and $C_{M_{LE}}$ at the stall with the slat in operation. It may be assumed that the pitching moment coefficient increment due to the slat is given by $\Delta C_{M_{LE}} = 0.9$ slat chord/wing chord. (Note that all coefficients are still referred to the same chord c).
   [2.140, –0.508]
8. For a rigid wing of root chord 2 m and semi-span 6 m with incidence 2°, write a MATLAB program to compare the different lift distributions obtained using strip theory and modified strip theory described in Equation (4.26). Determine the taper ratio that gives the closest strip theory lift distribution compared to the modified strip theories.

5

Introduction to Loads

In this book, the topics covered divide naturally into those related to (a) stability (e.g. flutter), (b) static deformation (e.g. static aeroelastic effects, steady flight manoeuvres) and (c) dynamic response (e.g. ground and flight manoeuvres, gusts/turbulence). The aircraft static and dynamic behavior needs to be calculated for a range of cases and once the response deformations, velocities and accelerations are obtained, the loads and stresses generated in the aircraft must be determined. Then the strength and fatigue/damage tolerance behavior may be assessed.

'Loads' is a general term that incorporates both forces and moments, discrete and distributed loads, external and internal loads. In this chapter, various basic concepts relevant to loads in general will be introduced, including Newton's laws of motion for a particle and their generalization to a body, d'Alembert's principle (leading to discrete inertia forces/couples and distributed inertia forces), externally applied/reactive loads, free body diagrams (FBDs), loads generated internally within a structure (i.e. internal loads) and inter-component loads. It will be shown how loads within an accelerating structure may be determined by introducing distributed inertia loads so as to bring the structure into an effective static equilibrium condition.

In this chapter, the way in which internal loads (or so-called 'stress resultants', namely bending moments, shear forces, torques and axial forces) are determined for slender members subject to uniformly or non-uniformly distributed loading will be explained; both continuous and discretized members will be considered. Then the classical way in which stresses are obtained from these internal loads for simple structures will be outlined. However, for complex aircraft structures, this classical approach is not always suitable; the different methodology for such structures, and the potential confusion in terminology, will be explained later in Chapters 16 and 23.

The treatment in this chapter aims to serve as a reminder of key concepts and an introduction to analyzing loads on structures that experience uniform or non-uniform distributions of acceleration; other texts should be referred to for more information and further explanation of basic concepts if required (Donaldson, 1993; Benham *et al.*, 1996; Megson, 1999).

The examples given in this chapter will be for simple continuous slender members under uniformly or non-uniformly distributed loading; this will allow a fundamental understanding to be gained of the kind of analyses relevant to slender wing and fuselage structures, represented in either a continuous or discretized manner. The application of these approaches to aircraft loads will be covered later in Chapter 16.

Introduction to Aircraft Aeroelasticity and Loads, Second Edition. Jan R. Wright and Jonathan E. Cooper.
© 2015 John Wiley & Sons, Ltd. Published 2015 by John Wiley & Sons, Ltd.

5.1   Laws of Motion

In Chapters 1 and 2, and indeed later on in the book, Lagrange's equations are normally used to set up the differential equations of motion. However, Newton's second law of motion could have been used instead to achieve the same end. In this chapter, Newton's laws will be introduced for completeness and because they are important for loads purposes.

5.1.1   Newton's Laws of Motion for a Particle

Newton's laws of motion (Meriam, 1980) are originally stated for a 'particle', a body of negligible dimensions but finite mass, such that acceleration occurs in translation, but without any rotation. The laws may be stated as follows:

1. Every 'particle' continues in a state of rest or of uniform motion unless acted upon by a net force.
2. When a net force acts upon a 'particle', it produces a rate of change of momentum equal to the force and in the same direction.
3. To every action, there is an equal and opposite reaction.

The second law is the most commonly used, and states mathematically that

$$F = \frac{d}{dt}(mv) = m\frac{dv}{dt} + v\frac{dm}{dt} \qquad (5.1)$$

where F is the net force, m is mass and v is velocity. If the rate of change of mass is insignificant, as is the case in most practical situations (except, for example, in rockets), then Equation (5.1) reduces to the better known form

$$F = m\frac{dv}{dt} = ma \qquad (5.2)$$

where a is the acceleration. When a 'particle' is not accelerating, there must be no net force, so $F = 0$ and the forces acting on it must be in equilibrium.

Strictly speaking, acceleration should be measured with respect to an inertial axes system (i.e. one fixed relative to the stars). However, for most engineering analyses, the motion of the earth may be ignored so that an axes system fixed to the earth may be used; this is obviously inappropriate for space flight. When expressed in two dimensions, Equation (5.2) is written for components aligned with two orthogonal axes. Also, the double dot notation seen in earlier chapters will be employed to represent acceleration in most of the book, but in this chapter the simple symbolic representation for acceleration often used when introducing basic dynamic principles will be retained.

5.1.2   Generalized Newton's Laws of Motion for a Body

When the applied forces cause both translation and rotation of a body of finite size, then Newton's second law as expressed for a particle no longer strictly applies. However, if the body is considered as an assembly of particles with equal and opposite forces acting between them, then it may be shown (Meriam, 1980) that this law can be extended to cover a body that is both translating and rotating.

5.1.2.1   Translation

For a body accelerating in translation, the generalized Newton's second law may be stated (in two dimensions) as

$$F_x = ma_x \quad F_y = ma_y \tag{5.3}$$

where the subscripts x and y refer to the components acting in the Oxy axes directions and the acceleration is that of the body centre of mass (CoM). Thus, Newton's second law for a particle may effectively be used for a finite sized body, but only provided that the accelerations are considered at its centre of mass.

Note that the term 'centre of mass' is used in this book to describe the point where the mass of a body may be assumed to act; it is a property of the body and remains the same whatever gravitational field the body is under. However, in aerospace applications, the term 'centre of gravity' (CG) is also in common usage; strictly speaking, this has a different definition to centre of mass in that the centre of gravity will change if the gravitational field is non-uniform, but the two are synonymous if a uniform gravitational field acts over the system.

5.1.2.2   Rotation

For a body accelerating in rotation, the generalized Newton's second law may be stated (in two dimensions) as

$$M_c = I_c \alpha \tag{5.4}$$

where M_c is the applied moment about the centre of mass, α is the angular acceleration (rad/s^2) and I_c is the moment of inertia about an axis Oz through the centre of mass. It is very important to recognize that for dynamic problems, this equation only applies for moments about the centre of mass (Meriam, 1980); the only exception is that moments may be taken about any fixed pivot point in the body (if one exists), with subscript 'o' replacing 'c' in Equation (5.4). If the body is not accelerating, then these equations reduce to the equations of *equilibrium* and moments may then be taken about *any* axis. Note that for a body in three dimensions, the net force may be expressed in three orthogonal directions and moments about the three orthogonal axes.

5.1.3   Units

In using the second law of motion as defined above, it is essential that consistent sets of units are used, given that both metric (SI) and Imperial units are still in use within the international aerospace industry. For Newton's equation in translation, the appropriate units of force, mass and acceleration are given in Table 5.1.

The term kilogram (or pound) force refers to the force produced by a kilogram (or pound) mass acting in the gravity field (the so-called 'weight' associated with the mass). However, for example, 'lb' is often used to denote force instead of 'lbf', though the authors consider it to be

Table 5.1   Units for Force, Mass and Acceleration

Unit set	A	B	C	D
A force of	1 N (Newton)	1 kgf (kilogram force)	1 lbf (pound force)	1 lbf
accelerates a mass of	1 kg (kilogram)	1 kg	1 lb (pound mass)	1 slug (=32.2 lb)
at	1 m/s^2	9.81 m/s^2 (1 g)	32.2 ft/s^2 (1 g)	1 ft/s^2

better practice to use 'lb' for mass only and to retain the symbol 'lbf' for pound force. It may be seen that the definition of 'slug' for mass in Imperial units derives from the second law. Clearly the unit sets A and D defined by kg m s N and slug ft s lbf are most suitable for use in Newton's law for translation (and indeed rotation); unit sets A and D also yield consistent results when used in natural frequency calculations involving mass and stiffness. However, the other two sets of units B and C are deemed unsuitable for dynamic calculations.

If the reader is ever faced with needing to convert between SI and Imperial units, then conversion factors are readily available for most common units. Alternatively, a procedure based on introducing conversion ratios of basic units may be employed as shown in the example following for a moment (or work done) unit, namely

$$1\,\text{lbf ft} = 1\left(\text{slug}\frac{\text{ft}}{\text{s}^2}\right)\text{ft} = 32.174\,\text{lb}\frac{\text{ft}^2}{\text{s}^2} = 32.174\,\text{kg}\left(\frac{\text{lb}}{\text{kg}}\right)\text{m}^2\left(\frac{\text{ft}}{\text{m}}\right)^2\frac{1}{\text{s}^2}$$

$$= 32.174\,\text{kg}\left(\frac{\text{lb}}{2.2046\,\text{lb}}\right)\text{m}^2\left(\frac{\text{ft}}{3.2808\,\text{ft}}\right)^2\frac{1}{\text{s}^2} = \left(\frac{32.174}{2.2046 \times 3.2808^2}\right)\frac{\text{kg}\,\text{m}^2}{\text{s}^2} = 1.3558\,\text{N}\,\text{m}$$

$$(5.5)$$

5.2   D'Alembert's Principle – Inertia Forces and Couples

In this section, d'Alembert's principle will be introduced to show how a dynamic problem may be reduced to an equivalent static one via so-called 'inertia forces (and couples)', even for a flexible aircraft in accelerated flight (see Chapter 16). This approach will allow the internal loads for a dynamic problem to be determined.

5.2.1   D'Alembert's Principle for a Particle

D'Alembert's principle (Meirovitch, 1986; Meriam, 1980) allows any *dynamic* problem solved using Newton's second law to be converted into an *equivalent static* problem by changing the reference axes system from an inertial set to one fixed in the particle (or 'body') and accelerating with it. The effect of the acceleration is handled by introducing a fictitious 'inertia force' equal to (mass × acceleration), acting in the *opposite* direction to the acceleration vector. The applied force and the inertia force are then simply in equilibrium ('dynamic equilibrium'), since the problem has been reduced to an equivalent static one. In essence, the observer accelerating with the particle/body considers it to be in equilibrium and concludes that an inertia force must be acting to balance the applied force.

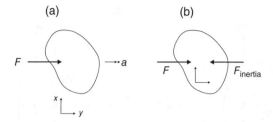

Figure 5.1 Comparison of (a) Newton's law and (b) d'Alembert's principle for a particle in translation.

The idea is illustrated for a particle, and compared with Newton's law, in Figure 5.1. Thus the equation of equilibrium associated with d'Alembert's principle becomes

$$F - F_{\text{Inertia}} = 0 \qquad (5.6)$$

where $F_{\text{Inertia}} = ma$. Note of course that this gives exactly the same result as may be found using Newton's second law. Newton's law and d'Alembert's principle are equivalent in outcome, and care should be taken not to use them together in a problem – otherwise the effective mass may end up being doubled.

The reader might wonder whether there is any point in using d'Alembert's principle; indeed Meriam (1980) recommends against using it. However, it is particularly powerful when determining the internal loads of a flexible accelerating body such as an aircraft. By employing d'Alembert's principle, a complex dynamic problem, such as an aircraft in manoeuvring flight, may be reduced to an equivalent static problem where simple static analysis methods may be employed. Indeed, so established is this approach in determining aircraft loads that the airworthiness requirements refer to the use of inertia forces and couples (CS-25 and FAR-25).

5.2.2 Application of d'Alembert's Principle to a Body

The approach described above for a particle may also be employed for a rigid body of finite size under the action of an applied force, except that the inertia forces must be introduced at the centre of mass. Also, a body subject to a net moment M_c about the centre of mass, and experiencing a rotational acceleration, is handled by introducing an *inertia couple* M_{Inertia} equal to the moment of inertia about the centre of mass $I_c \times$ angular acceleration α. The inertia couple acts at the centre of mass and in the opposite direction to the angular acceleration, so

$$M_c - M_{\text{Inertia}} = 0 \qquad (5.7)$$

where $M_{\text{Inertia}} = I_c \alpha$. The two approaches are illustrated in Figure 5.2.

Note that the term 'couple' refers to a pure moment, such as would be provided approximately by turning a screwdriver. It derives from the effect of two parallel and equal forces acting in opposite directions and has the effect of rotating a body but providing no tendency to translate.

5.2.3 Extension to Distributed Inertia Forces

The concept of a *discrete* inertia force (and/or couple) was introduced above for a body accelerating under an applied force (and/or couple). However, the idea of inertia forces is extremely

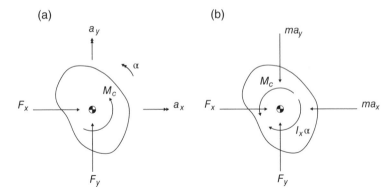

Figure 5.2 Comparison of (a) Newton's law and (b) d'Alembert's principle for a body.

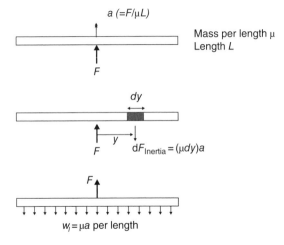

Figure 5.3 Distributed inertia forces on a body lying in the horizontal plane and experiencing a uniform horizontal acceleration.

powerful when considering the internal loads present in a body under an accelerated translational and rotational condition because inertia forces can be *distributed* over the body and the problem can be reduced to one that is in equivalent static equilibrium (see later). The use of distributed inertia forces will now be illustrated for the problem of a rigid *continuous* member for two different load cases.

5.2.3.1 Translation

Consider firstly the example of a uniform continuous *rigid* body (or slender member) of length L and mass per length μ, lying and accelerating in the horizontal plane (to avoid having to consider gravity) under the action of a force F applied at the centre of mass as shown in Figure 5.3. In the *steady* condition, the whole member, and thus every element dy of it, experiences a uniform acceleration of $a = F/(\mu L)$. Thus, using d'Alembert's principle for each element of mass μdy, then the inertia force dF_{Inertia} acting on each element of mass is given by

$$dF_{\text{Inertia}} = (\mu dy)a \qquad (5.8)$$

The elemental force acts in the opposite direction to the acceleration as shown in Figure 5.3. The inertia force per unit length is given by $w_I = \mu a$. The member is then effectively in static equilibrium because the applied force F and the total inertia force F_{Inertia} (obtained by integration of Equation (5.8)) are in balance. What is particularly useful, as a consequence of introducing distributed inertia forces, is that the internal loads for the member in the steady accelerating condition may be examined just as if the member was subjected to a central static force, balanced by a uniformly distributed load.

5.2.3.2   Rotation

A further example allows the ability of the approach to cater for a *non-uniformly distributed acceleration*. Consider the same uniform rigid member in the horizontal plane, but now accelerated in rotation by a moment M_c applied at the centre of mass as shown in Figure 5.4. Since the moment of inertia of a uniform slender member about its centre of mass is given by $I_c = \mu L^3/12$, the angular acceleration of the member is

$$\alpha = \frac{M_c}{I_c} \qquad (5.9)$$

Hence the acceleration of a typical element dy, at distance y from the centre of mass, will be given by

$$a = \alpha y \qquad (5.10)$$

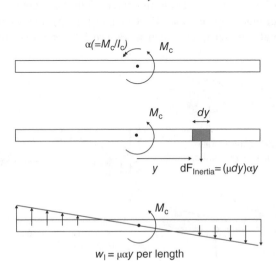

Figure 5.4   Distributed inertia forces on a member lying in the horizontal plane and experiencing a non-uniform horizontal acceleration.

and this may be seen to vary linearly with the distance from the centre of mass. Thus, when inertia forces are introduced for each element then, as shown in the figure, the inertia force per unit length varies linearly and is given by $w_I = \mu a y$. The applied couple is then in equilibrium with the net moment of the inertia forces acting on all the elements. Once again the internal loads could be determined by static methods, but this time an approach based on integrating elemental contributions will be required; this process is relevant to the real aircraft in dynamic response and will be illustrated in Chapter 16.

5.3   External Loads – Applied and Reactive

5.3.1   Applied Loads

Externally applied loads are defined as loads that may be considered to be acting *on* the whole body and may be constant or vary with time. They may be categorized as being:

a. *Discrete* (e.g. thrust),
b. *Distributed over a surface* (e.g. aerodynamic pressure loads) or
c. *Distributed over a volume* (e.g. weight and inertia forces, see later).

In practice, no force ever acts precisely at a 'point' but may often be represented as doing so for convenience of analysis, e.g. thrust will be considered as a discrete force for overall aircraft handling and load calculations. However, a discrete force representation is inappropriate for detailed engine load considerations because the load is actually distributed over the engine.

5.3.2   Reactive Loads (Reactions)

For an aircraft, many of the load cases involve the aircraft being airborne and so there would be no reactions to the ground. However, ground manoeuvres such as taxiing, landing, turning and braking involve the aircraft being in contact with the ground via the landing gear. Any such support arrangement has the effect of constraining one or more displacements and/or rotations, usually to zero, as shown in Figure 5.5 for a two-dimensional scenario

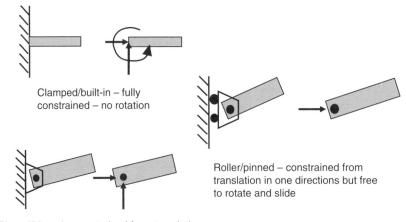

Figure 5.5   Constraints and reactions in two dimensions.

with several different constraint arrangements. This concept readily extends to the three-dimensional case.

When analyzing a body supported in some way, the body must be 'released' from the support (or ground) and equal and opposite reactions considered, corresponding to Newton's third law. The effect of the support on the body is replaced by an unknown reaction, shown in Figure 5.5 by the arrows; every constrained component of motion (translation and/or rotation) must be replaced by a reactive force and/or moment. A free body diagram (FBD) is then drawn (see the next section); the reactions are added to the FBD and their magnitudes determined using equilibrium considerations.

5.4 Free Body Diagrams

A free body diagram (FBD) is a diagrammatic representation of the forces acting on a whole body, or part of it. The body is isolated from its supports and all the applied and reactive forces are drawn on a diagram of the body, positioned at their effective points of action (e.g. centre of mass for inertia forces, centre of pressure for aerodynamic forces). This approach is used because only their total effect is required for *overall* considerations; however, when the internal loads are required (see later), then the *distributed* nature of the loads will need to be retained. Once an FBD is available, then the generalized Newton's laws of motion may be applied so as to yield relationships between the forces and any resulting accelerations; alternatively, d'Alembert's principle may be employed.

As an example of an FBD, consider an aircraft supported against vertical motion at the nose and main landing gear positions but free to roll horizontally, as shown in Figure 5.6; the aircraft is accelerating forward under thrust loading with friction and drag effects neglected. The mass is m, weight W $(=mg)$, thrust T and total support reactions R_N, R_M. The acceleration at the instant of interest is $a = T/m$. The dimensions are shown in the figure with the centre of mass being a distance d above the thrust line and h above the ground. The forces acting are shown on a FBD in the lower part of the figure. The arrows are the forces acting upon the aircraft if the generalized Newton's law were to be employed. On the other hand, were d'Alembert's principle to be applied, an additional horizontal inertia force ma would need to be present

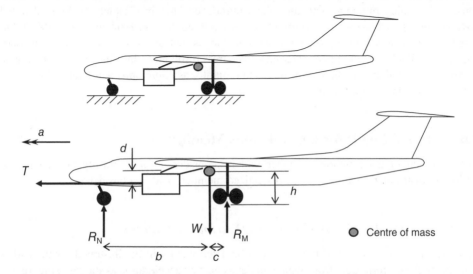

Figure 5.6   Free body diagram example when using Newton's law.

(not shown here). There is no pitch acceleration or inertia moment because both landing gears are assumed to remain in contact with the ground. Consider determining the nose gear reaction using the generalized Newton's law; the equations for translation and rotation *about the centre of mass* are

$$T = ma \quad R_N + R_M - W = 0 \quad R_N b - R_M c + Td = I_c \alpha = 0 \qquad (5.11)$$

and solving for the nose gear reaction yields

$$R_N = \frac{Wc - Td}{b + c}. \qquad (5.12)$$

However if, for example, moments were to be taken about the point of contact of the main landing gear with the ground instead of about the centre of mass, then the resulting reaction would be incorrect. This is because the basis of the generalized Newton's law is that moments must be taken about the centre of mass for a dynamic problem (this is a common error). However, if d'Alembert's principle were to be employed, the inertia force *ma* is included in the FBD and the problem is a static one, so it is immaterial where moments are taken.

5.5   Internal Loads

So far, only the external loads present on a body have been considered. However, in order to see whether the body can sustain the external loads applied to it, the so-called 'internal loads' present *within* the body must be determined. These internal loads will depend upon the *distribution* of the external loads and not simply on their net values. Internal loads may be determined for static or dynamic load cases, with the latter employing d'Alembert inertia forces/couples in order to create an effective static equilibrium condition.

Typical internal loads for a relatively slender body (or member/beam/shaft/rod/bar – see Chapter 3) are shear force, bending moment (hogging/sagging), axial force (tension/compression) and torque (or twisting moment) – a suitable acronym is 'MAST' loads (moment/axial/shear/torque). The examples considered in this chapter will only cover shear force and bending moment, with the others introduced later in Chapter 16. More detailed coverage of internal loads may be found in many references, e.g. Benham *et al.* (1996), but here the focus will be on members with distributed loads such as those induced by inertia loading. Internal loads are sometimes referred to as 'stress resultants' since they are the aggregate of the stresses acting. The body will be treated initially as *continuous* and later a *discretized* representation will be considered.

5.6   Internal Loads for a Continuous Member

The treatment of a continuous member will be considered for uniformly and then non-uniformly distributed loadings. Later, discretized members will be considered.

5.6.1   Internal Loads for Uniformly Distributed Loading

The idea of internal loads and how they can be determined will be illustrated for the earlier example of the uniform free–free member under a constant acceleration as shown in Figure 5.3.

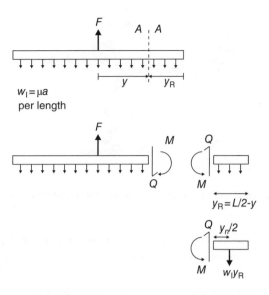

Figure 5.7   Free body diagrams showing internal loads on a member with uniformly distributed loading.

5.6.1.1   'Exposing' Internal Loads

To 'expose', and so determine the internal loads at a particular point on a member, then the analyst must 'imagine' the member as being 'cut' into two subsections at this point. However, when doing so, equal and opposite 'internal loads' must be introduced at the 'cut' to represent the effect of the missing structure (i.e. what has been 'cut' away); this is in effect a consequence of Newton's third law. The internal loads are then determined by considering the equilibrium of one or other subsection.

The FBDs for the two subsections generated by the 'cut' AA at distance y from the centre of mass are shown in Figure 5.7, with equal and opposite internal 'shear force' Q and 'bending moment' M (a pure couple) introduced. In order to simplify the resulting equilibrium expressions, a different coordinate measured from the right hand end will be used in place of y, namely $y_R = L/2 - y$. There are two versions of the FBD shown for the right-hand cut section, one with the distributed load per unit length w_I shown explicitly and the other showing the net force due to the distributed load acting on the relevant subsection; it is this latter version which is used for analysis and because the force is uniformly distributed, the net force $w_I y_R$ on the cut subsection may be placed at the centre of the subsection as shown.

5.6.1.2   Determining Internal Loads via Equilibrium of 'Cut' Sections

Each 'cut' subsection must be in equilibrium under the relevant external and internal loads acting on it. In this example, it is simpler to apply the equilibrium condition for the right-hand subsection, namely that there will be no net force or moment; moments are usually taken about the 'cut' to avoid including the shear force Q in the moment equation. Thus

$$Q - w_I y_R = 0 \quad \text{and} \quad M - w_I y_R \frac{y_R}{2} = 0 \tag{5.13}$$

Note the function of the two types of internal load introduced by the cut; the shear force Q balances the distributed external load $w_I y_R$ and the bending moment M resists the tendency

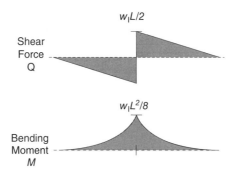

Figure 5.8 Internal load diagrams for a member accelerating under a central force.

of the member to rotate due to the offset of the net force from the cut. Rewriting the equations leads to expressions in terms of y or y_R, namely

$$Q = w_I y_R \quad \text{or} \quad w_I(L/2 - y) \qquad M = 1/2 w_I y_R^2 \quad \text{or} \quad 1/2 w_I(L/2 - y)^2 \qquad (5.14)$$

Exactly the same expressions would have been obtained had the equilibrium of the left-hand subsection in Figure 5.7 been considered instead, but the applied force $F(=\mu aL)$ would have been included. Note that the internal load expressions only apply for the subsection of the member covered by cut AA, namely $y > 0$ or $y_R < L/2$; the internal loads in the left-hand half of the member ($y < 0$) could be obtained by repeating the process for a cut in the left hand half, or more simply by exploiting symmetry.

The variation of these internal loads along the member is shown graphically in Figure 5.8. It may be seen that the shear force varies linearly and the bending moment quadratically with an increase in y_R; the maximum internal loads occur at the centre where the external load is applied and there is a step change in shear force. Note that the signs of these internal loads depend upon the sign convention chosen. What is important is to be consistent with the choice of sign convention within a particular problem.

5.6.1.3 Other Boundary Conditions

The above example was for a free–free member with no support boundaries. The determination of internal loads for members built-in at one end (i.e. cantilever) will be similar except that equilibrium of the section between the cut and tip will be used in order to avoid having to determine the reaction force and moment at the built-in end.

5.6.2 *Internal Loads for Non-uniformly Distributed Loading*

So far, internal loads have been shown for the dynamic load case of steady acceleration using a continuous, rigid and uniform member. In practice, an aircraft will be subject to transient dynamic loading and will be flexible; also the loads (inertia and aerodynamic) will be non-uniformly distributed. Resulting deformations and internal loads would then be required as a function of time and maximum values found.

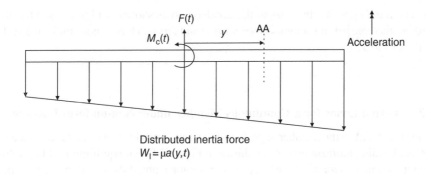

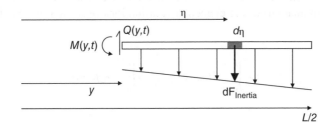

Figure 5.9   Internal loads for non-uniformly distributed loading – continuous member.

Such a problem may be treated as in the above example, but the effective static equilibrium condition needs to be considered at every instant of time. Also, the calculation of internal loads for a structure with non-uniform loading will require an approach involving integration, as shown later in this chapter. The calculation of loads for a rigid or flexible aircraft under steady/dynamic manoeuvres or gusts will be considered in Parts II and III of the book.

In this section, the analysis shown above will be extended to determining internal loads for a member with time-varying applied loads and non-uniformly distributed inertia loading. Consider a continuous, rigid and uniform member of length L, mass per length μ, mass m ($=\mu L$) and moment of inertia I_c, subject to centrally applied dynamic loads $F(t)$ and $M_c(t)$, as shown in Figure 5.9. The acceleration over the member varies non-uniformly and is defined by $a(\eta,\ t) = a_0(t) + \alpha(t)\eta$ where $a_0(t) = F(t)/m$ $\alpha(t) = M_c(t)/I_c$ and η is the distance from the centre of mass.

5.6.2.1   Distributed Inertia Forces for a Continuous Member

The acceleration of an element $d\eta$ of the member at position η from the centre is given by the function $a(\eta,\ t)$ for time t. Thus a distributed inertia force may be introduced to bring the member into effective static equilibrium at a chosen instant of time t. The inertia force acting on the element is

$$dF_{\text{Inertia}}(\eta,\ t) = dm\ a(\eta,t) = \mu d\eta a(\eta,\ t) \tag{5.15}$$

which acts in the opposite direction to the acceleration as shown in Figure 5.9. Thus the distributed inertia force per unit length is $w_I = \mu a(\eta, t)$ and this will have a non-uniform distribution along the member.

5.6.2.2 Internal Loads for a Continuous Member under Non-uniform Loading

Consider a 'cut' AA in the member at position y (>0) and introduce an instantaneous shear force $Q(y, t)$ and bending moment $M(y, t)$, as shown in Figure 5.9. For equilibrium of the right-hand subsection at time t, there will be no net force or moment about the cut. Previously, where the loading was uniformly distributed, the net force could be positioned at the centre of either cut section. However, the effect of the non-uniformly distributed inertia force needs to be included by integration (or summation) of the elemental contributions over the right-hand subsection, so that

$$Q(y,t) = \int_{\eta=y}^{L/2} dF_{\text{Inertia}}(\eta,t) = \mu \int_{\eta=y}^{L/2} a(\eta,t)d\eta$$

and   (5.16)

$$M(y,t) = \int_{\eta=y}^{L/2} (\eta-y)dF_{\text{Inertia}}(\eta,t) = \mu \int_{\eta=y}^{L/2} (\eta-y)a(\eta,t)d\eta$$

Knowing the time histories for the applied force and moment, and hence the acceleration along the member, the shear force and bending moment at every position along the member and at every instant of time may be calculated. Aircraft loads results are typically plotted against time for each internal load of interest at critical positions in order to determine maximum values for design (see Chapters 16 and 24).

Consider the case where only the moment $M_c(t)$ is applied, so $F(t) = 0$. The linearly varying acceleration of the member at position η is $a(\eta, t) = \eta\alpha(t)$, where $\alpha(t) = M_c(t)/I_c$ and $I_c = \mu L^3/12$ for a uniform member. If these values are substituted into Equation (5.16), then the time-varying internal loads are given by

$$Q(y,t) = \frac{M_c}{L} \left[\frac{3}{2} - 6\left(\frac{y}{L}\right)^2 \right] \qquad M(y,t) = M_c \left[\frac{1}{2} - \frac{3}{2}\left(\frac{y}{L}\right) + 2\left(\frac{y}{L}\right)^3 \right] \qquad (5.17)$$

The shear force and bending moment at every position along the member may then be calculated as a function of time, and antisymmetry may be employed for the left-hand half of the member. The variation of internal loads along the member at a particular instant of time is shown in Figure 5.10; note in this case that there is a step change in bending moment at the centre due to the presence of the applied moment.

Clearly, for an aircraft in flight, some applied forces will be aerodynamic in origin, will be a function of the response, and will also be distributed over the component in question. Also, the mass will not be uniformly distributed along the length. Thus the analysis for the aircraft case will be more complex than in these examples, but the principles will be the same.

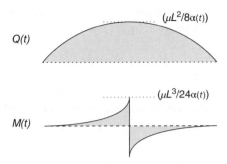

Figure 5.10   Internal loads for a member under rotational acceleration due to a central couple.

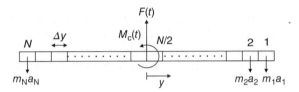

Figure 5.11   External loads for non-uniformly distributed loading – discretized member.

5.7   Internal Loads for a Discretized Member

The above analysis applies to continuous members, i.e. where both the structure and distributed loads are treated as continuous and the internal load expressions required integration over the length of the member. However, for a real structure such as an aircraft, while the structure and loading are still nominally continuous, they are far too complex to treat as analytic functions.

The structure is then idealized by dividing it into discrete sections, with the loads being applied to each section; only a finite set of internal loads will be determined. The analysis is approximate, with the analytic integral used earlier being replaced by a summation; however, provided an adequate number of sections are used, the accuracy should be satisfactory. The idea will be illustrated for the earlier example of a member under the action of a central force and moment, but this time the treatment is discretized. It should be recognized that the treatment of a discretized member is essentially the same whether the loading is uniformly or non-uniformly distributed, since in either case a summation approach would be required. Note that discretized structures were considered in Chapter 4, and will be considered further in Chapters 16, 17 and 19.

5.7.1   Distributed Inertia Forces for a Discretized Member

Consider the uniform rigid member of length L and mass per length μ, subject to a centrally applied dynamic load $F(t) = F_{\text{Applied}}$, as shown in Figure 5.11. The member is approximated by discretizing it into N sections, each of equal length Δy $(=L/N)$ and mass $m_k = \mu L/N$; the left-hand end of the kth section is at a distance y_k from the centre of the member. The mass is lumped at the centre of each section. d'Alembert's principle may then be applied to this discretized member by introducing, at the kth section, an inertia force equal to the product of the section mass m_k and the section acceleration $a_k(t)$ at a chosen instant of time t. The inertia force on the kth section at time t is then given by

$$F_{\text{Inertia\_k}}(t) = m_k a_k(t) \tag{5.18}$$

The inertia forces are shown in Figure 5.11 and the structure is in static equilibrium under the applied and inertia forces. The acceleration at each mass point can be found using the structure mass $m = \sum m_k$ and moment of inertia $I_c = \sum m_k r_k^2$, where r_k is the distance from the centre of mass to the kth mass point.

5.7.2 Internal Loads for a Discretized Member

For a discretized member, the internal loads are only determined at the interfaces between the sections. Consider a cut in the right-hand half of the member at the interface between the jth and $(j+1)$th sections, with internal loads $Q_j(t)$ and $M_j(t)$ introduced to the subsection, as shown in Figure 5.12. For equilibrium of the right-hand subsection, there is no net force or moment at the cut, so

$$Q_j(t) = \sum_{k=1}^{j} F_{\text{Inertia\_k}}(t) - F_{\text{Applied}}(t)$$

and   (5.19)

$$M_j(t) = \sum_{k=1}^{j} F_{\text{Inertia\_k}}(t) \left(y_k + \frac{\Delta y}{2} - y_j \right) - F_{\text{Applied}}(t) \left(\frac{L}{2} - y_j \right)$$

Thus the approach is the same as for the continuous system, except that integration is replaced by summation.

The above case will now be considered using numerical values. Consider the example of the member under uniform acceleration from an applied force alone, with $N = 10$ sections, $\mu = 100$ kg/m, $L = 1$ m (i.e. total mass of 100 kg) and $F_{\text{Applied}} = 1000$ N, so that $\Delta y = 0.1$ m, $m_{1-10} = \mu$, $\Delta y = 10$ kg, $a_{1-10} = F/\mu L = 10$ m/s^2 and $F_{\text{Inertia\_1-10}} = 100$ N. Applying Equation (5.19) yields the internal load diagram shown in Figure 5.13. The results may be compared to those in Figure 5.8 for the continuous member and there is good agreement.

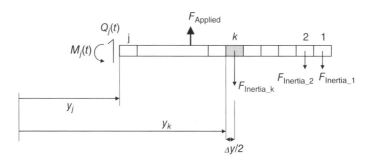

Figure 5.12 Internal loads for nonuniformly distributed loading – discretized member.

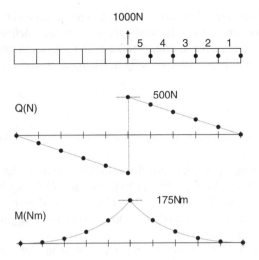

Figure 5.13   Internal load diagrams for a member accelerating under a central force.

5.8   Intercomponent Loads

An aircraft is often composed of several interconnected major components (e.g. wing, fuselage, tailplane, fin, landing gear and engines). It can be helpful to consider a particular component in isolation; this is achieved by 'cutting' it away and introducing equal and opposite 'intercomponent' loads at the interface. The internal loads for the isolated component may then be found; this concept will be illustrated for the wing and fuselage in Chapter 16.

5.9   Obtaining Stresses from Internal Loads – Structural Members with Simple Load Paths

The process of obtaining internal loads, described so far in this chapter, is essentially the first stage of a calculation aimed eventually at obtaining stresses. For a slender member with well-defined load paths (e.g. a circular tube or T-section member), once any moment, axial, shear and torsion ('MAST') internal loads present have been determined at any cross-section, it is then possible to determine the stresses at that cross-section from basic stress analysis theory using the cross-sectional properties (Benham *et al.*, 1996).

As an example, the direct stress σ for bending about the neutral axis of a member (i.e. the axis in the cross-section where this stress is zero) is given by

$$\sigma = \frac{My}{I} \tag{5.20}$$

where M is the bending moment, I is the cross-section second moment of area and y is the distance from the neutral axis (Benham *et al.*, 1996). Similar expressions apply for other loadings on simple structures (Young, 1989; Benham *et al.*, 1996) and for simple aerospace structures (Donaldson, 1993; Megson, 1999), but a treatment of stress analysis is beyond the scope of this book. Such methodologies may be employed for structures with well-defined load paths.

However, it should be noted that such classical formulaic approaches are not suitable for complex aerospace structures where the load paths are not well defined. Instead, additional analyses will need to be carried out to determine loads and stresses in structural elements using the internal loads; this will be explained briefly in Chapters 16 and 20.

5.10   Examples

Note that the signs of the internal loads depend upon the sign convention used.

1. Convert the following quantities into SI or Imperial units as appropriate: (a) $10\,\text{N}\,\text{m}$, (b) 5 lbf ft, (c) $10\,\text{N/m}^2$, (d) 5 lbf/ft^2, (e) $10\,\text{kg}\,\text{m}^2$ and (f) 5 slug ft^2.
   [7.376 lbf ft, 6.779 N m, 0.209 lbf/ft^2, 239.4 N/m^2, 7.376 slug ft^2, 6.779 $\text{kg}\,\text{m}^2$]
2. Draw the free body diagram and find the support reactions in magnitude and direction for the member shown in Figure 5.14.
   [Left-hand support 1.73 kN upwards and 5.50 kN to the left and right-hand support 0.87 kN upwards – do not forget the moment produced by the horizontal 5 kN force]

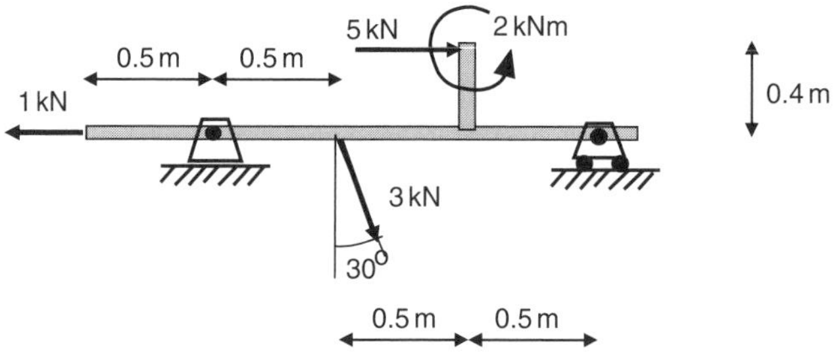

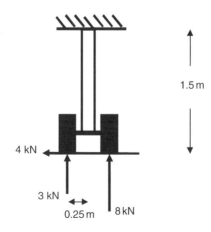

Figure 5.14

3. A nose landing gear is subject to a set of loads as shown in Figure 5.15. Find the force and moment reactions at the support.
   [11 kN down, 4 kN to the right and 4.75 kN m anticlockwise]

Figure 5.15

4. Replace the lift force L acting at the centre of pressure (CP) on the aerofoil shown in Figure 5.16 by an equivalent force and couple arrangement at the aerodynamic centre position (AC). Hint: place a pair of equal and opposite forces L at AC and replace one parallel force pair by a couple.
   [Force L and couple Lh nose down]

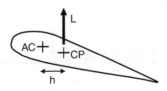

Figure 5.16

5. For the built-in slender member shown in Figure 5.17, draw each 'cut' part together with external and internal loads. Use equilibrium to determine expressions for the shear force and bending moment at each cut. Then draw the internal load diagrams and find the support reactions.
   [Maximum bending moment -10 kN m at 9 kN load position, reactions 4 kN and 2 kN m at root]

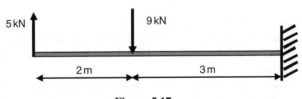

Figure 5.17

6. For the simply supported slender member shown in Figure 5.18, determine the reactions and draw each 'cut' part together with external reaction components and internal loads. Use equilibrium to determine expressions for the shear force and bending moment at each cut. Then draw the internal load diagrams and indicate the support reactions.
   [Maximum bending moment 12 kN m at the left-hand support, reactions 10 kN up at the left-hand support and 4 kN down at the right-hand support]

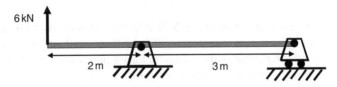

Figure 5.18

7. For the simply supported slender member shown in Figure 5.19, determine the reactions and draw each 'cut' part together with external and internal loads. Use equilibrium to determine expressions for the shear force and bending moment at each cut. Then draw the internal load diagrams and indicate the support reactions.
[Bending moment varying quadratically, with maximum 5.625 kN m at centre, reactions 7.5 kN at each end]

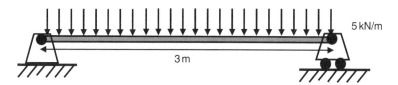

3 m

5 kN/m

Figure 5.19

8. A uniform free-free member of length L and mass per length μ is subject to applied forces $F/2$ acting normal to the member axis at a distance $b/2$ either side of its centre. Determine the inertia loading distribution and draw an FBD for the member in effective static equilibrium for an accelerated condition. Then determine the bending moments at the centre and at one load point. Sketch the bending moment variations for values of $b/L = 0$, $\frac{1}{3}$, $\frac{2}{3}$, 1 and note how the load position influences the behaviour.
$$\left[\text{Load point } (-FL/8)(1-b/L)^2 \text{ and centre } (FL/4)\left(b/L-\tfrac{1}{2}\right)\right]$$

9. For the built-in slender member shown below in Figure 5.20, draw each 'cut' part together with external and internal loads. Use equilibrium to determine expressions for the shear force and bending moment at each cut. Then draw the internal load diagrams and indicate the support reactions.
[Bending moment is constant at 5 kN m in the left-hand section, changing parabolically to –4 kN m at the support, where the other reaction is 6 kN]

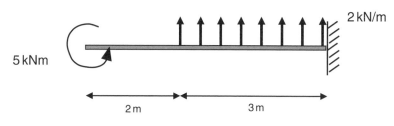

5 kNm

2 kN/m

2 m   3 m

Figure 5.20

10. The slender member shown in Figure 5.21 is built-in at point A and loaded by torques at points B and C. Draw each 'cut' part together with external and internal torsional loads. Use equilibrium to determine expressions for the torque at each cut. Then draw the internal load diagrams and indicate the support reaction.
[Torque is 4 kN m at root, essentially the same diagram as for the shear force in Example 5]

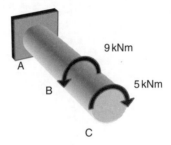

9 kNm

A

B   5 kNm

C

Figure 5.21

11. A uniform member of length 10 m and mass per length 200 kg/m is subject to an applied force of 10 kN acting normal to the member axis at its centre. Determine the inertia loading distribution and draw an FBD for the member in effective static equilibrium for this accelerated condition. Then determine the bending moment at the centre (a) assuming the member is continuous and (b) assuming the member is discretized into 10 sections.
   [Both 12.5 kN m]

12. A uniform member of length L and mass per length μ is subject to an applied force F acting normal to the member axis at one end of the member. Determine the inertia loading distribution and draw an FBD for the member in effective static equilibrium for this accelerated condition. Then determine the bending moment at the centre.
   [Centre $3FL/8$]

6

Introduction to Control

Control systems are used in a wide range of engineering applications and industries, to enable a system (e.g. an aircraft) to respond in some desired way when subjected to some form of external input. For example, a Gust Load Alleviation (GLA) system might use acceleration measurements on the aircraft to detect the motion due to turbulence and then employ the control surfaces in such a way as to reduce the loads acting on the aircraft structure. Other inputs could be provided by the pilot and the control system could limit the loads generated. The systems used on aircraft for control in flight may be electrical, mechanical, hydraulic or pneumatic and perform a widely differing range of tasks, e.g. provision of required stability and handling properties, carefree handling, Manoeuvre Load Alleviation system, etc. Modern aircraft have extremely sophisticated Flight Control Systems (FCS) (Pratt, 2000) that play an important role in the aeroelastic and loads behaviour of the aircraft, so it is important to understand key issues of control.

This chapter will examine some of the basic control tools and definitions that need to be understood before the application of control systems to aeroelastic systems, such as the science of aeroservoelasticity (or structural coupling), which will be tackled in Chapter 11. The aircraft Flight Control System will be considered further in Chapters 13 and 21. Many textbooks cover basic control theory, e.g. Raven (1994) and Dorf and Bishop (2004).

6.1 Open and Closed Loop Systems

Consider the system in Figure 6.1 that responds in some way, known as the output, to some given input. This representation of the resulting output due to an applied input could, for example, be used to describe the direction that an aircraft flies in, subject to the application of the control surfaces.

A controller can be added to the set-up, as shown in Figure 6.2, and is used to define the control inputs needed in order to manoeuvre the aircraft so that it flies on a particular heading. This is known as an *open loop* system. The controller may be designed using trial and error or past experience to dictate what control surface deflections are needed to change the direction of the aircraft. However, no account is made of any external influences, e.g. the wind direction and speed, and the consequent effect this has on the aircraft.

In order to steer the aircraft accurately, a continually updated comparison between the required direction and that actually being flown needs to be made. The aircraft's heading

---

Introduction to Aircraft Aeroelasticity and Loads, Second Edition. Jan R. Wright and Jonathan E. Cooper.
© 2015 John Wiley & Sons, Ltd. Published 2015 by John Wiley & Sons, Ltd.

Figure 6.1   Basic open loop system.

Figure 6.2   Open loop system with open loop controller.

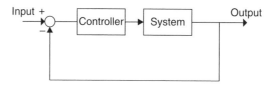

Figure 6.3   Closed loop system with a forward controller and a unity feedback loop.

can be continually changed until the difference between the required and actual direction is zero. Figure 6.3 shows this set-up, known as a *closed loop* system, where it can be seen that the output is fed back into the system and compared with the desired input value. The difference (error) between the two is then passed through the controller element which is designed to produce the required output from the system. Such a closed loop system is the basis of all control systems but there are many different ways of implementing the control.

The ratio between any two points in the system (often the output and input) in either the Laplace or frequency domains (Raven, 1994) is known as the Transfer Function (TF). Note however, that in vibration analysis the ratio between output and input signals is also commonly called the Frequency Response Function (FRF) (see Chapters 1 and 2).

6.2   Laplace Transforms

The Laplace Transform is one of the key mathematical tools used to model control systems. Essentially, it is a mathematical transformation that enables functions of time t to be reformulated in terms of the so-called Laplace operator s. One particular benefit of the transform is that differential equations in t can be expressed as algebraic expressions in s, which can then be used to determine the system Transfer Function with or without a control system present. The Transfer Function then enables the stability and dynamic characteristics of the system to be investigated. A further advantage of employing the Laplace approach is that, by using a transformation between s and ω, the Transfer Function can be expressed in the frequency domain.

The Laplace Transform $F(s)$ of a time function $f(t)$ is defined as

$$L\{f(t)\} = F(s) = \int_{0}^{\infty} f(t)e^{-st}dt \qquad (6.1)$$

Table 6.1 Example time functions and corresponding Laplace Transforms

Time domain function $f(t)$	Laplace domain function $F(s)$	Time domain function $f(t)$	Laplace domain function $F(s)$
Unit impulse	1	Unit step function	$\dfrac{1}{s}$
Exponential decay e^{-at}	$\dfrac{1}{s+a}$	$\sin \omega t$ and $\cos \omega t$	$\dfrac{\omega}{s^2+\omega^2}$ and $\dfrac{s}{s^2+\omega^2}$
$\dfrac{d}{dt}f(t)$	$sF(s)-f(0)$	$\dfrac{d^2}{dt^2}f(t)$	$s^2F(s)-sf(0)-\dfrac{df(0)}{dt}$
$e^{at}\sin \omega t$	$\dfrac{\omega}{(s-a)^2+\omega^2}$	$e^{at}\cos \omega t$	$\dfrac{s-a}{(s-a)^2+\omega^2}$

where the usual convention of time functions being written in lower case and Laplace operator functions written in upper case is followed. It is also possible to invert the process so that the corresponding time function for a given Laplace function can be found.

In general, the integral does not have to be solved since tables are available giving the Laplace Transforms of the most common functions, some of which are shown in Table 6.1. More complicated expressions can be tackled through the use of simple rules that apply to linear systems (Raven, 1994). Note in the table that $f(0)$, for example, is the initial condition of $f(t)$.

6.2.1 Solution of Differential Equations using Laplace Transforms

As an example of the use of Laplace Transforms, the solution of differential equations will be illustrated with a single degree of freedom (SDoF) system, subjected to a unit step function at time $t=0$ (with initial conditions $x(0)=0$, $\dot{x}(0)=0$). As determined in Chapter 1, the equation of motion is

$$m\ddot{x}+c\dot{x}+kx=f(t) \quad \text{or} \quad \ddot{x}+2\zeta\omega_n\dot{x}+\omega_n^2 x=\frac{f(t)}{m} \tag{6.2}$$

and for this example $f(t)$ is taken as the unit step function.

The idea of the analysis is to use the Laplace Transform to transform a differential equation in time into an algebraic equation in the Laplace variable. The algebraic equation may then be solved readily and transformed back to the time domain to yield the response solution.

Transforming each term in Equation (6.2) using the Laplace Transforms given in Table 6.1 leads to

$$m\left[s^2X(s)-sx(0)-\dot{x}(0)\right]+c\left[sX(s)-x(0)\right]+kX(s)=\frac{1}{s} \tag{6.3}$$

or, with zero initial conditions,

$$\left(ms^2+cs+k\right)X(s)=m\left(s^2+2\zeta\omega_n s+\omega_n^2\right)X(s)=\frac{1}{s} \tag{6.4}$$

This algebraic equation shows that a differential equation in time has been transformed into an algebraic expression in s. Rearranging this equation yields the response Laplace Transform

$$X(s) = \frac{1}{ms\left(s^2 + 2\zeta\omega_n s + \omega_n^2\right)} \tag{6.5}$$

This expression now needs inverse transformation to get back to the time domain, but the form of the expression needs modifying so as to be able to use the table. Rewriting Equation (6.5) in terms of a partial fraction expansion gives

$$X(s) = \frac{1}{m}\left(\frac{A}{s} + \frac{Bs + C}{s^2 + 2\zeta\omega_n s + \omega_n^2}\right) = \frac{1}{m}\left[\frac{A\left(s^2 + 2\zeta\omega_n s + \omega_n^2\right) + Bs^2 + Cs}{s\left(s^2 + 2\zeta\omega_n s + \omega_n^2\right)}\right] \tag{6.6}$$

where the unknown constants A, B and C are found by comparing coefficients of s in Equations (6.5) and (6.6). Thus the partial fraction expansion of $X(s)$ becomes

$$X(s) = \frac{1}{m\omega_n^2}\left(\frac{1}{s} - \frac{s + 2\zeta\omega_n}{s^2 + 2\zeta\omega_n s + \omega_n^2}\right) \tag{6.7}$$

The final step is to transform back into the time domain, using the relationships shown in Table 6.1. Thus

$$x(t) = \frac{1}{k}\left[1 - \frac{\omega_n}{\omega_d}e^{-\zeta\omega_n t}\sin\left(\omega_d t + \psi\right)\right] \tag{6.8}$$

which is the same answer for the Step Response Function as found using the analytical approach described in Chapter 1. Responses to other inputs may be obtained using a similar approach. Note that it is usual nowadays to use software such as MATLAB and SIMULINK to solve such systems rather than relying upon the solution using Laplace Transforms.

6.3 Modelling of Open and Closed Loop Systems using Laplace and Frequency Domains

The control of a system is achieved through either open or closed loop control systems. If the Laplace Transform of the open loop system is $G(s)$ in Figure 6.1, then the Transfer Function (TF) between the input $X(s)$ and output $Y(s)$ is given as

$$\text{TF}_{\text{System}} = G(s) = \frac{Y(s)}{X(s)} \tag{6.9}$$

The inclusion of a controller $H(s)$ as part of the *open loop* system as shown in Figure 6.2 results in the Transfer Function between the output and input

$$\text{TF}_{\text{Open Loop}} = \frac{Y(s)}{X(s)} = H(s)G(s) \tag{6.10}$$

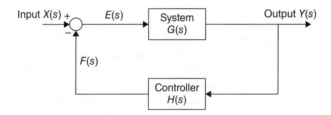

Figure 6.4   Closed loop system with controller in the feedback loop.

However, if the system is made *closed loop* so that the output to the system $G(s)$ is fed back into the input via a controller in the feedback path represented by $H(s)$, as shown in Figure 6.4, then it becomes possible to influence directly the input to the system in order to control the output. This system makes use of *negative feedback* shown by the minus sign on the feedback loop, and the error is input to the system.

There are two separate Transfer Functions to consider for each part of the system shown in Figure 6.4:

a. between the input $E(s)$ to the system and the output $Y(s)$ $G(s) = Y(s)/E(s)$
b. between the output $Y(s)$ and the feedback path output $F(s)$ $H(s) = F(s)/Y(s)$

In addition the relationship between the actual input $X(s)$, the error signal $E(s)$ and the feedback path output $F(s)$ can be written as

$$E(s) = X(s) - F(s) \qquad (6.11)$$

Rearranging these expressions gives the overall TF of the closed loop system with the controller placed in the *feedback* loop as

$$\text{TF}_{\text{CL feedback}} = \frac{Y(s)}{X(s)} = \frac{G(s)}{1 + G(s)H(s)} \qquad (6.12)$$

Using a similar approach for the closed loop system in Figure 6.3, where the controller is the *forward* path, yields the closed loop Transfer Function

$$\text{TF}_{\text{CL forward}} = \frac{Y(s)}{X(s)} = \frac{G(s)H(s)}{1 + G(s)H(s)}. \qquad (6.13)$$

Note that both the feedback and forward configurations of control, resulting in the closed loop Transfer Functions in Equations (6.12) and (6.13), have the same stability since the denominators are the same.

The Transfer Functions of far more complicated systems can be determined simply by following the above approach. It will be seen later how it can be useful to convert Transfer Functions based in the Laplace domain into the frequency domain simply through the use of the transformation $s = i\omega$, where $i = \sqrt{-1}$ is the complex variable used in this book (though the symbol j is normally used by control engineers).

6.4   Stability of Systems

Control systems are used to influence the behaviour of the system that is being controlled, in particular to manipulate the response to different inputs. Care must be taken that the characteristics of the closed loop system are favourable and that instabilities do not occur due to the interaction between the feedback control and the system. For example, the use of an incorrectly designed flight control system on an aircraft might result in flutter (refer to Chapter 11). Although time simulations of the system could be used to determine whether the application of a particular control loop results in a stable response, this is a very inefficient approach, particularly when the effect of changing many system and control parameters needs to be investigated.

   In this section, a range of commonly used tools are described that can be used to determine whether a system is stable or not, and also to determine what the critical conditions are that define the boundary between stable and unstable behaviour.

6.4.1   Poles and Zeros

Consider the following representation of the Transfer Function of a general closed loop system

$$TF_{CL}(s) = \mu \frac{s^m + b_{m-1}s^{m-1} + \cdots + b_1 s + b_0}{s^n + a_{n-1}s^{n-1} + \cdots + a_1 s + a_0} \tag{6.14}$$

where μ is a factor and a, b are the denominator and numerator polynomial coefficients. This equation can be written in terms of the roots of the denominator p_i, known as *poles* (TF reaches a peak at these roots), and the roots of the numerator z_i, known as *zeros* (TF reaches a minimum at these roots), such that

$$TF_{CL}(s) = \frac{\mu(s-z_1)(s-z_2)\ldots(s-z_m)}{(s-p_1)(s-p_2)\ldots(s-p_n)} \tag{6.15}$$

The poles are the roots of the *characteristic equation* of the system, which is found by setting the denominator of the TF to zero; the poles determine the stability of the closed loop system. For an oscillatory system (as usually encountered in aeroelastic or mechanical systems), the roots occur in complex conjugate pairs of the form $\sigma \pm i\theta = -\zeta\omega \pm i\omega\sqrt{1-\zeta^2}$.

   It can be shown that the system becomes unstable if the real part of any of the poles is positive ($\sigma > 0$). Figure 6.5 shows s plane plots (Argand diagrams) for three different cases of an SDoF system together with the corresponding time responses to an initial disturbance. Since this is a second order system, there are two poles that usually occur as a complex conjugate pair. When there are negative real parts of the poles ($\sigma < 0$), a stable time response occurs, as seen in Figure 6.5(a). Figure 6.5(b) shows the result for poles with a zero real part ($\sigma = 0$) and this leads to a critical response whereby the amplitude of the time history remains constant. Finally, in Figure 6.5(c), when the poles move to the right-hand side of the s plane the system response becomes unstable. If the imaginary part of a pole is zero ($\theta = 0$), then the motion relating to this pole cannot be oscillatory (but could still become unstable in a static sense if the real part is positive). For a multi-degree of freedom system (MDoF) (see Chapter 2), the motion involving all of the poles being damped is far more complicated than that shown in Figure 6.5(a). However, the motion of a single critically stable, or unstable, mode will dominate the response of an MDoF system and the resulting motion will be similar to that in Figures 6.5(a), (b) and (c) respectively.

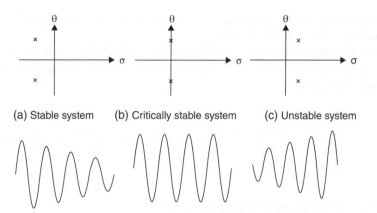

(a) Stable system   (b) Critically stable system   (c) Unstable system

Figure 6.5   Relation of position of poles to system stability.

6.4.2   Routh–Hurwitz Method

Although the roots of a polynomial can be determined using numerical software such as the ROOTS function in MATLAB, there can be occasions when the stability of a system needs to be determined without explicitly calculating the roots. The Routh–Hurwitz method (Bisplinghoff *et al.*, 1996) can be used to check whether a system is stable just by considering the coefficients of the characteristic polynomial. The technique will be used later on in Chapter 10 as a means to determine if and when flutter occurs. Only the method, and not the proof, will be described here.

Any *n*th-order polynomial

$$a_n s^n + a_{n-1} s^{n-1} + \cdots + a_1 s + a_0 = 0 \tag{6.16}$$

will have stable roots if all the coefficients $a_i > 0$ and the *n* determinants $T_1 \ldots T_n > 0$ where T_n is the $n \times n$ determinant that takes the form

$$
\begin{vmatrix}
a_{n-1} & a_{n-3} & \cdots & 0 & 0 & 0 \\
a_n & a_{n-2} & \cdots & a_0 & 0 & 0 \\
0 & a_{n-1} & \cdots & a_1 & 0 & 0 \\
\vdots & a_n & \cdots & a_2 & a_0 & 0 \\
\vdots & \vdots & & a_3 & a_1 & 0 \\
0 & 0 & \cdots & a_4 & a_2 & a_0
\end{vmatrix}
\tag{6.17}
$$

The subdeterminants $T_1 \ldots T_{n-1}$ are found as

$$
T_1 = |a_{n-1}| \quad
T_2 = \begin{vmatrix} a_{n-1} & a_{n-3} \\ a_n & a_{n-2} \end{vmatrix} \quad
T_3 = \begin{vmatrix} a_{n-1} & a_{n-3} & a_{n-5} \\ a_n & a_{n-2} & a_{n-4} \\ 0 & a_{n-1} & a_{n-3} \end{vmatrix} \quad \text{etc.}
\tag{6.18}
$$

A comparison of Equations (6.17) and (6.18) shows how these sub-determinants are formed. An increasing number of terms from the $n \times n$ determinant are taken, starting at its top left-hand corner. Note that if the subscript of any of the above terms is negative, then the term is taken as zero.

For example, consider the quartic equation $a_4s^4 + a_3 s^3 + a_2 s^2 + a_1s + a_0 = 0$ and set up the determinants up to 4×4 in size; therefore

The polynomial has stable roots if $a_i > 0$ for $i = 0,1,2,3,4$ (includes $T_1 > 0$), $a_3a_2 - a_1a_4 > 0$ (T_2) and $a_1a_2a_3 - a_0a_3^2 - a_1^2a_4 > 0(T_3)$. There is no need to calculate the largest determinant T_4 as this is equal to a_0T_3.

$$T_4 = \begin{vmatrix} a_3 & a_1 & 0 & 0 \\ a_4 & a_2 & a_0 & 0 \\ 0 & a_3 & a_1 & 0 \\ 0 & a_4 & a_2 & a_0 \end{vmatrix}, \quad T_3 = \begin{vmatrix} a_3 & a_1 & 0 \\ a_4 & a_2 & a_0 \\ 0 & a_3 & a_1 \end{vmatrix}, \quad T_2 = \begin{vmatrix} a_3 & a_1 \\ a_4 & a_2 \end{vmatrix}, \quad T_1 = a_3.$$

6.4.3 Frequency Domain Representation

When designing a system with a feedback or forward controller, it is often of interest to examine the effect of changing the gain and/or phase of the controller in order to determine if and when stability is lost. Such investigations can be carried out in the Laplace domain using root locus plots or in the frequency domain using Nyquist or Bode plots. The resulting figures can also be used to define how much the gain or phase can be increased before instability occurs.

6.4.3.1 Root Locus

Root locus plots are used to show the effect of changing the control system gain on the position of the closed loop poles (the so-called root loci); the real and imaginary parts of the poles are plotted on an Argand diagram. Instability occurs when any denominator root of the closed loop system crosses the imaginary axis and the real part becomes positive. The gain at which oscillatory roots become non-oscillatory (i.e. the imaginary part becomes zero) can also be determined. It is possible to use the root locus plot to adjust the open loop zeros and poles in order to affect the behaviour of the closed loop poles i.e. damping, frequencies and occurrence of instabilities.

There are a number of rules for drawing the root loci by hand (Raven, 1994). However, as it is usual nowadays to simply plot them out using numerical software, these rules will not be considered here.

As an example, consider the closed loop feedback system shown in Figure 6.4. The open loop Transfer Function is taken as $G(s) = 1/[s(s^2 + 4s + 8)]$ with a feedback loop term $H(s) = K$ where K is the gain. This gives a closed loop Transfer Function using the approach defined earlier as

$$\frac{Y(s)}{X(s)} = \frac{1}{s^3 + 4s^2 + 8s + K} \tag{6.19}$$

Figure 6.6 shows how the root loci change with values of gain K varying from 1 to 40. It can be seen that there is an oscillatory pair of (complex conjugate) poles and a single non-oscillatory pole (zero imaginary part). The system is stable for all values of $K < 32$ and this can be verified using the Routh–Hurwitz criteria. However, beyond this critical value the system becomes unstable and the real parts of the complex poles become positive. Note that as the poles change, the values of the corresponding frequency and damping ratio of the closed loop system alter as well.

It can be seen from the above example that the introduction of a feedback loop has allowed the system characteristics to be altered from those of the basic open loop system. In control design, the form and gain of the feedback controller $H(s)$ are chosen to achieve the desired

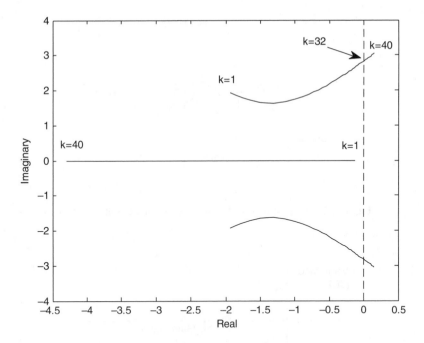

Figure 6.6   Root loci trends for different gain values.

closed loop characteristics, such as overshoot, rise time and settling time. These requirements vary for different types of system.

6.4.3.2   Stability Analysis using Nyquist and Bode Plots

By application of the transformation ($s = i\omega$) the Transfer Function in the Laplace domain is transformed into the frequency domain and thus Equation (6.12) becomes

$$\text{TF}_{\text{Closed Loop}}(\omega) = \frac{G(i\omega)}{1 + G(i\omega)H(i\omega)} \tag{6.20}$$

It is usual to display this type of representation in terms of the Bode plot (gain (dB) and phase angle versus frequency) or the Nyquist plot (real part versus imaginary part for different frequencies).

Considering the denominator of the expression in Equation (6.20) for the Transfer Function, then a system can be shown to be stable if the term $G(i\omega)H(i\omega)$ has an amplitude ratio on the Bode plot less than 0 dB when the phase angle is $-180°$. On the Nyquist diagram this is equivalent to an amplitude ratio of less than -1 at a phase angle of $-180°$ thus the Transfer Function must not enclose the point (-1) on the real axis. Figure 6.7 shows typical Nyquist plots for a stable and an unstable system.

It is useful for control law design to determine how far from instability the system is, and this can be defined by the gain and phase margins. The gain margin defines at the $-180°$ phase how much the magnitude is below 0 dB (Bode) or 1 (Nyquist), whereas the phase margin defines at 0 dB (Bode) or amplitude of 1 (Nyquist) how much the phase is greater than $-180°$. Schematic

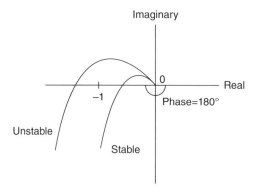

Figure 6.7 Nyquist representations of stable and unstable systems.

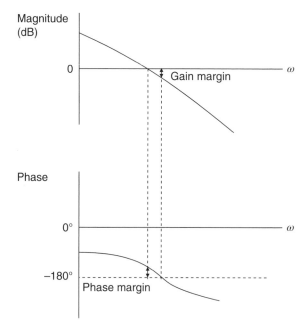

Figure 6.8 Bode plot showing gain and phase margins.

representations of the gain and phase margins for both Bode and Nyquist plots are shown in Figures 6.8 and 6.9.

Returning to the above example, defined by Equation (6.19), and making use of the substitution $s = i\omega$, then Bode and Nyquist plots of $G(i\omega)$ $H(i\omega)$ are shown for gain values of 20 and 32 in Figures 6.10 and 6.11. Note that they both have the same phase plot since the gain does not affect the phase in this case. It can be seen that when $K = 32$ the system is marginally stable, with the TF magnitude of 0 dB corresponding to the $-180°$ phase.

Airworthiness certification regulations define the amount of gain and phase margins that must be present when flight control systems are used for civil and military aircraft.

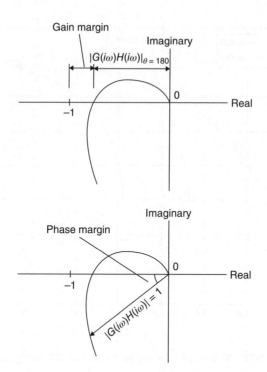

Figure 6.9 Nyquist plot showing gain and phase margins.

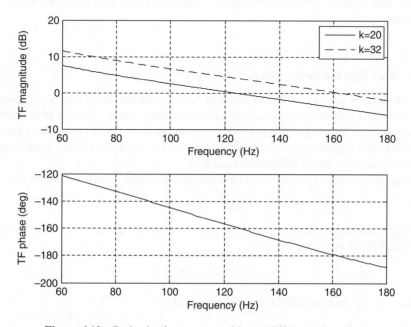

Figure 6.10 Bode plot for a system with two different gain values.

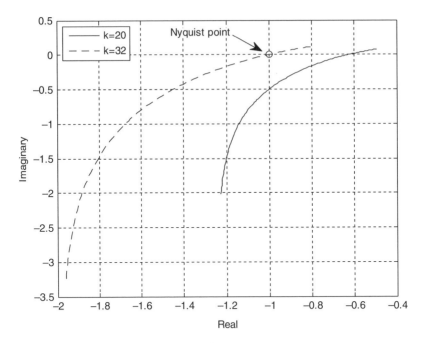

Figure 6.11   Nyquist plot for a system with two different gain values.

6.4.4   Time Domain Representation

The alternative to performing a control system analysis in the Laplace or frequency domains is to use the time domain, often deriving models expressed in terms of acceleration, velocity or displacement.

6.4.4.1   State Space Representation

For a time domain analysis, it is convenient to make use of the so-called *state space models*, which are based upon the so-called system *states*. The states are any sets of variables that must be linearly independent and sufficient in number to define the dynamic behaviour of the system, but cannot be the system inputs (or linear combinations of the inputs). State space models can be used to model any set of differential equations.

For example, consider a second-order mechanical system with input u and output y in the form

$$\ddot{y} + 2\zeta\omega_n\dot{y} + \omega_n^2 y = u \tag{6.21}$$

which can be rewritten in terms of two first-order differential equations with the states defined as $x_1 = y$ and $x_2 = \dot{y}$ such that $\dot{x}_1 = x_2$

$$\dot{x}_2 = u - \omega_n^2 x_1 - 2\zeta\omega_n x_2 \tag{6.22}$$

In matrix form, these equations may be expressed as

$$\left\{ \begin{array}{c} \dot{x}_1 \\ \dot{x}_2 \end{array} \right\} = \left[\begin{array}{cc} 0 & 1 \\ -\omega_n^2 & -2\zeta\omega_n \end{array} \right] \left\{ \begin{array}{c} x_1 \\ x_2 \end{array} \right\} + \left[\begin{array}{c} 0 \\ 1 \end{array} \right] \{u\} \quad \text{and} \quad \{y\} = \left[1 \ \ 0 \right] \left\{ \begin{array}{c} x_1 \\ x_2 \end{array} \right\} + [0]\{u\} \tag{6.23}$$

Equations (6.23) are in the form of the so-called *state space equations*

$$\dot{x} = \mathbf{A}_s x + \mathbf{B}_s u \quad \text{and} \quad y = \mathbf{C}_s x + \mathbf{D}_s u \tag{6.24}$$

where for an Nth-order system with N_I inputs and N_O outputs, x is the $N \times 1$ state vector, u the input vector $N_I \times 1$, y the $N_O \times 1$ output vector, $\mathbf{A}_s$ the $N \times N$ system matrix, $\mathbf{B}_s$ the $N \times N_I$ input matrix, $\mathbf{C}_s$ the $N_O \times N$ output matrix and $\mathbf{D}_s$ the $N_I \times N_O$ feedforward matrix (usually zero). The use of the subscript 's' for these matrices is nonstandard for state space analysis, but the subscript is added to avoid confusion with different matrices in the general aeroelastic and loads equations presented later.

The state space Equations (6.24) can then be solved in the time domain using numerical integration to obtain the response of the system to any input. For multivariable problems they are a succinct method for describing the dynamics and advanced matrix tools can be used to analyse the systems. Note that the eigenvalues of the system matrix $\mathbf{A}_s$ are the same as the poles of the Laplace domain Transfer Function. Also, the number of first-order state space equations is twice that of the second-order representation.

It is a straightforward operation to transform from a state space model to a Transfer Function model. Taking Laplace Transforms of Equation (6.24), with zero initial conditions, gives

$$sX(s) = \mathbf{A}_s X(s) + \mathbf{B}_s U(s) \quad \text{so} \quad X(s) = [s\mathbf{I} - \mathbf{A}_s]^{-1} \mathbf{B}_s U(s) \tag{6.25}$$

Hence

$$Y(s) = \left(\mathbf{C}_s [s\mathbf{I} - \mathbf{A}_s]^{-1} \mathbf{B}_s + \mathbf{D}_s \right) U(s) = \mathbf{G}(s) U(s) \tag{6.26}$$

where, of course, $\mathbf{G}(s)$ is now a matrix of Transfer Functions corresponding to the multiple inputs and outputs.

There is a large body of work devoted to the use of digital control systems (Kuo, 1995) as opposed to the continuous time model considered above. However, these are beyond the scope of this book.

6.5 PID Control

The simplest and most commonly used type of control strategy typically sets the controller of the system in Figure 6.3 as linear multiples of the error E (proportional) along with its integral (I) and derivative (D) multiplied by some gain values. Hence the proportional-integral-derivative (PID) controller can be written as

$$h_{PID}(t) = K_p E + K_i \int E \, dt + K_d \frac{dE}{dt}, \tag{6.27}$$

where K_p, K_i, K_d are the proportional, integral and derivative gains. In the Laplace domain this becomes

$$H(s) = K_p + \frac{K_i}{s} + K_d s \tag{6.28}$$

There are various empirical schemes that can be used for setting the three gain values, but tuning of the gains often still has to be executed in order to get optimal performance. The proportional term determines the speed of the response, the integral term improves the accuracy of the final steady state, while the derivative term helps to stabilize the response.

State feedback control typically used the set-up in Figure 6.4 with the controller feedback equal to $-K_x$. Such an approach leads to optimal control techniques (Whittle, 1996), which specify that the gain matrix $\mathbf{K}$ is such that some cost function is minimized.

6.6   Examples

1. For the system with a forward controller in Figure 6.3, determine the closed loop Transfer Function for the combinations of $G(s)$ and $H(s)$ given below.

$G(s)$	$H(s)$
$\dfrac{K}{s^2+3s+9}$	1
$\dfrac{K}{s(s^2+3s+9)}$	$(s+1)$
$\dfrac{K}{s(s+1)(s+3)}$	$(s+1)(s+2)$
$\dfrac{K(s+2)}{s(s^2+3s+9)}$	$K_p+\dfrac{K_i}{s}+K_ds$

2. Repeat Example 1 but this time using the feedback control system shown in Figure 6.4.
3. By plotting out the root locus, determine when the systems in Examples 1 and 2 become unstable.
4. By plotting Bode and Nyquist plots, determine the gain and phase margins of the systems in Examples 1 and 2.
5. Use the Routh–Hurwitz method to determine whether the roots of the following polynomials are stable:

$$(a)\ x^2+x+4=0, \quad (b)\ x^3+x^2+2x+1=0, \quad (c)\ 3x^3+x^2+2x+1=0$$

and $(d)\ \ x^4+3x^3+x^2+2x+1=0.$
6. Use the Routh–Hurwitz method to determine the values of p for which the roots of the following polynomials are stable:

$$(a)\ x^3+px^2+2x+1=0, \quad (b)\ x^3+x^2+px+1=0, \quad (c)\ x^3+2x^2+2x+p=0$$

and $(d)\ \ x^4+2x^3+x^2+x+p=0.$
7. Confirm the results of Example 3 using the Routh–Hurwitz method.

Part II

Introduction to Aeroelasticity and Loads

7

Static Aeroelasticity – Effect of Wing Flexibility on Lift Distribution and Divergence

Static aeroelasticity is the study of the deflection of flexible aircraft structures under aerodynamic loads, where the forces and motions are considered to be independent of time. Consider the aerodynamic lift and moment acting upon a wing to depend solely upon the incidence of each chordwise strip (i.e. strip theory, see Chapter 4). These loads cause the wing to bend and twist, so changing the incidence and consequently the aerodynamic flow, which in turn changes the loads acting on the wing and the consequent deflections, and so on until an equilibrium condition is usually reached. The interaction between the wing structural deflections and the aerodynamic loads determines the wing bending and twist at each flight condition, and must be considered in order to model the static aeroelastic behaviour. The static aeroelastic deformation in the steady flight condition is important since it governs the lift distribution, the drag forces, the effectiveness of the control surfaces, the aircraft trim behaviour and also the static stability and control characteristics. The aeroelastic wing shape in the cruise condition is of particular importance as this has a crucial effect on the drag and therefore the range.

Through the elimination of time-dependent forces and motion, the inertia forces can be ignored in the equilibrium equations as these are dependent upon acceleration. Also, only steady aerodynamic forces need to be included in the analysis. Consequently, the modelling of static phenomena is much easier than dynamic aeroelastic phenomena where unsteady aerodynamic effects must be considered (see Chapter 9).

There are two critical static aeroelastic phenomena that can be encountered, namely *divergence* and *control reversal*. The latter will be considered in Chapter 8. Divergence is the name given to the phenomenon that occurs when the moments due to aerodynamic forces overcome the restoring moments due to structural stiffness, so resulting in structural failure. The most common type is that of wing torsional divergence. On a historical note, it is thought that Langley's attempt to fly some months before the Wright brothers' successful flights in 1903 failed due to the onset of divergence (Collar, 1978; Garrick and Reid, 1981). When the Langley aircraft was rebuilt some years later by Curtis with a much stiffer wing structure, the aircraft

Introduction to Aircraft Aeroelasticity and Loads, Second Edition. Jan R. Wright and Jonathan E. Cooper.
© 2015 John Wiley & Sons, Ltd. Published 2015 by John Wiley & Sons, Ltd.

flew successfully. For aeroelastic considerations, the stiffness is of much greater importance than the strength.

In modern aircraft, the flutter speed (i.e. the air speed at which flutter, a dynamic aeroelastic instability, occurs: see Chapter 10) is usually reached before the divergence speed (the air speed at which divergence occurs) so divergence is not normally a problem. However, the divergence speed is a useful measure of the general stiffness of the aircraft structure and must be considered as part of the certification process (CS-25 and FAR-25).

In this chapter, the static aeroelastic behaviour of a rigid aerofoil supported upon a torsional spring will be examined. Then, the aerodynamic lift distribution on a flexible wing fixed at the root will be considered using a simple aeroelastic model involving wing twist, and the divergence condition will be shown. The influence of the aircraft trim on the divergence speed and lift distribution for a simple heave/pitch model combined with a flexible wing torsion branch mode will also be considered. Later, in Chapter 12, aircraft trim and the related issue of the equilibrium manoeuvre will be examined using a whole aircraft heave/pitch model with a free–free flexible mode, including other features such as the steady pitch rate, accelerated flight condition, wing camber, out-of-line thrust and drag, and downwash effects on the tailplane. Also, in Chapter 16, the internal loads (see Chapter 5) in the manoeuvre will be obtained.

7.1 Static Aeroelastic Behaviour of a Two-dimensional Rigid Aerofoil with a Torsional Spring Attachment

The static aeroelastic behaviour is considered initially using an iterative approach and then a direct (single step) approach.

7.1.1 Iterative Approach

As a first example of static aeroelastic behaviour, consider the two-dimensional aerofoil in Figure 7.1 with unit span and chord c. The rigid aerofoil section is symmetric (so has no inherent camber) and is attached to a torsional spring of stiffness K_θ at a distance ec aft of the aerodynamic centre on the quarter chord. The lift-curve slope is a_1. The aerofoil has an initial incidence of θ_0 and twists through an unknown angle θ (elastic twist) due to the aerodynamic loading.

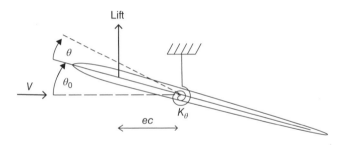

Figure 7.1 Two-dimensional aerofoil with a torsional spring.

The lift acting on the aerofoil at air speed V (true air speed, or TAS) and initial angle of incidence θ_0 causes a nose up pitching moment of

$$M = \left(\frac{1}{2}\rho V^2 c a_1 \theta_0\right) ec = \frac{1}{2}\rho V^2 ec^2 a_1 \theta_0 = qec^2 a_1 \theta_0 \tag{7.1}$$

to act about the elastic axis, where q is the dynamic pressure (not to be confused with the later usage of q for the pitch rate and q_e for the flexible mode generalized coordinate) and ρ is the air density at the relevant altitude. The equation for the aerofoil will be obtained using Lagrange's equations, introduced in Chapter 1. Since only static aeroelastic effects are being considered, the kinetic energy terms can be ignored. The elastic potential (or strain) energy U is found from the twist of the torsional spring, namely

$$U = \frac{1}{2}K_\theta \theta^2 \tag{7.2}$$

The generalized moment may be obtained from the incremental work done by the pitching moment acting through the incremental angle $\delta\theta$, and is given by

$$Q_\theta = \frac{\partial(\delta W)}{\partial(\delta\theta)} = \frac{\partial(qec^2 a_1 \theta_0 \delta\theta)}{\partial(\delta\theta)} = qec^2 a_1 \theta_0 \tag{7.3}$$

Then, application of Lagrange's equations for coordinate θ gives

$$K_\theta \theta = qec^2 a_1 \theta_0 \quad \text{so} \quad \theta = \frac{qec^2 a_1}{K_\theta}\theta_0 = qR\theta_0 \tag{7.4}$$

where $R = ec^2 a_1 / K_\theta$. Thus, having applied the initial aerodynamic loading, the aerofoil has twisted by angle θ, as determined in Equation (7.4). In performing this calculation, it has been assumed that the pitching moment has not changed due to the twist. However, as a consequence of the twist, the aerodynamic moment now changes due to the new angle of incidence. This new loading, in turn, causes the aerofoil twist to change again, leading to a further modification in the aerodynamic loading, and so on.

This stepping process, involving application of the aerodynamic loads on the aerofoil, changing the aerofoil twist, and then determining the new aerodynamic loading, illustrates the fundamental interaction between a flexible structure and aerodynamic forces that gives rise to aeroelastic phenomena.

7.1.1.1 First Iteration

The incidence of the aerofoil now includes the initial incidence and the estimate of twist, so the revised pitching moment becomes

$$M = qec^2 a_1 (\theta_0 + qR\theta_0) \tag{7.5}$$

and, since the elastic potential energy term remains the same as in Equation (7.2), application of Lagrange's equations gives a revised elastic twist angle of

$$\theta = qec^2 a_1 \left(\frac{1+qR}{K_\theta} \right) \theta_0 = qR(1+qR)\theta_0 \tag{7.6}$$

7.1.1.2   Further Iterations

The above process is then repeated by using the updated elastic twist value in the pitching moment and work expressions, leading to an infinite series expansion for the elastic twist in the form

$$\theta = qR \left[1 + qR + (qR)^2 + (qR)^3 + (qR)^4 + \dots \right] \theta_0 \tag{7.7}$$

Now, remembering that the binomial series is written as

$$(1-x)^{-1} = 1 + x + x^2 + x^3 + \dots \quad \text{with } |x| \le 1 \tag{7.8}$$

then in the limit, the aerofoil twist becomes

$$\theta = \frac{qR}{1-qR}\theta_0 \tag{7.9}$$

7.1.2   Direct (Single Step) Approach

Consider the same two-dimensional aerofoil used above, but now let the angle of incidence include the unknown elastic twist θ. The lift acting on the aerofoil at dynamic pressure q and total initial angle of incidence $(\theta_0 + \theta)$ causes a pitching moment

$$M = qec^2 a_1 (\theta_0 + \theta) \tag{7.10}$$

The elastic potential energy term is the same as in Equation (7.2).

The generalized moment, based on the incremental work done by the pitching moment acting through the incremental angle $\delta\theta$ is

$$Q_\theta = \frac{\partial(\delta W)}{\partial(\delta\theta)} = \frac{\partial[qec^2 a_1(\theta_0 + \theta)\delta\theta]}{\partial(\delta\theta)} = qec^2 a_1(\theta_0 + \theta) \tag{7.11}$$

Then, application of Lagrange's equations for coordinate θ gives

$$K_\theta\theta = qec^2 a_1 (\theta_0 + \theta) \quad \text{so} \quad \left(K_\theta - qec^2 a_1 \right)\theta = qec^2 a_1\theta_0 \tag{7.12}$$

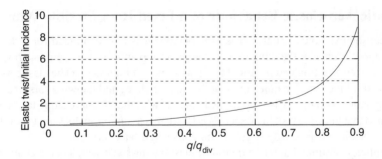

Figure 7.2   Typical twist behaviour for a two-dimensional aerofoil with torsional spring.

where it may be seen that the effective structural stiffness is reduced by the aerodynamic term as dynamic pressure increases. Solving this equation leads to the elastic twist

$$\theta = \frac{qec^2a_1}{K_\theta - qec^2a_1}\theta_0 = \frac{qR}{1-qR}\theta_0 \qquad (7.13)$$

Comparing Equations (7.9) and (7.13), it may be seen that both approaches give exactly the same value of elastic twist for a given dynamic pressure q. The elastic twist becomes infinite as q approaches $1/R$, and this defines the so-called divergence speed as

$$q_{div} = \frac{1}{R} = \frac{K_\theta}{ec^2a_1} \qquad (7.14)$$

and hence Equation (7.13) may be written

$$\theta = \frac{(q/q_{div})}{(1-q/q_{div})}\theta_0 \qquad (7.15)$$

This analysis demonstrates that the physical phenomenon of divergence occurs when the aerodynamic pitching moment overcomes the structural restoring moment. Since infinite deflections are not possible, then in practice the structure will fail. Figure 7.2 shows a plot of the ratio of elastic twist to initial angle of incidence against the ratio of dynamic pressure to that at divergence; it can be seen that, for this simple example, the elastic twist equals the initial incidence at $q = q_{div}/2$ and then increases markedly beyond this point.

It should be noted that the single step (often referred to as 'strongly coupled') approach introduced here is only feasible if there is a direct mathematical relationship between the aerodynamic forces and the deflections. For advanced static aeroelastic calculations involving the coupling of computational fluid dynamics (CFD) methods with Finite Element (FE) methods, the use of a 'loosely coupled' approach somewhat similar to the iterative process shown above is required. However, the more common and traditional methodology for static aeroelastic calculations is a single step approach and, in the following sections on the static aeroelastic behavior of more realistic wings, this is the one that will be used throughout.

7.2   Static Aeroelastic Behaviour of a Fixed Root Flexible Wing

A more realistic example of static aeroelastic behaviour is now examined for a flexible wing fixed at the root. Consider the wing to be rectangular in planform, with semi-span s, chord c, a symmetric aerofoil section and no initial twist, as shown in Figure 7.3. The elastic axis lies at a distance ec aft of the aerodynamic centre at the quarter chord and the wing torsional rigidity is GJ. Aerodynamic strip theory is used (see Chapter 4) with the lift curve slope taken as a_W. It is also assumed that the wing root incidence θ_0 is fixed; this assumption does not take into account the trim of the aircraft in steady flight (to be considered later on).

For simplicity, assume that the variation of elastic wing twist with span is characterized by the idealized linear relationship

$$\theta = \frac{y}{s}\theta_T \tag{7.16}$$

where θ_T is the twist at the wing tip (often called a generalized coordinate as it defines the amount of the assumed shape present). Thus the twist increases linearly away from the wing root. Using such an assumed shape enables an approximate analysis to be carried out using the Rayleigh–Ritz method introduced in Chapter 3.

7.2.1   Twist and Divergence of the Fixed Root Flexible Wing

The lift is taken to act at the aerodynamic centre and, because the section is symmetric, there is no pitching moment at zero incidence (see Chapter 4). Using an expression for the lift that takes both the root incidence and aeroelastic twist into account, then the lift on an elemental strip $\mathrm{d}y$ is given by

$$\mathrm{d}L = q c a_W \left(\theta_0 + \frac{y}{s}\theta_T\right)\mathrm{d}y \tag{7.17}$$

and thus the lift increases with distance from the wing root. The total lift on the wing is found by integrating over the semi-span, so that

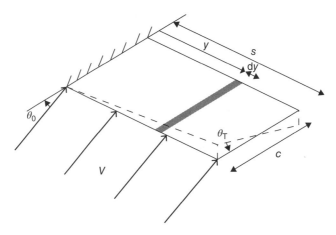

Figure 7.3   Flexible rectangular wing with a fixed root.

$$L = \int_0^s q c a_W \left(\theta_0 + \frac{y}{s} \theta_T \right) dy = q c a_W \left(s\theta_0 + \frac{s}{2}\theta_T \right) \tag{7.18}$$

Since there is no motion of the wing, the kinetic energy $T = 0$. The elastic potential energy corresponds to the strain energy due to twist (see Chapter 3), given by

$$U = \frac{1}{2} \int_0^s GJ \left(\frac{d\theta}{dy} \right)^2 dy = \frac{1}{2} \int_0^s GJ \left(\frac{\theta_T}{s} \right)^2 dy = \frac{GJ}{2s}\theta_T^2 \tag{7.19}$$

Now consider an incremental twist angle being expressed in terms of an incremental generalized coordinate, so that

$$\delta\theta = \frac{y}{s}\delta\theta_T \tag{7.20}$$

The work done by the aerodynamic forces is determined by considering the pitching moment acting upon each strip doing work through this incremental twist angle. The total incremental work done, δW, is obtained by integrating these work terms across the entire wing semi-span. Thus

$$\delta W = \int_0^s dL \, ec \, \delta\theta = \int_0^s q c a_W \left(\theta_0 + \frac{y}{s}\theta_T \right) ec \, \delta\theta dy$$
$$= \int_0^s q c^2 a_W \left(\theta_0 + \frac{y}{s}\theta_T \right) e \frac{y}{s} \delta\theta_T dy = q ec^2 a_W \left(\frac{s\theta_0}{2} + \frac{s\theta_T}{3} \right) \delta\theta_T \tag{7.21}$$

and for the generalized coordinate θ_T, Lagrange's equations yield

$$\frac{GJ\theta_T}{s} = q ec^2 a_W \left(\frac{s\theta_0}{2} + \frac{s\theta_T}{3} \right) \quad \text{so} \quad \left(\frac{GJ}{s} - q ec^2 a_W \frac{s}{3} \right) \theta_T = q ec^2 a_W \frac{s\theta_0}{2} \tag{7.22}$$

Note, for now, that equation (7.22) may be written in the general form

$$\rho V^2 \mathbf{C}(\theta + \theta_0) + \mathbf{E}\theta = 0 \tag{7.23}$$

and this formulation will be referred to in later chapters.

Here again the structural stiffness is seen to be reduced by the aerodynamic term, and hence the elastic tip twist is found from Equation (7.22) to be

$$\theta_T = \frac{3q ec^2 s^2 a_W}{6GJ - 2q ec^2 s^2 a_W}\theta_0 \tag{7.24}$$

The tip twist increases with dynamic pressure and behaves in a similar manner to that shown in Figure 7.2. When the divergence condition is reached, the twist tends to infinity although in reality structural failure will occur first. For this fixed root flexible wing, the dynamic pressure at divergence q_W is found as

$$q_W = \frac{3GJ}{ec^2 s^2 a_W} \tag{7.25}$$

Considering Equation (7.25), it is possible to make some deductions as to what design rules could be used to increase the divergence speed so that it does not occur within the desired flight envelope:

- The smaller the distance between the aerodynamic centre and the elastic axis, and/or the greater the flexural rigidity *GJ*, the greater the divergence speed becomes.
- If the elastic axis lies on the locus of aerodynamic centres (i.e. ¼ chord line) there is no twist due to aerodynamic loading and divergence will not occur.
- If the elastic axis is forward of the aerodynamic centre, the applied aerodynamic moment becomes negative, the tip twists nose down and divergence cannot occur.

Unfortunately, these last two design scenarios are not generally possible to implement in practical wing designs, so divergence must always be considered for aeroelastic design and adequate torsional stiffness is crucial. The first recognition of the effect of elastic axis position on the divergence phenomena is attributed to Anthony Fokker (Bisplinghoff et al, 1996) following modified strengthening of the rear spar of the Fokker D-VIII, shown on the front cover of this book. Repeated wing failures occurred in high-speed dives, although the wing had been made stronger, as the wing was now more prone to divergence due to the elastic axis being moved aft.

7.2.2 Variation of Lift Along the Fixed Root Flexible Wing

Having determined the wing twist, the corresponding lift distribution along the fixed root flexible wing may be determined. Combining Equations (7.17) and (7.24), the lift per unit span of the wing is found as

$$\frac{\mathrm{d}L}{\mathrm{d}y} = qca_W\left(\theta_0 + \frac{y}{s}\theta_T\right) = qca_W\left(1 + \frac{3qec^2s^2a_W}{6GJ - 2qec^2s^2a_W}\frac{y}{s}\right)\theta_0 \tag{7.26}$$

and this can be rewritten in terms of the dynamic pressure at divergence for the wing so that

$$\frac{\mathrm{d}L}{\mathrm{d}y} = qca_W\left(1 + \frac{3(q/q_W)}{2(1-(q/q_W))}\frac{y}{s}\right)\theta_0 \tag{7.27}$$

When the lift per unit span is plotted against the spanwise distance in Figure 7.4, it can be seen that the lift per unit span increases linearly along the wing span. This is due to the assumed linear twist shape and would differ if a more complicated shape were chosen, if the wing tapered or if modified strip theory were used. As the dynamic pressure increases, the spanwise slope of the lift distribution increases. The lift per unit span at the wing root depends solely upon the root incidence.

The total lift is found by integrating Equation (7.27) across the entire wing semi-span, giving

$$L = \int_0^s \frac{\mathrm{d}L}{\mathrm{d}y}\mathrm{d}y = qcsa_W\left[1 + \frac{3(q/q_W)}{4(1-q/q_W)}\right]\theta_0 \tag{7.28}$$

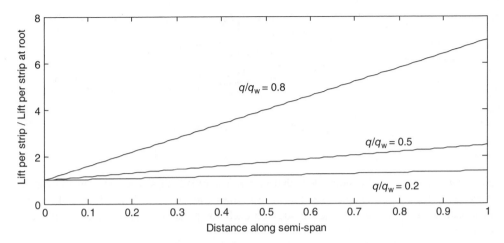

Figure 7.4   Lift per unit span for different dynamic pressures.

More lift is generated as the air speed (and hence dynamic pressure) increases. As the dynamic pressure q approaches the divergence condition for this fixed root wing model, the total lift becomes infinite.

7.3   Effect of Trim on Static Aeroelastic Behaviour

The above example shows that increasing the air speed leads to a greater wing twist, and thus to an increase in lift. However, in practice, a change in the air speed will require the trim of the aircraft to be adjusted via the elevator in order to maintain equilibrium of aerodynamic, inertia and propulsive forces. The following example of a simple flexible aircraft model illustrates how the divergence and load distribution behaviour is changed when the balance of overall lift and weight is preserved. The model has heave and pitch motions, together with the flexible wing effect, represented by adding a torsional wing branch mode (see Chapter 3).

Later in Chapter 12, where equilibrium manoeuvres are considered, a more advanced model, including out-of-line thrust and drag, wing camber and downwash effects at the tailplane, will be investigated for accelerated manoeuvres with a steady pitch rate for both rigid and flexible aircraft.

Note that because the content on static aeroelasticity and flutter in this book has employed the notation used in classical texts, there will be some differences in notation in the later parts on manoeuvres where a different notation is standard.

7.3.1   Effect of Trim on the Divergence and Lift Distribution for a Simple Aircraft Model

Consider the idealized flexible aircraft with weight W and wing lift L in steady level flight as shown in Figure 7.5. The fuselage is rigid, able to undergo heave and pitch motions. In order to simplify the analysis and highlight the key issues, only a balance in vertical forces is sought i.e. $L = W$ and tailplane effects are ignored. The balance of pitching moments is ignored by arranging for lift, weight, thrust and drag to all act through the aerodynamic centre of the

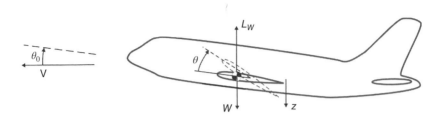

Figure 7.5   Aircraft in steady flight.

aircraft which is assumed to coincide with the aircraft centre of mass. The wings are the same as discussed in the above examples, i.e. they have a symmetric section and include flexible motion in torsion by employing a branch mode approach for a fixed wing root (as discussed in Chapter 3).

The equations of motion are determined once again using Lagrange's equations. As noted before for the flexible wing, the wing twist θ is taken about the elastic axis, distance ec aft of the wing aerodynamic centre; the corresponding generalized coordinate is the twist at the wing tip θ_T. However, in this whole aircraft case, the root incidence θ_0 and the heave displacement z (positive downwards) at the centre of mass also need to be included as generalized coordinates.

The kinetic energy is zero and the elastic potential energy for both wings is given as

$$U = 2\frac{1}{2}\int_0^s GJ\left(\frac{d\theta}{dy}\right)^2 dy = GJ\int_0^s \left(\frac{\theta_T}{s}\right)^2 dy = \frac{GJ}{s}\theta_T^2 \tag{7.29}$$

Consider the work done by the lift and weight acting through incremental displacements for wing tip twist $\delta\theta_T$ and heave δz, then

$$\delta W = W\delta z + 2\int_0^s (qca_W(\theta_0 + \theta))(-\delta z + ec\delta\theta)\,dy$$

$$= W\delta z - 2qcsa_W\left(\theta_0 + \frac{\theta_T}{2}\right)\delta z + 2qec^2 sa_W\left(\frac{\theta_0}{2} + \frac{\theta_T}{3}\right)\delta\theta_T \tag{7.30}$$

There is no generalized work done due to an incremental change in root incidence $\delta\theta_0$ so the problem has been reduced to two generalized coordinates. Application of Lagrange's equations for each of the two coordinates z and θ_T gives the following expressions. Firstly,

$$Q_z = \frac{\partial(\delta W)}{\partial(\delta z)} = W - 2qcsa_W\left(\theta_0 + \frac{\theta_T}{2}\right) = 0 \tag{7.31}$$

which is the equilibrium equation found by resolving forces in the vertical direction. It effectively imposes the constraint in steady level flight that the weight of the aircraft is equal to the wing lift. Secondly,

$$Q_{\theta_T} = 2qec^2 sa_W\left(\frac{\theta_0}{2} + \frac{\theta_T}{3}\right) \quad \Rightarrow \quad \frac{2GJ}{s}\theta_T - 2qec^2 sa_W\left(\frac{\theta_0}{2} + \frac{\theta_T}{3}\right) = 0 \tag{7.32}$$

which is the elastic mode equilibrium equation. The two unknowns in these equations are θ_0 and θ_T; note that z does not appear explicitly because the vertical position of the aircraft does not affect the steady lift. Rewriting Equations (7.31) and (7.32) in matrix form gives

$$
\begin{bmatrix}
2qcsa_W & qcsa_W \\
qec^2sa_W & \left(\dfrac{2}{3}qec^2sa_W - 2\dfrac{GJ}{s}\right)
\end{bmatrix}
\begin{Bmatrix} \theta_0 \\ \theta_T \end{Bmatrix} =
\begin{Bmatrix} W \\ 0 \end{Bmatrix}
\tag{7.33}
$$

For the trimmed aircraft, the simultaneous Equations (7.33) are solved to determine the combination of root incidence and wing twist that give rise to the equilibrium condition. The tip twist may be shown to be

$$
\theta_T = \frac{Wecs}{4GJ\left[1 - q/(4q_W)\right]}
\tag{7.34}
$$

whereas the root incidence is

$$
\theta_0 = \frac{W\left[1 - q/q_W\right]}{2qcsa_W\left[1 - q/(4q_W)\right]}
\tag{7.35}
$$

The variation of normalized values of θ_T and θ_0 versus normalized dynamic pressure are shown in Figure 7.6. As before, an increase in the air speed leads to an increase in the wing tip twist. However, it can be seen that the root incidence decreases with air speed, and beyond

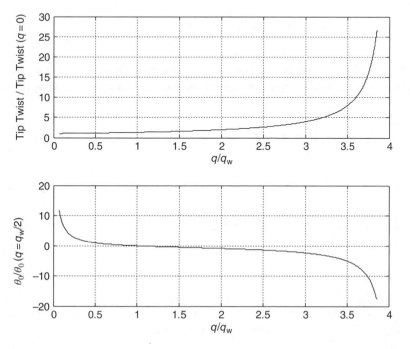

Figure 7.6 Wing tip twist and incidence for an aircraft with a flexible wing in trim.

the *fixed root wing divergence speed* the incidence becomes negative and the inboard sections of the wing are at negative angles of incidence.

Considering the denominator of Equation (7.35), it can be seen that the *trimmed aircraft divergence speed* occurs when

$$q = q_A = \frac{12GJ}{ec^2s^2a_W} = 4q_W \tag{7.36}$$

where q_A is the trimmed aircraft divergence speed and q_W is the divergence speed for the fixed root flexible wing (this latter value can be obtained from Equations (7.33) by constraining θ_0 to be zero.) Thus, for this case where trim is maintained at increasing air speed, the dynamic pressure at divergence q_A is four times that of the fixed root flexible wing case (i.e. double the divergence speed). At this air speed, both the tip twist and root incidence tend towards infinity, so again structural failure will occur. However, in practice, it is unlikely that the divergence speed will be achieved as the aircraft will run out of trim at a lower air speed, i.e. more and more elevator will need to be applied to maintain trim.

Note that the effects of introducing centre of mass position, wing camber, tailplane downwash, out-of-line thrust and drag, steady pitch rate, accelerated flight condition etc. to the model will alter the detail of the trim condition, but not the essence of the result.

7.3.2 Effect of Trim on the Variation of Lift along the Wing

Substituting the above expressions for θ_0 and θ_T as functions of the dynamic pressure into Equation (7.26) for the lift distribution (lift per unit span) leads to

$$\frac{dL}{dy} = \frac{W\left[2 + q/q_W(3y/s - 2)\right]}{4s\left[1 - q/(4q_W)\right]} \tag{7.37}$$

Figure 7.7 shows the lift distribution along the span for the trimmed aircraft at several normalized dynamic pressures. The lift may be seen to increase linearly from the root to the tip. It can

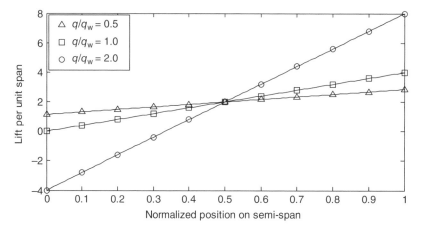

Figure 7.7 Lift per unit span for different dynamic pressures on the trimmed aircraft.

be seen that above the fixed root wing divergence speed, negative lift occurs close to the root because the root incidence required for equilibrium in trim is negative. The spanwise position of the net wing lift is seen to shift outboard as the air speed increases. Since trim has to be maintained, then the total wing lift must remain constant and therefore the area under each line in this figure remains constant. However, as the lift shifts outboard, the corresponding wing root bending moment (see Chapters 5 and 16) will increase.

7.3.3   Effect of Trim on the Wing and Tailplane Lift

So far, only the wing lift for the aircraft has been considered. If the relative positions of the wing aerodynamic centre and the centre of mass were changed such that the tailplane lift needed to be included, then to maintain trim, both the vertical force and moment equilibrium equations must hold. Provided that the wing has a symmetric section (i.e. no camber), it may be shown that the wing and tailplane lift forces remain constant with changes in air speed. However, wing camber introduces a pitching moment at zero lift; this means that the wing and tailplane lift forces will actually change with air speed, though their sum will still equal the weight. The presence of camber will be considered later in Chapter 12.

Also, when camber is present it may be shown that the magnitude of the elevator angle must be increased at higher air speeds in order to maintain trim but, in practice, the aircraft is likely to run out of available elevator trim angle before divergence can occur.

7.4   Effect of Wing Sweep on Static Aeroelastic Behaviour

Most aircraft are designed with swept-back wings. The reasons for this are mainly aerodynamic, since for subsonic aircraft sweep-back increases the air speed at which shock waves are formed on the wings, so delaying the associated increase in drag. The sweep also reduces the effective thickness to chord ratio and moves the centre of mass aft.

Similar reductions in drag could be obtained through the use of swept-forward wings, which is also beneficial as the flow separation then occurs initially near the wing root. Such a configuration preserves the aileron control at the wing tip, whereas for swept-back wings flow separation occurs first towards the wing tips. However, very few aircraft (e.g. X-29, Sukhoi-47) have been built with swept-forward wings. The main reason for this is the static aeroelastic behaviour of swept-forward wings, in particular, the detrimental effect that forward sweep has on the divergence speed.

This section introduces a simple aeroelastic wing model that demonstrates how wing sweep changes the static aerodynamic lift and aeroelastic behaviour. The differences between wings with forward or backward sweep are emphasized.

7.4.1   Effect of Wing Sweep on Effective Angle of Incidence

In order to illustrate the effect of wing sweep on a flexible wing, consider the rectangular wing shown in Figure 7.8 subjected to an upwards bending displacement along the mid-chord line (for simplicity); the wing can be unswept, swept-backward or swept-forward. Wing bending has a much greater effect on the effective angle of incidence compared to wing twist, but the actual mode of divergence is still torsion. Of particular importance is the effective angle of incidence of the streamwise strips when the sweep angle is changed (Broadbent, 1954).

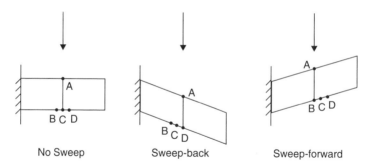

Figure 7.8   Streamwise strips for wings with no sweep, sweep-back and sweep-forward.

The streamwise sections in Figure 7.8 are denoted as AC, AD and AB for the unswept, swept-back and swept-forward cases respectively. When the wing is considered to bend upwards, the following occurs:

a. For the unswept case (AC), the incidence is unchanged due to bending.
b. For the swept-back case (AD), the effective streamwise angle of incidence reduces since point D moves upwards more under bending than point A.
c. For the swept-forward case (AB), the effective incidence increases (point A moves upwards more than point B).

Consequently, swept-forward wings have a decreased divergence speed compared to unswept wings due to increased effective incidence, whereas swept-back wings have an increased divergence speed.

7.4.2   Effective Streamwise Angle of Incidence due to Bending/Twisting

Consider now the untapered wing with chord c, semi-span s and sweep angle Λ (sweep-back positive) moving at speed V with an initial root angle of incidence θ_0, as shown in Figure 7.9. Both bending (downwards positive) and twisting (nose up positive) type deflections need to be included in order to achieve a realistic behaviour; here they will be taken as

$$z = \left(\frac{y_0}{s_0}\right)^2 q_b \quad \text{and} \quad \theta = \left(\frac{y_0}{s_0}\right) q_t \qquad (7.38)$$

Respectively, where q_b and q_t are generalized coordinates for the bending and twisting assumed shapes (see Chapter 3). The bending displacement and twist are defined with respect to a reference axis lying along the centre of the wing. Any bending-torsion coupling, typically occurring due to wing sweep on full scale aircraft, will be ignored.

Consider the flow over an elemental streamwise strip of width dy at spanwise distance y. The effective angle of incidence depends upon the difference in the deflection of the two ends of the strip, and also upon the geometry of points A, B and D. Including the root angle of incidence θ_0, the variation of incidence can be found as (Niu; 1988)

$$\alpha = \theta_0 + \theta \cos\Lambda + \frac{\partial z}{\partial y_0} \sin\Lambda = \theta_0 + \left(\frac{y_0}{s_0}\right) \cos\Lambda \; q_t + 2\left(\frac{y_0}{s_0^2}\right) \sin\Lambda \; q_b \qquad (7.39)$$

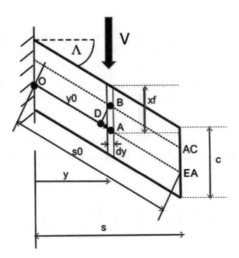

Figure 7.9   Swept untapered wing.

Note that the inclusion of the bending term in this model means that the angle of incidence along each aerodynamic strip is now not constant and therefore gives some further approximation to the analysis.

Examination of Equation (7.39) shows that twisting the wing increases the effective angle of incidence for both forward and backward sweep. A positive bending deflection increases the effective incidence for swept-back and reduces it for swept-forward wings; however, these effects are reversed in steady flight as the wing deflection is upwards (negative deflection). The larger bending deflections of a wing dominate the changes in the effective angle of incidence compared to those from wing twist; however, the twisting degree of freedom must also be included in any mathematical model as this is still part of the mechanism that gives rise to divergence.

7.4.3   Effect of Sweep Angle on Divergence Speed

For the uniform swept wing introduced above, the lift acting at the quarter chord of the elemental strip of area $c\,\mathrm{d}y$ (i.e. point B) when bending and torsion are considered together is

$$\mathrm{d}L = \frac{1}{2}\rho V^2 c\,\mathrm{d}y\,a_{\mathrm{W}}\alpha = q a_{\mathrm{W}} c\alpha\,\mathrm{d}y \qquad (7.40)$$

Also, the downwards displacement of point B is found to be

$$z_{\mathrm{B}} = \frac{(y_0 - x_{\mathrm{A}}\sin\Lambda)^2}{s_0} q_b - \left(\frac{y_0 - x_{\mathrm{A}}\sin\Lambda}{s_0}\right) q_t x_{\mathrm{A}}\cos\Lambda \quad \text{where } \left(x_{\mathrm{f}} - \frac{c}{4}\right) = x_{\mathrm{A}} \qquad (7.41)$$

Here x_{A} is the distance that the reference axis lies aft of the quarter chord. Thus the work done by this lift acting through the incremental displacement, when integrated along the span of the wing, is

$$\delta W = -\int_0^s dL \delta z_B = -qca_W \int_0^s \left\{ \theta_0 + \left(\frac{y_0}{s_0}\right) \cos\Lambda \ q_t + 2\left(\frac{y_0}{s_0^2}\right) \sin\Lambda \ \delta q_b \right\}$$

$$\left\{ \frac{(y_0 - x_A \sin\Lambda)^2}{s_0} q_b - \left(\frac{(y_0 - x_A \sin\Lambda)}{s_0}\right) x_A \cos\Lambda \delta q_t \right\} dy \tag{7.42}$$

The structural behaviour of the wing is represented by a beam lying along the reference axis. The elastic potential energy is derived from the strain energy for both bending and torsion and is given by

$$U = \frac{1}{2}\int_0^{s_0} GJ \left(\frac{\partial\theta}{\partial y_0}\right)^2 dy_0 + \frac{1}{2}\int_0^{s_0} EI \left(\frac{\partial^2 z}{\partial y_0^2}\right)^2 dy_0 = \frac{GJ}{2s_0} q_t^2 + \frac{2EI}{s_0^3} q_b^2 \tag{7.43}$$

Then application of Lagrange's equation for generalized coordinates q_b and q_t leads to the linear simultaneous equations

$$\begin{bmatrix} \frac{4EI}{s_0^3} + 2qca_W c_\Lambda s_\Lambda \left(\frac{1}{4} - \frac{2x_A s_\Lambda}{3s_0} + \frac{x_A^2 s_\Lambda^2}{2s_0^2}\right) & qca_W c_\Lambda^2 \left(\frac{s_0}{4} - \frac{2x_A s_\Lambda}{3} + \frac{x_A^2 s_\Lambda^2}{2s_0}\right) \\ -2qca_W x_A c_\Lambda^2 s_\Lambda \left(\frac{1}{3} - \frac{x_A s_\Lambda}{2s_0}\right) & \frac{GJ}{s_0} - qca_W x_A c_\Lambda^3 \left(\frac{s_0}{3} - \frac{x_A s_\Lambda}{2}\right) \end{bmatrix} \begin{Bmatrix} q_b \\ q_t \end{Bmatrix}$$

$$\tag{7.44}$$

$$= \begin{Bmatrix} -qca_W c_\Lambda \left(\frac{s_0}{3} - x_A s_\Lambda + \frac{x_A^2 s_\Lambda^2}{s_0}\right) \\ qcx_A a_W c_\Lambda^2 \left(\frac{s_0}{2} - x_A s_\Lambda\right) \end{Bmatrix} \theta_0$$

where $s_\Lambda = \sin\Lambda$ and $c_\Lambda = \cos\Lambda$.

These equations can be solved to give the bending and torsion deflections, via the generalized coordinates, for a given air speed and root angle of incidence. Also, divergence for the swept wing occurs when the determinant of the left-hand side square matrix becomes zero.

Figure 7.10 shows how the divergence speed increases with sweepback ($\Lambda > 0$) and decreases for the sweep-forward case ($\Lambda < 0$) for different aspect ratios; as expected the lower aspect ratio has a higher divergence speed. Figure 7.11 shows a similar plot, but this time varying the position of the reference axis; the further aft this axis is from the quarter chord, the lower is the divergence speed.

When there is no sweep ($\Lambda = 0$), then the divergence speed becomes

$$V_{div} = \frac{3GJ}{cs_0^2 a_W x_A} \tag{7.45}$$

and this corresponds to the divergence speed obtained earlier for the unswept wing. This result is to be expected as the bending degree of freedom has no effect on the steady aerodynamic lift of a streamwise section for the unswept case.

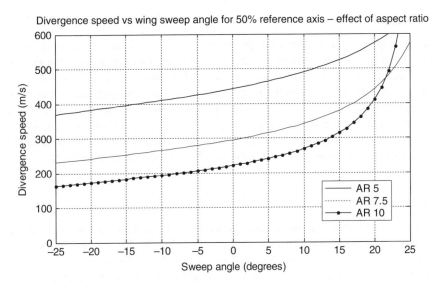

Figure 7.10   Effect of sweep angle on divergence speed for different aspect ratios.

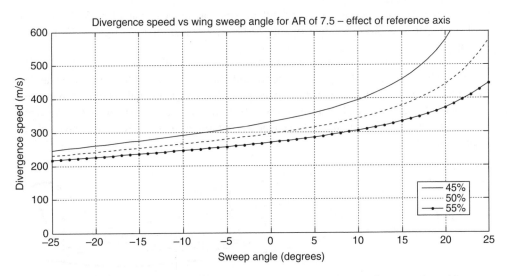

Figure 7.11   Effect of sweep angle on divergence speed for different reference axis positions.

The reduction of the divergence speed becomes the limiting case for swept-forward designs and consequently very few have been built. Experimental aircraft such as the X-29 were only able to have swept-forward wings by using aeroelastic tailoring, where the wing characteristics were altered using composite laminates oriented in such a manner that an upwards bending deflection resulted in a nose down twist.

It should be remembered that earlier in this chapter it was shown that by allowing for the trim of the aircraft the divergence speed changed, and a similar effect is found for the swept wing case.

7.4.5 Comments

There are a number of significant assumptions in the above analysis. Sweep-back (or sweep-forward) will increase the aerodynamic interactions between different parts of the wing, which will make the strip theory aerodynamics more inaccurate, and also the inclusion of a bending shape means that the slope of each streamwise strip is not now constant. It has also been assumed that the wing behaves as a beam-like structure, and consequently that the reference axis remains parallel to the axis of sweep along or near to the mid-chord line. In cases where the wing behaves more like a plate, such as for low aspect ratio tapered swept wings, the structural bending/torsion coupling effects for the swept wing must also be included.

7.5 Examples

1. Determine the lift distribution and divergence speed for a fixed root rectangular wing of semi-span s, chord c and torsional rigidity GJ using modified strip theory such that $a_W(y) = a_W(1 - y^2/s^2)$. Evaluate the effect of the modified strip theory.
2. Determine the lift distribution and divergence speed of a fixed root tapered rectangular wing of semi-span s and chord $c = c_0(1 - y^2/s^2)$ and torsional rigidity $GJ = GJ_0(1 - y^2/s^2)$.
3. Repeat Examples 1 and 2 but this time for the case where whole aircraft trim in heave is taken into account.
4. Determine the divergence speed and lift distribution for a rectangular fixed root wing of semi-span s, chord c, bending rigidity $EI_0(1 - y^2/s^2)$ and torsional rigidity GJ for wing sweep Λ.
5. Consider the idealized aircraft of weight W in steady level flight as shown in Figure 7.12. The fuselage is rigid, able to undergo heave and pitch motions, and the wings include flexible motion in torsion as in the above examples. The wings and tailplane have symmetric sections. Also, thrust and drag are assumed to be in-line and so do not contribute to the pitching moment terms. Determine the full equations of motion for this system and find the divergence speed.

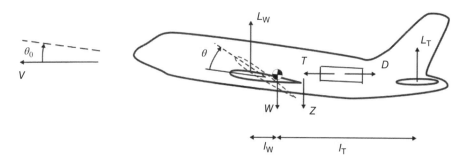

Figure 7.12 Example 5: Aircraft in steady flight.

8

Static Aeroelasticity – Effect of Wing Flexibility on Control Effectiveness

It was shown in Chapter 4 how the use of control surfaces changed the effective camber of an aerofoil and that this could be used to change the lift. Thus control surfaces are used to manoeuvre an aircraft in-flight and their sizing is an important issue when the aircraft is designed. It is important to know how sensitive an aircraft is to application of the control surfaces and what loads are generated; this is of particular significance for military aircraft where the need to manoeuvre rapidly is essential, but is also important of course for the performance of commercial aircraft.

This chapter will consider the effect that aeroelastic deflections of the flexible wing have on the aerodynamic influence, or *effectiveness*, of the control surfaces in comparison to the rigid wing. It will be shown that as the speed increases the effectiveness reduces until at some critical speed – the *reversal speed* – there is no response to application of the control surface. At speeds greater than the reversal speed, the action of the controls reverses, a phenomenon known as *control reversal*. Although not necessarily disastrous, it is unacceptable that, at speeds near to the reversal speed, the aircraft responds either very slowly or not at all to application of the controls, and that the opposite response to that demanded occurs beyond the reversal speed.

There are two basic ways that the aircraft industry considers these static aeroelastic phenomena (although there is motion, it is considered to be steady). These will be illustrated (i) by considering a wing with a fixed root experiencing a control deflection, and (ii) for a wing rolling at a constant rate; in both cases a simple rectangular wing plus aileron is used. In Chapter 12, where equilibrium (or so-called *bookcase*) manoeuvres are considered, the steady application of the elevator to a whole aircraft flexible model is considered. Also, in Chapter 13, the dynamic (or so-called *rational*) manoeuvre using the flight mechanics model allows the flexible aircraft dynamic response to a transient application of a roll, yaw or pitch control to be considered, and non-linear effects can be included.

Static aeroelastic calculations are employed fairly early in the design process to size the control surfaces. However, later on when the Flight Control System (FCS) has been designed, the flight mechanics model allows rational calculations to be performed in order to evaluate the

Introduction to Aircraft Aeroelasticity and Loads, Second Edition. Jan R. Wright and Jonathan E. Cooper.
© 2015 John Wiley & Sons, Ltd. Published 2015 by John Wiley & Sons, Ltd.

control effectiveness with non-linear effects included. The deployment and performance of the controls is then fine-tuned via the FCS to obtain the characteristics required; clearly it would then be too late to change the control surface sizes at that stage.

8.1 Rolling Effectiveness of a Flexible Wing – Fixed Wing Root Case

Consider a flexible uniform wing shown in Figure 8.1, with semi-span s, chord c, a symmetric section (i.e. no camber), a root incidence θ_0 and a rigid full span aileron whose rotation angle is β (trailing edge down); this latter symbol is used as standard in the classical flutter books (e.g. Fung, 1969), but is not to be confused with the usage later in Chapter 13 for the sideslip angle in the flight mechanics model.

As in Chapter 7, assume that the wing is flexible in twist, taken to vary linearly as

$$\theta = \left(\frac{y}{s}\right)\theta_T \tag{8.1}$$

The twist is defined as nose up about the elastic axis, taken at distance ec aft of the aerodynamic centre on the quarter chord. Also, using the results in Chapter 4 for any section of the wing plus control, the lift and moment (defined positive nose upwards and referred to the elastic axis) coefficients are

$$C_L = a_0 + a_W(\theta_0 + \theta) + a_C\beta \quad \text{and} \quad C_M = b_0 + b_W(\theta_0 + \theta) + b_C\beta \tag{8.2}$$

where $a_0 = b_0 = 0$ for the symmetric aerofoil and $b_W = a_W e$. Figure 8.2 shows the effect of applying a downwards aileron deflection on a flexible (or elastic) wing; it can be seen that the

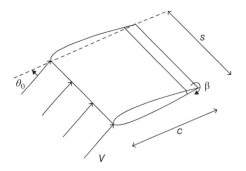

Figure 8.1   Cantilevered wing with a full span aileron.

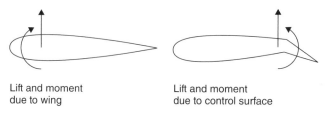

Lift and moment due to wing

Lift and moment due to control surface

Figure 8.2   Effect on lift distribution of applying a control surface rotation.

incremental lift due to the control rotation acts towards the aileron hinge line, around the 2/3 to 3/4 chord. Thus, any applied control rotation provides not only a lift force introducing roll but also a nose down pitching moment, leading to nose down twist and therefore a reduction in the angle of incidence for an elastic wing.

8.1.1 Determination of Reversal Speed

Since the wing is fixed then the roll rate and roll angle are zero so $\dot{\phi} = \phi = 0$. The lift and pitching moment on the elemental strip dy due to the fixed root incidence θ_0, twist θ and control rotation β are

$$dL = qc\,dy\left[a_W\left(\theta_0 + \frac{y}{s}\theta_T\right) + a_C\beta\right] \quad \text{and} \quad dM = qc^2\,dy\left[b_W\left(\theta_0 + \frac{y}{s}\theta_T\right) + b_C\beta\right] \quad (8.3)$$

The work done by the lift is zero since the wing is not allowed to roll. However, the incremental work done by the pitching moment acting through an incremental twist $\delta\theta$ for the single wing is

$$\delta W = \int_{wing} dM\,\delta\theta = qc^2\int_0^s\left[b_W\left(\theta_0 + \frac{y}{s}\theta_T\right) + b_C\beta\right]\frac{y}{s}\delta\theta_T\,dy \quad (8.4)$$

and thus the generalized force in twist is

$$Q_{\theta_T} = \frac{\partial(\delta W)}{\partial(\delta\theta_T)} = qc^2s\left(\frac{b_W}{2}\theta_0 + \frac{b_W}{3}\theta_T + \frac{b_C}{2}\beta\right) \quad (8.5)$$

The elastic potential (or strain) energy for a single wing is the same as used in Chapters 3 and 7 so that

$$U = \frac{GJ}{2s}\theta_T^2 \quad (8.6)$$

and so applying Lagrange's equation gives the expression

$$\left(\frac{GJ}{s} - qc^2s\frac{b_W}{3}\right)\theta_T = qc^2s\left(\frac{b_W}{2}\theta_0 + \frac{b_C}{2}\beta\right) \quad (8.7)$$

The wing tip twist is then given by

$$\theta_T = \frac{qc^2s}{(2GJ/s - 2qc^2sb_W/3)}(b_W\theta_0 + b_C\beta) = \frac{qc^2s^2}{2GJ\left(1 - \dfrac{q}{q_W}\right)}(b_W\theta_0 + b_C\beta) \quad (8.8)$$

Now consider the effect due to control rotation, in isolation from the root incidence effect (examined in Chapter 7). Since b_C is negative, a nose down twist will result from an increased control angle or dynamic pressure. The lift per unit span due to control rotation alone can be determined as

$$\frac{dL}{dy} = qc\left[a_W\frac{y}{s}\theta_T + a_C\beta\right] \quad (8.9)$$

and substituting in the relevant part of the expression for the wing tip twist in Equation (8.8) gives

$$\frac{dL}{dy} = qc\left[\frac{qc^2s^2a_W}{2GJ(1-q/q_W)}\frac{y}{s}b_C + a_C\right]\beta \tag{8.10}$$

The total root bending moment (see internal loads in Chapters 5 and 16) due to application of control rotation is

$$M_{root} = \int_0^s y\frac{dL}{dy}dy = qcs^2\left[\frac{qc^2s^2a_W}{6GJ(1-q/q_W)}b_C + \frac{a_C}{2}\right]\beta \tag{8.11}$$

At the reversal speed there will be zero total bending moment at the wing root, so reversal occurs when $q \rightarrow q_{rev}$, and

$$\frac{q_{rev}c^2s^2a_W}{6GJ(1-q_{rev}/q_W)}b_C + \frac{a_C}{2} = 0 \tag{8.12}$$

Then, following some algebraic manipulation, the dynamic pressure at reversal can be found as

$$\frac{q_{rev}}{q_W} = \frac{ea_C}{ea_C - b_C} \tag{8.13}$$

and as b_c is negative, the dynamic pressure at reversal is less than that at wing divergence. Note that the root bending moment due to control rotation is equivalent to the total rolling moment, but the former term is more appropriate for a fixed root wing.

8.1.2 Rolling Effectiveness – Rigid Fixed Wing Root Case

A measure of the influence of the wing flexibility is provided by the so-called *effectiveness* which compares the behaviour for flexible and rigid wings. For a rigid fixed root wing, the lift per unit span due to control rotation is

$$\frac{dL}{dy} = qca_C\beta \tag{8.14}$$

and thus the total root bending moment due to control rotation for the rigid wing is given as

$$M_{rigid} = \int_0^s qca_C\beta y\,dy = \frac{qcs^2a_C}{2}\beta \tag{8.15}$$

The *root bending moment effectiveness* is then found by combining Equations (8.11) and (8.15) to give (after some algebraic manipulation)

$$\mathfrak{I}_{static\ moment} = \frac{\text{root bending moment (flexible)}}{\text{root bending moment (rigid)}} = \frac{1-q/q_{rev}}{1-q/q_W} \tag{8.16}$$

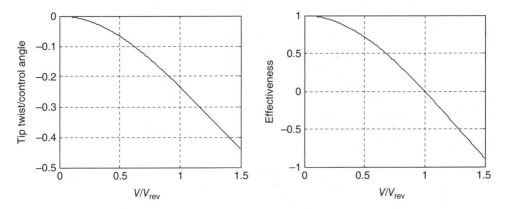

Figure 8.3 Fixed root case: roll effectiveness and tip twist/control angle against velocity normalized to the reversal speed.

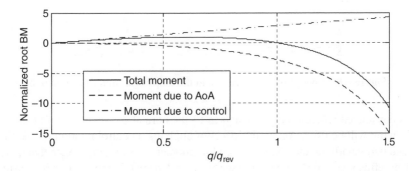

Figure 8.4 Contribution to wing root bending moment by the control surface and angle of attack.

Typical plots of the control effectiveness and tip twist per control angle are shown in Figure 8.3 as a function of velocity normalized to the reversal speed; the figures demonstrate that the control effectiveness reduces from a value of unity with increasing air speed, reaches zero at the reversal speed and then becomes increasingly negative. Military aircraft are sometimes designed to take advantage of this effect to achieve high manoeuvrability through the use of an active control system that takes into account the opposite effect of the controls beyond reversal. The tip twist also becomes increasingly negative (nose down, since b_C is negative) for increases in control angle and dynamic pressure.

Figure 8.4 shows why control reversal occurs; the root bending moment contribution due application of the control surface increases linearly with dynamic pressure; however, this gets cancelled out at the reversal speed by an ever increasing nose-down root bending moment due to the downwards twist of the wing.

8.2   Rolling Effectiveness of a Flexible Wing – Steady Roll Case

Consider the same flexible wing model with a full span aileron as used earlier in Section 8.1, but now with the wing root allowed to undergo a steady roll rate $\dot{\phi}$ about an axis at the root, as shown in Figure 8.5. To avoid problems with signs at this early stage, the port wing with down

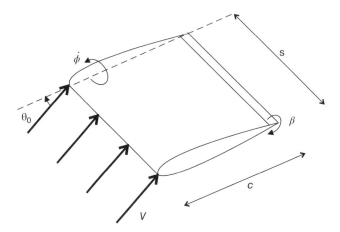

Figure 8.5   Rolling cantilevered wing with a full span aileron.

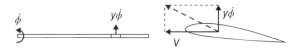

Figure 8.6   Change of incidence due to the downwash from rolling motion.

aileron is considered whereas later in Chapter 13 the normal convention of y positive on the starboard wing will be used. Again, the effect of applying a control rotation on the lift distribution, and particularly on the wing root bending moment, is of interest. Note that no fuselage or tailplane effects are considered. In essence, as in Chapter 7, the aircraft is considered to have a rigid body roll motion, together with a wing torsion branch mode (behaving antisymmetrically on the two sides of the aircraft).

8.2.1   Determination of Reversal Speed for Steady Roll Case

The lift force and pitching moment acting upon an elemental strip dy for a down aileron application on the port wing are

$$dL = qc\,dy\left[a_W\left(\theta_0 + \frac{y}{s}\theta_T - \frac{\dot{\phi}y}{V}\right) + a_C\beta\right] \quad \text{and} \quad dM = qc^2\,dy\left[b_W\left(\theta_0 + \frac{y}{s}\theta_T - \frac{\dot{\phi}y}{V}\right) + b_C\beta\right] \quad (8.17)$$

where q is again the dynamic pressure and $\dot{\phi}y/V$ is the reduction of incidence associated with the downwash due to the rate of roll (port wing moving upwards), as seen in Figure 8.6.

The total incremental work done associated with incremental twist $\delta\theta$ and roll $\delta\phi$ angles is

$$\delta W = \int_{\text{wing}} (dL y\,\delta\phi + dM\,\delta\theta)$$

$$= 2qc\int_0^s \left[a_W\left(\frac{y}{s}\theta_T - \frac{\dot{\phi}y}{V}\right) + a_C\beta\right] y\,\delta\phi\,dy + 2qc^2\int_0^s \left[b_W\left(\frac{y}{s}\theta_T - \frac{\dot{\phi}y}{V}\right) + b_C\beta\right]\delta\theta\,dy \qquad (8.18)$$

Here a factor of two has been applied to include the effect of the starboard wing and the θ_0 terms cancel out on the two sides of the aircraft. Thus the generalized forces in the ϕ and θ_T generalized coordinates are

$$Q_\phi = \frac{\partial(\delta W)}{\partial(\delta\phi)} = 2qc \int_0^s \left[a_W \left(\frac{y^2}{s}\theta_T - \frac{\dot{\phi}y^2}{V} \right) + a_C \beta y \right] dy = 2qc \left[a_W \left(\frac{s^2}{3}\theta_T - \frac{\dot{\phi}s^3}{3V} \right) + \frac{a_C \beta s^2}{2} \right]$$

(8.19)

and

$$Q_{\theta_T} = \frac{\partial(\delta W)}{\partial(\delta\theta_T)} = 2qc^2 \int_0^s \left[b_W \left(\frac{y^2}{s^2}\theta_T - \frac{\dot{\phi}y^2}{sV} \right) + \frac{b_C \beta y}{s} \right] dy = 2qc^2 \left[b_W \left(\frac{s}{3}\theta_T - \frac{\dot{\phi}s^2}{3V} \right) + \frac{b_C \beta s}{2} \right]$$

(8.20)

Since the roll motion of the aircraft is steady, the kinetic energy terms associated with roll ϕ and twist θ are constant and do not contribute inertia terms to the roll or twist equations. Thus, using Lagrange's equations and evaluating the elastic potential (or strain) energy for both wings, the relationship between the rate of roll, tip twist and aileron angle is found as

$$\begin{bmatrix} \dfrac{2qcs^3 a_W}{3V} & \dfrac{-2qcs^2 a_W}{3} \\[3mm] \dfrac{2qc^2 s^2 b_W}{3V} & \left(\dfrac{2GJ}{s} - \dfrac{2qc^2 sb_W}{3} \right) \end{bmatrix} \begin{Bmatrix} \dot{\phi} \\ \theta_T \end{Bmatrix} = \begin{Bmatrix} qcs^2 a_C \\ qc^2 sb_C \end{Bmatrix} \beta$$

(8.21)

or

$$\begin{bmatrix} 1 & -1 \\ e & (\mu-e) \end{bmatrix} \begin{Bmatrix} \dfrac{s\dot{\phi}}{V} \\ \theta_T \end{Bmatrix} = \begin{Bmatrix} \dfrac{3a_C}{2a_W} \\[3mm] \dfrac{3b_C}{2a_W} \end{Bmatrix} \beta$$

(8.22)

where

$$\mu = \frac{3GJ}{qc^2 s^2 a_W}$$

(8.23)

Solving Equations (8.22) leads to the expressions for the roll rate per control angle and also for the tip twist per control angle

$$\frac{\dot{\phi}}{\beta} = \frac{3V}{2\mu s a_W}[a_C(\mu-e) + b_C] \quad \text{and} \quad \frac{\theta_T}{\beta} = \frac{3(b_C - ea_C)}{2a_W\mu} = \frac{qc^2 s^2 (b_C - ea_C)}{2GJ}$$

(8.24)

Remembering that b_C is negative, the second expression in Equation (8.24) is always negative and there is an increasing nose-down twist with increasing dynamic pressure.

When the roll rate per control angle is compared to the rigid aircraft result, then the *control effectiveness in roll* $\Im$ may be defined as

$$\Im = \frac{(\dot{\phi}/\beta)_{\text{flexible}}}{(\dot{\phi}/\beta)_{\text{rigid}}} \tag{8.25}$$

and since for the rigid wing $GJ \rightarrow \infty$ and $\mu \rightarrow \infty$, then

$$\Im = \frac{(\dot{\phi}/\beta)_{\text{flexible}}}{(\dot{\phi}/\beta)_{\text{rigid}}} = \frac{[3V/(2\mu s a_W)][a_C(\mu - e) + b_C]}{3Va_C/2sa_W} = \frac{a_C(\mu - e) + b_C}{\mu a_C} \tag{8.26}$$

where it should be noted that $\mu > 0$, $a_C > 0$ and $b_C < 0$.

At the reversal speed, there is no change of roll rate with respect to control angle, i.e. $\dot{\phi}/\beta = 0$, which occurs when $[a_C(\mu - e) + b_C] = 0$. Thus the dynamic pressure at the reversal speed, q_{rev}, is found as

$$q_{\text{rev}} = \frac{3GJa_C}{c^2 s^2 a_W(ea_C - b_C)} \tag{8.27}$$

with a corresponding tip twist per aileron angle of

$$\frac{\theta_{T_{\text{rev}}}}{\beta} = -\frac{3a_C}{2a_W} \tag{8.28}$$

At the reversal speed, the rolling moment due to the angle of incidence of the wing is exactly cancelled out by the rolling moment generated by the control angle. Combining Equations (8.26) and (8.27) leads to a different expression for the control effectiveness purely in terms of the dynamic pressure, so that

$$\Im = 1 - \frac{q}{q_{\text{rev}}} \tag{8.29}$$

Finally, comparing the reversal speed to the fixed wing divergence speed gives

$$\frac{q_{\text{rev}}}{q_W} = \frac{3GJa_C/[c^2 s^2 a_W(ea_C - b_C)]}{3GJ/(ec^2 s^2 a_W)} = \frac{ea_C}{ea_C - b_C} \tag{8.30}$$

which is the same value as found for the fixed wing root case considered earlier.

Figure 8.7 compares the constant roll rate and static moment effectiveness expressions in Equations (8.16) and (8.29), for the case where the reversal speed is 80% of the divergence speed. It can be seen that although the reversal speed is the same for both approaches, there is a significant difference between the two effectiveness curves in the subcritical region.

The actual dynamic roll performance of the aircraft may be examined further when the flight mechanics model is used with the Flight Control System (FCS) and flexible modes included, or with relevant aerodynamic terms corrected for flexible effects (see Chapter 13). The scheduling of different controls (e.g. inboard and outboard ailerons and spoilers) may be adjusted as necessary.

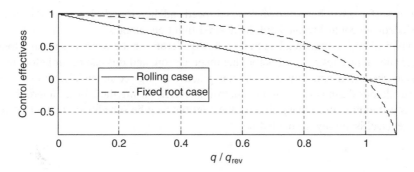

Figure 8.7 Control effectiveness for constant roll rate and fixed root cases.

8.2.2   Lift Distribution for the Steady Roll Case

Consider the lift acting upon a chordwise elemental strip at distance y from the root. The incremental lift on each chordwise strip dy due to the twist, roll rate and control rotation (i.e. ignoring the steady contribution from the root incidence θ_0) is given by

$$dL = qc\left[a_W\left(\frac{y}{s}\theta_T - \frac{\dot{\phi}y}{V}\right) + a_C\beta\right]dy \qquad (8.31)$$

Substituting Equations (8.22) for the roll rate and wing twist at the reversal condition into this equation leads to the lift per unit span expression at the reversal speed

$$\frac{dL}{dy} = qca_C\left(1 - \frac{3y}{2s}\right)\beta \qquad (8.32)$$

Although the lift per unit span increases linearly with dynamic pressure and control angle, the lift per unit span reduces with distance from the wing root due to the negative (nose down) twist caused by the control action. In the outboard wing, beyond two-thirds of the semi-span, the lift per unit span is negative and this counteracts the positive lift inboard to give no rolling moment at reversal.

The total wing root rolling moment on one wing at reversal is found by integrating the moment of lift on each strip across the entire wing, such that

$$\text{Wing root rolling moment} = \int_0^s qca_C\left(1 - \frac{3y}{2s}\right)\beta y\,dy = qca_C\left(\frac{s^2}{2} - \frac{s^2}{2}\right)\beta = 0 \qquad (8.33)$$

This is to be expected since the achieved rate of roll is zero at reversal.

8.3   Effect of Spanwise Position of the Control Surface

In practice there are usually a number of ailerons along the wings of commercial and military aircraft. They can be used in combination, often to reduce the root bending moment in manoeuvres, and also when a Gust Load Alleviation (or suppression) system is installed. Each of the control surfaces will have its own reversal speed and therefore designers need to ensure that there is adequate control throughout the desired flight envelope.

For a typical large commercial aircraft with two ailerons on the wing, it is usually found that the outer aileron reaches reversal before the inner aileron, and this can occur before the cruise speed is reached. Consequently, it is common practice in large commercial aircraft to 'lock' the outboard ailerons during cruise, and to control the aircraft using the inboard ailerons. Alternatively, as is common in military aircraft, the controls may be scheduled via the FCS and any control reversal is simply accounted for.

8.4   Full Aircraft Model – Control Effectiveness

In practice, when dealing with the general form of the full aircraft model for static aeroelastic considerations, the generalized (or modal) coordinate equations are used to represent the rigid body and flexible modes (see Chapter 3). In this section, a more general approach that will allow a full aircraft to be considered is described.

For the constant roll rate example given above, consider the equations for roll rate and tip twist developed earlier

$$
\begin{bmatrix}
\dfrac{2qcs^3 a_{\mathrm{W}}}{3V} & \dfrac{-2qcs^2 a_{\mathrm{W}}}{3} \\[2mm]
\dfrac{2qc^2 s^2 b_{\mathrm{W}}}{3V} & \left(\dfrac{2GJ}{s} - \dfrac{2qc^2 s b_{\mathrm{W}}}{3}\right)
\end{bmatrix}
\left\{ \begin{array}{c} \dot{\phi} \\ \theta_T \end{array} \right\}
=
\left\{ \begin{array}{c} qcs^2 a_C \\ qc^2 s b_C \end{array} \right\} \beta
\tag{8.34}
$$

Now, a constraining moment M may be added in the roll degree of freedom such that there is no roll rate allowed; then a dummy equation for roll is added and the roll angle set to zero. Equation (8.34) becomes

$$
\begin{bmatrix}
0 & \dfrac{-2qcs^2 a_{\mathrm{W}}}{3} \\[2mm]
0 & \dfrac{2GJ}{s} - \dfrac{2qc^2 s b_{\mathrm{W}}}{3}
\end{bmatrix}
\left\{ \begin{array}{c} \phi \\ \theta_T \end{array} \right\}
+
\left\{ \begin{array}{c} 1 \\ 0 \end{array} \right\} M
=
\left\{ \begin{array}{c} qcs^2 a_C \\ qc^2 s b_C \end{array} \right\} \beta
\tag{8.35}
$$

and this may be rewritten as

$$
\begin{bmatrix}
0 & \dfrac{-2qcs^2 a_{\mathrm{W}}}{3} & 1 \\[2mm]
0 & \dfrac{2GJ}{s} - \dfrac{2qc^2 s b_{\mathrm{W}}}{3} & 0 \\[2mm]
1 & 0 & 0
\end{bmatrix}
\left\{ \begin{array}{c} \phi \\ \theta_T \\ M \end{array} \right\}
=
\left\{ \begin{array}{c} qcs^2 a_C \\ qc^2 s b_C \\ 0 \end{array} \right\} \beta
\tag{8.36}
$$

This matrix equation may be solved for the constraining moment

$$M = qcs^2 \left[\left(\frac{qc^2 s b_C}{3GJ/s - qc^2 s b_W} \right) a_W + a_C \right] \beta \qquad (8.37)$$

Once the constraining moment is determined, the ratio of the flexible to rigid wing moments yields the static effectiveness and the same reversal speed as for the previous approaches.

Thus, for a flexible aircraft, the constraining Equation (8.36) takes the general matrix form

$$\begin{bmatrix} \mathbf{E} + \rho V^2 \mathbf{C} & \mathbf{Z}_c^T \\ \mathbf{Z}_c & 0 \end{bmatrix} \begin{bmatrix} \mathbf{p} \\ \mathbf{F} \end{bmatrix} = \begin{bmatrix} \mathbf{R}_{con} \\ 0 \end{bmatrix} \qquad (8.38)$$

where $\mathbf{E}$ is the generalized stiffness matrix, $\mathbf{C}$ is the aerodynamic generalized stiffness matrix (see Chapter 9), $\mathbf{p}$ is a vector of generalized coordinates representing rigid body and flexible modes, $\mathbf{F}$ are the forces/moments at some reference point required to constrain the relevant displacements/rotations to zero, $\mathbf{R}_{con}$ is the vector of control generalized forces and $\mathbf{Z}_c$ (equal to $[1\ 0]$ in the above example) is the physical constraint matrix setting the relevant displacements/rotations in the rigid body motions to zero via $\mathbf{Z}_c \mathbf{p} = \mathbf{0}$. The upper equation is the overall force/moment balance and the lower equation constrains the defined displacement/rotation to be zero.

The effectiveness may be obtained from any of the constraint forces or moments. The same approach may be employed for elevator reversal by constraining the aircraft in the centre fuselage.

8.5   Effect of Trim on Reversal Speed

So far it has been assumed that the aircraft can reach and exceed the calculated reversal speed if the wing is sufficiently flexible. However, in practice the trim of the aircraft needs to be maintained by slight adjustments of the controls, i.e. ailerons/rudder for lateral trim and elevator for longitudinal trim. As the reversal condition is approached, the relevant control becomes less effective, not only at controlling the aircraft but also at adjusting trim. Thus the aircraft may 'run out of trim' prior to reversal being reached.

8.6   Examples

1. Investigate the effect that varying the position of the elastic axis (ec) and the chord/aileron ratio (Ec) have upon the reversal and divergence speeds.
2. For the fixed root case, determine the combinations of e and E that produce the best effectiveness values below the reversal speed.
3. For the fixed root case, explore the effect of the ratio between the dynamic pressures at divergence and reversal on the aileron effectiveness.
4. For a wing containing two ailerons that together total the entire semi-span of the wing, determine the size of the two parts such that they have the same reversal speed.
5. For a wing containing two ailerons that together total the entire semi-span of the wing, determine the size of the two parts such that they each give the same root bending moment.

·

9

Introduction to Unsteady Aerodynamics

So far, when considering static aeroelastic effects in Chapters 7 and 8, the aerodynamic surfaces (such as wings) have been in a steady condition and so the resulting forces and moments have been *steady* (i.e. constant with time). However, for flutter, manoeuvre and gust response analyses the behaviour of aerodynamic surfaces under dynamic motion is required and it is necessary to include the effect of the aerodynamic surface motion upon the resulting forces and moments. These so-called *unsteady* effects are an outcome of the changing circulation and wake acting upon a moving aerofoil, and can have a considerable influence upon the resulting aerodynamic forces and moments. Consequently, a more sophisticated analysis is required than simply considering the angle of incidence. Most aeroelasticity textbooks cover unsteady aerodynamic effects (Scanlan and Rosenbaum, 1960; Fung, 1969; Bisplinghoff *et al.*, 1996; Hodges and Pierce, 2011; Dowell *et al.*, 2004).

In this chapter, the two-dimensional inviscid, incompressible flow over a thin, rigid section aerofoil undergoing small amplitude heave and pitch motions will be considered. Starting with the effect of a sudden step change in incidence on the lift acting on an aerofoil, the lift and moment resulting from a harmonically oscillating aerofoil in a steady flow will be investigated, followed by consideration of how a general motion would be dealt with. Analytical models, using the so-called oscillatory aerodynamic derivatives, will be developed to show how the aerodynamic forces and moments can be expressed via aerodynamic damping and stiffness terms. The results will be used in Chapter 10 on flutter, albeit using a highly simplified version of the derivatives.

The related issue of unsteady aerodynamic effects for the aerofoil encountering a sharp-edged or harmonic gust will also be considered, since this will be required for the discrete gust and continuous turbulence response analysis in Chapter 14. Unsteady aerodynamics will not be considered for flight and ground manoeuvres in Chapters 12, 13 and 15.

Simple examples will be used to illustrate the underlying principles rather than addressing the most up-to-date aerodynamic methods. There is a wide range of more advanced methods available for computing unsteady aerodynamics for more general three-dimensional geometries, and these will be addressed briefly in Chapters 18 and 19.

Introduction to Aircraft Aeroelasticity and Loads, Second Edition. Jan R. Wright and Jonathan E. Cooper.
© 2015 John Wiley & Sons, Ltd. Published 2015 by John Wiley & Sons, Ltd.

9.1 Quasi-steady Aerodynamics

So far in this book, the static aeroelastic cases considered in Chapters 7 and 8 have been for aerofoils fixed relative to the air flow and where the aerodynamic forces and moments are constant with time, i.e. so-called *steady* aerodynamics (see Chapter 4).

Where the aerofoil is undergoing a general motion in heave and/or pitch relative to the upstream flow, then the forces and moments vary with time. One simple approach for the calculation of such forces and moments is to assume that at any instant of time the aerofoil behaves with the characteristics of the same aerofoil with instantaneous values of displacement and velocity. This is known as the *quasi-steady* assumption and implies that there are no frequency-dependent effects.

9.2 Unsteady Aerodynamics related to Motion

The *quasi-steady* assumption, while attractive in its simplicity, is not sufficiently accurate for flutter and gust response calculations and a more advanced *unsteady* aerodynamic analysis must be used in order to predict accurately the dependency of aerodynamic forces and moments on the frequency content of dynamic motions.

In order to understand the effect of aerofoil heave and/or pitch motions on the aerodynamic loads and moments generated, the result of instantaneous changes in the angle of incidence and harmonic motion of the aerofoil need to be considered; the key tools to analyze these effects are the Wagner and Theodorsen functions respectively (Fung, 1969; Bisplinghoff *et al.*, 1996). The Wagner function can be used to consider the case of general motion in the *time* domain, whereas the related Theodorsen function is an important component in predicting the onset of flutter in the *frequency* domain and in the analysis of the response to continuous turbulence.

9.2.1 Instantaneous Change in Angle of Incidence – Wagner Function

Consider a two-dimensional aerofoil of chord c, initially at some small angle of incidence α and moving at air speed V in still air; assume inviscid and incompressible flow. The aerofoil is then subjected to an instantaneous change in angle of incidence of $\Delta\alpha$. If a quasi-steady aerodynamic model were used, the lift would increase instantaneously to the new quasi-steady value for the new angle of incidence, but this does not occur in practice.

Figure 9.1 shows how the unsteady lift changes instantaneously to half of the difference between the initial and final steady values, and then increases asymptotically towards the final steady value. Approximately 90% of the change in lift is achieved after 15 semi-chords have been travelled by the aerofoil; however, there is clearly a considerable delay after the change in incidence before the quasi-steady value is reached. The delay in achieving the new steady lift value occurs due to the time taken for the circulation around the aerofoil to change to that of the new steady flow condition and for changes in the wake to reach a steady state. Since the results are expressed in terms of semi-chords travelled by the aerofoil, a non-dimensional measure, this behaviour is independent of chord size or air speed. Note also that the same effect would be found if the air speed were to change suddenly instead.

The *Wagner function* (Fung, 1969; Bisplinghoff *et al.*, 1996) is used to model this behaviour, i.e. how the lift acting at the quarter chord on the aerofoil builds up following the step change of incidence (or air speed), by obtaining the effective downwash at the 3/4 chord point (downwash is the velocity component normal to the airflow). In terms of non-dimensional time

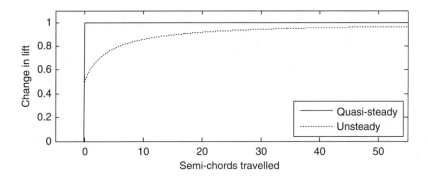

Figure 9.1 Change in aerofoil lift due to a sudden change in the angle of incidence.

$\tau = 2Vt/c = Vt/b$ (i.e. the time taken for the flow to cross a semi-chord b of the aerodynamic surface), the *increase* in lift per unit span following the step change in incidence $\Delta\alpha$ is expressed as

$$\Delta L = \frac{1}{2}\rho\, V^2 ca_1 \Delta\alpha\Phi(\tau) = \frac{1}{2}\rho\, Vca_1 w\Phi(\tau) \qquad (9.1)$$

where $w = V\sin\Delta\alpha \approx V\Delta\alpha$ is the change in downwash on the aerofoil (the symbol w is not to be confused with its later usage for the flight mechanics model in Chapter 13) and $\Phi(\tau)$ is the Wagner function, defined approximately for the incompressible case as (Fung, 1969; Bisplinghoff *et al.*, 1996)

$$\Phi(\tau) = 0 \quad \tau \le 0 \quad \text{and} \quad \Phi(\tau) = \frac{\tau+2}{\tau+4} \quad \tau > 0 \qquad (9.2)$$

The Wagner function is often defined using exponential functions, as they are easier to manipulate using Laplace transforms compared to the simple expression in Equation (9.2), but they will not be considered here. The exact function may be determined using Bessel functions (Fung, 1969).

9.2.2 Harmonic Motion – Convolution using the Wagner Function

For a general heave and pitch motion of the aerofoil, the Wagner function may be used to find the lift by obtaining the effective downwash w at the 3/4 chord point and using a convolution integral approach (see Chapter 1). This approach is analogous to determining the response of a system to a general excitation expressed as a superposition of a series of steps, knowing the step response function. The downwash is then represented by a series of step changes that follow the motion of the aerofoil.

By considering the step change in downwash dw over time $d\tau_0$, the lift may be written as

$$L(\tau) = \frac{1}{2}\rho Vca_1 \left[w_0 + \int_{\tau_0=0}^{\tau} \Phi(\tau-\tau_0)\frac{dw}{d\tau_0}d\tau_0 \right] \qquad (9.3)$$

where $\Phi(\tau-\tau_0)(dw/d\tau_0)d\tau_0$ defines the lift at τ due to the step change in downwash at τ_0. The overall lift time history is obtained by summing (or integrating) the lift obtained from each step.

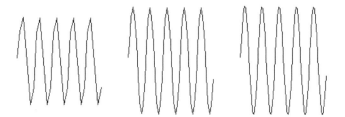

Figure 9.2 Resultant lift for oscillating aerofoil using the convolution process with decreasing time steps.

Figure 9.2 shows the result of the convolution process, for decreasing time steps, of the aerofoil oscillating sinusoidally in pitch with an angle of incidence varying as $\alpha = \alpha_0 \sin \omega t$. When the time step is large, the predicted lift time history is somewhat uneven and in error. However, if the time step is reduced then, in the limit, the resulting lift is *sinusoidal* and of the *same frequency* as the oscillation of the aerofoil.

9.2.3 Harmonic Motion using the Theodorsen Function

For flutter calculations, the general unsteady aerodynamic behaviour in the time domain is rarely used, since the motion at a single oscillation frequency is of more interest (note, however, that the general motion in the time domain is of interest for the gust response; see Chapter 14). Returning to the example in the previous section, with an aerofoil oscillating in pitch at frequency ω, and applying the convolution approach using the Wagner function to obtain the lift time history, the effect of varying the frequency of the oscillation in pitch is now examined. Figure 9.3 shows that, compared to the quasi-steady lift values, there is a *reduction in the magnitude* of the lift and an *introduction of a phase lag* between the aerofoil motion and the unsteady forces (the quasi-steady values are always in-phase by definition). As the frequency increases, the unsteady force amplitude decreases and the phase lag changes.

Further investigation shows that the amplitude attenuation and phase lag are a function of the dimensionless *frequency parameter* ν, defined as

$$\nu = \frac{\omega c}{V} \tag{9.4}$$

which can be interpreted as the number of oscillations undergone by the aerofoil during the time taken for the airflow to travel across the chord of the aerofoil, multiplied by 2π (the frequency is defined in radians per second). However, often the so-called *reduced frequency k* is used, as in Figure 9.3, and this is defined in terms of the semi-chord $b = c/2$ such that

$$k = \frac{\omega b}{V} = \frac{\omega c}{2V} = \frac{\nu}{2} \tag{9.5}$$

Historically, the fundamental work on unsteady aerodynamics and aeroelasticity in the UK (Frazer and Duncan, 1928; Collar, 1978) used the frequency parameter ν, whereas the equivalent research in the USA (Theodorsen, 1935) was based upon the reduced frequency, k. For most of this book the reduced frequency k will be employed as this has been used in the classic

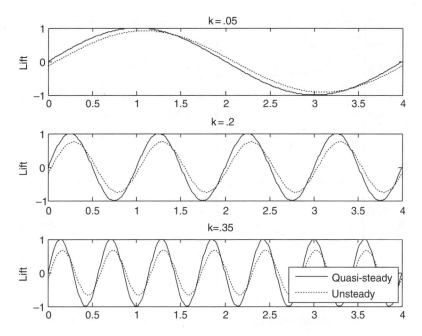

Figure 9.3   Lift for an aerofoil oscillating in pitch at different reduced frequencies.

aeroelasticity textbooks (Fung, 1969; Bisplinghoff *et al.*, 1996). Every time that the reduced frequency is mentioned in a descriptive part of the text, it can simply be replaced with the frequency parameter.

The *Theodorsen function* is used to model the changes in amplitude and phase of the sinusoidal unsteady aerodynamic forces relative to the quasi-steady forces for different reduced frequencies (or frequency parameter). The function behaves, effectively, as the Fourier Transform of the Wagner function, and can be thought of as a filter that modifies the input to a system (i.e. the quasi-steady lift for aerofoil oscillations at some frequency) to give an output (i.e. the unsteady air forces) depending upon the reduced frequency. The Theodorsen function $C(k) = F(k) + iG(k)$, where $C(k)$ is a complex quantity (required since both the amplitude and phase need to change), is expressed (Fung, 1969) as a function of reduced frequency such that

$$C(k) = F(k) + iG(k) = \frac{H_1^{(2)}(k)}{H_1^{(2)}(k) + iH_0^{(2)}(k)} = \frac{K_1(ik)}{K_0(ik) + K_1(ik)} \tag{9.6}$$

where the K_j (ik) $(j = 0,1, \ldots)$ terms are modified Bessel functions of the second kind and $H_n^{(2)}(k)$ are Hankel functions of the second kind. Although an explanation of Bessel and Hankel functions is beyond the scope of this book, these functions are included in many software libraries and are easy to calculate.

Figure 9.4 shows the real and imaginary parts, and amplitude and phase, of the Theodorsen function in graphical form. Note that as k increases the magnitude decreases, and the phase lag increases up to a value of around $k = 0.3$ and then reduces again. The complex plane representation is shown in Figure 9.5, with the function following the curve in a clockwise direction for increasing frequency.

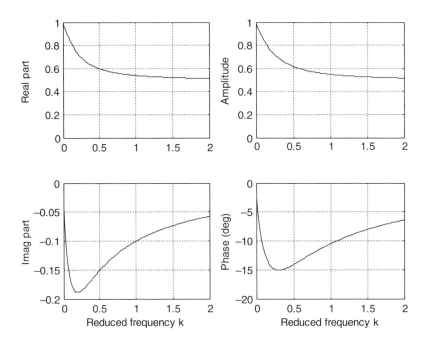

Figure 9.4   Theodorsen function.

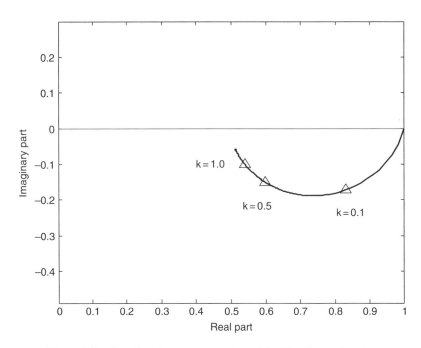

Figure 9.5   Complex plane representation of the Theodorsen function.

For the *quasi-steady aerodynamics* case $\omega = 0$, thus $k = \nu = 0$ and hence $F = 1$ and $G = 0$, so the unsteady lift may be seen to tend towards the quasi-steady values. In the limit as $k \to \infty$, then $F \to 0.5$ and $G \to 0$, but typically for full size aircraft, k has a maximum value of the order of unity.

9.3 Aerodynamic Lift and Moment for an Aerofoil Oscillating Harmonically in Heave and Pitch

The solution of the flow around the aerofoil undergoing harmonic oscillations in heave and pitch can be divided into two parts:

a. *Circulatory terms.* Lift and moment terms occurring due to the vorticity in the flow (related to the Theodorsen function).
b. *Non-circulatory terms.* 'Apparent inertia' forces whose creation is not related to vorticity; as the aerofoil moves, a cylindrical mass of air accelerates with the aerofoil and introduces a reactive force and moment upon the aerofoil. These terms are of minor importance for bending/torsion type flutter of cantilever wings at low reduced frequencies, but are more important for flutter of control surfaces at higher reduced frequencies.

Consider a symmetric two-dimensional aerofoil $(C_{M_0} = 0)$ of chord c, with the elastic axis positioned at distance $ab(=ac/2)$ aft of the mid-chord as shown in Figure 9.6. The aerofoil undergoes oscillatory harmonic motion in heave $z = z_0 e^{i\omega t}$ (positive downwards) and pitch $\theta = \theta_0 e^{i\omega t}$ (positive nose up). The classical solution for the lift and moment about the elastic axis, both expressed per unit span, may be written (Theodorsen, 1935; Fung, 1969; Bisplinghoff *et al.*, 1996) as

$$L = \pi \rho b^2 \left[\ddot{z} + V\dot{\theta} - ba\ddot{\theta} \right] + 2\pi \rho VbC(k) \left[\dot{z} + V\theta + b\left(\frac{1}{2} - a\right)\dot{\theta} \right] \tag{9.7}$$

$$M = \pi \rho b^2 \left[ba\ddot{z} - Vb\left(\frac{1}{2} - a\right)\dot{\theta} - b^2\left(\frac{1}{8} + a^2\right)\ddot{\theta} \right]$$
$$+ 2\pi \rho Vb^2 \left(a + \frac{1}{2}\right)C(k)\left[\dot{z} + V\theta + b\left(\frac{1}{2} - a\right)\dot{\theta}\right] \tag{9.8}$$

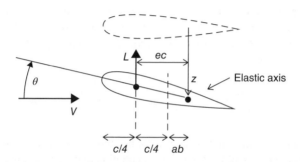

Figure 9.6 Two-dimensional aerofoil undergoing heave and pitch motion.

The derivation of these two equations is beyond the scope of this book, but they are included as they help to describe how the aerodynamic lift and moment vary with reduced frequency. The first part of each expression shows the non-circulatory terms, and the second part shows the circulatory terms which are dependent upon the value of the Theodorsen function. There are terms dependent upon the displacement, velocity and acceleration of both heave and pitch motions, except for the heave displacement term (the vertical aerofoil position does not affect the lift and moment). Note that the Theodorsen function is applied by a multiplication in the frequency domain. Here, the two-dimensional lift curve slope has been taken as $a_1 = 2\pi$.

9.4   Oscillatory Aerodynamic Derivatives

Taking the above expressions for the lift and moment about the elastic axis of the oscillating aerofoil and substituting for the complex form of the Theodorsen function and the heave and pitch motions in complex algebra form (see Chapter 1), then Equations (9.7) and (9.8) become

$$L = \left\{ \pi\rho b^2 \left[-\omega^2 z_0 + i\omega V\theta_0 + \omega^2 ba\theta_0 \right] + 2\pi\rho Vb(F+iG) \left[i\omega z_0 + V\theta_0 + i\omega b \left(\frac{1}{2} - a \right)\theta_0 \right] \right\} e^{i\omega t}$$

(9.9)

$$M = \left\{ \pi\rho b^2 \left[-\omega^2 ba z_0 - i\omega Vb \left(\frac{1}{2} - a \right)\theta_0 + b^2\omega^2 \left(\frac{1}{8} + a^2 \right)\theta_0 \right] \right.$$

$$\left. + 2\pi\rho Vb^2 \left(a + \frac{1}{2} \right)(F+iG) \left[i\omega z_0 + V\theta_0 + i\omega b \left(\frac{1}{2} - a \right)\theta_0 \right] \right\} e^{i\omega t}$$

(9.10)

These equations can then be written in the oscillatory derivative form

$$L = \rho V^2 b \left[(L_z + ikL_{\dot{z}}) \frac{z_0}{b} + (L_\theta + ikL_{\dot{\theta}})\theta_0 \right] e^{i\omega t}$$

(9.11)

$$M = \rho V^2 b^2 \left[(M_z + ikM_{\dot{z}}) \frac{z_0}{b} + (M_\theta + ikM_{\dot{\theta}})\theta_0 \right] e^{i\omega t}$$

where L_z, M_z etc., are the non-dimensional *oscillatory aerodynamic derivatives* (not to be confused with classical aerodynamic or stability and control derivatives; see Chapters 12 and 13). These derivatives are expressed in terms of the normalized displacement and velocity for heave and pitch, so, for example,

$$L_z = \frac{\partial C_L}{\partial(z/b)} \qquad L_{\dot{z}} = \frac{\partial C_L}{\partial(\dot{z}/V)} \qquad L_{\dot{\theta}} = \frac{\partial C_L}{\partial(\dot{\theta}c/V)} \qquad \text{etc.}$$

(9.12)

Note that there are no acceleration-based terms as they have now been included in the displacement terms via the conversion of the double differentiation to the frequency domain. In terms of the Theodorsen function, comparison of Equations (9.9) and (9.11) leads to the lift derivatives being expressed as

$$L_z = 2\pi\left(-\frac{k^2}{2} - Gk\right) \quad L_{\dot{z}} = 2\pi F$$

$$L_\theta = 2\pi\left[\frac{k^2 a}{2} + F - Gk\left(\frac{1}{2} - a\right)\right] \quad L_{\dot{\theta}} = 2\pi\left[\frac{1}{2} + F\left(\frac{1}{2} - a\right) + \frac{G}{k}\right] \tag{9.13}$$

and, from comparison of Equations (9.10) and (9.11), the relevant moment derivatives are

$$M_z = 2\pi\left[-\frac{k^2 a}{2} - k\left(a + \frac{1}{2}\right)G\right] \quad M_{\dot{z}} = 2\pi\left(a + \frac{1}{2}\right)F$$

$$M_\theta = 2\pi\left[\frac{k^2}{2}\left(\frac{1}{8} + a^2\right) + F\left(a + \frac{1}{2}\right) - kG\left(a + \frac{1}{2}\right)\left(\frac{1}{2} - a\right)\right] \tag{9.14}$$

$$M_{\dot{\theta}} = 2\pi\left[-\frac{k}{2}\left(\frac{1}{2} - a\right) + kF\left(a + \frac{1}{2}\right)\left(\frac{1}{2} - a\right) + \frac{G}{k}\left(a + \frac{1}{2}\right)\right]$$

Apart from L_z and $L_{\dot{z}}$, these derivative values depend upon where the elastic axis is located on the chord.

The quasi-steady values of the aerodynamic derivatives ($k \to 0$, $F \to 1$, $G \to 0$) can be found as

$$L_z = 0 \quad L_{\dot{z}} = 2\pi \quad L_\theta = 2\pi \quad kL_{\dot{\theta}} = 0 \quad M_z = 0$$

$$M_{\dot{z}} = 2\pi\left(a + \frac{1}{2}\right) \quad M_\theta = 2\pi\left(a + \frac{1}{2}\right) \quad kM_{\dot{\theta}} = 0 \tag{9.15}$$

Note the singularity in the expressions for $M_{\dot{\theta}}$ and $L_{\dot{\theta}}$ as $k \to 0$. However, since both $kL_{\dot{\theta}}$ and $kM_{\dot{\theta}}$ tend to zero, then the contribution to the lift and moment from these derivatives is also zero as $k \to 0$. Therefore, the concept of quasi-steady derivatives does not apply to the $\dot{\theta}$ derivatives (Hancock et al., 1985). The other derivatives agree with the expressions found earlier for the quasi-steady forces and moments.

9.5 Aerodynamic Damping and Stiffness

Further insight into the effect of the unsteady aerodynamic forces can be obtained by considering

$$k = \frac{\omega b}{V}, \quad z = z_0 e^{i\omega t}, \quad \dot{z} = i\omega z_0 e^{i\omega t} \quad \theta = \theta_0 e^{i\omega t} \quad \text{and} \quad \dot{\theta} = i\omega\theta_0 e^{i\omega t} \tag{9.16}$$

Substituting these expressions into the lift and moment in Equation (9.11) gives

$$L = \rho V^2\left(L_z z + L_{\dot{z}}\frac{b\dot{z}}{V} + L_\theta b\theta + L_{\dot{\theta}}\frac{b^2\dot{\theta}}{V}\right)$$

$$M = \rho V^2\left(M_z bz + M_{\dot{z}}\frac{b^2\dot{z}}{V} + M_\theta b^2\theta + M_{\dot{\theta}}\frac{b^3\dot{\theta}}{V}\right) \tag{9.17}$$

and this can be written in the matrix form

$$\begin{Bmatrix} L \\ M \end{Bmatrix} = \rho V \begin{bmatrix} bL_{\dot{z}} & b^2 L_{\dot{\theta}} \\ b^2 M_{\dot{z}} & b^3 M_{\dot{\theta}} \end{bmatrix} \begin{Bmatrix} \dot{z} \\ \dot{\theta} \end{Bmatrix} + \rho V^2 \begin{bmatrix} L_z & bL_{\theta} \\ bM_z & b^2 M_{\theta} \end{bmatrix} \begin{Bmatrix} z \\ \theta \end{Bmatrix} = \rho V \mathbf{B} \begin{Bmatrix} \dot{z} \\ \dot{\theta} \end{Bmatrix} + \rho V^2 \mathbf{C} \begin{Bmatrix} z \\ \theta \end{Bmatrix}$$

$$(9.18)$$

It can be seen that one term is proportional to the heave and pitch velocities, while the other term is proportional to the heave and pitch displacements. Thus, the aerodynamic forces acting on an aerofoil undergoing oscillatory motion can be considered to behave in a similar way to that of damping and stiffness in a structure; hence $\mathbf{B}$ and $\mathbf{C}$ are termed the aerodynamic damping and stiffness matrices respectively. A key difference to structural damping and stiffness matrices is that the aerodynamic matrices are non-symmetric, aiding coupling between the modes which is a key element of the flutter instability (see Chapter 10). Note also that the aerodynamic damping and stiffness matrices depend upon the reduced frequency and Mach number.

When applied to aeroelastic systems, as will be shown in the next chapter, the aerodynamic forces are considered together with the structural equations, leading to the equations of motion in the classical form of

$$\mathbf{A}\ddot{q} + (\rho V \mathbf{B} + \mathbf{D})\dot{q} + (\rho V^2 \mathbf{C} + \mathbf{E})q = 0 \qquad (9.19)$$

where $\mathbf{A}$, $\mathbf{B}$, $\mathbf{C}$, $\mathbf{D}$, $\mathbf{E}$ are the structural inertia, aerodynamic damping, aerodynamic stiffness, structural damping and structural stiffness matrices respectively, and q are the generalized coordinates (typically modal coordinates). It is important to note that the $\mathbf{B}$, $\mathbf{C}$ matrices only apply for the reduced frequency for which they are defined; this can cause some difficulty for flutter calculations, discussed later in Chapter 10.

Equation (9.19) is one of the most important equations in this book and describes the fundamental interaction between the flexible structure and the aerodynamic forces. Note that it is usual when considering aeroelastic systems to write the structural inertia, damping and stiffness matrices as $\mathbf{A}$, $\mathbf{D}$, $\mathbf{E}$ respectively, rather than the $\mathbf{M}$, $\mathbf{C}$, $\mathbf{K}$ notation often used in classical structural dynamics (see Chapter 2).

9.6 Approximation of Unsteady Aerodynamic Terms

It has been shown (Hancock *et al.*, 1985) that the most important unsteady aerodynamic term is the pitch damping derivative $M_{\dot{\theta}}$, i.e. pitch damping per pitch rate; it adds damping to the moment equation. The relevant Equation (9.14) is plotted for different values of reduced frequency and elastic axis position (i.e. x_f aft of the leading edge) in Figure 9.7. Although there is some variation with reduced frequency, a constant value of $M_{\dot{\theta}} = -1.2$ provides a very good fit through the data, with the added bonus that it has no frequency dependency. This approximation will be used initially in Chapter 10 for the flutter calculations.

9.7 Unsteady Aerodynamics related to Gusts

Similar changes in the aerodynamic forces, as shown above for aerofoil heave and pitch motions, occur when the aerofoil encounters a gust field, with the aerodynamic forces also taking time to build up. Here, the gust analysis equivalent of the Wagner and Theodorsen

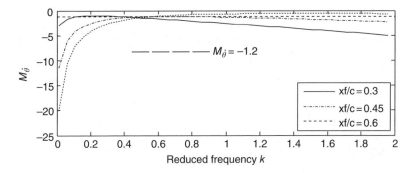

Figure 9.7 Effect of reduced frequency and elastic axis position on $M_{\dot{\theta}}$.

functions will be considered briefly. The response to 'sharp-edged' and 'sinusoidal' gusts will provide the unsteady aerodynamic tools used for the cases of a rigid or flexible aircraft encountering a discrete gust or continuous turbulence, considered later in Chapter 14.

9.7.1 Lift due to a Sharp-Edged Gust – Küssner Function

Consider a rigid aerofoil of chord c and unit span moving at air speed V in still air and suddenly encountering a vertical sharp-edged gust of velocity w_g. The increment in lift acting upon the aerofoil is due to the effective change in the angle of incidence caused by the vertical gust velocity, namely

$$\tan \Delta\alpha \approx \Delta\alpha = \frac{w_g}{V} \tag{9.20}$$

A quasi-steady analysis of this situation assumes that the lift per unit span is developed, as soon as the aircraft enters the gust, according to the expression

$$L = \frac{1}{2}\rho V^2 c a_1 \frac{w_g}{V} = \frac{1}{2}\rho V c a_1 w_g \tag{9.21}$$

However, in practice, the lift takes time to build up as the wing penetrates the gust and this effect can be modelled by rewriting the lift as

$$L = \frac{1}{2}\rho V a_1 c w_g \Psi(\tau) \tag{9.22}$$

where $\Psi(\tau)$ is the Küssner function which describes how the aerodynamic forces build up upon entering a gust. The function (Bisplinghoff et al., 1996) is defined approximately in terms of non-dimensional time τ (= distance travelled in semi-chords) as

$$\Psi(\tau) = \frac{\tau^2 + \tau}{\tau^2 + 2.82\tau + 0.80} \tag{9.23}$$

Figure 9.8 shows how this function builds up from zero, when the aerofoil starts to enter the gust, and asymptotically tends towards unity. As with the Wagner function, there is a

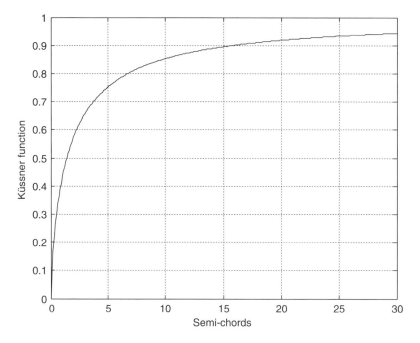

Figure 9.8   Küssner function.

significant delay before the quasi-steady value is reached. The response to any arbitrary gust field can be determined using a convolution approach (see Chapter 14) similar to that demonstrated earlier with the Wagner function.

9.7.2   *Lift due to a Sinusoidal Gust – Sears Function*

Clearly, the response to a general time-varying gust velocity field may be obtained using the Küssner function and a convolution approach. However, an alternative way of modelling the response of an aerofoil subjected to continuous turbulence is to perform a frequency domain Power Spectral Density (PSD) based analysis (see Chapters 1, 14 and 16), which is the approach adopted in industry (see Chapter 23). In order to include the unsteady aerodynamic effects in a frequency domain analysis, the effect at each frequency must be evaluated. The resulting force and moment acting on an aerofoil encountering a sinusoidal gust will be attenuated and delayed in-phase with respect to the quasi-steady result, in much the same way as was seen earlier for an oscillating aerofoil in a steady flow field. These effects are dependent on the reduced frequency.

Consider an aerofoil of chord c moving at air speed V within a sinusoidal gust field having a vertical velocity expressed as a function of time as

$$w_g = w_{g0}e^{i\omega t} \tag{9.24}$$

The lift acting at the quarter chord of the aerofoil can be written as a function of the reduced frequency such that

$$L = \frac{1}{2}\rho V c a_1 w_{g0}e^{i\omega t}\phi(k) \tag{9.25}$$

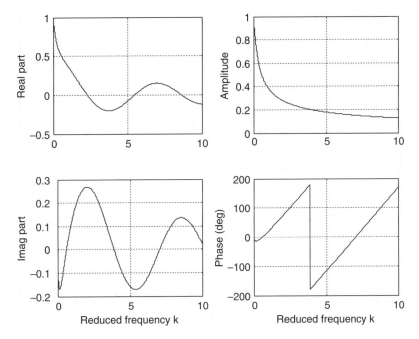

Figure 9.9   Sears function.

Here $\phi(k)$ is the Sears function, which is defined as

$$\phi(k) = [J_0(k) - iJ_1(k)]\, C(k) + iJ_1(k) \tag{9.26}$$

where $C(k)$ is the Theodorsen function and $J_j(k)(j=0,1,\dots)$ are Bessel functions of the first kind. Again, these expressions can be easily computed nowadays without resorting to numerical approximations.

The variation of the real and imaginary parts, and amplitude and phase, of the Sears function for different k values is shown in Figure 9.9, and the complex plane form is shown in Figure 9.10 to have a characteristic spiral shape. If the temporal variation of the sinusoidal gust velocity is transformed into a spatial variation using $k = \omega b/V$ and $x_g = Vt$ (see Chapter 14), then

$$w_g = w_{g0}\exp\left(i\frac{kV}{b}t\right) = w_{g0}\exp\left(i\frac{k}{b}x_g\right) \tag{9.27}$$

The wavelength of the sinusoidal gust is then given by $\lambda_g = 2\pi b/k$, which may be compared to the chord length $c = 2b$ for different values of k.

At first sight, it might be thought that the forces developed on an aerofoil that is not moving in heave due to the harmonic gust would be the same as those developed for the aerofoil itself moving with a harmonic heave velocity in a steady air stream. In fact, this is more or less true for gusts where the wavelength is large in comparison with the chord ($\lambda_g \gg c$, i.e. a small reduced frequency) and so the gust velocity is almost constant across the chord, as seen in Figure 9.11. However, where the wavelength is small compared to the chord ($\lambda_g \ll c$, i.e. a high reduced frequency) there will be a significant difference between the two results because the downwash

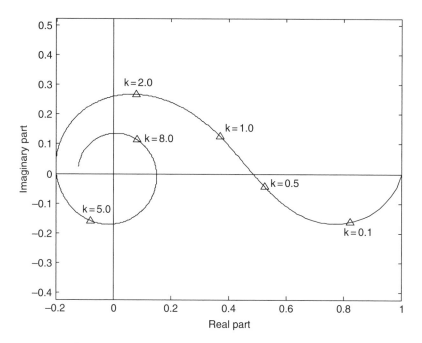

Figure 9.10   Sears function – complex plane representation.

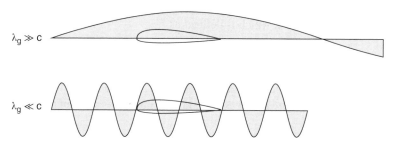

Figure 9.11   Effect of the gust wavelength compared to the chord.

due to the gust velocity will vary significantly across the chord. Consequently the Theodorsen and Sears functions are very similar for a small reduced frequency, as seen by comparing Figure 9.5 with Figure 9.10.

9.8   Examples

1. Write a MATLAB program to model the change in lift for a step change in incidence for a two-dimensional aerofoil making use of the Wagner function. Apply this approach for a harmonically oscillating aerofoil and explore the effect of different frequency parameters and convolution time steps.
2. Making use of the *besselk* function, write a MATLAB program to determine the Theodorsen function for $0 < k < 10$.

3. Write a MATLAB program to determine how the oscillatory aerodynamic derivatives vary with reduced frequency. How great is the difference from the quasi-steady values?
4. Write a MATLAB program to model the change in lift for a two-dimensional aerofoil entering a sharp-edged gust making use of the Küssner function. Extend this to develop a convolution approach for a harmonically oscillating aerofoil and explore the effect of different frequency parameters and convolution time steps.
5. Making use of the *besselj* function, write a MATLAB program to determine the Sears function for $0 < k < 10$.
6. Compare the Theodorsen and Sears functions. For what values of gust wavelengths and reduced frequencies can they be considered to be the same?

10

Dynamic Aeroelasticity – Flutter

Flutter is arguably the most important of all the aeroelastic phenomena (Collar, 1978; Garrick and Reid, 1981) and is the most difficult to predict. It is an unstable self-excited vibration in which the structure extracts energy from the air stream and often results in catastrophic structural failure. Classical binary flutter (Scanlan and Rosenbaum, 1960; Fung, 1969; Hancock *et al.*, 1985; Niblett, 1998; Bisplinghoff *et al.*, 1996; Hodges and Pierce, 2014; Dowell *et al.*, 2004; Rodden, 2011) occurs when the aerodynamic forces associated with motion in two modes of vibration cause the modes to couple in an unfavourable manner, although there are cases where more than two modes have combined to cause flutter and in industry the mathematical models employ many modes (see Chapters 21 and 22 where a typical industry approach is described).

At some critical speed, known as the flutter speed, the structure sustains oscillations following some initial disturbance. Below this speed the oscillations are damped, whereas above it one of the modes becomes negatively damped and unstable (often violent) oscillations occur, unless some form of non-linearity (not considered in detail here) bounds the motion. Flutter can take various forms involving different pairs of interacting modes, e.g. wing bending/torsion, wing torsion/control surface, wing/engine, T-tail etc.

In this chapter, a simple binary flutter model is developed, making use of strip theory with simplified unsteady aerodynamic terms; the model is then used to illustrate the dynamic characteristics of aeroelastic systems, considering the effect of varying the position of the elastic axis, the mass distribution and the frequency spacing between the two modes. Various methods for determining the critical flutter speeds and associated flutter frequencies are examined, including the realistic case where the aerodynamic terms are reduced frequency-dependent. The final part of the chapter considers the phenomenon of control surface flutter and the effect of rigid body modes, it also briefly explores flutter in the transonic flight regime, and introduces some effects of non-linearities. It will be shown that it is important to include unsteady aerodynamic effects in the dynamic models that are used to predict the subcritical aeroelastic behaviour and the onset of flutter. A number of MATLAB codes related to this chapter are included in the companion website.

Introduction to Aircraft Aeroelasticity and Loads, Second Edition. Jan R. Wright and Jonathan E. Cooper.
© 2015 John Wiley & Sons, Ltd. Published 2015 by John Wiley & Sons, Ltd.

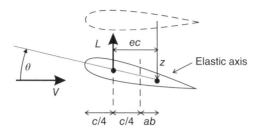

Figure 10.1   Two-dimensional aerofoil.

10.1   Simplified Unsteady Aerodynamic Model

The full two-dimensional unsteady aerodynamic model was described in Chapter 9 for a harmonically oscillating aerofoil, but here a simplified unsteady aerodynamic model will be introduced. Consider the two-dimensional aerofoil shown in Figure 10.1 with the elastic axis positioned a distance ec aft of the aerodynamic centre and ab aft of the mid chord, so

$$ec = \frac{c}{4} + ab = \frac{c}{4} + \frac{ac}{2} \qquad (10.1)$$

It was shown in Chapter 9 that the *unsteady* lift and moment per unit span for an aerofoil may be expressed, for a particular reduced frequency, as

$$L = \rho V^2 \left(L_z z + L_{\dot z} \frac{b \dot z}{V} + L_\theta b \theta + L_{\dot\theta} \frac{b^2 \dot\theta}{V} \right)$$

$$M = \rho V^2 \left(M_z b z + M_{\dot z} \frac{b^2 \dot z}{V} + M_\theta b^2 \theta + M_{\dot\theta} \frac{b^3 \dot\theta}{V} \right) \qquad (10.2)$$

Taking the *quasi-steady* assumption ($k \to 0$, $F \to 1$, $G \to 0$) and the quasi-steady derivatives quoted in Section 9.4, then the lift and pitching moment per unit span about the elastic axis become

$$L = \rho V^2 \left(L_{\dot z} \frac{b \dot z}{V} + L_\theta b \theta \right) = \frac{1}{2} \rho V^2 c a_1 \left(\theta + \frac{\dot z}{V} \right)$$

$$M = \rho V^2 \left(M_{\dot z} \frac{b^2 \dot z}{V} + M_\theta b^2 \theta \right) = \frac{1}{2} \rho V^2 e c^2 a_1 \left(\theta + \frac{\dot z}{V} \right) \qquad (10.3)$$

Compared to the lift force used in the static aeroelastic case, there is now an extra term in each expression due to the effective incidence associated with the aerofoil moving downwards with constant heave velocity $\dot z$, so causing an effective 'upwash'. The quasi-steady assumption implies that the aerodynamic loads acting on an aerofoil undergoing variable heave and pitch motions are equal, at any moment in time, to the characteristics of the same aerofoil with constant position and rate values.

   As seen in Chapter 9, the major drawback in using quasi-steady aerodynamics is that no account is made for the time that it takes for changes in the wake associated with the aerofoil motion to develop (as defined by the Wagner function) and this can lead to serious aeroelastic modelling errors. Consequently, the $M_{\dot{\theta}}$ unsteady aerodynamic derivative term in Equation (10.2) will be retained in the flutter analysis as it has been shown (Hancock *et al.*, 1985) that this has an important effect on the unsteady aerodynamic behaviour. It adds a pitch damping term to the pitching moment Equation (10.3) and so the model then becomes

$$L=\frac{1}{2}\rho V^2 ca_1\left(\theta+\frac{\dot{z}}{V}\right), \quad M=\frac{1}{2}\rho V^2 c^2\left[ea_1\left(\theta+\frac{\dot{z}}{V}\right)+M_{\dot{\theta}}\frac{\dot{\theta}c}{4V}\right] \tag{10.4}$$

where $M_{\dot{\theta}}$ is negative and will initially be assumed to have a constant value, as discussed in Chapter 9. This 'simplified unsteady aerodynamic' model will now be used to develop a binary aeroelastic model. Note that the pitch damping term here differs numerically from that in Hancock *et al.* (1985) by a factor of four, which occurs because the unsteady aerodynamic derivatives are derived here in terms of semi-chord b instead of the chord c.

10.2   Binary Aeroelastic Model

10.2.1   Aeroelastic Equations of Motion

The simple unswept/untapered (i.e. rectangular) cantilevered wing model shown in Figure 10.2 is used throughout this chapter to illustrate classical binary flutter. The rectangular wing of span s and chord c has flexural rigidity EI and torsional rigidity GJ. Note that there is no stiffness coupling between the two motions. The elastic axis is positioned at a distance ec behind the aerodynamic centre (on the quarter chord). The wing is assumed to have a uniform mass distribution and thus the mass axis lies on the mid-chord.

The displacement z (downwards positive) of a general point on the wing is

$$z(x,y,t)=\left(\frac{y}{s}\right)^2 q_b(t)+\left(\frac{y}{s}\right)(x-x_f)q_t(t)=\phi_b q_b+\phi_t q_t \quad \text{so}\quad \theta=\left(\frac{y}{s}\right)q_t \tag{10.5}$$

where q_b and q_t are generalized coordinates and ϕ_b and ϕ_t are the simple assumed shapes. They are actually normal mode shapes (i.e. pure bending and torsion) if there is no inertia coupling about the elastic axis.

The equations of motion can be found using Lagrange's equations. The kinetic energy now exists due to the dynamic motion and is

$$T=\int\limits_{\text{Wing}}\frac{1}{2}dm\dot{z}^2=\frac{m}{2}\int\limits_0^s\int\limits_0^c\left(\left(\frac{y}{s}\right)^2\dot{q}_b+\left(\frac{y}{s}\right)(x-x_f)\dot{q}_t\right)^2 dxdy \tag{10.6}$$

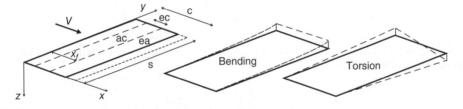

Figure 10.2   Binary aeroelastic model showing bending and torsion modes.

where m is the mass per unit area of the wing. The elastic potential energy corresponds to the strain energy in bending and torsion, such that

$$U = \frac{1}{2}\int_0^s EI\left(\frac{2q_b}{s^2}\right)^2 dy + \frac{1}{2}\int_0^s GJ\left(\frac{q_t}{s}\right)^2 dy \tag{10.7}$$

Note that for a dynamic analysis, any initial angle of incidence can be ignored as vibrations about the trim condition are considered. Applying Lagrange's equations for both generalized coordinates gives

$$\left(\frac{\partial T}{\partial \dot{q}_b}\right) = m \int_0^s\int_0^c \left(\left(\frac{y}{s}\right)^4 \dot{q}_b + \left(\frac{y}{s}\right)^3 (x-x_f)\dot{q}_t\right) dxdy \rightarrow \frac{d}{dt}\left(\frac{\partial T}{\partial \dot{q}_b}\right) = m\left[\frac{sc}{5}\ddot{q}_b + \frac{s}{4}\left(\frac{c^2}{2}-cx_f\right)\ddot{q}_t\right]$$

$$\left(\frac{\partial T}{\partial \dot{q}_t}\right) = m\int_0^s\int_0^c\left(\left(\frac{y}{s}\right)^3(x-x_f)\dot{q}_b + \left(\frac{y}{s}\right)^2(x-x_f)^2\dot{q}_t\right)dxdy \rightarrow \frac{d}{dt}\left(\frac{\partial T}{\partial \dot{q}_t}\right) \tag{10.8}$$

$$= m\left[\frac{s}{4}\left(\frac{c^2}{2}-cx_f\right)\ddot{q}_b + \frac{s}{3}\left(\frac{c^3}{3}-c^2x_f+x_f^2\right)\ddot{q}_t\right]$$

and

$$\frac{\partial U}{\partial q_b} = \int_0^s EI\left(\frac{4q_b}{s^4}\right)dy = \frac{4EI}{s^3}q_b \qquad \frac{\partial U}{\partial q_t} = \int_0^s GJ\left(\frac{q_t}{s^2}\right)dy = \frac{GJ}{s}q_t \tag{10.9}$$

These expressions lead to the equations of motion for the wing, without any aerodynamic forces acting, as

$$m\begin{bmatrix} \dfrac{sc}{5} & \dfrac{s}{4}\left(\dfrac{c^2}{2}-cx_f\right) \\ \dfrac{s}{4}\left(\dfrac{c^2}{2}-cx_f\right) & \dfrac{s}{3}\left(\dfrac{c^3}{3}-c^2x_f+x_f^2c\right) \end{bmatrix}\begin{Bmatrix}\ddot{q}_b \\ \ddot{q}_t\end{Bmatrix} + \begin{bmatrix} \dfrac{4EI}{s^3} & 0 \\ 0 & \dfrac{GJ}{s} \end{bmatrix}\begin{Bmatrix}q_b \\ q_t\end{Bmatrix} = \begin{Bmatrix}0 \\ 0\end{Bmatrix} \tag{10.10}$$

The inertia matrix takes the form

$$\begin{bmatrix} A_{bb} & A_{bt} \\ A_{tb} & A_{tt} \end{bmatrix}$$

so if there is no inertial coupling ($A_{bt} = 0$, i.e. $x_f = c/2$ for this model) then the bending and torsion natural frequencies are

$$\omega_b = \sqrt{\frac{4EI}{A_{bb}s^3}} \qquad \omega_t = \sqrt{\frac{GJ}{A_{tt}s}} \tag{10.11}$$

However, the presence of a non-zero value for $A_{bt}(=A_{tb})$ couples the two motions in the mode shapes and the natural frequencies differ.

Generalized forces Q_b and Q_t act on the system in the form of unsteady aerodynamic forces; in general for an oscillatory motion, they may be written in terms of the aerodynamic derivatives for a particular reduced frequency $k = \omega c/(2V)$. As shown in Chapter 9, these forces are complex but can be expressed in terms of displacements and velocities, bearing in mind that the result only applies for the relevant reduced frequency value. Applying strip theory, together with the simplified unsteady aerodynamics representation, leads to expressions for lift and nose up pitching moment (about the elastic axis) for each elemental strip dy of

$$dL = \frac{1}{2}\rho V^2 c\,dy\,a_W\left(\frac{y^2\dot{q}_b}{s^2V} + \frac{y}{s}q_t\right) \qquad dM = \frac{1}{2}\rho V^2 c^2 dy\left[ea_W\left(\frac{y^2\dot{q}_b}{s^2V} + \frac{y}{s}q_t\right) + M_{\dot\theta}c\frac{(y\dot{q}_t)}{4sV}\right] \quad (10.12)$$

where $\dfrac{y^2}{s^2}\dot{q}_b$ is the effective heave velocity (positive downwards), $b_W = ea_W$ and $M_{\dot\theta} < 0$.

The incremental work done over the wing, corresponding to the aerodynamic force/moment doing work through incremental deflections δq_b, δq_t is

$$\delta W = \int_{\text{Wing}}\left[dL\left(-\left(\frac{y}{s}\right)^2\delta q_b\right) + dM\left(\left(\frac{y}{s}\right)\delta q_t\right)\right] \qquad (10.13)$$

remembering that z is positive downwards. Thus, the generalized forces are

$$Q_{q_b} = \frac{\partial(\delta W)}{\partial(\delta q_b)} = -\int_0^s\left(\frac{y}{s}\right)^2 dL = -\frac{1}{2}\rho V^2 ca_W\int_0^s\left(\left(\frac{y}{s}\right)^4\frac{\dot{q}_b}{V} + \left(\frac{y}{s}\right)^3 q_t\right)dy$$

$$= -\frac{1}{2}\rho V^2 ca_W\left(\frac{s}{5V}\dot{q}_b + \frac{s}{4}q_t\right)$$

$$Q_{q_t} = \frac{\partial(\delta W)}{\partial(\delta q_t)} = \int_0^s\left(\frac{y}{s}\right)dM = \frac{1}{2}\rho V^2 c^2\int_0^s\left[ea_W\left(\left(\frac{y}{s}\right)^3\frac{\dot{q}_b}{V} + \left(\frac{y}{s}\right)^2 q_t\right) + M_{\dot\theta}c\left(\frac{y}{s}\right)^2\frac{\dot{q}_t}{4V}\right]dy \qquad (10.14)$$

$$= \frac{1}{2}\rho V^2 c^2\left[ea_W\left(\frac{s}{4V}\dot{q}_b + \frac{s}{3}q_t\right) + M_{\dot\theta}c\left(\frac{s}{12V}\dot{q}_t\right)\right]$$

The full aeroelastic equations of motion now become

$$m\begin{bmatrix}\dfrac{sc}{5} & \dfrac{s}{4}\left(\dfrac{c^2}{2}-cx_f\right) \\[2mm] \dfrac{s}{4}\left(\dfrac{c^2}{2}-cx_f\right) & \dfrac{s}{3}\left(\dfrac{c^3}{3}-c^2x_f+cx_f^2\right)\end{bmatrix}\begin{Bmatrix}\ddot{q}_b \\ \ddot{q}_t\end{Bmatrix} + \rho V\begin{bmatrix}\dfrac{cs}{10}a_W & 0 \\[2mm] -\dfrac{c^2 s}{8}ea_W & -\dfrac{c^3 s}{24}M_{\dot\theta}\end{bmatrix}\begin{Bmatrix}\dot{q}_b \\ \dot{q}_t\end{Bmatrix}$$

$$+ \left\{\rho V^2\begin{bmatrix}0 & \dfrac{cs}{8}a_W \\[2mm] 0 & -\dfrac{c^2 s}{6}ea_W\end{bmatrix} + \begin{bmatrix}\dfrac{4EI}{s^3} & 0 \\[2mm] 0 & \dfrac{GJ}{s}\end{bmatrix}\right\}\begin{Bmatrix}q_b \\ q_t\end{Bmatrix} = \begin{Bmatrix}0 \\ 0\end{Bmatrix}$$

$$(10.15)$$

and it may be seen that the mass and stiffness matrices are symmetric whilst the aerodynamic matrices are non-symmetric. Hence the two degrees of freedom are coupled and it is this coupling that can give rise to flutter.

10.3 General Form of the Aeroelastic Equations

Equation (10.15) is in the classical second-order form for N DoF discussed earlier in Chapter 9, namely

$$\mathbf{A}\ddot{q} + (\rho V\mathbf{B} + \mathbf{D})\dot{q} + (\rho V^2\mathbf{C} + \mathbf{E})q = \mathbf{0}. \tag{10.16}$$

As is often the case, structural damping has been ignored in the model developed here (i.e. $\mathbf{D} = \mathbf{0}$). An alternative representation that is sometimes used is to reform Equation (10.16) in terms of the equivalent air speed V_{EAS} such that

$$\mathbf{A}\ddot{q} + (\rho_0 \sqrt{\sigma} V_{\mathrm{EAS}}\mathbf{B} + \mathbf{D})\dot{q} + (\rho_0 V_{\mathrm{EAS}}^2 \mathbf{C} + \mathbf{E})q = \mathbf{0} \tag{10.17}$$

where $\sigma = \rho/\rho_0$ is the ratio of air densities at altitude and sea level, defined in Chapter 4.

Since these aeroelastic equations have a zero right-hand side (and so are homogeneous), it is not possible to determine the absolute values of the model response. Instead, the stability of the system needs to be explored using an eigenvalue approach.

10.4 Eigenvalue Solution of the Flutter Equations

The aeroelastic Equation (10.16) can be solved efficiently for an N DoF system using an eigenvalue solution to determine the system frequencies and damping ratios at a particular flight condition (air speed and altitude). Introducing the (trivial) expression

$$\mathbf{I}\dot{q} - \mathbf{I}\dot{q} = \mathbf{0} \tag{10.18}$$

where $\mathbf{I}$ is the $N \times N$ identity matrix, and combining it with Equation (10.16) in partitioned form gives the formulation

$$\begin{bmatrix} \mathbf{I} & 0 \\ 0 & \mathbf{A} \end{bmatrix} \begin{Bmatrix} \dot{q} \\ \ddot{q} \end{Bmatrix} - \begin{bmatrix} 0 & \mathbf{I} \\ -(\rho V^2\mathbf{C} + \mathbf{E}) & -(\rho V\,\mathbf{B} + \mathbf{D}) \end{bmatrix} \begin{Bmatrix} q \\ \dot{q} \end{Bmatrix} = \begin{Bmatrix} \mathbf{0} \\ \mathbf{0} \end{Bmatrix} \tag{10.19}$$

This equation may be rewritten as

$$\begin{Bmatrix} \dot{q} \\ \ddot{q} \end{Bmatrix} - \begin{bmatrix} 0 & \mathbf{I} \\ -\mathbf{A}^{-1}(\rho V^2\mathbf{C} + \mathbf{E}) & -\mathbf{A}^{-1}(\rho V\,\mathbf{B} + \mathbf{D}) \end{bmatrix} \begin{Bmatrix} q \\ \dot{q} \end{Bmatrix} = \mathbf{0} \quad \text{or} \quad \dot{x} - \mathbf{Q}x = \mathbf{0} \tag{10.20}$$

Equation (10.20) are now in first-order form, but note that the $\mathbf{Q}$ matrix is $2N \times 2N$, double the size of the matrices in the aeroelastic Equation (10.16). The equation can be solved by assuming $x = x_0 e^{\lambda t}$ and thus Equation (10.20) becomes

$$(\mathbf{I}\lambda - \mathbf{Q})\mathbf{x}_0 = \mathbf{0} \quad \text{or} \quad (\mathbf{Q} - \mathbf{I}\lambda)\mathbf{x}_0 = \mathbf{0} \tag{10.21}$$

which is in the classical eigensolution form $(\mathbf{A} - \mathbf{I}\lambda)\mathbf{x} = \mathbf{0}$.

For an oscillatory system, such as the aeroelastic system considered here, the eigenvalues λ of the system matrix $\mathbf{Q}$ occur in complex conjugate pairs and are in the form (Frazer *et al.*, 1938; Collar and Simpson, 1987)

$$\lambda_j = -\zeta_j \omega_j \pm i\omega_j \sqrt{1 - \zeta_j^2}, \quad j = 1, 2, ..., N \tag{10.22}$$

where $\omega_j, j = 1, 2, ..., N$ are the natural frequencies and $\zeta_j, j = 1, 2, ..., N$ are the damping ratios. The corresponding eigenvectors appear in complex conjugate columns and take the form

$$\mathbf{x}_j = \left\{ \begin{array}{c} \mathbf{q}_j \\ \lambda_j \mathbf{q}_j \end{array} \right\}, \quad j = 1, 2, ..., N \tag{10.23}$$

Thus the upper (or lower) halves of the eigenvectors yield the mode shapes in terms of generalized coordinates. Note that due to the influence of the non-symmetric aerodynamic terms these modes are complex (see Chapter 2). Also note that the same solution can be found by solving the matrix polynomial directly by letting $q = q\,\mathrm{e}_{\lambda t}$ i.e.

$$\left[\lambda^2 \mathbf{A} + \lambda(\rho V \mathbf{B} + \mathbf{D}) + (\rho V^2 \mathbf{C} + \mathbf{E}) \right] \mathbf{q}_0 = \mathbf{0} \tag{10.24}$$

If the real part of the complex eigenvalues is positive then the system becomes unstable. However, if the eigenvalues are real, then the roots are non-oscillatory and do not occur in complex conjugate pairs, although, if the real part becomes positive, the system becomes statically unstable (i.e. divergent; see Chapter 7).

It is possible to improve the model further by introducing more terms into the definition of the deflection shape; e.g. for two bending and two torsion assumed modes, the deflection could be modelled as

$$z = \left(\frac{y}{s}\right)^2 q_1 + \left(\frac{y}{s}\right)^3 q_2 + \left(\frac{y}{s}\right)(x - x_\mathrm{f})q_3 + \left(\frac{y}{s}\right)^2 (x - x_\mathrm{f})q_4 \tag{10.25}$$

The more terms that are included in the model, the more accurate the results will be (see Chapter 3).

10.5 Aeroelastic Behaviour of the Binary Model

The dynamic aeroelastic behaviour for the binary wing can now be determined at different air speeds and altitudes by forming the eigensolution of matrix $\mathbf{Q}$ in Equation (10.20) for each flight condition and then calculating the corresponding frequencies and damping ratios. In the following section, the effect of varying different structural and aerodynamic parameters on the frequency and damping trends (so-called $V\omega$ and Vg plots) is investigated.

The baseline system parameters considered are shown in Table 10.1, noting that the mass axis is at the semi-chord ($x_\mathrm{m} = 0.5c$) and the elastic axis is at $x_\mathrm{f} = 0.48c$ from the leading edge.

Table 10.1 Baseline parameters for the binary flutter model

Semi-span (s)	7.5 m	Bending Rigidity (EI)	2×10^7 Nm^2
Chord (c)	2 m	Torsional Rigidity (GJ)	2×10^6 Nm^2
Elastic axis (x_f)	0.48c	Lift curve slope (a_W)	2π
Mass axis	0.5c	Non-dimensional pitch damping derivative ($M_{\dot\theta}$)	-1.2
Mass per unit area	200 kg/m^2	Air density (ρ)	1.225 kg/m^3

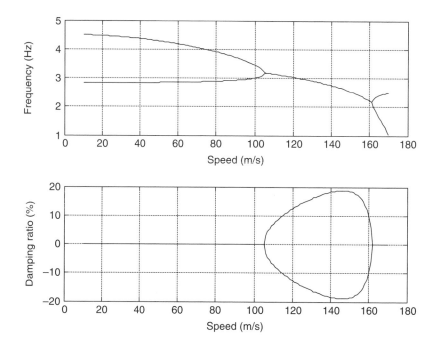

Figure 10.3 Frequency and damping trends for the baseline system with zero aerodynamic and structural damping ($x_f = 0.48c$, $x_m = 0.5c$).

10.5.1 Zero Aerodynamic Damping

If the aerodynamic and structural damping related terms in Equation (10.16) are ignored (i.e.
$\mathbf{B} = \mathbf{D} = \mathbf{0}$) then the Vg and $V\omega$ trends in Figure 10.3 for the baseline system show that as the air
speed increases, the two frequencies move closer to each other; however, the damping of both
modes remains at zero. Once the two frequencies become equal at around 105 m/s, the frequencies are said to 'coalesce'; one of the damping ratios becomes positive and the other negative.
Hence the system becomes unstable, which is the flutter condition. Beyond $V = 162$ m/s, the
frequency coalescence stops and both modes become undamped once again.

It is often stated that the frequencies of an aeroelastic system must coalesce for flutter
to occur, as in this case of zero aerodynamic damping; however, this is not true for general
aeroelastic systems, as will be seen in the later examples when aerodynamic damping is
included.

10.5.2   Aerodynamic Damping with Quasi-steady Aerodynamics

When the quasi-steady aerodynamic damping terms in Equation (10.12) are included in the computations, but the $M_{\dot{\theta}}$ term is still set to zero, then the frequency and damping behaviour become markedly different. The frequencies in Figure 10.4 start to converge gradually with an increase in air speed but do not coalesce. One of the damping ratios rises with increasing air speed; however, the other one is marginally stable until the flutter speed of 40 m/s.

Further investigation of the flutter speeds for this system with $M_{\dot{\theta}} = 0$ for different positions of the elastic axis is shown in Figure 10.5. Starting from a position on the quarter chord, the

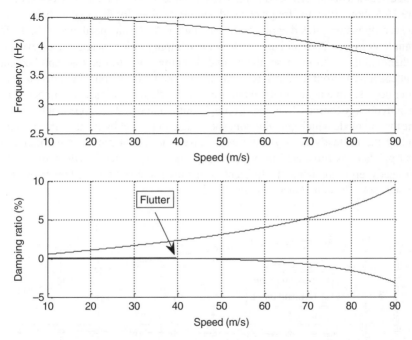

Figure 10.4   Frequency and damping trends for the baseline system with quasi-steady aerodynamic damping included $\left(M_{\dot{\theta}} = 0, x_{\mathrm{f}} = 0.48c, x_{\mathrm{m}} = 0.5c\right)$.

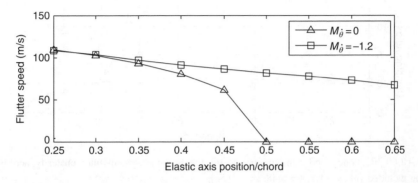

Figure 10.5   Flutter speed values for baseline binary system for different elastic axis positions.

flutter speed reduces significantly until it becomes zero with the elastic axis on the mid-chord, and then remains at zero as the elastic axis moves further aft. It will be shown in the next section that this is not the case if the unsteady aerodynamics terms are included.

10.5.3 Aerodynamic Damping with Unsteady Aerodynamics

The full aeroelastic equations defined in Equation (10.15) are now considered, again with the elastic axis at $x_f = 0.48c$ but now with the unsteady aerodynamic term $M_{\dot{\theta}}$ included (taken here as -1.2 as discussed in Chapter 9, equivalent to the value of -0.3 used in Hancock *et al.* (1985)).

Figure 10.6 shows how the frequency and damping values for this system behave when the unsteady aerodynamics pitch damping term is included. As the air speed increases, the frequencies begin to converge and, initially, both of the damping ratios increase. However, whereas one of them continues to increase, the second damping starts to decrease and becomes zero at the flutter speed of around 82 m/s; beyond this air speed this damping ratio becomes negative and flutter occurs. Note again, that the two frequencies do not coalesce, but rather move close enough in frequency for the two modes to couple unfavourably. The mode that becomes unstable is the torsion mode which contains some bending component by the time that flutter occurs.

Including the unsteady aerodynamic term makes the model more representative of what occurs in practice, as illustrated by Figure 10.5 where the flutter speeds with the inclusion of the unsteady terms are markedly different from those for the system without them. This issue is discussed more fully in Hancock *et al.* (1985).

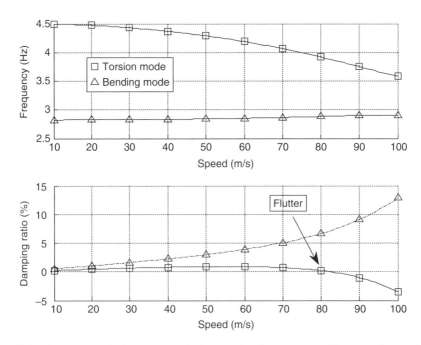

Figure 10.6 Frequency and damping trends for the baseline system with unsteady aerodynamic damping included $\left(M_{\dot{\theta}} = -1.2, x_f = 0.48c, x_m = 0.5c \right)$.

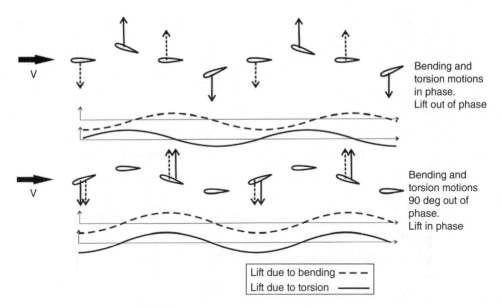

Figure 10.7   Lift due to bending and torsion components of the wing tip motion with different phasing between motions.

10.5.4   Illustration of Phasing for Flutter

A simplistic illustration as to how the phasing between the bending and torsion motions enables their corresponding lift components to work together to extract energy from the airflow at the flutter condition is shown in Figure 10.7. Imagine the aerofoil shape representing the tip section of the simple wing under consideration, undergoing bending and torsion motion; the quasi-steady aerodynamic components of the bending and torsion motions are considered. In the upper plot, the motions of the bending and torsion components are in phase with each other, and the maximum and minimum values of the resulting lift forces are 90° out-of-phase. However, in the lower plot the bending and torsion motions are 90° out-of-phase, and thus the maximum and minimum lift components are in-phase.

This characteristic behaviour is illustrated in Figure 10.8 where is can be seen that there is nearly a 90° phase difference between the two modes at the flutter speed. This example also illustrates that, in practice, the presence of unsteady aerodynamic terms means that the phasing of the critical motions at flutter is not exactly 90°.

10.5.5   Soft and Hard Flutter

If the critical damping ratio trend approaches the critical speed with a shallow gradient, this is known as a *soft flutter*. Figure 10.9a shows a case where the system parameters were modified such that the system becomes unstable around 120 m/s, but stable again at about 270 m/s. Of course, in practice it would not be possible to fly to this second stable region without flutter occurring first. Such a characteristic shape is sometimes referred to as a *hump mode*. The presence of structural damping may prevent such a flutter mechanism from occurring by

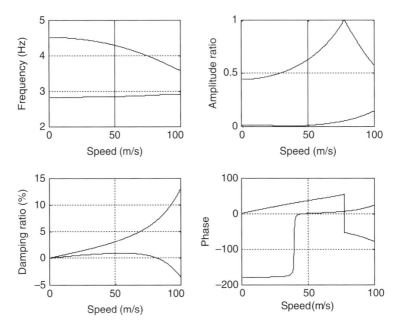

Figure 10.8 Frequency, damping ratio and mode shape amplitude and phase trends for baseline system.

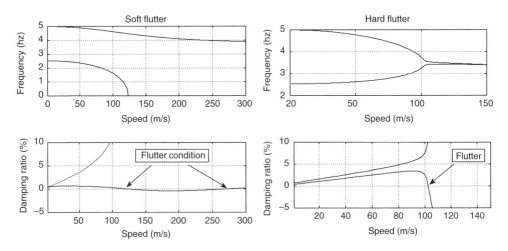

Figure 10.9 Illustration of (a) soft and (b) hard flutter.

moving the lowest point of the dip in the damping curve above the stability boundary (see next section).

However, if there is a very sudden drop in the damping values with increasing speed as the flutter condition is approached, as shown in Figure 10.9b, then this is known as a *hard flutter*. This latter case is of greatest concern during Flight Flutter Test (see Chapter 25) since a stable system may suddenly become unstable with a relatively small increase in air speed.

10.5.6 Inclusion of Structural Damping

The aeroelastic behaviour will be altered somewhat if structural damping is present. Consider the general aeroelastic model, but now with the addition of a proportional structural damping matrix, defined as a linear combination of the structural mass and stiffness matrices (see Chapter 2), namely

$$\mathbf{D} = \alpha \mathbf{A} + \beta \mathbf{E} \qquad (10.26)$$

In order to obtain values of the Rayleigh coefficients α, β for the system, a range of frequencies with bounds ω_a and ω_b must be chosen. It can be shown (NAFEMS, 1987) that in order to achieve damping ratios ζ_a, ζ_b at these frequencies, the Rayleigh coefficients must be defined as

$$\alpha = \frac{2\omega_a\omega_b(\zeta_b\omega_a - \zeta_a\omega_b)}{\omega_a^2 - \omega_b^2} \qquad \beta = \frac{2(\zeta_a\omega_a - \zeta_b\omega_b)}{\omega_a^2 - \omega_b^2} \qquad (10.27)$$

However, the damping ratios do not remain constant over the specified range of frequencies. If there are only two modes considered, then the values of ω_a and ω_b may be taken as being equal to the natural frequencies and the damping ratios for both modes have to be defined.

For the baseline system, the two natural frequencies were chosen in this way so as to yield damping ratios of 1% critical per mode. Figure 10.10 shows the trends of the frequency and damping ratios versus air speed for the baseline system considered in Figure 10.6 overlaid with the results obtained when damping ratios of 1% critical are present for each of the modes. Similar trends are obtained as before, with very little change to the frequency behaviour;

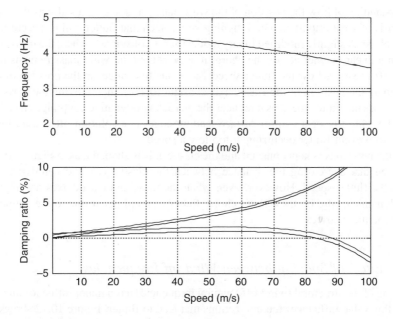

Figure 10.10 Effect of viscous structural damping (---) on frequency and damping trends.

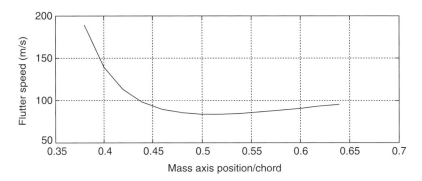

Figure 10.11 Effect of mass axis position on flutter speed of baseline model.

however, there is a change in the damping trends and the onset of flutter is delayed from 80 to 88 m/s. In general, the presence of structural damping is always beneficial in terms of flutter.

A more usual approach to modelling damping is not to use a Rayleigh model but rather to transform the equations into modal form and then to add a diagonal modal damping matrix where the damping terms are of the form $2\zeta_j\omega_j m_j, j = 1, 2, ..., N$. The damping ratios ζ_j may be defined initially based on experience and later using values from the Ground Vibration Test (GVT; see Chapter 25) once available. It is explained in Part III how the certification requirements allow some damping to be included, although it is often ignored in aeroelastic models since it can only be determined from measurement; aerodynamic damping tends to dominate anyhow and the flutter predictions without structural damping will be conservative.

10.5.7   Effect of Changes in Position of the Elastic and Mass Axes

An understanding of how the position of the wing elastic and mass axes affects the aeroelastic behaviour is of great importance in designing wings such that flutter will not occur inside the flight envelope. A slightly modified binary model is now considered, where the mass is allowed to vary in a linear manner across the chord; however, the total wing mass remains the same.

Figure 10.11 shows how the flutter speed varies due to changes in the position of the mass axis. Keeping the position of the elastic axis constant ($0.48c$), then moving the mass axis forward increases the flutter speed as it reduces the inertial/aerodynamic coupling. The addition of mass at the wing tip, or a control surface leading edge, is a solution often used by aircraft designers to prevent flutter occurring at too low a speed.

When the mass axis is kept constant and the elastic axis is changed, as in Figure 10.5, then it can be seen that decreasing the distance between the elastic axis and aerodynamic centre increases the flutter speed. However, even when the mass and elastic axes are aligned with the aerodynamic centre on the quarter chord, flutter can still occur due to the coupling from the aerodynamic matrices.

10.5.8   Effect of Spacing between Wind-off Frequencies

The spacing of the structural (wind-off) natural frequencies has a major influence on the flutter speed as this value influences the interactions that lead to flutter. Figure 10.12 shows the aeroelastic behaviour of the baseline system compared to that when the two structural frequencies

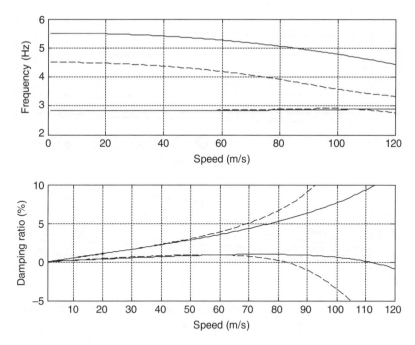

Figure 10.12   Frequency and damping ratio trends for systems with different wind-off frequencies.

are further apart (wind-off torsion mode increased to around 5.5 Hz); in this latter case the onset of flutter is delayed. In general, the closer the wind-off frequencies become, the more rapidly the critical interaction between the modes occurs and the flutter speed reduces. Designers often try to increase the frequency gap between modes (by changing the mass distribution or increasing stiffness) to increase the flutter speed; however, for a complete aircraft, care must be taken to ensure that in solving one problem a different critical flutter mechanism is not created.

10.6   Aeroelastic Behaviour of a Multiple Mode System

If the aerodynamic and structural matrices are known, then the process for plotting the $V\omega$ and Vg plots for a full-scale aircraft model is exactly the same as that described above, making use of the eigensolution approach in Equation (10.21). Figure 10.13 shows a typical set of the first 10 symmetric modes for an aircraft model with unsteady aerodynamics evaluated at a single reduced frequency. The behaviour is much more complicated than for the binary systems considered previously; however, it can be seen that the flutter speed occurs at around 190 m/s, is due to the interaction of the first and third modes, and is a moderately hard flutter. A further multiple mode example is shown in Chapter 19.

10.7   Flutter Speed Prediction for Binary Systems

Instead of plotting the damping trends for different speeds and determining the flutter speed by eye, or by trial and error from the air speed at which zero damping ratio occurs, it is more accurate to calculate the flutter condition directly. This process is straightforward when

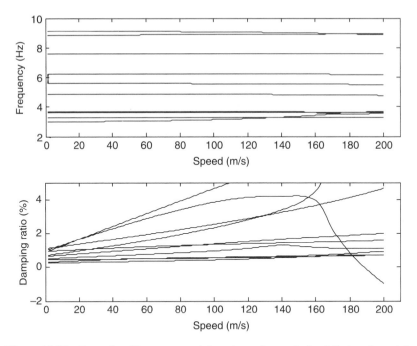

Figure 10.13   Example of frequency and damping ratio trends for full aircraft model.

frequency-independent aerodynamics is considered, but not for frequency-dependent aerodynamics (see later). The approach shown here is based upon the Routh–Hurwitz method introduced in Chapter 6.

Consider a binary aeroelastic system with frequency-independent aerodynamics, whose
equation of motion is

$$\begin{bmatrix} a_{11} & a_{12} \\ a_{21} & a_{22} \end{bmatrix} \begin{Bmatrix} \ddot{q}_1 \\ \ddot{q}_2 \end{Bmatrix} + V \begin{bmatrix} b_{11} & b_{12} \\ b_{21} & b_{22} \end{bmatrix} \begin{Bmatrix} \dot{q}_1 \\ \dot{q}_2 \end{Bmatrix} + \left(V^2 \begin{bmatrix} c_{11} & c_{12} \\ c_{21} & c_{22} \end{bmatrix} + \begin{bmatrix} e_{11} & 0 \\ 0 & e_{22} \end{bmatrix} \right) \begin{Bmatrix} q_1 \\ q_2 \end{Bmatrix} = \begin{Bmatrix} 0 \\ 0 \end{Bmatrix}$$

$$(10.28)$$

Assume a solution of the form

$$\begin{Bmatrix} q_1 \\ q_2 \end{Bmatrix} = \begin{Bmatrix} q_1 \\ q_2 \end{Bmatrix}_0 e^{\lambda t}$$

and also make the substitutions

$$e_{11} = xV^2 \qquad e_{22} = \mu e_{11} = \mu x V^2 \qquad\qquad (10.29)$$

where μ is the ratio between the two spring stiffness values and x is an unknown that has to be
found. The non-trivial solution of the equations is defined by

$$\begin{vmatrix} a_{11}\lambda^2 + b_{11}V\lambda + (c_{11}+x)V^2 & a_{12}\lambda^2 + b_{12}V\lambda + c_{12}V^2 \\ a_{21}\lambda^2 + b_{21}V\lambda + c_{21}V^2 & a_{22}\lambda^2 + b_{22}V\lambda + (c_{22}+\mu x)V^2 \end{vmatrix} = 0 \qquad (10.30)$$

and solving the determinant yields the quartic equation

$$b_4\lambda^4 + b_3\lambda^3 + b_2\lambda^2 + b_1\lambda + b_0 = 0 \tag{10.31}$$

where $b_0, \ldots, b_4$ are functions of the parameters in Equation (10.30). The roots of the equation are in two complex conjugate pairs, namely

$$\lambda_{1,2} = -\zeta_1\omega_1 \pm i\omega_1\sqrt{1-\zeta_1^2} \quad \lambda_{3,4} = -\zeta_2\omega_2 \pm i\omega_2\sqrt{1-\zeta_2^2} \tag{10.32}$$

and at the flutter speed, since one of the damping ratios becomes zero, one of the root pairs becomes

$$\lambda = \pm i\omega \tag{10.33}$$

Substituting both solutions from Equation (10.33) into the quartic Equation (10.31) gives

$$b_4\omega^4 - ib_3\omega^3 - b_2\omega^2 + ib_1\omega + b_0 = 0 \quad \text{and} \quad b_4\omega^4 + ib_3\omega^3 - b_2\omega^2 - ib_1\omega + b_0 = 0 \tag{10.34}$$

Now, adding and subtracting Equation (10.34) gives

$$b_4\omega^4 - b_2\omega^2 + b_0 = 0 \quad \text{and} \quad ib_3\omega^3 - ib_1\omega = 0 \tag{10.35}$$

Hence the frequency at the flutter condition is given by the second expression, such that

$$\omega = \sqrt{\frac{b_1}{b_3}} \tag{10.36}$$

Equation (10.36) can be substituted into the first part of Equation (10.35) to give the expression

$$b_4b_1^2 - b_1b_2b_3 + b_0b_3^2 = 0 \tag{10.37}$$

and this equation allows the flutter speed to be obtained. The same result can be found by considering the Routh–Hurwitz stability criteria (see Chapter 6) for the quartic Equation (10.31), leading to the condition for stability

$$b_1b_2b_3 - b_4b_1^2 - b_0b_3^2 > 0 \tag{10.38}$$

Knowing the matrix terms in Equation (10.28) it is possible to determine directly the critical flutter speeds and frequencies of a binary aeroelastic system using the following procedure. On a historical note, this was the procedure (Bairstow and Fage, 1916) used to investigate the flutter incident of the Handley Page O-400 bomber (see the book front cover of the first edition), the first documented flutter analysis.

Expanding the determinant in Equation (10.30) leads to a fourth-order characteristic polynomial equation. Substituting of these equations into Equation (10.37) for the critical condition, and eliminating a factor of V^6, gives a quadratic equation in terms of the unknown variable x. By substituting the two roots of this equation into Equation (10.29), the two critical flutter speeds

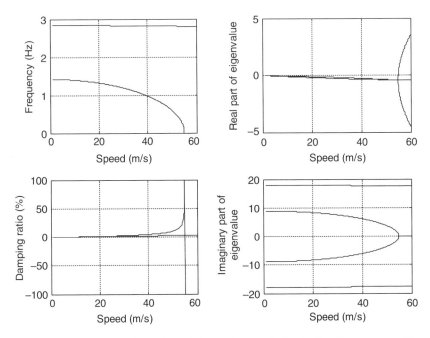

Figure 10.14 Frequency, damping ratio and eigenvalue trends for the baseline system with reduced torsional stiffness.

between which the system is unstable may be obtained. Obviously the lowest speed is the one that is of interest, since any aircraft will probably have been destroyed long before the second critical condition has been reached. The corresponding flutter frequencies are then found using Equation (10.36).

10.8 Divergence of Dynamic Aeroelastic Systems

So far, this chapter has only been concerned with determining the flutter speed but although flutter normally occurs before divergence, the latter condition does have to be checked. Now consider again the baseline aeroelastic system of Section 10.5, but this time with the torsional stiffness reduced to 10% of its earlier value. Figure 10.14 shows how the frequency, damping ratio and real and imaginary parts of the eigenvalues change with air speed. The torsion mode reduces in frequency and splits into two non-oscillatory solutions at 54.8 m/s; this is where the imaginary part of the eigenvalue becomes zero and the corresponding frequency on the $V\omega$ plot becomes zero. With a further (slight) increase in air speed, the system undergoes divergence at 54.9 m/s when the real part of one of the non-oscillatory eigenvalues becomes positive. The solution for the divergence speed is considered further in Chapter 22.

It is instructive to examine the eigenvalue behaviour on a root locus plot (see Chapter 6), shown in Figure 10.15. The complex conjugate pair of eigenvalues corresponding to the mode that remains oscillatory moves from A to B throughout the speed range. The second mode starts off as a complex conjugate pair at point C; however, at point D the imaginary part becomes zero and the oscillatory motion ceases. The eigenvalues then split and move along the real axis in both directions towards points E and G. Once the eigenvalue heading for point G crosses the imaginary axis at point F the system becomes statically unstable and divergence occurs.

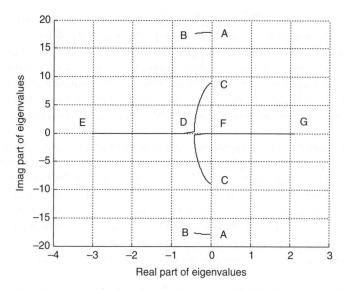

Figure 10.15 Root locus of eigenvalues for the baseline system with reduced torsional stiffness.

An exact calculation of the divergence speed for the general aeroelastic system in Equation (10.16) can be found by considering only the displacement related terms, leading to

$$\left(\rho V^2 \mathbf{C} + \mathbf{E}\right)\mathbf{q} = \mathbf{0} \tag{10.39}$$

Divergence then occurs at the non-trivial solution defined by

$$\left|\rho V^2 \mathbf{C} + \mathbf{E}\right| = 0 \tag{10.40}$$

or this can also be expressed as an eigenvalue problem. When applied to the system considered in this section, a divergence speed of 54.9 m/s is found; this is exactly the same as that obtained from the $V\omega$ plots above.

In order to determine the static aeroelastic deflections, the initial deformed shape (e.g. wing root incidence) must be included as a right-hand side term (see Chapters 7, 12 and 23). Equation (10.39) then becomes

$$\left(\rho V^2 \mathbf{C} + \mathbf{E}\right)\mathbf{q} = -\rho V^2 \mathbf{C} \mathbf{q}_0 \tag{10.41}$$

where $\mathbf{q}_0$ defines the initial deflection at zero air speed. Note that the airframe starts off with a jig shape which is stress free (see Chapter 23), deforms statically under 1 g inertia loading (i.e. gravity) and then experiences a static aeroelastic deformation in flight as described here.

10.9   Inclusion of Unsteady Reduced Frequency Effects

It has been seen that in order to model aeroelastic systems correctly, the unsteady aerodynamics described in Chapter 9 need to be accounted for. However, so far only the very simplistic approach of including a constant unsteady $M_{\dot{\theta}}$ term has been used. In practice, as shown in

Chapter 9, the aerodynamic stiffness and damping matrices are reduced frequency-dependent, which leads to the so-called 'frequency matching' problem. If the **B** and **C** matrices in the aeroelastic Equation (10.16) are known, then the eigenproblem posed in Equation (10.20) can be solved. However, the **B** and **C** matrices cannot be formed unless the reduced frequency of interest is known, and this cannot be determined until the eigensolution of the system matrix, **Q**, involving both the **B** and **C** matrices, has been solved. There is no direct way of solving this 'chicken and egg' problem and some form of iterative approach, known as 'frequency matching', must be used.

There are a number of ad hoc approaches that have been developed to solve the frequency matching problem. Here, simplified versions of two commonly used approaches, the so-called 'k' and 'p–k' methods (Hassig, 1971), will be illustrated on the binary system considered earlier. Both methods are based upon the assumption that the aerodynamics behaviour is dependent upon a harmonic response. This is fine at the flutter condition, but is not true below (and above) this speed; consequently the methods give the same critical flutter speed and frequency but predict different subcritical behaviour. The methods tend to be fairly robust in their use, although there are concerns about the damping ratio trend predictions for the 'k' method.

Here, in order to illustrate the two methods, the baseline binary aeroelastic system will be used with the frequency dependent expression for $M_{\dot\theta}$ given in Equation (9.14).

10.9.1 Frequency Matching: 'k' Method

Consider the classical form of the aeroelastic Equation (10.16), with **B** and **C** now being functions of reduced frequency $k = \omega b/V$, and also include a structural (or hysteretic) damping term in the form $\mathbf{D} = ig\mathbf{E}$ (see Chapter 1), where g is the symbol commonly used for the structural damping coefficient in flutter calculations (not to be confused with the symbol for acceleration due to gravity). Assuming a harmonic solution in the form $q = q_0 e^{i\omega t}$ and dividing throughout by $-\omega^2$, then Equation (10.16) becomes

$$\left[\mathbf{A} - ip\left(\frac{b}{k}\right)\mathbf{B} - p\left(\frac{b}{k}\right)^2\mathbf{C} - \frac{1+ig}{\omega^2}\mathbf{E}\right]q_0 = 0 \tag{10.42}$$

This equation has been written solely in terms of the reduced frequency k, and is a generalized eigenvalue problem such that

$$(\mathbf{F} - \lambda\mathbf{E})q_0 = 0 \quad \text{where} \quad \mathbf{F} = \left[\mathbf{A} - ip\left(\frac{b}{k}\right)\mathbf{B} - p\left(\frac{b}{k}\right)^2\mathbf{C}\right] \quad \text{and} \quad \lambda = \frac{1+ig}{\omega^2} \tag{10.43}$$

Thus it may be seen that

$$\omega = \frac{1}{\sqrt{\text{Re}(\lambda)}} \qquad g = 2\zeta = \frac{\text{Im}(\lambda)}{\text{Re}(\lambda)} \quad \text{and} \quad V = \frac{\omega c}{2k} \tag{10.44}$$

where ζ is the equivalent viscous damping ratio for motion at the natural frequency (see Chapter 1).

The 'k' method is applied in the following manner. For each reduced frequency of interest:

a. Calculate the corresponding **B** and **C** matrices.
b. Solve the complex eigenproblem in Equation (10.43) to yield complex eigenvalues λ.
c. Determine the frequencies and damping coefficients (or ratios) from the eigenvalues using Equation (10.44).
d. Relate these values to the air speed via the definition of the reduced frequency k in Equation (10.44).

Consider the next reduced frequency and keep repeating the process until all k values have been investigated.

Then join up the frequencies and corresponding damping coefficients (or ratios) to form $V\omega$ and Vg plots. Care must be taken in interpreting the results since it is possible for the damping values to 'fold back', i.e. to have more than one solution for a particular mode at some flight condition.

A typical result is shown in Figure 10.16 where three different reduced frequencies have been considered for a 3DOF aeroelastic system. At each reduced frequency k, there are three eigenvalues λ that correspond to natural frequencies ω and speeds V determined from Equation (10.44). The $V\omega$ and corresponding Vg trends (the latter not shown here) can then be formed, giving an estimate of the aeroelastic behaviour of the system.

The addition of the structural damping terms in this solution is somewhat artificial; the eigenvalues that are being found actually allow determination of the structural damping required to give zero *overall* damping for that mode at that flight condition (Garrick and Reid, 1981). Consequently, for a stable condition, the damping values that are determined are negative, and vice versa for an unstable fluttering system. It is often the convention that flutter plots use these damping coefficients g plotted against air speed and so flutter occurs for positive 'required' damping; this is why the classical flutter plot is known as a Vg plot. Care must be taken to confirm what is meant by damping in the interpretation of such plots. In this book, true damping ratio values are presented and damping ratio results are a factor of -2 different from the approach involving the required structural damping.

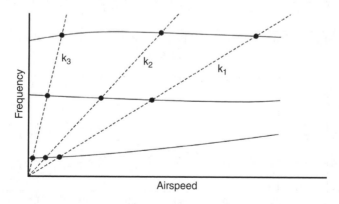

Figure 10.16   The 'k' method – determination of frequency points for lines of constant k.

10.9.2 Frequency Matching: 'p–k' Method

The 'p–k' method is a widely used frequency matching method and is applied in the following manner. For each air speed of interest in the flight envelope and for each mode of interest:

a. Make an initial guess of the frequency for the mode (often the previous air speed or wind-off results are used) and calculate the corresponding reduced frequency for the air speed/frequency combination.
b. Determine the aerodynamic stiffness and damping matrices **B**, **C** using this reduced frequency.
c. Determine the frequencies for the system at this flight condition using the eigenvalue solution of the matrix shown in Equation (10.20) for the first-order form.
d. Take the frequency solution closest to the initial guess and repeat the process.
e. Continue until the frequency converges (usually after four or five iterations) and note the corresponding damping ratio.

Consider the next mode of interest and repeat until all modes of interest have been investigated. Then consider the next flight speed and repeat until all air speeds of interest have been explored.

A set of frequency, damping ratio and air speed values will then be assembled and plotted, each corresponding to the correct reduced frequency, and the flutter speed may be found where the damping is zero.

10.9.3 Comparison of Results for 'k' and 'p–k' Methods

Using the baseline binary aeroelastic system, but with the frequency-dependent $M_{\dot{\theta}}$ term described in Equation (9.14), Figure 10.17 shows the frequency and true damping ratio (not

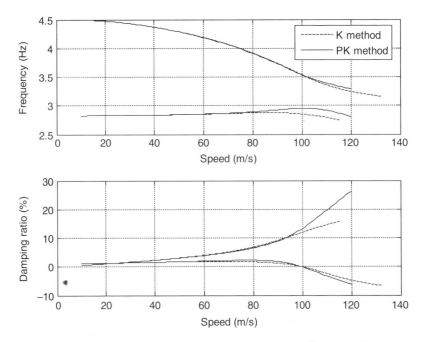

Figure 10.17 Frequency and damping ratio trends from the 'k' and 'p–k' methods.

the 'required' structural damping coefficient) trends for the 'k' and 'p–k' methods, where it can be seen that there are some differences in the subcritical behaviour but that, as expected, the same solution is found at the flutter speed.

One major difficulty with these approaches is that, whereas at the flutter condition the critical mode dominates the motion, at subcritical speeds the motion is made up of a number of different modes. The frequency matching process must be undertaken for each mode at a particular flight condition. However, since the aerodynamic forces are assumed to occur at a single frequency, there will be some error in the estimates. Consequently, the various frequency matching methods give different frequency and damping values at subcritical speeds (or beyond flutter), but give the same estimate at the flutter condition.

More sophisticated frequency matching methods exist (Chen, 2000; Edwards and Weiseman, 2003) but are beyond the scope of this book.

10.10 Control Surface Flutter

Historically, flutter involving the control surfaces has occurred more frequently than classical wing bending/torsion flutter. This has often resulted in the loss of control surfaces and/or part of the wing/tail structure but in many cases the aircraft has survived. Usually the flutter mechanism still occurs due to the interaction of two modes.

To illustrate some of the characteristics of control surface flutter, consider the 3DoF aeroelastic system shown in Figure 10.18. The binary bending/torsion aeroelastic model has now been altered so that a full span control surface is attached to the wing by a distributed torsional spring of stiffness k_β per unit span. As before, the non-dimensional pitch damping derivative $M_{\dot{\theta}}$ is included in order to approximate the unsteady aerodynamic behaviour, but now a control aerodynamic damping derivative $M_{\dot{\beta}}$ is also included (Wright et $al.$, 2003).

The vertical deflection (positive downwards) is expressed as

$$z = \left(\frac{y}{s}\right)^2 q_b + \left(\frac{y}{s}\right)(x - x_f)q_t + [x - x_h]\beta \qquad (10.45)$$

where β is the control rotation (trailing edge down and the same at every spanwise position) and $[X]$ is the Heaviside function defined by $[X] = 0$ if $X < 0$ and $[X] = X$ if $X \geq 0$. The downwards displacement at the elastic axis and the nose up twist are given by

$$z_f = \left(\frac{y}{s}\right)^2 q_b \quad \text{and} \quad \theta = \left(\frac{y}{s}\right)q_t \qquad (10.46)$$

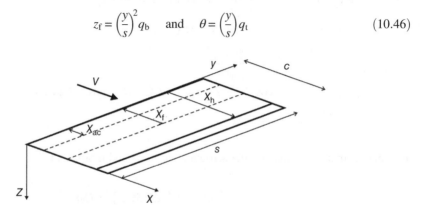

Figure 10.18 Aeroelastic model including a full span control surface.

The kinetic energy terms for the wing (w) and control (c) are given by

$$T_w = \frac{m_w}{2} \int_0^s \int_0^{x_h} \left\{ \left(\frac{y}{s}\right)^2 \dot{q}_b + (x - x_f)\left(\frac{y}{s}\right)\dot{q}_t \right\}^2 dxdy \tag{10.47}$$

and

$$T_c = \frac{m_c}{2} \int_0^s \int_{x_h}^c \left\{ \left(\frac{y}{s}\right)^2 \dot{q}_b + (x - x_f)\left(\frac{y}{s}\right)\dot{q}_t + (x - x_h)\dot{\beta} \right\}^2 dxdy \tag{10.48}$$

where m_w, m_c are the mass per unit area for the wing and control surface. The elastic potential (or strain) energy terms are the same for bending and torsion as earlier but with an additional term due to the distributed control spring, so

$$U = \frac{1}{2} \int_0^s EI \left(\frac{2q_b}{s^2}\right)^2 dy + \frac{1}{2} \int_0^s GJ \left(\frac{q_t}{s}\right)^2 dy + \frac{1}{2} k_\beta s \beta^2 \tag{10.49}$$

The approach used earlier to obtaining the aerodynamic matrices is adopted here. The lift, pitching moment about the elastic axis and the hinge moment about the hinge line are written for a strip of width dyas follows

$$dL = \frac{1}{2}\rho V^2 cdy \left[a_w \left(\frac{y^2 \dot{q}_b}{s^2 V} + \frac{y}{s} q_t\right) + a_c \beta \right]$$

$$dM = \frac{1}{2}\rho V^2 c^2 dy \left[b_w \left(\frac{y^2 \dot{q}_b}{s^2 V} + \frac{y}{s} q_t\right) + M_{\dot{\theta}} c \frac{(y\dot{q}_t)}{4sV} + b_c \beta \right] \tag{10.50}$$

$$dH = \frac{1}{2}\rho V^2 c^2 dy \left[c_w \left(\frac{y^2 \dot{q}_b}{s^2 V} + \frac{y}{s} q_t\right) + M_{\dot{\beta}} \frac{c\dot{\beta}}{4V} + c_c \beta \right]$$

Here the lift, pitching moment and hinge moment coefficients may be estimated as in Chapter 4 (Scanlan *et al.*, 1960), namely

$$a_w = 2\pi \quad a_c = \frac{a_w}{\pi} \left[\cos^{-1}(1 - 2E) + 2\sqrt{E(1 - E)} \right]$$

$$b_w = ea_w \quad b_c = -\frac{a_w}{\pi}(1 - E)\sqrt{E(1 - E)}$$

$$c_w = -\frac{T_{12}}{2} \quad c_c = -\frac{T_{12}T_{10}}{2\pi} \tag{10.51}$$

$$T_{10} = \sqrt{1 - d^2} + \cos^{-1} d \quad T_{12} = \sqrt{1 - d^2}(2 + d) + \cos^{-1} d(2d + 1)$$

$$d = \frac{2x_h}{c} - 1 \quad Ec = c - x_h$$

The incremental work done by the aerodynamic force and moments is

$$\delta W = - \int_0^s dL\delta z_f + \int_0^s dM\delta\theta + \int_0^s dH\delta\beta \tag{10.52}$$

Using Lagrange's energy equations then yields the 3DoF aeroelastic equations that involve interaction of the wing bending, torsion and control rotation, namely

$$
\begin{bmatrix} A_{bb} & A_{bt} & A_{b\beta} \\ A_{tb} & A_{tt} & A_{t\beta} \\ A_{\beta b} & A_{\beta t} & A_{\beta\beta} \end{bmatrix} \begin{Bmatrix} \ddot{q}_b \\ \ddot{q}_t \\ \ddot{\beta} \end{Bmatrix} + \rho V \begin{bmatrix} \dfrac{cs}{10} a_{\mathrm{W}} & 0 & 0 \\ -\dfrac{c^2 s}{8} b_{\mathrm{W}} & -\dfrac{c^3 s}{24} M_{\dot{\theta}} & 0 \\ -\dfrac{c^2 s}{6} c_{\mathrm{W}} & 0 & -\dfrac{c^3 s}{8} M_{\dot{\beta}} \end{bmatrix} \begin{Bmatrix} \dot{q}_b \\ \dot{q}_t \\ \dot{\beta} \end{Bmatrix}
$$

$$
+ \left(\rho V^2 \begin{bmatrix} 0 & \dfrac{cs}{8} a_{\mathrm{W}} & \dfrac{cs}{6} a_{C} \\ 0 & -\dfrac{c^2 s}{6} b_{\mathrm{W}} & -\dfrac{c^2 s}{4} b_{C} \\ 0 & -\dfrac{c^2 s}{4} c_{\mathrm{W}} & -\dfrac{c^2 s}{2} c_{C} \end{bmatrix} + \begin{bmatrix} \dfrac{4EI}{s^3} & 0 & 0 \\ 0 & \dfrac{GJ}{s} & 0 \\ 0 & 0 & k_{\beta}s \end{bmatrix} \right) \begin{Bmatrix} q_b \\ q_t \\ \beta \end{Bmatrix} = \begin{Bmatrix} 0 \\ 0 \\ 0 \end{Bmatrix}
$$

$$(10.53)$$

where **A** is a symmetric matrix with the elements (for $m_{\mathrm{W}} = m_{C} = m$)

$$
A_{bb} = m\frac{sc}{5} \quad A_{tt} = m\frac{s}{3}\left(\frac{c^3}{3} - x_f c^2 + x_f^2 c\right) \quad A_{bt} = m\frac{s}{4}\left(\frac{c^2}{2} - x_f c\right) \quad A_{b\beta} = m\frac{s}{3}\left(\frac{c^2 - x_h^2}{2} - x_h(c - x_h)\right)
$$

$$
A_{t\beta} = m\frac{s}{2}\left(\frac{c^3 - x_h^3}{3} - (x_f + x_h)\frac{c^2 - x_h^2}{2} + x_f x_h(c - x_h)\right) \quad \text{and} \quad A_{\beta\beta} = ms\left(\frac{c^3 - x_h^3}{3} + x_h^2 c - x_h c^2\right)
$$

The flutter characteristics of this system can be obtained using the same methodology as earlier for the bending/torsion flutter. In order to assess the effect of the control stiffness, a nominal control frequency is calculated based on the control rotating with respect to a fixed wing; however, the wind-off normal modes will actually be a mixture of bending, torsion and control rotation. In practice, the control stiffness will be low if a mechanical linkage is employed and high if a hydraulic power control unit is used. The variation of the flutter speed with nominal control frequency is shown in Figure 10.19; the values used are hinge line at 80%, elastic axis

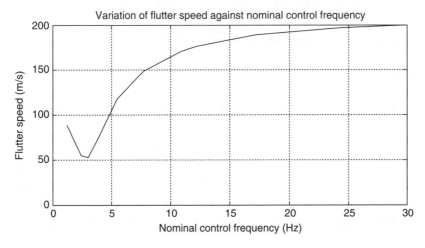

Figure 10.19   Variation of flutter speed with nominal control frequency.

at 40%, mass axis at 50%, $m_W = m_C = 400$ kg/m^2, $EI = 4 \times 10^7$, $GJ = 8 \times 10^6$ N m^2, $M_{\dot\theta} = -1.2$, $M_{\dot\beta} = -0.1$, $s = 7.5$ m and $c = 2$ m. For this example and data set, the flutter speed reduces significantly as the control stiffness decreases, except at very low frequency where it starts to rise again.

In order to see how the flutter mechanism changes with nominal control frequency, then three values of control stiffness are examined, namely low, medium and high ($k_\beta = 1 \times 10^3$, 1×10^4 and 1×10^5 (Nm/m)/rad, leading to nominal control frequencies of 1.72, 5.45 and 17.2 Hz. In each case the wind-off mode shapes were calculated and the results shown in Figures 10.20a to 10.22a; the mode shapes are presented in terms of the motion of the main wing and the control at the wing tip chord. The corresponding $V\omega$ and Vg plots are shown in Figures 10.20b to 10.22b. It may be seen that as the control frequency increases, the flutter mechanism changes from (bending + control) to (bending + torsion + control) and then finally to (bending + torsion, rather as for the wing alone); the lowest two control frequencies demonstrate quite a 'hard' flutter. Clearly, it may be seen that one way to delay the onset of control surface flutter is to increase the control stiffness. The above results also highlight the risks of control surface failure.

The key approach used classically to eliminate control surface flutter is to add extra mass to the control surface in order to change the inertia characteristics (so-called 'mass balancing'). The use of a 'horn balance' to move the control surface centre of mass on to the hinge line by adding mass ahead of the hinge line is sometimes seen. However, whereas adding mass can improve the critical speed of one flutter mechanism, if too great a mass is added it is possible for a different flutter mechanism to occur instead.

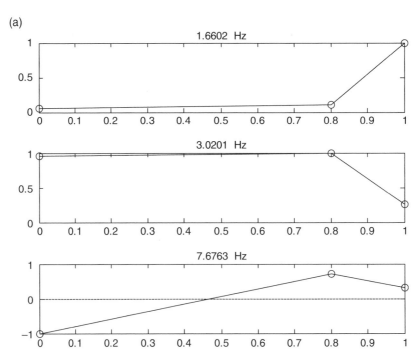

Figure 10.20a Mode shapes for low control stiffness.

(b)

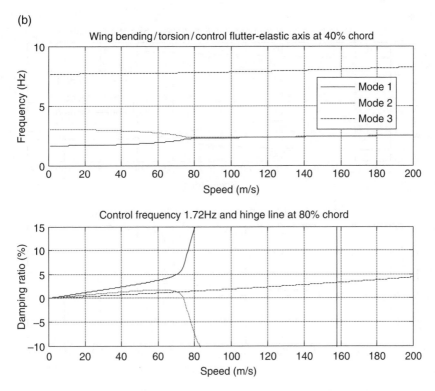

Figure 10.20b   *Vω* and *Vg* plots for low control stiffness.

(a)

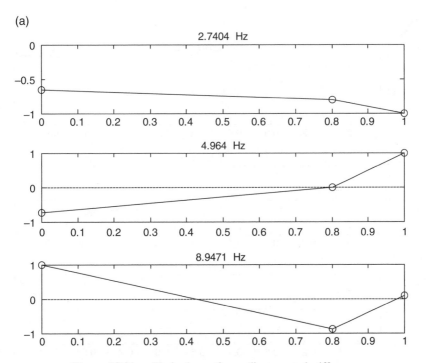

Figure 10.21a   Mode shapes for medium control stiffness.

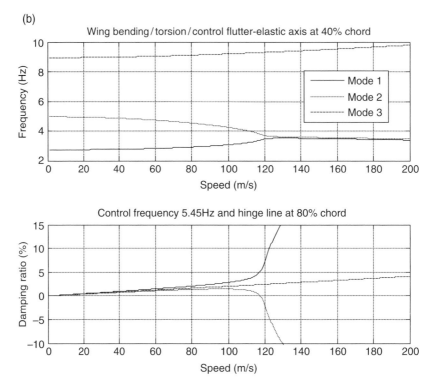

Figure 10.21b $V\omega$ and Vg plots for medium control stiffness.

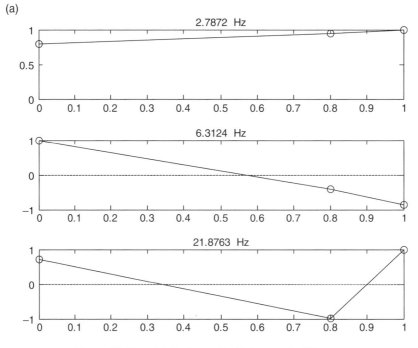

Figure 10.22a   Mode shapes for high control stiffness.

(b)

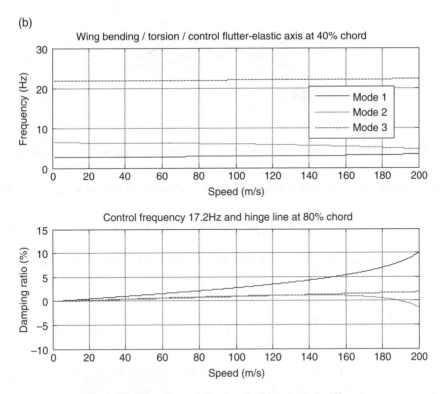

Figure 10.22b $V\omega$ and Vg plots for high control stiffness.

10.11   Whole Aircraft Model – Inclusion of Rigid Body Modes

So far in this chapter, the behaviour of a simple 'wing-alone' with a built-in root has been considered; the equations were derived in terms of displacements and/or rotations relative to inertial axes. Historically, flutter calculations were often performed for individual lifting surfaces built in at their root because the natural frequencies were significantly higher than the rigid body frequencies and the aim was for the calculation to be kept at a low model order. Also, it was assumed that there was no aeroelastic coupling between different components of the aircraft and this allowed smaller mathematical models to be used.

However, in recent years it has been more normal practice to carry out whole aircraft flutter calculations including both flexible and rigid body modes since the available computational power is now much greater. Also, rigid body and flexible mode frequencies are often sufficiently close for coupling terms to be relevant and for the flutter behaviour to be influenced somewhat by the rigid body modes.

10.11.1   Binary Aeroelastic Model with Free–Free Heave Motion

The same simple unswept/untapered (i.e. rectangular) cantilevered wing model used previously is now defined with a rigid lumped mass of M positioned at the elastic axis position at the root of the wing, as shown in Figure 10.23; also, there is now a rigid body heave DOF q_h (positive downwards). The combined system only has one half wing and is not allowed to roll.

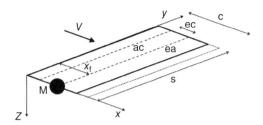

Figure 10.23 Combination of binary aeroelastic model with rigid body heave motion.

The displacement z (downwards positive) of a general point (x,y) on the wing now becomes

$$z = \left(\frac{y}{s}\right)^2 q_b + \left(\frac{y}{s}\right)(x-x_f)q_t + q_h \;\; \text{so} \;\; \theta = \left(\frac{y}{s}\right)q_t \qquad (10.54)$$

where q_b, q_t and q_h are the generalized coordinates. The kinetic energy due to the dynamic motion is

$$T = \int_{\text{Wing}} \frac{1}{2} dm \dot{z}^2 = \frac{1}{2} M \dot{q}_h{}^2 + \frac{1}{2} \int_0^s \int_0^c m \left(\left(\frac{y}{s}\right)^2 \dot{q}_b + \left(\frac{y}{s}\right)(x-x_f)\dot{q}_t + \dot{q}_h \right)^2 dxdy \qquad (10.55)$$

where the total mass of the wing is $m_{\text{wing}} = mcs$. The elastic potential energy due to bending and torsion remains the same as earlier since there is no constraint in the rigid body motion.

Applying Lagrange's equations for all three generalized coordinates, and considering the aerodynamic forces and moments acting on each streamwise strip leads to the aeroelastic equations of motion

$$\begin{bmatrix} m\dfrac{sc}{5} & m\dfrac{s}{4}\left(\dfrac{c^2}{2}-cx_f\right) & m\dfrac{sc}{3} \\[2ex] m\dfrac{s}{4}\left(\dfrac{c^2}{2}-cx_f\right) & m\dfrac{s}{3}\left(\dfrac{c^3}{3}-c^2x_f+cx_f^2\right) & m\dfrac{s}{2}\left(\dfrac{c^2}{2}-cx_f\right) \\[2ex] m\dfrac{sc}{3} & m\dfrac{s}{2}\left(\dfrac{c^2}{2}-cx_f\right) & (msc+M) \end{bmatrix} \begin{Bmatrix} \ddot{q}_b \\ \ddot{q}_t \\ \ddot{q}_h \end{Bmatrix}$$

$$+ \rho V \begin{bmatrix} \dfrac{cs}{10}a_w & 0 & \dfrac{cs}{6}a_w \\[2ex] -\dfrac{c^2s}{8}b_w & -\dfrac{c^3s}{24}M_{\dot{\theta}} & -\dfrac{c^2s}{4}b_w \\[2ex] \dfrac{cs}{6}a_w & 0 & \dfrac{cs}{2}a_w \end{bmatrix} \begin{Bmatrix} \dot{q}_b \\ \dot{q}_t \\ \dot{q}_h \end{Bmatrix} + \begin{bmatrix} \dfrac{4EI}{s^3} & \rho V^2 \dfrac{cs}{8}a_w & 0 \\[2ex] 0 & \left(\dfrac{GJ}{s}-\rho V^2\dfrac{c^2s}{6}b_w\right) & 0 \\[2ex] 0 & \rho V^2\dfrac{cs}{4}a_w & 0 \end{bmatrix} \begin{Bmatrix} q_b \\ q_t \\ q_h \end{Bmatrix} = 0$$

$$(10.56)$$

Note the presence of inertia and aerodynamic cross-couplings between the rigid body and flexible motions. Now consider the aeroelastic system with the same baseline parameters

Table 10.2   Flutter characteristics of a binary wing with a rigid body heave mode.

Fuselage mass	Flutter speed (m/s)	Flutter frequency (Hz)	Normalized magnitude of DoF at flutter		
			q_b	q_t	q_h
m_{wing}	57	4.21	0.70	1	0.20
$10\, m_{wing}$	75	3.98	0.60	1	0.08
$100\, m_{wing}$	80	3.92	0.45	1	0.01

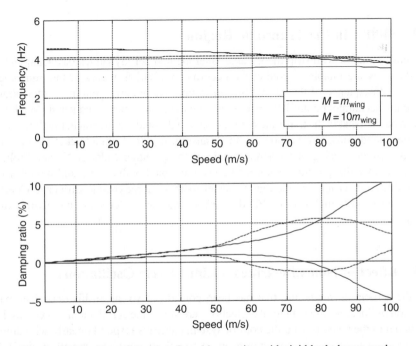

Figure 10.24 $V\omega$ and Vg plots for a binary wing with rigid body heave mode.

for the wing, but now using three different values for the fuselage mass M (expressed in terms of the wing mass). The resulting flutter behaviour is shown in Table 10.2 and Figure 10.24. For an extremely large fuselage mass ($M = 100m_{wing}$), the flutter plots are exactly the same as for the fixed root case except that there is now a zero frequency rigid body mode at all speeds; the flutter speed is 80 m/s.

When the fuselage mass reduces, but is still much larger than that of the wing ($M = 10m_{wing}$), the flutter speed reduces to 75 m/s and now the flutter mode contains some rigid body motion. Finally, when the fuselage mass reduces to that of the wing ($M = m_{wing}$), the flutter speed reduces further to 57 m/s, and the flutter mechanism now contains a significant amount of rigid body motion as well as the torsion and bending DOF. Inspection of Figure 10.24 shows how the $M = 10m_{wing}$ and $M = m_{wing}$ cases compare, in the latter case the bending frequency is greater and the flutter mode is a 'hump mode'.

10.11.2   *Relevance of Rigid Body Motions to Loads*

In the later chapters on loads, further examples of simple models with both rigid body and flexible modes are also presented. The inertial axes representation is employed for the equilibrium manoeuvre, gusts and ground loads (see Chapters 12, 14 and 15) whereas the body fixed axes (or so-called flight mechanics) model is used for dynamic manoeuvres (see Chapter 13). This flight mechanics model with flexible effects included is also used to assess the impact of flexibility upon the aircraft handling, but could also provide another view of any particular flutter mechanism encountered where rigid body effects were important. The inclusion of unsteady aerodynamic effects on the rigid body and flexible modes is considered in Chapter 19.

10.12   Flutter in the Transonic Regime

One major limitation with strip theory and panel method aerodynamics (see Chapters 18 and 19) is that they are unable to predict the occurrence of shock waves in the transonic flight regime. A consequence of this is that the prediction of the corresponding flutter boundaries can become inaccurate. Figure 10.25 shows a typical plot of flutter speed versus Mach number, and it can be seen that in the transonic region there is a dramatic reduction in the flutter speed for certain flow conditions. This is known as the 'transonic dip' (or 'flutter bucket') but it cannot be predicted accurately using linear aerodynamic methods; either Euler or Navier–Stokes CFD aerodynamic modelling techniques must be used, coupled with a structural model. However it is possible to determine corrections to panel method Aerodynamic Influence Coefficients (AICs) using wind tunnel test or CFD data where transonic effects can be incorporated (see Part III).

10.13   Effect of Non-Linearities – Limit Cycle Oscillations

All of the aeroelastic modelling that has been considered so far in this book has made the assumption of linearity; the structural deflections are small, the aerodynamic forces are linearly proportional to the response and the control system elements respond linearly with amplitude. In practice, non-linearities can be present in an aeroelastic system (Dowell *et al.*, 2003) via

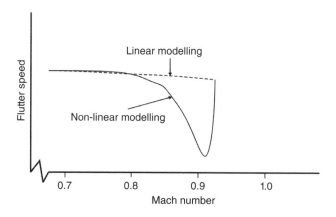

Figure 10.25   Typical flutter speed behaviour in a transonic regime.

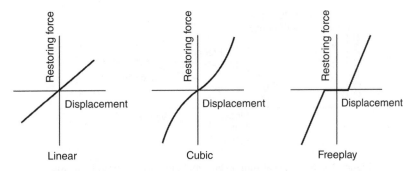

Figure 10.26 Typical structural non-linearities.

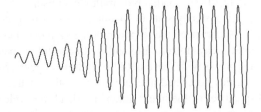

Figure 10.27 Typical limit cycle oscillation.

structural, aerodynamic and control system phenomena. These non-linearities affect the aero-elastic behaviour and cannot be predicted using linear analysis methods.

Structural non-linearities occur primarily as non-uniform stiffness effects, including cubic stiffening root attachments of engine pylons, bilinear stiffness of structural joints, freeplay of control surface attachments and geometric stiffening due to large deflections. Figure 10.26 shows some typical restoring force versus displacement plots for different stiffness non-linearities. Very flexible aircraft exhibit geometric stiffness non-linearities due to the large deflections that can occur.

Aerodynamic non-linearities occur primarily in the transonic flight regime, where shock waves are present upon the wing or control surfaces, and the position of the shock waves changes in response to motion of the wings; the interaction of control surfaces with shock waves is sometimes referred to as 'buzz'. Further aerodynamic non-linearities are 'stall flutter' when stall occurs at the wing tips and lift is lost on the outer part of the wing, and 'buffeting' where the separated flow from one part of the structure impinges onto another part.

Control non-linearities include control surface deflection and rate limits where the control surfaces cannot respond in the manner that is required by control laws. Also, the control surface actuation mechanism tends to be non-linear as well as the control laws that are used. The use of non-linear, or multiple, control laws and time delays in their application also leads to non-linear aeroelastic behaviour.

The main non-linear aeroelastic response phenomena are Limit Cycle Oscillations (LCOs) which can be considered as bounded flutter and an example of which is shown in Figure 10.27. Sometimes this is referred to as non-linear flutter. If an aeroelastic system is considered that includes the cubic stiffness as shown in Figure 10.26, then at some air speed, depending upon the stiffness at zero deflection, flutter will start to occur and an unstable motion results.

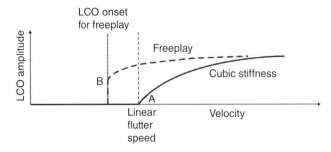

Figure 10.28 Typical limit cycle amplitude versus speed behaviour.

However, as the deflections get larger, the stiffness will become greater and the motion will be limited. In some cases the LCOs are made up of multiple sinusoids.

Figure 10.28 shows a typical steady-state LCO amplitude versus speed plot, where it can be seen that below the linear flutter speed the response of the system containing a cubic stiffness following some initial input decays to zero. Beyond the linear flutter speed at point A, an LCO develops and its amplitude grows with speed until finally flutter occurs. Such a response occurs for most of the non-linearities described above. One exception is the freeplay non-linearity where, at some critical speed below the linear flutter speed, a limit cycle suddenly occurs with a jump to point B. Much research is currently being undertaken to investigate accurate and efficient ways of predicting LCOs, including non-linear FE and aerodynamic models.

10.14 Examples

1. Using the MATLAB code given in the companion website, generate $V\omega$ and Vg plots for the binary flutter system and explore the effect of the following on the aeroelastic behaviour:
   a. Effect of ratio and spacing between wind-off torsion and bending natural frequencies.
   b. Effect of position of elastic and mass axes.
   c. Inclusion of structural damping.
   d. Altitude and hence plot Mach number vs. speed at flutter.
2. Develop a MATLAB code to determine the aeroelastic behaviour of the assumed mode representation described in Section 10.6 and explore the effect of including more modes in the mathematical model. Take $EI = 10^6 \, \text{N m}^2$, $GJ = 10^6 \, \text{N m}^2$, $c = 1.5 \, \text{m}$ and $s = 7 \, \text{m}$.
3. A wing bending–torsion system (in SI units) is described in terms of coordinates q_1 and q_2 as

$$\begin{bmatrix} 14D^2 + 6VD + \sigma - 6V^2 & -2D^2 + VD + V^2 \\ -2D^2 - 2VD - 5V^2 & D^2 + VD + V^2 \end{bmatrix} \begin{Bmatrix} q_1 \\ q_2 \end{Bmatrix} q_1 = 0$$

where $D = d/dt$ and $\sigma = 1 \times 10^5 \, \text{N/rad}$. Determine the critical flutter speeds and corresponding frequencies using the Routh–Hurwitz approach.
[73.35 m/s 13.92 Hz, 132.89 m/s 17.74 Hz]
4. A wing bending–torsion system (in SI units) is modelled in terms of coordinates α and θ:

$$12\ddot{\alpha} + 6V\dot{\alpha} + \left(4 \times 10^5 - 9V^2\right)\alpha + 3V\dot{\theta} + 3V^2\theta = 0$$
$$-3V^2\alpha + \ddot{\theta} + V\dot{\theta} + V^2\theta = 0$$

Determine the critical flutter speeds and corresponding frequencies using the Routh–Hurwitz approach.

[115.5 m/s 26 Hz, 365 m/s 41 Hz]

5. A binary aeroelastic system (in SI units) takes the form

$$\begin{bmatrix} 130 & 0 \\ 0 & 10 \end{bmatrix} \begin{Bmatrix} \ddot{\theta} \\ \ddot{\gamma} \end{Bmatrix} + \begin{bmatrix} 6V & 0 \\ -3V & V \end{bmatrix} \begin{Bmatrix} \dot{\theta} \\ \dot{\gamma} \end{Bmatrix} + \begin{bmatrix} k & 3V^2 \\ 0 & 2k-3V^2 \end{bmatrix} \begin{Bmatrix} \theta \\ \gamma \end{Bmatrix} = 0$$

Find the stiffness value k that gives a critical flutter speed of $V = 250$ m/s and the corresponding flutter frequency. Also find the divergence speed.

[1.173×10^5 N/m, 10.33 Hz, 279.6 m/s]

6. A binary aeroelastic system (in SI units) takes the form

$$\begin{bmatrix} 120 & 0 \\ 0 & 9 \end{bmatrix} \begin{Bmatrix} \ddot{\theta} \\ \ddot{\gamma} \end{Bmatrix} + \begin{bmatrix} 6V & 0 \\ -3V & V \end{bmatrix} \begin{Bmatrix} \dot{\theta} \\ \dot{\gamma} \end{Bmatrix} + \begin{bmatrix} k_1 & 4V^2 \\ 0 & k_2-3V^2 \end{bmatrix} \begin{Bmatrix} \theta \\ \gamma \end{Bmatrix} = 0$$

where $k_1 = 5 \times 10^4$ Nm/rad and $k_2 = 7 \times 10^4$ N m/rad. Determine the critical flutter speeds and corresponding frequencies using the Routh–Hurwitz approach. Also, obtain the divergence speed.

[256.8 m/s, 3.29 Hz; 131.9 m/s, 7.30 Hz; 152.8 m/s]

11

Aeroservoelasticity

The science of aeroservoelasticity (ASE) extends the aeroelastic interactions between aerodynamic forces and a flexible structure, discussed in Chapters 7, 8 and 10, to include a control system, introduced in Chapter 6. The classic Collar aeroelastic triangle can be extended to form the aeroservoelastic pyramid shown in Figure 11.1, where there are now forces resulting from the control system as well as the aerodynamic, elastic and inertial forces. ASE effects (Zimmermann, 1991; Pratt, 2000; Librescu, 2005) are becoming of increasing importance in modern aircraft design as it is usual nowadays to employ some form of Flight Control System (FCS) (Pratt, 2000; see also Chapter 13) to improve the handling and stability, flight performance and ride quality throughout the flight envelope, and also to reduce loads and improve service life. For commercial aircraft, the FCS might include a Gust and/or Manoeuvre Load Alleviation System in addition to a control system that meets the basic handling requirements. Modern military aircraft are often designed for carefree handling and the ability to fly with reduced, or unstable, open loop static stability so as to improve their manoeuvrability; however, they can only stay airborne through the use of the FCS. All control implementations involve the use of sensors, usually accelerometers and rate gyros placed at the aircraft centre of mass and air data sensors (e.g. angle of incidence, air speed). Some form of control input (defined by a control law; see Chapter 6) is then applied via the control surfaces. It is feasible to develop flutter suppression systems that enable aircraft to fly beyond the flutter speed; however, such an approach has a very high risk and would not be contemplated for a commercial aircraft.

ASE effects, sometimes referred to as 'structural coupling', can potentially cause a major structural failure due to a flutter mechanism involving coupling of the aeroelastic and control systems. However, there is also the possibility of causing fatigue damage and reducing control surface actuator performance. Most structural coupling problems occur when the motion sensors detect not only the aircraft rigid body motion but also the motion in the flexible modes, and these vibrations are fed back into the FCS. In this case, the movement of the control surfaces is then likely to excite the flexible modes, so causing the aircraft to vibrate further. Notch filters are often used to remedy this problem by introducing significant attenuation of the response in the region of critical frequencies.

In this chapter, the use of feedback control on a simple binary aeroelastic system with a control surface is considered and the effects of the control law on the stability and response investigated.

Introduction to Aircraft Aeroelasticity and Loads, Second Edition. Jan R. Wright and Jonathan E. Cooper.
© 2015 John Wiley & Sons, Ltd. Published 2015 by John Wiley & Sons, Ltd.

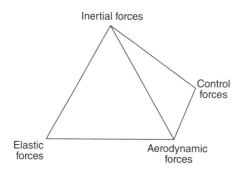

Figure 11.1   Aeroservoelastic pyramid.

A simple proportional-integral (PI) controller (see Chapter 6) is used to demonstrate how the gust response can be reduced and the flutter speed increased. Modelling of ASE systems in the time or frequency domain is also considered, including representation of reduced frequency-dependent aerodynamics and control effects along with the use of state space models. A number of MATLAB codes related to this chapter are included in the companion website.

11.1   Mathematical Modelling of a Simple Aeroelastic System with a Control Surface

Consider the binary aeroelastic model examined for its flutter behaviour in Chapter 10; it is composed of a cantilevered uniform rectangular wing with bending and torsion degrees of freedom. As can be seen in Figure 11.2, a full span control surface is now included which has an infinite stiffness attachment to the wing but can be moved to any angle β that is demanded. The inertial effects of the control surface are ignored. Thus the control surface is not involved in the basic dynamics of the wing, but simply acts as an excitation device.

The lift and pitching moment acting upon an elemental strip of wing can be written using the same assumptions and notation as considered before in Chapters 4 and 10, with the $M_{\dot\theta}$ term being included to allow for simple unsteady aerodynamic effects on the aerofoil (see Chapter 9), such that

$$dL = \frac{1}{2}\rho V^2 c\,dy\left[a_W\left(\frac{y^2}{s^2 V}\dot{q}_b + \frac{y}{s}q_t\right) + a_C\beta\right]$$

$$dM = \frac{1}{2}\rho V^2 c^2\,dy\left[ea_W\left(\frac{y^2}{s^2 V}\dot{q}_b + \frac{y}{s}q_t\right) + M_{\dot\theta}\frac{cy}{4sV}\dot{q}_t + b_C\beta\right]$$

(11.1)

where there are now lift and moment components present due to the application of the control surface through angle β (Fung, 1969).

The kinetic energy and incremental work done terms are evaluated across the entire semi-span of the wing (as in Chapter 10) and Lagrange's equations applied. Then, adding in the prescribed motion of the control surface gives the expression for the open loop system as

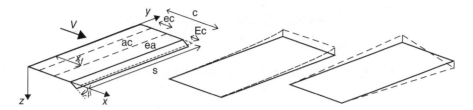

Figure 11.2   Binary flutter system with a control surface with bending and torsion modes.

$$\begin{bmatrix} A_{bb} & A_{bt} \\ A_{tb} & A_{tt} \end{bmatrix} \begin{Bmatrix} \ddot{q}_b \\ \ddot{q}_t \end{Bmatrix} + \rho V \begin{bmatrix} \dfrac{ca_W s}{10} & 0 \\ -\dfrac{c^2 ea_W s}{8} & -\dfrac{c^3 M_{\dot\theta} s}{24} \end{bmatrix} \begin{Bmatrix} \dot{q}_b \\ \dot{q}_t \end{Bmatrix}$$

$$+ \left(\rho V^2 \begin{bmatrix} 0 & \dfrac{ca_W s}{8} \\ 0 & -\dfrac{c^2 ea_W s}{6} \end{bmatrix} + \begin{bmatrix} \dfrac{4EI}{s^3} & 0 \\ 0 & \dfrac{GJ}{s} \end{bmatrix} \right) \begin{Bmatrix} q_b \\ q_t \end{Bmatrix} = \rho V^2 \begin{Bmatrix} -\dfrac{ca_C s}{6} \\ \dfrac{c^2 b_C s}{4} \end{Bmatrix} \beta$$

$$(11.2)$$

where the **A** matrix was quoted fully in Chapter 10. It can be seen that there is now a forcing vector on the right-hand side of the equations due to the control surface deflection. Equation (11.2) can be written more compactly in the matrix form

$$\mathbf{A}\ddot{q} + \rho V \mathbf{B}\dot{q} + \left(\rho V^2 \mathbf{C} + \mathbf{E} \right) q = \mathbf{g}\beta \qquad (11.3)$$

where for convenience the term ρV^2 is embedded in the excitation vector $\mathbf{g}$. Then, using the approach of Chapter 2, a harmonic excitation $\beta = \beta_0 e^{i\omega t}$ and response $q = q_0 e^{i\omega t}$ leads to the Frequency Response Function between the response degrees of freedom and the control surface rotation. A frequency domain approach may then be used to determine the response of the system due to the input of the control system at any point in the flight envelope. Alternatively, an approach based on time domain numerical integration could be employed for any general excitation input.

For the complete aircraft, the free–free rigid body and flexible modes need to be incorporated, together with the basic FCS. Although some explanation of the FCS model is given in Chapters 13 and 21, highlighting its importance, a detailed consideration of the FCS is beyond the scope of this book.

11.2   Inclusion of Gust Terms

The effect of gusts and turbulence will be considered in much greater detail in Chapter 14. However, it is useful here to include the effect of a uniform vertical gust of velocity w_g encountered along the whole span of the wing in order to provide a disturbance; this simplified approach contains a number of assumptions. Figure 11.3 shows that a gust gives rise to an effective instantaneous change of incidence $\Delta\theta$ of

$$\Delta\theta = \frac{w_g}{V} \qquad (11.4)$$

Figure 11.3 Effective angle of incidence due to the vertical gust.

The additional lift and pitching moment acting on an elemental streamwise strip on the wing due to the gust now become

$$dL_{\text{Gust}} = \frac{1}{2}\rho V^2 c \, dy a_{\text{W}} \frac{w_g}{V} \qquad dM_{\text{Gust}} = \frac{1}{2}\rho V^2 c^2 \, dy e a_{\text{W}} \frac{w_g}{V} \qquad (11.5)$$

so the additional right hand side excitation term due to the gust input is

$$\rho V \left\{ \begin{array}{c} -\dfrac{c a_{\text{W}} s}{6} \\[2mm] \dfrac{c^2 e a_{\text{W}} s}{4} \end{array} \right\} w_g = \left\{ \begin{array}{c} h_1 \\ h_2 \end{array} \right\} w_g = \boldsymbol{h} w_g \qquad (11.6)$$

Thus the system with both control and gust excitation is of the following general form

$$\mathbf{A}\ddot{\boldsymbol{q}} + \rho V \mathbf{B}\dot{\boldsymbol{q}} + \left(\rho V^2 \mathbf{C} + \mathbf{E}\right)\boldsymbol{q} = \boldsymbol{g}\beta + \boldsymbol{h} w_g \qquad (11.7)$$

Again, for convenience, the term ρV is embedded in the gust excitation vector $\boldsymbol{h}$. Now the gust disturbance term is seen to appear on the right-hand side of the equations together with the control surface input. Clearly, the response due to a known gust time history may now be calculated.

11.3 Implementation of a Control System

One of the simplest forms of control system is the PI approach (see Chapter 6), and when implemented here the control surface demand angle is linearly proportional to the velocity and displacement of the system. For simplicity, consider a transducer positioned at the wing leading edge ($x = 0$) a distance s_0 from the root, and the control surface deflection is taken as being proportional to its displacement and velocity such that

$$\begin{aligned} \beta &= K_v \dot{z}_{\text{Wing}} + K_d z_{\text{Wing}} = K_v \left(\left(\frac{s_0}{s}\right)^2 \dot{q}_b - x_{\text{f}} \left(\frac{s_0}{s}\right) \dot{q}_t \right) + K_d \left(\left(\frac{s_0}{s}\right)^2 q_b - x_{\text{f}} \left(\frac{s_0}{s}\right) q_t \right) \\ &= K_v \left\{ \left(\frac{s_0}{s}\right)^2 \ -x_{\text{f}} \left(\frac{s_0}{s}\right) \right\} \left\{ \begin{array}{c} \dot{q}_b \\ \dot{q}_t \end{array} \right\} + K_d \left\{ \left(\frac{s_0}{s}\right)^2 \ -x_{\text{f}} \left(\frac{s_0}{s}\right) \right\} \left\{ \begin{array}{c} q_b \\ q_t \end{array} \right\} \end{aligned} \qquad (11.8)$$

where K_v and K_d are weightings (commonly called feedback 'gains'; see Chapter 6) applied to the velocity and displacement terms respectively. Thus the aircraft response is fed back via the control surface to modify the aircraft characteristics. This mathematical model of the wing with gust and control input, modified by the feedback law, can be represented by the block diagram

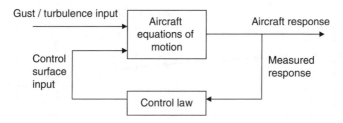

Figure 11.4   Block diagram for the aeroservoelastic system.

shown in Figure 11.4, where the system is generalized to that of an aircraft (not just a wing); see also Chapters 13 and 21.

11.4   Determination of Closed Loop System Stability

In order to examine the stability of the closed loop system, the feedback law in Equation (11.8) must be combined with the basic wing equations in Equation (11.3). In order to simplify the expressions, the transducer will be placed at the tip leading edge ($s_0 = s$) so that $s_0/s = 1$. Thus the closed loop Equation (11.3) of the wing plus control surface (but in the absence of a gust disturbance) becomes

$$A\ddot{q} + \rho V B \dot{q} + (\rho V^2 C + E)q = g\beta = K_v \begin{Bmatrix} g_1 \\ g_2 \end{Bmatrix} \{1 \quad -x_f\} \begin{Bmatrix} \dot{q}_b \\ \dot{q}_t \end{Bmatrix} + K_d \begin{Bmatrix} g_1 \\ g_2 \end{Bmatrix} \{1 \quad -x_f\} \begin{Bmatrix} q_b \\ q_t \end{Bmatrix}$$

$$= K_v \begin{bmatrix} g_1 & -g_1 x_f \\ g_2 & -g_2 x_f \end{bmatrix} \begin{Bmatrix} \dot{q}_b \\ \dot{q}_t \end{Bmatrix} + K_d \begin{bmatrix} g_1 & -g_1 x_f \\ g_2 & -g_2 x_f \end{bmatrix} \begin{Bmatrix} q_b \\ q_t \end{Bmatrix} = F\dot{q} + Gq \qquad (11.9)$$

where F, G are composite feedback matrices with coefficients that are a function of the control gains, density and air speed. This closed loop equation can be rearranged as

$$A\ddot{q} + (\rho V B - F)\dot{q} + (\rho V^2 C + E - G)q = 0 \qquad (11.10)$$

and the closed loop system can now be solved to examine its stability. Clearly, the dynamics of the system have now been altered since there are extra stiffness and damping matrices present due to the control system, and this will affect the aeroelastic behaviour including the flutter speed. Equation (11.10) is still in the same general form of the aeroelastic equations and the analysis can be carried out in exactly the same way as in Chapter 10 for flutter, determining the natural frequencies and damping ratios at different flight conditions for different combinations of constant feedback gains K_v, K_d.

As an example, the effect of using gain K_v alone ($K_d = 0$) for the baseline binary aeroelastic system can be seen in Figures 11.5 and 11.6. In Figure 11.5 the frequency and damping ratio trends are plotted against speed and it can be seen that whereas the open loop systems flutters at an air speed of 145 m/s, for the closed loop system the flutter is delayed until 165 m/s. Similar characteristics can be seen in the root locus plot of the system eigenvalues in Figure 11.6 where the path of the critical mode is clearly delayed in crossing the imaginary axis.

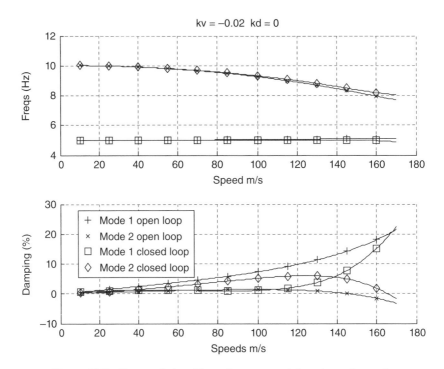

Figure 11.5 Open and closed loop frequency and damping ratio trends.

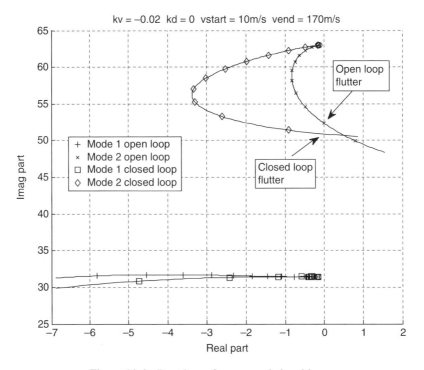

Figure 11.6 Root locus for open and closed loop.

11.5 Gust Response of the Closed Loop System

Inclusion of the gust excitation terms on the right-hand side of Equation (11.10) for the closed loop system gives

$$\mathbf{A}\ddot{q} + (\rho V\mathbf{B} - \mathbf{F})\dot{q} + (\rho V^2\mathbf{C} + \mathbf{E} - \mathbf{G})q = hw_g \qquad (11.11)$$

This equation allows the response of the system to gusts with the inclusion of the feedback control to be calculated. Having determined the responses, the required control angle deflection can be found using Equation (11.8).

Figure 11.7 shows the wing leading edge tip displacement for a short '1-cosine' gust input to the open loop system and also to the closed loop system employing a different control law ($K_d = 1.0$, $K_v = 0$). The results show that there is a 15% reduction in the maximum amplitude of the gust response. The control surface demand angle is also shown in Figure 11.8 with the control law applied. It can be seen that the control can be used to reduce the time that it takes for the response to decay.

One problem with this implementation of control is that the control surface deflection is taken to be linearly related to the wing displacement and velocity. There will be limits to both the control deflection (e.g. ± 15°) and rate (e.g. 30°/s) that can be realized. Unsteady aerodynamic effects also reduce the control surface effectiveness as the application frequency increases. In practice, more sophisticated models need to be developed to allow for the unsteady aerodynamic behaviour, including reduced frequency effects.

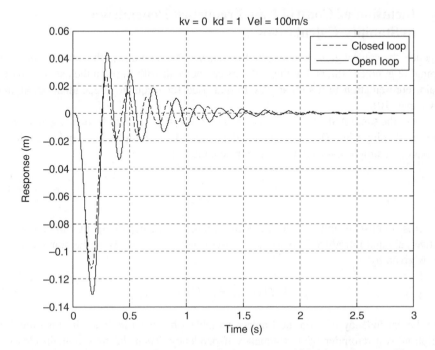

Figure 11.7 Leading edge tip response of wing to gust with/without the control law.

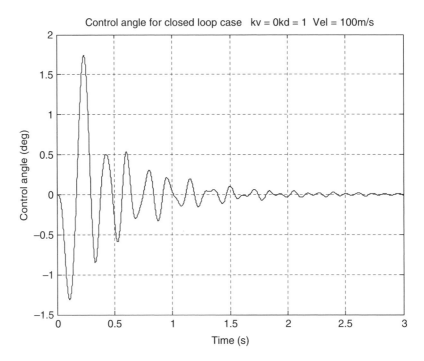

Figure 11.8 Control surface deflection for closed loop gust response.

11.6 Inclusion of Control Law Frequency Dependency in Stability Calculations

In practice, the control gains will also be frequency-dependent due to the presence of so-called 'shaping' (e.g. notch) filters, but these effects can be dealt with through the use of a frequency domain representation in a similar way to that used for frequency-dependent aerodynamics (see Chapter 10).

Consider the same ASE system as above, but now with a velocity feedback gain of $K_v\,T(s)$, where the shaping filter $T(s)$ is frequency dependent. For example, $T(s)$ could be a low pass filter having a simple Laplace domain representation of

$$T(s) = \frac{a}{s+a} \tag{11.12}$$

where a is a constant and s is the Laplace variable (see Chapter 6); this filter reduces the effective gain and introduces a phase lag as frequency increases. Then the demanded control angle is given by

$$\beta = K_v T(s) \dot{z}_{Wing} = K_v T(s) \left(\dot{q}_b - x_f \dot{q}_t \right) \tag{11.13}$$

where the dependency of T on the Laplace variable s has been included in this time domain description as a reminder of its frequency dependency. Then the time domain closed loop equations become

$$\mathbf{A}\ddot{\boldsymbol{q}} + \rho V \mathbf{B}\dot{\boldsymbol{q}} + \left(\rho V^2 \mathbf{C} + \mathbf{E}\right)\boldsymbol{q} = \boldsymbol{g}\beta = K_v T(s)\begin{bmatrix} g_1 & -g_1 x_{\mathrm{f}} \\ g_2 & -g_2 x_{\mathrm{f}} \end{bmatrix}\begin{Bmatrix} \dot{q}_b \\ \dot{q}_t \end{Bmatrix} = T(s)\mathbf{F}\dot{\boldsymbol{q}} \qquad (11.14)$$

Transformation to the frequency domain using $\boldsymbol{q} = \boldsymbol{q}_0 e^{i\omega t}$ and $s = i\omega$ gives

$$\left\{-\omega^2 \mathbf{A} + i\omega(\rho V \mathbf{B} - T(i\omega)\mathbf{F}) + \left(\rho V^2 \mathbf{C} + \mathbf{E}\right)\right\}\boldsymbol{q}_0 = 0 \qquad (11.15)$$

where matrices $\mathbf{B}$, $\mathbf{C}$ are reduced frequency (k) dependent and the filter T is frequency (ω) dependent.

In Chapter 10, the reduced frequency dependency of the $\mathbf{B}$, $\mathbf{C}$ matrices was handled by seeking a matched frequency solution; this approach determined the correct aerodynamic terms at each flight condition and thus the flutter speed was determined. The presence of the shaping filter in this equation may be dealt with in exactly the same way since, for a particular reduced frequency k and air speed V, the frequency $\omega = 2\,kV/c$ in T will be known and the shaping filter contribution to the equation can be evaluated. More complicated control laws can be utilized using this approach as long as they can be approximated in the form shown above; nonlinear control laws must be approximated by a linear representation.

11.7 Response Determination via the Frequency Domain

Since the aerodynamic and control terms may be determined for particular reduced frequencies (or frequencies) for a given air speed, it is possible to determine the response to turbulence using a frequency domain representation (see Chapter 14 for a full explanation). Assuming that the ASE system considered above with velocity feedback control including a shaping filter behaves in a linear manner, then encountering a harmonic gust of the form $w_{\mathrm{g}} = w_{\mathrm{g}0} e^{i\omega t}$ results in a harmonic response $\boldsymbol{q}_0 = \boldsymbol{q}_0 e^{i\omega t}$, and Equation (11.11) then becomes

$$\left(-\omega^2 \mathbf{A} + i\omega(\rho V \mathbf{B} - T(i\omega)\mathbf{F}) + \left(\rho V^2 \mathbf{C} + \mathbf{E}\right)\right)\boldsymbol{q}_0 = \boldsymbol{h}w_{\mathrm{g}0} \qquad (11.16)$$

where $\mathbf{B}$, $\mathbf{C}$ and T are evaluated at the appropriate k and ω. Hence, the vector of closed loop transfer functions between the generalized coordinates and the gust excitation (see Chapter 14) is defined by

$$\boldsymbol{q}_0 = \mathbf{H}_{\mathrm{qg}}(\omega)w_{\mathrm{g}0} \quad \text{where} \quad \mathbf{H}_{\mathrm{qg}}(\omega) = \left(-\omega^2 \mathbf{A} + i\omega(\rho V \mathbf{B} - T(i\omega)\mathbf{F}) + \left(\rho V^2 \mathbf{C} + \mathbf{E}\right)\right)^{-1}\boldsymbol{h} \quad (11.17)$$

The transfer function between the deflection at the wing tip leading edge and the gust excitation for this closed loop system is

$$z_{Wing} = \left\{1 - x_f\right\}\boldsymbol{q}_0 = \left\{1 - x_f\right\}\mathbf{H}_{\mathrm{qg}}(\omega)w_{\mathrm{g}0} = H_{zg}(\omega)w_{\mathrm{g}0} \qquad (11.18)$$

The von Karman frequency representation of turbulence (see Chapter 14) can be used to provide the Power Spectral Density (PSD) input to the system from which the response PSD for the closed loop system can be calculated using Equation (11.17). The control system gains can then be designed such that the resulting deflections and loads are reduced to required levels and required gain and phase margins are achieved.

11.8   State Space Modelling

An alternative approach to using second-order models is to employ the first-order state space representation introduced in Chapter 6. The state space formulation is particularly useful for the application of many control design techniques (e.g. optimal control theory).

Equation (11.7) for the open loop system with control and gust input can be reformulated into the first-order state space form, such that

$$\begin{Bmatrix} \dot{q} \\ \ddot{q} \end{Bmatrix} = \begin{bmatrix} 0 & \mathbf{I} \\ -\mathbf{A}^{-1}(\rho V^2 \mathbf{C} + \mathbf{E}) & -\mathbf{A}^{-1}(\rho V \mathbf{B}) \end{bmatrix} \begin{Bmatrix} q \\ \dot{q} \end{Bmatrix} + \begin{Bmatrix} 0 \\ \mathbf{A}^{-1}g \end{Bmatrix} \beta + \begin{Bmatrix} 0 \\ \mathbf{A}^{-1}h \end{Bmatrix} w_g \qquad (11.19)$$

or as shown in Chapter 6,

$$\dot{x} = \mathbf{A}_s x + \mathbf{B}_s u + \mathbf{E}_s w_g \qquad y = \mathbf{C}_s x + \mathbf{D}_s u \qquad (11.20)$$

Here the control input $u = \{\beta\}$ and a gust disturbance term are now present and, when velocity-only measurement at the leading edge is considered, the measured output y is given by

$$y = \{\dot{z}_{\text{Wing}}\} = \begin{bmatrix} 1 & -x_f & 0 & 0 \end{bmatrix} \{\dot{q}_b \quad \dot{q}_t \quad q_b \quad q_t\}^{\mathrm{T}} = \mathbf{C}_s x \qquad (11.21)$$

since $\mathbf{D}_s = \mathbf{0}$ is normally assumed. In order to avoid confusion with the notation used earlier in the flutter equation, the subscript 's' has been used here to denote the state space matrices.

To introduce the feedback effect, the control input is written in terms of the measured response y (in this case a velocity), namely

$$u = \{\beta\} = [K_v] \{\dot{z}_{\text{Wing}}\} = \mathbf{K}_s y \qquad (11.22)$$

where $\mathbf{K}_s$ is the state space gain matrix. Then substituting Equation (11.22) into Equation (11.20) and simplifying leads to the revised state equation for the closed loop system

$$\dot{x} = [\mathbf{A}_s + \mathbf{B}_s \mathbf{K}_s \mathbf{C}_s] x + \mathbf{E}_s w_g \qquad (11.23)$$

In the same way that a frequency domain representation was used with the second order form of the equations, it is possible to determine the frequencies and damping ratios for the closed loop system at a particular flight condition and gain value using the eigensolution of the system matrix $[\mathbf{A}_s + \mathbf{B}_s \, \mathbf{K}_s \, \mathbf{C}_s]$ that has been revised to account for control feedback effects. As with the frequency domain approach, the effects of frequency-dependent unsteady aerodynamics and control shaping filters need to be taken into account if accurate models are required. The frequency domain gust response transfer function may also be derived from the state space equations.

These state space equations, and indeed the earlier second-order equations, can be solved in the time domain to give the state response x, and hence the measured output y, for the closed loop system due to any gust input. In this case, representation of frequency-dependent aerodynamic effects is achieved through the use of so-called Rational Function Approximation, which is described in Chapter 19. The effects of structural and control law nonlinearities can be included when a simulation is carried out in the time domain.

11.9   Examples

Make use of the MATLAB and SIMULINK routines shown in the companion website for the baseline binary aeroelastic system with feedback control input and gust excitation.

1. Determine the open and closed loop response due to '1-cosine' gusts and explore the effect of different gust wavelengths.
2. Determine the response to a 'chirp' (fast sinusoidal sweep) control input of linearly varying frequency and explore the effect of changing the start and end frequencies of the 'chirp'.
3. Explore the effect of varying the gains K_v and K_d on the flutter and divergence speed of the system.
4. Determine the range of gains K_v and K_d that will enable the flutter speed of the baseline system with the control system to be increased by 30 m/s.
5. For '1-cosine' gusts of duration 0.005, 0.01 and 0.05 s, explore the effect of varying the gains K_v and K_d on the closed loop system response.

12

Equilibrium Manoeuvres

Aircraft are controlled when the pilot uses the control surfaces (namely aileron/spoiler for roll, rudder for yaw and elevator for pitch) singly or in combination for a range of different manoeuvres; of course, thrust may also be varied. The structure must be designed to withstand these manoeuvres and the load calculations required are critical in the aircraft clearance process, often involving many tens of thousands of cases. A useful background to meeting some of the loads requirements in the certification specifications (CS-25 and FAR-25) is given in (Howe, 2004; Lomax, 1996).

There is a difference between manoeuvres performed by commercial and military aircraft. Military aircraft (excepting transport and bomber aircraft) are subject to far more severe manoeuvres, involving higher g levels, control angles and rates. However, military combat aircraft are generally stiffer than commercial aircraft, with natural frequencies usually greater than 5 Hz, so the manoeuvre loads calculations are sometimes carried out using a rigid aircraft model, though this is changing for more highly flexible combat aircraft and unmanned air vehicles (UAVs). In contrast, although their manoeuvres are less severe, large commercial aircraft are generally significantly more flexible, some (e.g. Airbus A380) with modes of vibration lower than 1 Hz; thus it is becoming more essential to perform loads calculations using a flexible aircraft model that incorporates the rigid aircraft characteristics. This means that the aeroelastic and loads domains are becoming more interdependent, a key reason why this book seeks to balance these two aspects.

There are two types of *flight manoeuvre* that have to be considered in the design of an aircraft, often referred to as:

- equilibrium (or balanced/steady/trimmed) manoeuvres and
- dynamic manoeuvres.

The calculation methodology is different in each case and will be addressed in this and the next chapter.

The term *equilibrium (or balanced/steady/trimmed) manoeuvre* refers to the case where the aircraft is in a steady manoeuvre. In the symmetric case, where *pitching* at a steady pitch rate is usually involved (i.e. zero pitch acceleration), the aircraft will experience accelerations normal

---

Introduction to Aircraft Aeroelasticity and Loads, Second Edition. Jan R. Wright and Jonathan E. Cooper.
© 2015 John Wiley & Sons, Ltd. Published 2015 by John Wiley & Sons, Ltd.

to the flight path. Such manoeuvres are intended to represent the aircraft in an emergency pull-up or push-down situation, with the wings horizontal; this load case is important for the design of inboard parts of aerodynamic surfaces as well as for engine pylons and fuselage components. A steady banked turn is also an equilibrium manoeuvre. In such accelerated conditions, the aircraft is in effective static equilibrium once d'Alembert's principle has been used to introduce inertia forces. The symmetric equilibrium manoeuvre is considered in Howe (2004), Lomax (1996), Megson (1999) and in ESDU Data Sheets 94009, 97032 and 99033.

In this chapter, the process of determining the balanced response (deformation and component loads) in a number of *symmetric* equilibrium manoeuvres will be considered, using a progression of fairly basic mathematical models for both rigid and simple flexible aircraft. The flexible aircraft needs to be considered since flexibility can affect the loads distribution. CS-25 states: 'If deflections under load would significantly change the distribution of internal or external loads, this redistribution must be taken into account'. The accounting for aeroelastic effects in loads calculations may be done by retaining a rigid aircraft model, but modifying it for flexible effects (e.g. with modified aerodynamic derivatives), or else by including flexible modes and deformations in the model. Note that the axes system used in this chapter will be inertial (earth fixed) and the unknowns will be displacements, angles and generalized coordinates.

However, there are also a number of *asymmetric* manoeuvres involving *rolling* and *yawing* that may be classed as equilibrium manoeuvres, but these make use of more simplified aircraft representations. Such manoeuvres will be considered briefly in Chapter 13 once the lateral aircraft dynamic model has been introduced.

In a *dynamic manoeuvre*, the variation with time of the response and loads of the aircraft when subject to a transient input is determined using a representative dynamic model of the aircraft; both symmetric and asymmetric manoeuvres are possible and the types of case will be considered later in Chapters 13 and 23. Often, the non-linear (large angle) flight mechanics model of the aircraft is used, sometimes with flexible modes included especially for highly flexible aircraft; alternatively, a linear (small angle) model may be employed but this is less accurate for manoeuvres with substantial changes of altitude and angle. Therefore, in Chapter 13, the non-linear flight mechanics model of the rigid aircraft will be introduced, where axes fixed in the aircraft are used, the unknowns are velocities, and large angle manoeuvres may be handled. The extension to the flexible aircraft will be considered since flexibility effects are important. In Chapter 13, the application of this flight mechanics model to simple dynamic heave/pitch examples will be considered for both rigid and simple flexible aircraft. In addition, simple roll/yaw manoeuvres will be considered for a rigid aircraft model.

Balanced manoeuvres for pitching, rolling and yawing are sometimes known as *bookcase* manoeuvres, where the aircraft model and load cases are often somewhat artificial but will yield load estimates at a relatively early stage in the aircraft life. Dynamic manoeuvres are sometimes referred to as *rational* manoeuvres where full dynamic simulations of the aircraft behaviour in time are carried out using a more realistic aircraft model and load case.

Later, in Chapter 14, the related issues of the response to *gusts* and *continuous turbulence* will be considered, though of course these are not deliberate flight manoeuvres. Then, in Chapter 15, the treatment of *ground manoeuvres* (e.g. taxiing, landing, braking, etc.) will be introduced. In both these cases, inertial axes will be used where unknowns are displacements, rotations and generalized coordinates.

Because a whole range of concepts will be introduced in the next few chapters, the aerodynamics will be kept as simple as possible and strip theory will continue to be used. Later in Chapters 18 and 19, the use of more realistic aerodynamics will be considered. Also, the

Rayleigh–Ritz approach for incorporating a flexible free–free mode together with the rigid body motions will be used, thus allowing the whole flexible aircraft to be studied in concept for symmetric flight cases using a total of only 3DoF. In Chapter 7, some of the concepts of static aeroelastic deformation for a simple wing model in level flight were introduced.

In the next few chapters on manoeuvre and gust/turbulence inputs, the focus will be on determining the aircraft response, steady or time-varying. In all the cases covered, the aerodynamic and inertia loads distributed over the aircraft will give rise to *internal 'MAST' loads* (i.e. moment, axial, shear and torque - see Chapter 5). The calculation of internal loads will be covered in Chapter 16, since there are many common factors in determination of internal loads for manoeuvre and gust/turbulence inputs.

12.1   Equilibrium Manoeuvre – Rigid Aircraft under Normal Acceleration

In this section, the aircraft is treated effectively as a 'particle' (see Chapter 5) under heave acceleration, i.e. no rotational effects are present. Different manoeuvres are considered and the *load factor* concept is introduced.

12.1.1   Steady Level Flight

Firstly, consider an aircraft of mass m in steady level flight. There are two ways to define the aerodynamic forces (Lomax, 1996). Firstly, as shown in Figure 12.1a, the overall force may be defined by the chordwise force C and normal force N acting along the aircraft x and z axes Alternatively, as shown in Figure 12.1b, the aerodynamic forces are defined by the drag force D and the lift force L which act parallel and perpendicular to the freestream direction. The two sets of forces are related by the wing incidence. It is possible for either force representation to be employed but in this book the lift and drag combination will be used and for convenience, recognising that the wing incidence is small and seeking to make the equations as simple as possible, the thrust is taken to act parallel to the drag.

Then in steady level flight, there will be no net vertical force and total lift L will balance weight W ($=mg$) so $L = W$. Also, for this simple 'particle' representation, thrust T will balance drag D so $T = D$.

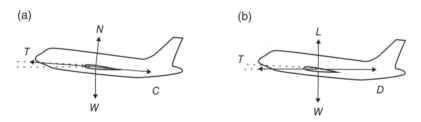

(a)   (b)

Figure 12.1   Rigid aircraft in level flight: (a) normal/chordwise and (b) lift/drag forces.

12.1.2 *Accelerated Flight Manoeuvre – Load Factor*

Now consider the aircraft accelerating upwards at a constant acceleration of a (m/s^2), as shown via the Free Body Diagrams (FBDs; see Chapter 5) in Figure 12.2, but ignoring any velocity component normal to the flight path; a steady turn or a pull-up from a dive fall into this category. Velocity components normal to the flight path introduce a change in incidence and become relevant later for dynamic manoeuvres and gust encounters.

Using d'Alembert's principle (see Chapter 5), the aircraft dynamics may be converted into an equivalent static equilibrium problem by introducing an inertia force equal to ma in the opposite direction to the acceleration, as shown in Figure 12.2. Thus, for equilibrium of aerodynamic and inertia forces normal to the flight path

$$L - W - ma = 0 \quad \text{or} \quad L = W\left(1 + \frac{a}{g}\right) \quad \text{or} \quad L = nW \tag{12.1}$$

where n is the so-called *load factor*, providing a measure of the severity of a manoeuvre and defined by

$$\text{Load factor } n = \frac{\text{Total Lift}}{\text{Weight}} \tag{12.2}$$

It is common practice in accelerated flight manoeuvres to use the load factor concept and therefore to balance the lift by the force nW (a combination of weight and inertia force), as shown in Figure 12.2. Thus, wherever W appears in equations for steady level flight, it may be replaced by nW for accelerated flight; the usefulness of this concept will be illustrated later in the chapter. Clearly, steady level flight is characterized by a load factor value of $n = 1$. The thrust/drag equation along the flight path remains the same as for steady level flight though changes in lift will affect the drag and therefore the thrust required to maintain steady flight.

Next, some simple symmetric manoeuvres will be considered to see how load factor relates to flight conditions. The term 'symmetric' is used whenever the response and loads on both sides of the aircraft are the same.

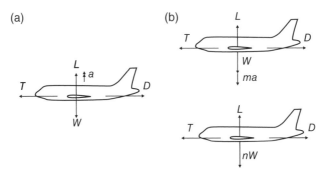

Figure 12.2 Rigid aircraft in accelerated flight: (a) Newton's law and (b) d'Alembert's principle.

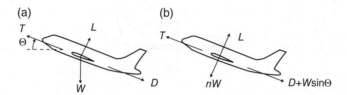

Figure 12.3 Rigid aircraft in steady climb: (a) basic forces and (b) load factor form.

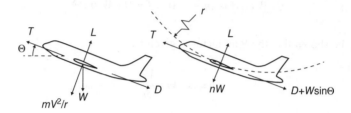

Figure 12.4 Rigid aircraft in steady pull-out.

12.1.3   Steady Climb/Descent

Consider an aircraft in a steady climb with its flight path at an angle of inclination Θ to the horizontal; the FBD is shown in Figure 12.3. In this case there will be no velocity of the aircraft normal to the flight path, no pitch rate and no accelerations. For equilibrium *normal* to and *along* the flight path, then

$$L = W \cos\Theta, \quad T = D + W \sin\Theta \tag{12.3}$$

and so thrust will need to change when the aircraft climbs. The load factor is therefore given by

$$n = \cos\Theta \quad (n \leq 1) \tag{12.4}$$

12.1.4   Steady Pull-Up and Push-Down

Now consider the aircraft in a steady pull-up from a dive. Assume that the aircraft follows a circular path of radius r (not to be confused with yaw velocity in Chapter 13) in the vertical plane at true air speed (TAS) V (see Chapter 4 for air speed definitions), as shown in Figure 12.4. Again, in this case there will be no velocity of the aircraft normal to the flight path and no acceleration along the flight path. However, because the aircraft is following a circular path, it must experience a centripetal acceleration V^2/r (i.e. acting towards the centre).

It is essential to use true air speed TAS when evaluating a centripetal acceleration and not equivalent air speed EAS. Introducing an inertia force and considering equilibrium in a direction *normal* to and *along* the flight path when the aircraft is at an inclination Θ to the horizontal yields

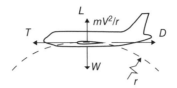

Figure 12.5   Rigid aircraft in steady push-down.

$$L - W \cos\Theta - m\frac{V^2}{r} = 0 \quad T = D + W \sin\Theta \tag{12.5}$$

and so it may be shown that the load factor is given by

$$n = \cos\Theta + \frac{V^2}{gr} \tag{12.6}$$

It may be seen from this expression that the load factor increases for higher air speeds and tighter pull-ups (i.e. smaller radius). In the pull-up, the aircraft velocity is $V = rq$, where $q = V/r$ is the steady pitch rate (nose up positive); the symbol q is traditionally used in this way in flight mechanics (see Chapter 13) and should not be confused with its use elsewhere in this book as a generalized coordinate or dynamic pressure.

When an aircraft 'loops the loop', the load factor varies continuously because of the change in angle to the horizontal; the highest load factor occurs at the bottom of the loop where the inertia force and weight act in the same direction, whereas the lowest load factor occurs at the top of the loop. Where the pitch rate is negative (nose down), the manoeuvre is push-down (sometimes known as a 'bunt') and can give rise to a zero load factor, or the 'weightless' (zero g) flight condition. Consider the push-down manoeuvre shown in Figure 12.5 ($n < 1$); here both the lift and inertia force act upwards. It may be shown that the load factor is zero (i.e. the weightless condition) when $V = \sqrt{rg}$. In the $n = -1$ manoeuvre, seen later to be a possible extreme push-down manoeuvre, the lift actually acts downwards i.e. it flies with a negative incidence.

12.1.5   Example: Steady Pull-up

An aircraft is pulling up from a dive at velocity 200 m/s TAS at 10000 ft altitude, following a flight path with a radius of 2000 m. Determine the load factor and pitch rate at the bottom of the pull-up. Note that altitude is irrelevant when TAS is used but would be if EAS had been specified because it is the absolute air speed that is important.

$$\text{Load factor } n = 1 + \frac{V^2}{gr} = 1 + \frac{200^2}{9.81 \times 2000} = 3.04$$

$$\text{Pitch rate } q = \frac{V}{r} = \frac{200}{2000} = 0.1\,\text{rad/s} = 5.73\,\text{deg/s}$$

Figure 12.6   Rigid aircraft in steady banked turn.

12.1.6   Steady Turn

A symmetric steady turn is an equilibrium manoeuvre and is characterized by the aircraft flying at a constant true air speed V around a circular path of radius r in the horizontal plane. The aircraft is oriented at a bank angle Φ as shown in Figure 12.6, such that there is no sideslip (i.e. no net lateral force or acceleration), no roll rate and therefore the lift on each wing is the same; this is known as a banked turn. Again there is no velocity normal to the flight path.

Recognizing that the aircraft experiences a centripetal acceleration V^2/r and employing d'Alembert's principle, yields the FBD shown in Figure 12.6. The equilibrium equations in the vertical and horizontal directions are

$$L\cos\Phi = W \quad L\sin\Phi - m\frac{V^2}{r} = 0 \tag{12.7}$$

Solving these equations yields the expressions for the bank angle and load factor, namely

$$\Phi = \tan^{-1}\left(\frac{V^2}{gr}\right) \quad \text{and} \quad n = \frac{1}{\cos\Phi} \quad (n \geq 1) \tag{12.8}$$

The rate of turn is given by $\omega_{Turn} = {}^{V}/_{r}$ and so the aircraft experiences steady yaw and pitch rates

$$n_{Yaw} = \omega_{Turn}\,\cos\Phi = \frac{V}{rn} \quad q = \omega_{Turn}\,\sin\Phi = \frac{V}{r}\frac{\sqrt{n^2-1}}{n} \tag{12.9}$$

where n_{Yaw} is the yaw rate (n is the standard flight mechanics symbol for yaw rate but the subscript is added here to avoid the confusion with the load factor within the same equation). Thus the pitch rate is related non-linearly to the load factor and linearly to air speed. Because the lift increases in a steady turn, then so will the drag and therefore the thrust must be increased to maintain the same air speed in the turn.

12.1.7   Example: Steady Banked Turn

An aircraft is in a level steady banked turn at velocity 100 m/s EAS at 10 000 ft (3148 m) altitude ($\sigma = 0.738$), following a flight path with a radius of 1000 m. Determine the bank angle and load factor. Note that altitude is important when EAS is used because load factors depend upon the true air speed $V_{TAS} = V_{EAS}/\sqrt{\sigma} = 116.4\,\text{m/s}$.

$$\text{Bank angle} \quad \Phi = \tan^{-1}\left(\frac{V^2}{gr}\right) = \tan^{-1}\left(\frac{116.4^2}{9.81 \times 1000}\right) = 54.1^{\circ}$$

$$\text{Load factor} \quad n = \frac{1}{\cos \Phi} = 1.70$$

12.2 Manoeuvre Envelope

Clearly, there are an infinite number of combinations of manoeuvre, load factor and true air speed that could be considered for loads clearance purposes. However, in order to simplify the position, a *manoeuvre (or manoeuvring) envelope* can be defined to show boundaries of n and V_{EAS} within which the aircraft must withstand the loads for symmetric equilibrium manoeuvres. In many cases (e.g. pull-up, turn), a steady pitch rate is involved.

The envelope is defined by the design cruise speed V_C, design dive speed V_D (to account for a prescribed 'upset' manoeuvre), the positive and negative stall curves (given by the maximum normal force coefficient $\pm C_{N_{Max}}$), the maximum manoeuvring limit load factor n_1 (typically between 2.5 and 3.8 for a commercial aircraft, depending upon the maximum take-off weight) and the minimum manoeuvring load factor $n_3 = -1$. The stall curves define how the load factor developed is limited by stall so, for example, on the positive stall boundary

$$L = nW = \frac{1}{2}\rho_0 V_{EAS}^2 S_W C_{N_{Max}} \tag{12.10}$$

where ρ_0 is the sea level air density and S_W is the wing area. It may be seen that the equivalent air speed (EAS) is used to provide a unified envelope and so avoid needing to construct envelopes for a series of different altitudes. The normal force coefficient $C_{N_{Max}}$ is normally taken to equal the lift coefficient $C_{L_{Max}}$. The relationship between n and V_{EAS} along the positive stall curve is then given by the quadratic expression

$$n = \frac{\rho_0 S_W C_{L_{Max}}}{2W} V_{EAS}^2 \tag{12.11}$$

The manoeuvre envelope shown in Figure 12.7 is bounded by the stall curves, the design dive speed and the maximum and minimum load factors, noting that the most severe negative

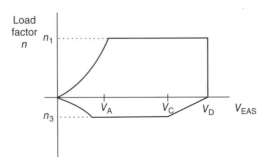

Figure 12.7 Manoeuvre envelope.

load factor is not achievable at V_D because it would imply that the aircraft is attempting to go even faster than it was designed for. Boundaries for both flaps up and flaps down cases would be considered. The design manoeuvring speed V_A is defined by the intersection of the positive stall curve and the maximum load factor.

The aircraft must be designed to withstand loads arising from symmetric (pitching) manoeuvres defined by values of n and V_{EAS}, both on and within the envelope for a range of weights, centre of mass positions, altitudes etc. Usually the corner points are the most critical, but checks along the boundaries would also be needed to ensure that the maximum loads in each part of the aircraft are found (CS-25). Note that the load calculations need to be carried out for the flexible aircraft and with any non-linearity due to control system constraints and aerodynamics represented.

12.3   Equilibrium Manoeuvre – Rigid Aircraft Pitching

In Section 12.1, the balance of lift, drag, weight, inertia and thrust forces was considered for the aircraft represented as a 'particle' in heave, so no pitch rate effects were included. In this section, the aircraft will be treated as a 'body' (see Chapter 5) and pitch rate effects will be included to show how the tailplane load balances the aircraft in pitch. The elevator angle and incidence required to trim the aircraft in the manoeuvre may be determined. At this stage, the aircraft is considered as rigid and the analysis approach applies for both unswept and swept wings.

12.3.1   Inertial Axes System

Before carrying out an analysis on the rigid aircraft undergoing an equilibrium manoeuvre in heave and pitch, it is worth making some comments about the axes system to be employed. For equilibrium manoeuvres, flutter, ground manoeuvres and gust/turbulence encounters, the aircraft behaviour may be referenced to a datum position defined for inertial axes fixed in space because the aircraft excursion from its datum position is considered to be small; the unknowns will be displacements and rotations. In Chapter 13, on the other hand, axes will be considered to be fixed in the aircraft, and move with it, because for a dynamic manoeuvre the excursion from the initial datum position may be significant and a non-linear flight mechanics model may be used; the unknowns will then be velocities. When aerodynamic derivatives are considered to represent the aerodynamic forces and moments in a compact notation, it is important to recognize which axes system the derivatives are evaluated for.

12.3.2   Determination of External Forces to Balance the Aircraft

In the analysis of the aircraft behaviour in heave and pitch, the tailplane now needs to be included and it is assumed to be symmetric. For completeness the thrust and drag lines of action are allowed to be different (though it is assumed for simplicity that they are parallel). The FBD for the aircraft flying horizontally ($\Theta = 0$), but under an accelerated flight condition defined by the vertical load factor n, is shown in Figure 12.8. There is no horizontal load factor as the air speed is assumed to be constant during the manoeuvre. nW is the combined weight/inertia force, T is the thrust, L_W is the wing lift, L_T is the tailplane lift, D is the drag and M_{0W} is the wing zero lift pitching moment (nose up positive). Since a pitch rate is permitted but no pitch acceleration, then there will be no inertia moment.

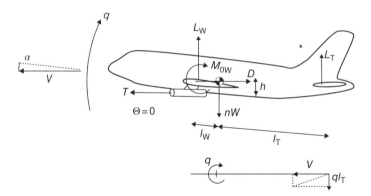

Figure 12.8 Rigid aircraft in an equilibrium manoeuvre, with heave and pitch effects included.

The definitions of pitching moment, lift and drag will involve air speed and density (see Chapter 4); these will be taken as true air speed V and corresponding density ρ (at whatever altitude the aircraft is flying) for most of the remainder of the book, unless indicated otherwise. Thus the wing zero lift pitching moment is given by the expression

$$M_{0W} = \frac{1}{2}\rho V^2 S_W \bar{c} C_{M_{0W}} \tag{12.12}$$

where $\bar{c}$ is the wing mean chord (though in the simple examples considered in this book, wings are usually untapered so the 'bar' will be dropped from now on) and $C_{M_{0W}}$ is the zero lift pitching moment coefficient for the wing (usually negative because of camber). This constant pitching moment coefficient arises when the lift is considered to act at the aerodynamic centre (see Chapter 4). The aircraft dimensions are shown in Figure 12.8. In order to simplify the equations, consider the drag force to act through the aircraft centre of mass.

Taking a view from the side of the aircraft, this FBD is seen to cover all the manoeuvres considered in Section 12.1 for $\Theta = 0$. To cater for inclined flight, the weight component along the flight path also needs to be included. The whole process of introducing inertia forces via d'Alembert's principle is highlighted in the Certification Specifications (CS-25) by the statement 'the linear inertia forces must be considered in equilibrium with the thrust and all aerodynamic loads'.

As the aircraft shown must be in equilibrium, there must be no net vertical or horizontal forces and no net pitching moment (about the centre of mass, say). Thus the equations defining equilibrium, assuming that the incidence angle is small (so $\cos \alpha = 1$) and that the lift and drag forces act essentially normal to and along the aircraft axis, are

$$L_W + L_T - nW = 0 \quad T - D = 0 \quad M_{0W} + l_W L_W - l_T L_T + hT = 0 \tag{12.13}$$

For a given flight condition, nW and M_{0W} are known and L_W, L_T, D and T are unknown. The tailplane is normally designed such that the tailplane lift (might be up or down) required for trim is relatively small in cruise and that maximum up and down loads are approximately equal; since the tailplane has been assumed to have a symmetric section, its zero lift pitching moment M_{0T} is zero. Thus it may be seen that the overall loads on the aircraft may be determined for a given accelerated flight condition by solving Equations (12.13). There are two different conditions to consider for the solution (see below). A more complete analysis is shown in Lomax (1996) and ESDU Data Sheet 94009.

12.3.3   Thrust and Drag In-line

When the thrust and drag are *in-line*, then $h = 0$ so the drag is not involved in the moment equation for vertical loads and the tailplane and wing lift forces may be solved directly as

$$L_T = \frac{M_{0W} + nWl_W}{l_W + l_T} \qquad L_W = nW - L_T \qquad (12.14)$$

These forces express the 'balance' or trim of the aircraft in heave and pitch. It may be seen that the required tailplane lift varies linearly with the load factor. Note that because the zero lift pitching moment for the wing is usually negative, the tailplane lift can act downwards or upwards, depending upon the load factor.

12.3.4   Example: Thrust and Drag In-line

An aircraft flying at 175 m/s (sea level) has the following characteristics: $m = 20\,000$ kg, $S_W = 80$ m^2, $c = 3$ m, $l_W = 1$ m, $l_T = 16$ m and $C_{M0W} = -0.06$. Obtain the tailplane lift for load factors of 1 and 3.

$$M_{0W} = \frac{1}{2}\rho V^2 S_W \bar{c} C_{M0W} = \frac{1}{2} \times 1.225 \times 175^2 \times 80 \times 3 \times -0.06 = -270000\,\mathrm{N\,m}$$

$$W = mg = 20000 \times 9.81 = 196000\,\mathrm{N}$$

Calculating the tailplane lift using Equation (12.14) yields values of –4.3 kN (download) and +18.7 kN (upload) respectively for the two load factors. Corresponding wing lift values are 191.7 kN and 569.3 kN.

12.3.5   Thrust and Drag Out-of-line

When the thrust and drag are *out-of-line* (but parallel), then $h \neq 0$ and so the lift forces will depend upon the thrust. However, the thrust is equal to the drag which in turn depends upon the wing lift (see Chapter 4). Thus an iterative approach must be adopted. Initially, the tailplane lift is set to zero so the wing lift is equal to nW; thus the wing lift coefficient, drag coefficient, drag and hence thrust may be estimated. Then a new estimate for the tailplane lift force may be obtained using Equations (12.14). The wing lift, drag, thrust and tailplane lift are then re-estimated and this process repeated until the solution converges, usually quite rapidly.

12.3.6   Example: Thrust and Drag Out-of-line

An aircraft flying at 250 m/s EAS has the following characteristics: $m = 44\,200$ kg, $S_W = 145$ m^2, $c = 5$ m, $l_W = 0.3$ m, $l_T = 14.9$ m, $C_{M0W} = -0.07$ and drag coefficient $C_D = 0.02 + 0.072C_{L_W}^2$, where C_{L_W} is the wing lift coefficient. The drag may be assumed to act through the centre of mass and the thrust acts a distance $h = 1.5$ m below the drag. Obtain the tailplane lift for a load factor of 2.5:

$$M_{0W} = \frac{1}{2} \times 1.225 \times 250^2 \times 145 \times 5 \times -0.07 = -1943 \, kNm$$

$$W = mg = 44\,200 \times 981 = 433.6 \, kN$$

Iteration 1: $L_W = nW = 2.5 \times 433.6 \, kN = 1084 \, kN$ $C_{Lw} = L_W / \frac{1}{2} \rho V^2 S_W = 0.1953$

$C_D = 0.02 + 0.072 \times 0.1953^2 = 0.0227$ $T = D = \frac{1}{2} \rho V_0^2 S_W C_D = 126.3 \, kN$

$L_T = (M_{0W} + l_W L_W + hT)/l_T = -95.8 \, kN$

Iteration 2: $L_W = nW - L_T = 1084 - (-958) = 1179.8 \, kN$ $C_{Lw} = 0.2126$

$C_D = 0.0233 \quad T = D = 129.1 \, kN$ $L_T = -93.6 \, kN$

The process may then be continued until the required level of accuracy is obtained; after three iterations the tailplane lift is $L_T = -93.7$ kN.

12.3.7 Determination of Balanced Condition – Thrust/Drag In-line

Having obtained the wing and tailplane lift forces in a steady manoeuvre with a load factor n, the next step is to find the corresponding incidence and elevator angles required to generate the trim lift forces; the analysis is shown for an elevator, but a similar approach may be used for an all-moving tailplane. It will be assumed that the aircraft is pitching steadily (nose up) with a rate q; this is true for the pull-up and turn cases but not the descent or climb cases where the pitch rate is zero. The pitch rate effect on wing lift will be ignored as being an unsteady term (see Chapter 9), but the effect of pitch rate on the tailplane lift has a significant steady component and is included.

The elevator angle required for an equilibrium manoeuvre (or elevator angle to trim) may be found by expressing the wing and tailplane lift forces in Equations (12.13) in terms of incidence, elevator angle, pitch rate and downwash. The wing lift is given by

$$L_W = -Z_W = \frac{1}{2} \rho V^2 S_W a_W (\alpha - \alpha_0) \qquad (12.15)$$

where α is the aircraft incidence (assumed small) and α_0 is the incidence for zero wing lift (see Chapter 4); the setting angles defining the inclination of the wing and tailplane aerofoil sections to the fuselage are neglected for simplicity. Z_W is the downwards force on the wing (in the opposite direction to lift), used because the aircraft z axis points downwards.

To write an expression for tailplane lift in terms of incidence and elevator angle, it should be recognized that there is an effective steady incidence of ql_T/V due to the pitch rate effect (i.e. the tail moving forwards at velocity V and downwards with velocity ql_T), as shown in Figure 12.8. A further effect that needs to be taken into account for the tailplane is the mean downwash angle ε, i.e. the reduction in the effective incidence at the tailplane due to the downward flow associated with the wing trailing vortices (see Chapter 4). The tailplane incidence is then

$$\alpha_T = \alpha + \frac{q l_T}{V} - \varepsilon \tag{12.16}$$

and remembering that the tailplane has a symmetric section, the tailplane lift is

$$L_T = -Z_T = \frac{1}{2}\rho V^2 S_T \left[a_T \left(\alpha + \frac{q l_T}{V} - \varepsilon \right) + a_E \eta \right] \tag{12.17}$$

Here S_T is the tailplane area, a_T is the tailplane lift curve slope (defined with respect to incidence), η is the elevator angle and a_E is the tailplane lift curve slope defined with respect to the elevator angle. The mean downwash angle ε may be assumed to be proportional to the effective wing incidence (and therefore wing lift), namely

$$\varepsilon = \frac{d\varepsilon}{d\alpha}(\alpha - \alpha_0) = k_\varepsilon(\alpha - \alpha_0) = \varepsilon_0 + k_\varepsilon \alpha \tag{12.18}$$

where $k_\varepsilon = d\varepsilon/d\alpha$ is typically of the order of 0.35–0.4 and $\varepsilon_0 \,(= -k_\varepsilon \alpha_0)$ is the downwash at zero incidence.

The total aerodynamic lift force required in Equations (12.13) is then given by

$$L_W + L_T (= nW) = -Z = \frac{1}{2}\rho V^2 \left(S_W a_W (\alpha - \alpha_0) + S_T \left\{ a_T \left[(1 - k_\varepsilon)\alpha + k_\varepsilon \alpha_0 + \frac{q l_T}{V} \right] + a_E \eta \right\} \right) \tag{12.19}$$

where it may be seen that the angle of zero lift for the wing influences the tailplane lift via the downwash. It is convenient to write Equation (12.19) in terms of so-called *aerodynamic derivatives*. For example, the total normal force (in the opposite direction to lift) for the whole aircraft may be expressed in derivative form as

$$Z = -L = Z_0 + Z_\alpha \alpha + Z_q q + Z_\eta \eta \tag{12.20}$$

where Z_0 is the constant force term due to α_0 and, for example, $Z_\alpha = \partial Z/\partial \alpha$. For equilibrium manoeuvres, these derivatives are effectively defined with respect to the inertial axes system; other rate dependent derivatives will be introduced in Chapter 14 for gust response analysis. The derivatives obtained with respect to inertial axes for symmetric manoeuvres are tabulated in Appendix B. Also, note that these derivatives for inertial axes are different to the stability derivatives for wind axes in Chapter 13 and also to the unsteady aerodynamic derivatives introduced in Chapter 9.

The normal force derivatives in Equation (12.20) are given from inspection of Equations (12.19) and (12.20) as

$$Z_0 = -\frac{1}{2}\rho V^2 [-S_W a_W + S_T a_T k_\varepsilon]\alpha_0 \quad Z_q = -\frac{1}{2}\rho V S_T a_T l_T$$

$$Z_\alpha = -\frac{1}{2}\rho V^2 [S_W a_W + S_T a_T (1 - k_\varepsilon)] \quad Z_\eta = -\frac{1}{2}\rho V^2 S_T a_E \tag{12.21}$$

Similarly, the total aerodynamic pitching moment (nose up positive) about the aircraft centre of mass required in Equations (12.13) is given in aerodynamic moment derivative form as

$$M_{CoM}(=0) = M_{0W} + l_W L_W - l_T L_T = M_0 + M_\alpha \alpha + M_q q + M_\eta \eta \quad (12.22)$$

where the moment derivative values may be seen by inspection of the lift expressions as

$$M_0 = M_{0W} - \frac{1}{2}\rho V^2 [S_W a_W l_W + S_T a_T k_\varepsilon l_T]\alpha_0 \quad M_q = -\frac{1}{2}\rho V S_T a_T l_T^2$$

$$M_\alpha = -\frac{1}{2}\rho V^2 [S_W a_W l_W + S_T a_T (1 - k_\varepsilon) l_T] \quad M_\eta = -\frac{1}{2}\rho V^2 S_T a_E l_T$$

$$(12.23)$$

It should be noted that the constant derivative term M_0 is for the whole aircraft and is a combination of the wing zero lift pitching moment M_{0W} and the moment terms associated with the wing zero lift angle α_0.

When the equilibrium heave and pitch Equations (12.19) and (12.22) are combined and expressed in matrix form (assuming that thrust and drag are in line and may be omitted from the pitch equation) then it may be shown that

$$-\begin{bmatrix} Z_\eta & Z_\alpha \\ M_\eta & M_\alpha \end{bmatrix}\begin{Bmatrix} \eta \\ \alpha \end{Bmatrix} = \begin{Bmatrix} 1 \\ 0 \end{Bmatrix}nW + \begin{Bmatrix} Z_q \\ M_q \end{Bmatrix}q + \begin{Bmatrix} Z_0 \\ M_0 \end{Bmatrix} \quad (12.24)$$

The ordering of the unknowns in the left-hand side vector is chosen for convenience to compare with the result for the flexible aircraft determined later in this chapter. The three input terms on the right-hand side are respectively the *inertia input* and *pitch rate input* (both related to the load factor) and the *aerodynamic input at zero incidence* related to the presence of wing camber and the angle of zero lift. Solving these simultaneous equations for the unknown left hand side vector leads to the elevator angle η required to trim the aircraft in the manoeuvre and the corresponding trim incidence α. The analysis may be extended to include the balance of fore-and-aft motion via the thrust and drag. In practice, the aerodynamic forces on the fuselage and nacelles would also be included.

If the aircraft is flying near to stall where the lift varies non-linearly with incidence, then the derivative Z_α will be a non-linear function and so the solution of Equation (12.24) will be non-linear (typically found using the Newton–Raphson method). The non-linear trim condition may then be found. Also, the equations may be modified for large angles of incidence. The same issue applies if a non-linear Flight Control System (FCS) is present.

12.3.8 Determination of Balanced Condition – Thrust/Drag Out-of-line

In the above analysis, the solution was kept relatively simple by assuming that the thrust and drag forces were in line. In the case where the thrust and drag were out of line, it was shown earlier that this led to an iterative solution. The solution may perhaps be seen more clearly by adding the $T - D = 0$ equation to the above analysis and including thrust in the moment equation, with X derivatives used in place of drag (with X defined as positive forward). Thus, Equation (12.24) becomes

$$
-\begin{bmatrix} Z_\eta & Z_\alpha & Z_T \\ M_\eta & M_\alpha & M_T \\ X_\eta & X_\alpha & 1 \end{bmatrix} \begin{Bmatrix} \eta \\ \alpha \\ T \end{Bmatrix} = \begin{Bmatrix} 1 \\ 0 \\ 0 \end{Bmatrix} nW + \begin{Bmatrix} Z_q \\ M_q \\ X_q \end{Bmatrix} q + \begin{Bmatrix} Z_0 \\ M_0 \\ X_0 \end{Bmatrix} \tag{12.25}
$$

where $M_T = h$ and $Z_T = 0$ (assuming thrust is perpendicular to lift and parallel to drag). The drag term due to incidence will be $-X_\alpha \alpha$ and since the induced drag (see Chapter 4) is proportional to the square of the lift (and thus the incidence) then the derivative X_α will be a linear function of incidence. Thus the solution of Equation (12.25) will be non-linear, yielding the incidence, elevator angle and thrust required for the balanced condition involving the appropriate load factor. Such a trimmed or balanced condition (usually at $1\,g$, $n = 1$) will be the starting point for a dynamic manoeuvre or landing using the flight mechanics model (see Chapters 13 and 15).

12.3.9 Aerodynamic Derivatives

Before moving on, it should be pointed out that in Chapter 13, other derivatives (actually very similar to the inertially defined derivatives) will be defined (see Appendix B) with respect to wind (i.e. body fixed) axes for stability and control purposes (Bryan, 1911; Babister, 1980; Cook, 1997; ESDU Data Sheet). In Cook (1997), these *dimensional* derivatives are indicated by using the 'over-o' symbol, e.g. Z_q^o, and the *non-dimensional* derivatives by Z_q. However, since non-dimensional derivatives are not used at all in this book, this distinction will not be used and Z_q will indicate a dimensional derivative throughout (see also Chapter 13). The context (i.e. which equations and chapter they are included in) will define which axes the derivatives apply to, i.e. inertial or body fixed.

12.3.10 Static Stability (Stick Fixed)

One important issue related to the equilibrium manoeuvre is the static stability of the aircraft when encountering a disturbance, i.e. the aircraft should return to its balanced state without pilot intervention (stick fixed). The condition for static stability is that $\partial C_M / \partial \alpha < 0$, where C_M is the overall nose up pitching moment coefficient for the aircraft related to the centre of mass position. This condition may be evaluated and rearranged to give a criterion for the location of the centre of mass for stability, namely

$$
l_W < l_T \frac{a_T}{a_W} \frac{S_T}{S_W} (1 - k_\varepsilon) \tag{12.26}
$$

Thus the centre of mass must be ahead of the so-called neutral point where the aircraft is neutrally stable (i.e. where a change in incidence does not introduce a pitching moment) or else, if it is behind it, the aircraft will be unstable. The static margin is the distance that the centre of mass is ahead of the neutral point.

12.3.11 Example: Steady Equilibrium Manoeuvre – Rigid Aircraft Pitching

An example using the simple tailplane and wing lift expressions for an equilibrium manoeuvre was given earlier. In this section, the calculation of trim incidence and elevator angle for manoeuvres with different load factors is considered using the simultaneous Equations (12.24)

(a)

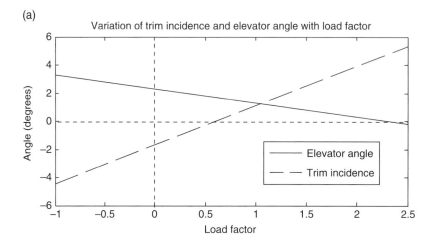

(b)

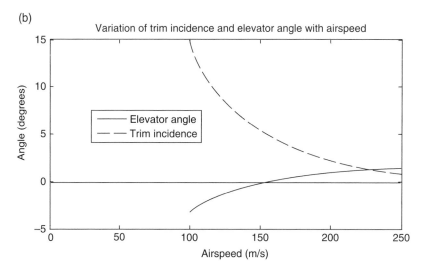

Figure 12.9 Results for the example of a rigid aircraft in an equilibrium manoeuvre – trim incidence and elevator angle as a function of (a) load factor (150 m/s EAS) and (b) air speed ($n = 2.5$).

developed for the thrust and drag in-line case. The analysis and example will be repeated later for a simple flexible aircraft.

Consider a rigid aircraft with the following data: $m = 10\,000$ kg, $S_W = 30$ m^2, $S_T = 7.5$ m^2, $c = 2.0$ m, $l_W = 0.6$ m, $l_T = 7$ m, $a_W = 4.5$ per rad, $a_T = 3.2$ per rad, $a_E = 1.5$ per rad, $k_\varepsilon = 0.35$, $\alpha_0 = -0.03$ rad and $C_{M0W} = -0.03$. Further parameters will be defined later for the flexible aircraft. It should be noted that the data used corresponds to a static margin of 0.21 m (10% c) and a flaps-up stall speed of 60 m/s EAS ($C_{L_{Max}} = 1.5$).

The aim of the example is to determine the trim incidence and elevator angle as (a) a function of load factor between -1.0 and 2.5 at an air speed of 150 m/s EAS and also as (b) a function of air speed in the range 120 to 200 m/s EAS for a load factor of 2.5. A MATLAB program set up to solve the two simultaneous equations is included in the companion website. The analysis used in these examples is based upon the solution of Equation (12.24).

Results are shown in Figures 12.9(a) and (b). It may be noted that: (a) the elevator angle to trim is proportional to the load factor, (b) the trim incidence increases with the load factor since a larger wing lift needs to be developed, (c) the elevator angle to trim may be positive or negative depending upon the air speed and (d) the trim incidence increases at lower air speeds to maintain lift (heading towards stall). The elevator angle acts together with the trim incidence to produce the required tailplane lift force for trim; the trim elevator angle will also depend upon the centre of mass position and whether flaps are deployed (since $C_{M_{0W}}$ is more negative).

12.4   Equilibrium Manoeuvre – Flexible Aircraft Pitching

So far, the equilibrium manoeuvres have only been considered for the rigid aircraft, either in heave alone or in combined heave and pitch; also, in the analysis so far, the wing could be unswept/swept and untapered/tapered. However, in practice, flexible effects can make a significant difference to the balanced/trimmed condition and also to the magnitude and distribution of loads developed on the aircraft in any manoeuvre.

For an unswept wing, if it is assumed that the wing and tailplane aerodynamic centre positions and zero lift pitching moment coefficient are unaffected by flexible effects, then the wing and tailplane lift forces required for balance will not change due to flexible effects; what will change is the trim incidence and elevator angle needed to generate these forces, and also the distribution of the lift loading along the wing. The situation for a swept wing is different and will be discussed later without formal analysis.

In this section, the analysis of a simple aircraft with a rigid tailplane and a flexible fuselage or wing will be carried out. To make the treatment more manageable, it is assumed that the wing is unswept and untapered. A single whole aircraft free–free symmetric flexible mode involving fuselage bending, wing bending and wing twist will be included in the two DoF heave/pitch type analysis covered in the previous section. Solving this three DoF problem, and including the elevator, allows the equilibrium manoeuvre for a simple flexible aircraft to be considered when it experiences a load factor n and a steady pitch rate q; the analysis will lead to the balanced/trimmed deformation of the flexible aircraft and revised values for the incidence and elevator angle to trim.

Cases with different mode shapes may be considered after the model has been developed e.g. fuselage bending may be chosen to be dominant (see Appendix C). It should be noted that the same flexible model will be used in Chapters 13 to 16 and that much of the analysis undertaken in this chapter will be applicable later on.

12.4.1   Definition of the Flexible Aircraft with Unswept Wings

The simplified flexible aircraft consists of uniform, untapered, unswept but flexible wings of chord c and semi-span s, plus a flexible fuselage and a rigid tailplane, as shown in Figure 12.10. The wings are assumed to have a uniform mass distribution with mass μ_W and pitch moment of inertia χ_W per unit span. The wing mass axis (WM) lies a distance l_{WM} ahead of the aircraft centre of mass (CM) and is typically around the mid-chord. The mass and pitch moment of inertia of the aircraft fuselage will be represented by discretizing it into three 'lumps' of mass m_F, m_C and m_T; these discrete masses are positioned at the front fuselage (a distance l_F forward

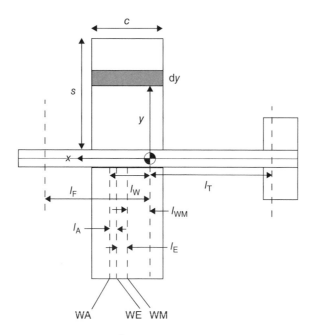

Figure 12.10 Flexible aircraft in an equilibrium manoeuvre – heave and pitch effects included.

of the centre of mass), at the whole aircraft CM, and at the tailplane aerodynamic centre (a distance l_T aft of the centre of mass), respectively.

The wing elastic axis WE (typically around one-third chord) is assumed to lie a distance l_E ahead of the wing mass axis (WM) so that wing bending/torsion coupling can occur. The wing aerodynamic centre axis (WA) (typically at the wing quarter chord) is at a distance l_W ahead of the centre of mass and a distance l_A ahead of the elastic axis. Thus the relationship between the wing chord dimensions is $l_W = l_A + l_E + l_{WM}$.

If the elastic stiffness distributions along the wing and fuselage were to be specified, then it would be possible to calculate the aircraft modal characteristics using the Rayleigh–Ritz or Finite Element approaches (see Chapters 3 and 17); different stiffness values would lead to different mode shapes and natural frequencies for the flexible mode. However, to allow control and modification of the mode shapes and natural frequencies, and to allow an analytical approach to be used, the philosophy adopted will be as follows (see Appendix C):

- define the rigid body heave and pitch modes;
- define the type of flexible mode required (e.g. wing bending, fuselage bending);
- obtain mode shape parameters and modal mass by constraining the chosen flexible mode to be orthogonal to the rigid body heave and pitch modes; and
- examine the effects of varying natural frequency by simply altering the modal stiffness while retaining the same modal mass and mode shape.

The analysis in this chapter will be carried out from first principles using Lagrange's equations for the flexible mode equation; both rigid and elastic aerodynamic derivatives will emerge. The same concepts are seen in later chapters.

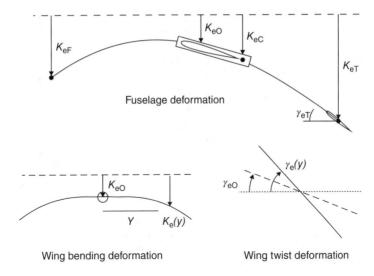

Figure 12.11 Free–free mode shape for flexible aircraft.

12.4.2 Definition of the Flexible Mode Shape

Selecting a suitable symmetric flexible mode (subscript e) that allows the illustration of key features in an equilibrium manoeuvre, while keeping the analysis as simple as possible, is not straightforward. In practice, many modes are required to represent the true behaviour of the flexible aircraft, but in the interest of keeping the mathematical model to only 3DoF, a single flexible mode, having the possibility of including wing twist, wing bending or fuselage bending effects by simply changing a set of mode shape parameters, will be employed. In Chapter 3, the analysis of a free–free flexible aircraft was considered, with rigid body heave and pitch modes combined with a flexible mode; it may be helpful for the reader to refer back to that chapter if not familiar with this kind of analysis.

The symmetric free–free flexible mode shape to be used, illustrated in Figure 12.11 for positive values of all parameters, has the following features:

- The wing twists (nose up) and bends (downwards) according to the assumed shapes, $\gamma_e(y)$ and $\kappa_e(y)$ respectively, both defined relative to the wing elastic axis.
- The fuselage centre section, where the wing is attached and incorporating the aircraft centre of mass location, is treated as rigid and so experiences a slope (nose up) of $\gamma_e(0) = \gamma_{e0}$ and a (downwards) displacement of $\kappa_e(0) = \kappa_{e0}$, both defined at the wing elastic axis location.
- The fuselage displacements (downwards) at the front fuselage, wing elastic and mass axes, aircraft centre of mass and tailplane positions are $\kappa_{eF}, \kappa_{e0}, \kappa_{eW}, \kappa_{eC}, \kappa_{eT}$ respectively. The fuselage bends such that the (nose up) pitch in the mode shape at the rigid tailplane is γ_{eT}.

The generalized coordinate describing the absolute magnitude of the modal deformation is q_e (not to be confused with symbol q used elsewhere for pitch rate or dynamic pressure). Thus, for example, the absolute displacement at the tailplane in the flexible mode is $\kappa_{eT}q_e$. The parameters defining the mode shape are considered in more detail in Appendix C, which also defines the modal parameter values to be used later. Flexible modes will be defined that involve wing bending, wing twist or fuselage bending as being dominant.

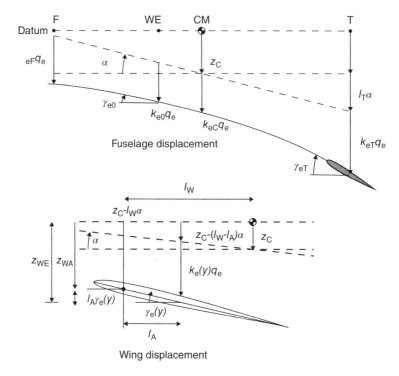

Figure 12.12 Aircraft displacements and rotations.

12.4.3 Expressions for Displacement and Angles over the Aircraft

Now that the flexible mode shape has been defined in general terms, expressions for the variation of (downwards) displacement and (nose up) twist along the wing (and rotation at the tailplane) may be written for the combination of rigid body and flexible motions shown in Figure 12.12. The rigid body motions are defined by a (downward) displacement z_C at the centre of mass and angle α (nose up) whereas the flexible mode is defined by the generalized coordinate q_e.

The displacements (downwards positive) at the *front fuselage* (subscript F), *aircraft centre of mass* (CM) and *tailplane* (T), are

$$z_F = z_C - l_F\alpha + \kappa_{eF}q_e \quad z_{CM} = z_C + \kappa_{ec}q_e \quad z_T = z_C + l_T\alpha + \kappa_{eT}q_e \quad (12.27)$$

where the displacements are a summation of the heave, pitch and flexible mode effects. The symbol z is commonly used for downwards displacement in aeroelasticity and gusts, and its use here should not be confused with its other use on some occasions as a position coordinate (see also Chapter 3).

Along the axis of *wing aerodynamic centres* (subscript WA), the *wing elastic axis* (WE) and *wing mass axis* (WM) for the starboard wing, the displacement variations are

$$z_{WA}(y) = z_C - l_W\alpha + [\kappa_e(y) - l_A\gamma_e(y)]q_e$$
$$z_{WE}(y) = z_C - (l_W - l_A)\alpha + \kappa_e(y)q_e \quad (12.28)$$
$$z_{WM}(y) = z_C - l_{WM}\alpha + [\kappa_e(y) + l_E\gamma_e(y)]q_e$$

where the elastic deformation is a combination of wing bending and twist. The variation of the *effective wing incidence* along the starboard wing, including flexible effects, involves a combination of the trim incidence and the flexible mode twist and is given by

$$\alpha_W(y) = \alpha + \gamma_e(y)q_e \tag{12.29}$$

These expressions will be required when Lagrange's equations are used to develop the equations of motion.

12.4.4 Aerodynamic Terms

Firstly, the aerodynamic terms due to the wing and tailplane will be determined. The tailplane may be considered as a whole (since it is rigid) whereas evaluating the wing contribution will involve integration using a strip dy because of the flexible twist present (see Chapter 7). The *wing* incidence α_W on a strip dy at position y is given by Equation (12.29) so the steady lift on the strip is

$$dL_W = \frac{1}{2}\rho V^2 c\, dy a_W [\alpha - \alpha_0 + \gamma_e(y)q_e] \tag{12.30}$$

where a_W is the sectional wing lift curve slope and the additional flexible term may be seen when compared to the earlier analysis. The total wing lift is then given by the integration of this equation over the semi-span and doubling it for the two wings, namely

$$L_W = -Z_W = 2\frac{1}{2}\rho V^2 c \int_{y=0}^{s} a_W(\alpha - \alpha_0 + \gamma_e q_e)\, dy \tag{12.31}$$

where it may be noted that the dependency of γ_e, κ_e on y under the integrals will not always be explicitly stated from now on. It may be seen, as expected, that there is no steady aerodynamic contribution from the wing bending. There is also a zero lift pitching moment for the wing as seen before, namely

$$M_{0W} = \frac{1}{2}\rho V^2 S_W c C_{M0W} \tag{12.32}$$

The *tailplane* incidence is given by

$$\alpha_T = \alpha - \varepsilon + \frac{q l_T}{V} + \gamma_{eT} q_e = \alpha - k_\varepsilon(\alpha - \alpha_0) + \frac{q l_T}{V} + \gamma_{eT} q_e \tag{12.33}$$

using the expression for mean downwash in Equation (12.18). Note that a term is present to account for the fuselage slope γ_{eT} at the tailplane (nose up positive) in the flexible mode. The effect of the flexible deformation of the wing on downwash is neglected; more advanced aerodynamic methods would account for this effect. The tailplane lift is then given by

$$L_T = -Z_T = \frac{1}{2}\rho V^2 S_T \left\{ a_T \left[k_\varepsilon \alpha_0 + (1-k_\varepsilon)\alpha + \frac{q l_T}{V} + \gamma_{eT} q_e \right] + a_E \eta \right\} \tag{12.34}$$

where the effect of the elevator angle η is included to provide trim. Note that the effective incidence due to the nose up pitch rate q is included again.

12.4.5   Inertia Terms

Because the aircraft is in a steady accelerated manoeuvre with a load factor n, the inertia forces (subscript I) may be considered using d'Alembert's principle. The (downwards) inertia force on the strip of wing will be

$$dF_{WI} = n(\mu_W dy)g \qquad (12.35)$$

and the inertia forces for the front fuselage (F), centre fuselage (C) and tailplane (T) are

$$F_{FI} = nm_F g \qquad F_{CI} = nm_C g \qquad F_{TI} = nm_T g \qquad (12.36)$$

No inertia moments need to be included in this chapter as the aircraft is not accelerating in pitch. Note that there will be no kinetic energy term T because the manoeuvre is steady. When this model is used later for gusts and turbulence, for example, then kinetic energy terms will need to be included.

12.4.6   Stiffness Term

Only the flexible mode will introduce any structural stiffness effects. However, rather than evaluate the elastic potential energy by spatial integration over the wing and fuselage using assumed bending and torsional rigidity expressions, it is simpler to write the elastic potential energy using modal quantities, namely

$$U = \frac{1}{2}k_e q_e^2 \qquad (12.37)$$

where k_e is the modal stiffness value (see Chapter 3). Adjustment of this modal stiffness will allow the natural frequency to be controlled readily for a given modal mass.

12.4.7   Incremental Work Done Terms

Incremental displacements along the wing and at the centre of mass, tailplane and fuselage are required in order to determine the incremental work done, and therefore the generalized forces for aerodynamic and inertia contributions. As an example, the incremental displacement corresponding to the wing aerodynamic centre axis may be written from Equation (12.28) as

$$\delta z_{WA}(y) = \delta z_C - l_W \delta\alpha + (\kappa_e - l_A \gamma_e)\delta q_e \qquad (12.38)$$

The incremental work done for the wing total lift and zero lift pitching moment will then be given by

$$\delta W_{WA} = -2\int_{y=0}^{s} dL_W \delta z_{WA} + M_{0W}\delta\alpha$$

$$\delta W_{WA} = -2\int_{y=0}^{s} \frac{1}{2}\rho V^2 c a_W (\alpha - \alpha_0 + \gamma_e q_e)[\delta z_C - l_W \delta\alpha + (\kappa_e - l_A \gamma_e)\delta q_e]\,dy + M_{0W}\delta\alpha \tag{12.39}$$

where the minus sign in the work expression occurs because lift is positive upwards and displacement is positive downwards. The zero lift pitching moment does work through an incremental incidence angle $\delta\alpha$. Similarly, the incremental work done for the tailplane lift is

$$\delta W_{TA} = -L_T \delta z_T$$

$$= -\frac{1}{2}\rho V^2 S_T \left\{ a_T \left[k_\varepsilon \alpha_0 + (1 - k_\varepsilon)\alpha + \frac{q l_T}{V} + \gamma_{eT} q_e \right] + a_E \eta \right\} (\delta z_C + l_T \delta\alpha + \kappa_{eT}\delta q_e) \tag{12.40}$$

Finally, the net incremental work done by the wing and fuselage/tailplane inertia forces is

$$\delta W_I = F_{FI}\delta z_F + F_{CI}\delta z_{CM} + 2\int_0^s dF_{WI}\delta z_{WM} + F_{TI}\delta z_T$$

$$= nm_F g(\delta z_C - l_F \delta\alpha + \kappa_{eF}\delta q_e) + nm_C g(\delta z_C + \kappa_{ec}\delta q_e)$$

$$+ 2\int_0^s n\mu_W g\{\delta z_C - l_{WM}\delta\alpha + [\kappa_e(y) + l_E \gamma_e(y)]\delta q_e\}\,dy + nm_T g(\delta z_C + l_T \delta\alpha + \kappa_{eT}\delta q_e) \tag{12.41}$$

In this work expression for inertia forces, the overall term in $\delta\alpha$ will be zero (since masses of fuselage, wing and tailplane have to balance about the overall aircraft centre of mass). The term in δq_e is also zero, since there must be no net inertia force in a free–free mode. Therefore, this work done term can be simplified to

$$\delta W_I = nmg\delta z_C = nW\delta z_C \tag{12.42}$$

The right-hand side applied aerodynamic and inertia forces in both the rigid body and flexible modes can now be determined by employing the differentials of these incremental work terms in Lagrange's equations.

12.4.8   *Aerodynamic Derivatives – Rigid Body and Flexible*

As an example of how the aerodynamic contribution converts to derivative form when applying Lagrange's equation, considering the term in the heave equation for the wing aerodynamics only, then

$$\frac{\partial(\delta W_{WA})}{\partial(\delta z_C)} = -2\int_{y=0}^{s} \frac{1}{2}\rho V^2 c a_W (\alpha - \alpha_0 + \gamma_e(y)q_e)\,dy = Z_W = Z_{0w} + Z_{\alpha w}\alpha + Z_{ew}q_e \tag{12.43}$$

where each of the derivatives for the wing may be seen by inspection and it may be noted that there is no aerodynamic derivative term associated with the coordinate z_C. It may be seen that there is a flexible derivative Z_{eW} denoting the normal force on the wing per unit flexible mode deformation. When tailplane terms are added and the equivalent pitching moment expression is derived, it may be shown that the derivatives for the flexible mode (wing + tailplane) appearing in the force and moment equations are

$$Z_e = \frac{\partial Z}{\partial q_e} = \frac{1}{2}\rho V^2(-S_W a_W J_1 - S_T a_T \gamma_{eT})$$

$$M_e = \frac{\partial M}{\partial q_e} = \frac{1}{2}\rho V^2(S_W a_W l_W J_1 - S_T a_T l_T \gamma_{eT})$$

(12.44)

where the wing area is $S_W = 2cs$ and $J_1 = 1/s \int_{y=0}^{s} \gamma_e dy$ is a constant that depends upon the wing twist shape. The basic lift and moment derivatives associated with zero lift, incidence, pitch rate and elevator angle are the same as for the rigid aircraft in the previous section.

As an example of finding derivatives in the flexible mode equation, consider the flexible mode term for the wing aerodynamics

$$\frac{\partial(\delta W_{WA})}{\partial(\delta q_e)} = -2\int_{y=0}^{s} \frac{1}{2}\rho V^2 c a_W (\alpha - \alpha_0 + \gamma_e q_e)(\kappa_e - l_A \gamma_e) dy = Q_{0w} + Q_{\alpha w}\alpha + Q_{ew} q_e \quad (12.45)$$

This term yields flexible derivatives such as $Q_\alpha = \partial Q/\partial \alpha$ in the generalized equation for the flexible mode. When tailplane terms are added, the derivatives (wing + tailplane) in the flexible mode equation are given by

$$Q_0 = \frac{1}{2}\rho V^2(S_W a_W J_2 - S_T a_T k_\varepsilon \kappa_{eT})\alpha_0 \qquad Q_q = -\frac{1}{2}\rho V S_T a_T l_T \kappa_{eT}$$

$$Q_\alpha = \frac{1}{2}\rho V^2(-S_W a_W J_2 - S_T a_T (1-k_\varepsilon)\kappa_{eT}) \qquad Q_\eta = -\frac{1}{2}\rho V^2 S_T a_E \kappa_{eT} \qquad (12.46)$$

$$Q_e = \frac{1}{2}\rho V^2(-S_W a_W J_3 - S_T a_T \gamma_{eT}\kappa_{eT})$$

where $J_2 = 1/s \int_{y=0}^{s} (\kappa_e - l_A \gamma e) dy$ and $J_3 = 1/s \int_{y=0}^{s} (\kappa l_e - l_A \gamma_e)\gamma_e dy$ are further constants that depend upon the bending and twist shapes. The wing bending deformation dictated by $\kappa_e(y)$ only affects the Q_0, Q_α, Q_e derivatives where J_2 and J_3 appear and wing bending should only affect the deformed shape in trim but not the trimmed incidence and elevator angle derivatives. The derivatives are tabulated in Appendix B.

12.4.9 Equations of Motion for Flexible Aircraft Pitching

When Lagrange's equations are used, it may be shown after simplification that the equations may be written in derivative form (defined with respect to inertial axes) such that

$$
\left[\begin{bmatrix} 0 & Z_\alpha & Z_e \\ \hline 0 & M_\alpha & M_e \\ \hline 0 & Q_\alpha & Q_e \end{bmatrix} + \begin{bmatrix} 0 & 0 & 0 \\ \hline 0 & 0 & 0 \\ \hline 0 & 0 & k_e \end{bmatrix} \right] \begin{Bmatrix} z_C \\ \alpha \\ q_e \end{Bmatrix} = \begin{Bmatrix} Z_\eta \\ M_\eta \\ Q_\eta \end{Bmatrix} \eta + \begin{Bmatrix} 1 \\ 0 \\ 0 \end{Bmatrix} nW + \begin{Bmatrix} Z_q \\ M_q \\ Q_q \end{Bmatrix} q + \begin{Bmatrix} Z_0 \\ M_0 \\ Q_0 \end{Bmatrix} \qquad (12.47)
$$

where the rigid and flexible terms are partitioned. Additional flexible modes would simply increase the partition dimensions corresponding to the flexible modal coordinates. If Equation (12.47) is examined carefully, it can be seen that the square matrix on the left hand side is singular, so these equations cannot apparently be solved in this form. In fact, the rigid body heave displacement z_C would not even appear if the equations were written out in long-hand form, and so it is therefore not possible to identify its value. This is because any arbitrary value of z_C could satisfy the equation, since the model used does not have any means of defining a unique vertical position for the aircraft. However, on more careful observation the three equations may be seen to have three unknowns, namely α, q_e and η. Rewriting the equations in terms of these unknowns yields

$$
\left[\begin{bmatrix} Z_\eta & Z_\alpha & Z_e \\ \hline M_\eta & M_\alpha & M_e \\ \hline Q_\eta & Q_\alpha & Q_e \end{bmatrix} + \begin{bmatrix} 0 & 0 & 0 \\ \hline 0 & 0 & 0 \\ \hline 0 & 0 & k_e \end{bmatrix} \right] \begin{Bmatrix} \eta \\ \alpha \\ q_e \end{Bmatrix} = \begin{Bmatrix} 1 \\ 0 \\ 0 \end{Bmatrix} nW + \begin{Bmatrix} Z_q \\ M_q \\ Q_q \end{Bmatrix} q + \begin{Bmatrix} Z_0 \\ M_0 \\ Q_0 \end{Bmatrix} \qquad (12.48)
$$

The equations may now be solved for the elevator angle to trim, trim incidence and flexible deformation in the manoeuvre. This form of the equations also applies for models with more flexible modes. The effect of introducing aircraft flexibility may be seen by comparing this equation with that for the rigid aircraft in Equation (12.24); an additional flexible mode equation, coupled aerodynamically to the rigid body motions, is present. The effect of out-of-line thrust and drag may be included as before by adding a drag equation.

12.4.10 General Form of Equilibrium Manoeuvre Equations

The equations of motion (12.48) for the trimmed/balanced state may now be written in the more general form (c.f. Chapter 10 for flutter and divergence)

$$
\left[\rho V^2 \mathbf{C} + \mathbf{E} \right] \begin{Bmatrix} \eta \\ \alpha \\ q_e \end{Bmatrix} = \mathbf{F_I} nW + \rho V \mathbf{F_q} q + \rho V^2 \mathbf{F_0} \alpha_0 \qquad (12.49)
$$

where $\mathbf{C}$ is an aerodynamic stiffness matrix, $\mathbf{E}$ is the structural stiffness matrix and the right-hand side force vectors $\mathbf{F}$ correspond to the inertia, pitch rate and zero incidence effects. The dependency upon density and air speed has been highlighted. Clearly, by comparing Equations (12.48) and (12.49), these matrices and vectors may be related to the aerodynamic derivatives.

The determinant of the stiffness matrix $[\rho V^2 \mathbf{C} + \mathbf{E}]$ in Equation (12.48) will become zero at the whole flexible aircraft divergence speed for whatever flexible mode is chosen.

12.4.11 Values for the Flexible Mode Parameters

In Appendix C, three versions of the whole aircraft symmetric free–free flexible mode are considered, with either fuselage bending, wing bending or wing twist being the dominant feature of the mode shape; the fuselage bending and wing bending/twist contributions are controlled by the bending and torsion rigidities EI_{Fuselage}, EI_{Wing} and GJ_{Wing}, and an individual contribution (e.g. wing bending) becomes 'dominant' when the other two rigidities are allowed to tend to infinity. It is then simple to change an example from one mode type to another. Also, a mode may have all three components if desired, but that case is not considered. Example mode shapes and modal masses are shown in Appendix C for the set of aircraft parameters used in the example to follow.

12.4.12 Lift Distribution and Deformed Shape in the Manoeuvre

Having found the elevator angle, incidence and flexible mode generalized coordinate for trim in the balanced manoeuvre, the lift distribution along the wing and the deformed shape of the aircraft may be determined. The lift per unit span, or lift distribution, may be found by rewriting Equation (12.30) as

$$\frac{\mathrm{d}L_{\text{W}}}{\mathrm{d}y} = \frac{1}{2}\rho V^2 c a_{\text{W}} [\alpha - \alpha_0 + \gamma_{\text{e}}(y) q_{\text{e}}] \qquad (12.50)$$

Clearly, the lift distribution is only affected by the aircraft flexibility for a mode involving wing torsion; the effect will be to shift the lift outboard (see Chapter 7). Also, the deformation of the wing and fuselage may be found using the mode shape definition shown in Figure 12.11 and the value of the generalized coordinate q_{e}.

12.4.13 Example: Equilibrium Manoeuvre – Flexible Aircraft Pitching

This example corresponds to the rigid aircraft in Section 12.3.11 but additional parameters are specified to cater for the introduction of flexible effects, namely, fuselage mass terms $m_{\text{F}} = 1500$ kg, $m_{\text{C}} = 4000$ kg, $m_{\text{T}} = 1500$ kg, wing mass terms $m_{\text{W}} = 3000 \ (=2 \ \mu_{\text{W}} s)$ kg, wing inertia term $I_{\text{W}} = 1330 \ (=2 \chi_{\text{W}} s)$ kg m^2, aircraft pitch moment of inertia $I_y = 144 \ 000$ kg m^2 and dimensions $s = 7.5$ m, $l_{\text{A}} = 0.25$ m, $l_{\text{E}} = 0.25$ m, $l_{\text{WM}} = 0.1$ m and $l_{\text{F}} = 6.8$ m. The modal mass and mode shape parameters for dominant (a) fuselage bending, (b) wing bending and (c) wing twist modes are shown in Appendix C. Then, by varying the value of modal stiffness, the natural frequency of the flexible mode can be altered without changing anything else.

The aim of the example is to determine the effect of the flexible mode natural frequency for the three different mode types on the flexible deformation and on the trim incidence/elevator angle. Only a load factor of 1.0 and air speed of 150 m/s EAS are considered; the results can then be compared to those for the rigid aircraft. The analysis used in these examples is based upon the solution of Equation (12.48).

12.4.13.1 Fuselage Bending Mode

Results for the trim incidence/elevator angle and the fuselage deformation are shown plotted against the natural frequency for this fuselage bending mode in Figure 12.13(a) and (b);

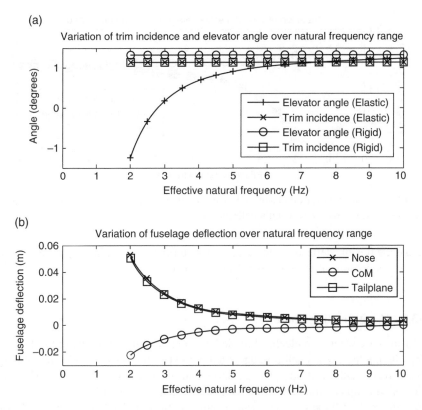

Figure 12.13 Results for the example of a flexible aircraft in an equilibrium manoeuvre – fuselage bending mode at 150 m/s EAS and $n = 1$: (a) trim incidence and elevator angle and (b) fuselage deflection.

the idealized mode shape involves bending of the fuselage, but not wing bending or twist. The wings behave rigidly and heave the same amount as the centre of mass. The trim incidence is unaffected by flexible effects (so rigid and flexible incidence curves overlay each other), but the elevator angle becomes more negative as the aircraft becomes more flexible (i.e. at lower natural frequencies). An explanation of what is happening physically now follows.

The fuselage deformation in the balanced state due to flexibility effects may be seen to involve the centre fuselage (and wings) deflecting upwards and the nose/tailplane bending downwards; this corresponds to a negative value of the modal generalized coordinate. This sense of the modal deflection may be understood by considering the modal force generated by the aerodynamic and inertia forces present for the balanced aircraft. It may be shown that the wing lift force and nose/tailplane inertia forces cause a negative modal force that exceeds the positive modal force contribution due to the tailplane lift and centre fuselage/wing inertia forces; the result is the negative modal deformation seen in the results. This deformed shape will not influence the wing lift because there is no wing twist, but it will tend to increase the tailplane lift due to the nose up flexible pitch angle at the tailplane. Therefore a more negative elevator angle is required to compensate for this effect and so maintain the same tailplane lift force for trim.

For the most flexible 2 Hz natural frequency case (excessively low for this size of aircraft), the fuselage deflections are of the order of 0.05 m and as the effective natural frequency increases, this deformation reduces rapidly. For a higher load factor of 2.5, the behaviour is

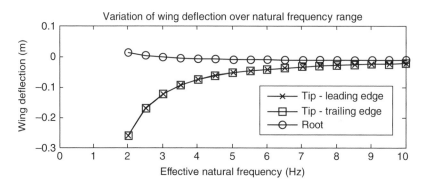

Figure 12.14   Results for the example of a flexible aircraft in an equilibrium manoeuvre – wing bending mode at 150 m/s EAS and $n = 1$: wingtip deflection.

similar but the flexible deflection increases to around 0.12 m. At higher air speeds, the flexible effect changes because of the revised balance of forces.

12.4.13.2   Wing Bending Mode

Consider the case where the idealized mode has wing bending but no wing twist or fuselage bending (though there is a small fuselage pitch present). The flexible mode natural frequency only has a very small effect upon the incidence and the elevator angle to trim. However, the variation of wing tip and root deflection with natural frequency for 150 m/s air speed and $n = 1$ is seen in Figure 12.14; there is some wing bending deflection due to the wing lift distribution (around 0.25 m upwards at the tip and 0.028 m downwards at the fuselage for the most flexible 2 Hz mode case) and also a very small nose down pitch (around 0.01°). These values reduce quickly with an increase in natural frequency. It is expected that the inclusion of this mode would have no effect on the trimmed state but the small pitch deflection explains the very small flexible effect on trim incidence and elevator angle found (typically there is only about a 1% change in the angles due to the flexible effect, which is not worth plotting).

12.4.13.3   Wing Twist Mode

Here the idealized mode has wing twist but no wing or fuselage bending (though again there is a small fuselage pitch). The variation of the trim incidence, elevator angle and the wing twist deflection with natural frequency is shown in Figures 12.15(a) and (b); since the deflections at 2 Hz were found to be unrealistic (16° steady twist), the lowest frequency considered was 4 Hz (about 2° twist), but even that frequency is low for a torsion mode.

For this mode, Figure 12.15(a) shows that both the trim incidence and elevator angle are changed noticeably by flexible effects. The wing twists nose up because the aerodynamic lift force acts ahead of the elastic axis and so will tend to increase the wing lift further; in order to maintain the same total wing lift for trim, the trim incidence has to reduce. As a result, the tail-plane lift reduces and less up elevator is required to maintain the same total lift force. As mentioned earlier (see Chapter 7), the lift will shift outboard due to the flexible effect and so increase the wing root shear force and bending moment when compared to the rigid aircraft (see also Chapter 16).

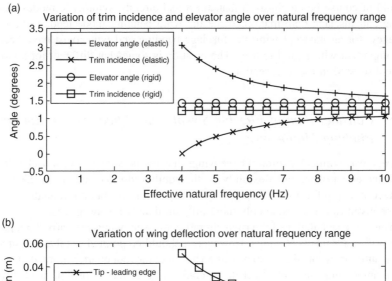

Figure 12.15 Results for the example of a flexible aircraft in an equilibrium manoeuvre – wing twist mode at 150 m/s EAS and $n = 1$: (a) trim incidence and elevator angle and (b) flexible mode deflection.

12.4.14   Summary of Flexible Effects in an Equilibrium Pitching Manoeuvre for an Unswept Wing

The above analysis and examples have demonstrated a number of features for the equilibrium manoeuvre of a simple flexible aircraft with unswept wings, summarized as follows:

- the wing and tailplane lift forces required for trim are the same whether the aircraft is rigid or flexible, but the incidence and elevator angle required for trim will change when flexibility is present;
- the flexible aircraft will deform under the aerodynamic and inertia loading for all mode types;
- the presence of a flexible mode involving fuselage bending and/or wing twist can alter the elevator angle to trim and hence affect the control requirements;
- the presence of a flexible mode involving wing bending will not affect the overall trim forces and angles but will cause wing bending deformation; and
- the presence of a flexible mode involving wing twist will modify the spanwise aerodynamic loading distribution such that the loading reduces at the root and increases at the tip, so shifting the lift towards the tip and increasing the wing root shear force and bending moment in the manoeuvre.

It should of course be emphasized that, for a real aircraft, symmetric modes will involve both wing and fuselage deformation: while the wing bending mode tends to be the lowest in frequency, for an unswept wing the fuselage bending and wing twist have been seen to be more important with regard to trim. The position is somewhat different for a swept wing, as will be discussed in the next section.

12.4.15   Consideration of Flexible Swept Wing Effects on an Equilibrium Pitching Manoeuvre

In practice, most commercial aircraft have wings that are moderately swept back (20–30°) and this alters the effect of aircraft flexibility on the equilibrium manoeuvre. In this section, the likely effect of sweep for a flexible aircraft will be considered, but a full analytical treatment will not be given as it is considerably more difficult than the unswept case.

The manoeuvre calculations for a rigid aircraft with swept wings are carried out in precisely the same way as for unswept wings. However, it should be pointed out that the aircraft aerodynamic centre for swept wings will be somewhat aft of the root quarter chord, depending upon the sweep angle, and this will affect the aircraft centre of mass position needed to achieve trimmed flight. Also, the swept wing mass will have a much greater contribution to the aircraft pitch moment of inertia than for an unswept wing, but this is only relevant for dynamic manoeuvres when the aircraft experiences pitch acceleration.

There are other key differences between swept and unswept wings when flexible effects are considered:

- the aerodynamic behaviour is significantly more three-dimensional so strip theory is less appropriate;
- the elastic axis is forward of the mid-chord over most of the span (typically around a two-thirds chord for a 30° sweep) but moves further aft as the wing root is approached;
- the simple treatment of a high aspect ratio wing with moderate sweep back has traditionally considered the wing to be a swept beam with bending rigidity EI and torsional rigidity GJ aligned with the sweep axis (see Chapter 7), although of course nowadays the Finite Element method is normally used to model the entire swept box structure;
- the angle of incidence is not only affected by wing twist but also by wing bending (see Chapter 7); and
- bending and torsion motions are coupled.

The impact of these points on the equilibrium manoeuvre will be discussed once the effect of flexibility on the incidence has been considered.

12.4.15.1   Effects of a Flexible Swept Wing on Incidence

This issue was raised earlier in Chapter 7 when considering the effect of forward or backward sweep on the aircraft divergence speed using the model of a simple flexible swept wing fixed at the root. The flexible swept wing allowed the wing to twist (nose up positive) and bend (downwards positive) according to the assumed deformation functions. However, in this case the shapes were defined with respect to a reference axis, neither the elastic or flexural axis, and the functions would vary with the distance from the root along the reference axis. The approach used earlier in Chapter 7 for a swept back wing, and also given in Niu (1988), indicated that nose up flexible twist increases the effective incidence whereas upwards bending (i.e. a negative slope) will decrease it (for a backward sweep). Clearly, for an unswept wing, only the twist has an effect on incidence.

12.4.15.2   Effects of a Flexible Swept Wing on Equilibrium Manoeuvre

In regard to the balanced/trimmed condition for an aircraft with swept wings undergoing an equilibrium manoeuvre, the key difference to the unswept case is that any spanwise change of incidence can alter the fore-and-aft position of the net wing lift force and hence the magnitude of the wing and tailplane lift forces required for balance; this was not the case for the unswept wing. Note that when the aircraft is flying with a positive angle of incidence (in a sufficiently positive load factor manoeuvre), the positive lift distribution will act ahead of the reference axis and thus have the potential to cause nose up twist and upwards bending of the wing. The different types of flexible mode will now be considered briefly with regard to their impact on trim in such a manoeuvre.

Fuselage bending mode
The special case of a mode where only the fuselage bends and the wing is rigid in bending and twist is straightforward. The positions and therefore magnitudes of the lift forces required for balance will not change and therefore the elevator angle and incidence will alter to maintain the same tailplane lift force once the fuselage has deformed elastically; thus there is little difference to the unswept wing case.

Wing bending mode
In this idealized case, where the wing is rigid in twist about the reference axis and the fuselage rigid in bending, the sweep effect will mean that there will be a greater fuselage pitch in the free–free mode shape and therefore some influence from the tailplane pitch. However, the main effect is that the wing lift distribution will give rise to a bending deflection (upwards for positive lift) and this will reduce the effective incidence, more towards the tip where the bending slope is greater. What this means is that the effective centre of pressure on each wing moves towards the root and therefore forward on the aircraft. Then, in order to maintain balance, the incidence and elevator angle will need to be adjusted but the final balance forces will not be the same as for the rigid aircraft. The inboard shift in the lift will decrease the wing root bending moment, for example, and so reduce the internal loads.

Wing twist mode
In this idealized case, where the wing twist about the reference axis is assumed to be dominant in the mode, there will be some fuselage pitch as before but no fuselage or wing bending. The wing lift distribution will give rise to a nose up twist and so increase the lift further, as well as moving the centre of pressure on each wing outboard towards the tip and therefore aft on the aircraft. In order to maintain trim, the incidence and elevator angle will need to be adjusted but the final balanced state will not be the same as for the rigid aircraft. The outboard shift in the lift will increase the wing root bending moment.

Wing mode with bending and twist

Having considered the idealized wing twist and bending cases separately, it is important to point out that the nature of the aircraft structure is such that, for most wings, the wing bending mode will be at a lower frequency than the twist mode (though of course the deformations in any given mode tend to be a combination of both motions, particularly when sweep is involved). Wing bending will be present in most modes and the reality for a swept wing aircraft is that the first free–free mode will be dominated by wing bending with some twist. Unlike the unswept case, this mode will have the most impact on trim. The bending effect will tend to dominate the twist effect on incidence and so bending would be the most important term; therefore the comments made for wing bending are likely to be most relevant for the equilibrium manoeuvre.

Finally, it should be noted that in the ESDU series, the static aeroelastic behaviour of a flexible aircraft with a swept wing is considered in some detail using assumed modes (ESDU Data Sheet 97032) and normal modes (ESDU Data Sheet 99033).

12.5 Representation of the Flight Control System (FCS)

The Flight Control System (FCS) has a significant impact upon the dynamic response and loads in a commercial aircraft and will be introduced in Chapter 13. When equilibrium (or steady) manoeuvres are considered, only the steady effect of the FCS is important and so a full FCS model is not required. Instead, the effect of the FCS is represented by including the limits imposed by the FCS on the aircraft behaviour e.g. in providing a maximum incidence (for high angle of incidence protection) or restricting pitch rate.

12.6 Examples

1. Draw a manoeuvre envelope for an aircraft with the following characteristics: mass 10 000 kg, wing area 30 m², maximum lift coefficient 1.6, maximum and minimum load factors 3 and −1, cruise speed 220 m/s EAS and design dive speed 250 m/s EAS. Air density at sea level is 1.225 kg/m³.
   [Stall speed at $n = 3$ is 100 m/s EAS].
2. For the following manoeuvres of the aircraft above in Example 1, determine the load factors and indicate the manoeuvres on the manoeuvre envelope:
   a. in a level turn (sea level) at 130 m/s, turning through 180° in 50 s;
      [$n = 1.30$, $R = 2069$ m, $\phi = 39.8°$]
   b. in a level turn with bank angle 30° at 210 m/s TAS (at 6000 m where $\sigma = 0.53$);
      [$n = 1.15$, $R = 7786$ m]
   c. in a dive at 150 m/s TAS and rate of descent 30 m/s TAS (at 6000 m), pulling out to a rate of climb of 30 m/s TAS in 10 s at constant air speed following a circular path.
      [$n = 0.98 \rightarrow 1.58 \rightarrow 1.61 \rightarrow 1.58 \rightarrow 0.98$, $R = 3737$ m].
3. An aircraft flying at 150 m/s (sea level) has the following characteristics: $m = 100\,000$ kg, $S_W = 95$ m², $c = 3.8$ m, $l_W = 0.4$ m, $l_T = 13$ m, $C_{M_{0W}} = -0.028$. Obtain the wing and tailplane lift forces for $n = 2.5$.
   [2390 kN and 63 kN].
4. An aircraft has the following characteristics: $m = 9000$ kg, $S_W = 46$ m², $c = 3$ m, $C_{M_{0W}} = -0.035$ and $l_W + l_T = 8$ m (distance between tailplane and wing aerodynamic centres).

If the maximum allowable *download* on the tailplane is 9000 N, what limitations must be imposed upon the centre of mass position to meet this constraint in steady level flight at 180 m/s (sea level)?

$[l_W > 0.27 \text{ m}]$.

5. An aircraft flying at 87.3 m/s EAS is at the point of stalling for a load factor of 2.5. It has the following characteristics: $m = 44\,200$ kg, $S_W = 145$ m^2, $c = 5$ m, $l_W = 0.3$ m, $l_T = 14.9$ m, $C_{Mow} = -0.07$, drag coefficient $C_D = 0.02 + 0.072C_{LW}^2$ where C_{Lw} is the wing lift coefficient. The drag may be assumed to act through the wing aerodynamic centre and the thrust acts a distance $h = 1.5$ m below the wing. Estimate the tailplane lift using three iterations.

[19.0 kN].

6. Consider a rigid aircraft with: $m = 10\,000$ kg, $S_W = 30$ m^2, $S_T = 7.5$ m^2, $c = 2.0$ m, $l_W = 0.6$ m, $l_T = 7$ m, $a_W = 4.5$ per rad, $a_T = 3.2$ per rad, $a_E = 1.5$ per rad, $k_\varepsilon = 0.35$, $\alpha_0 = -0.03$ rad, $C_{Mow} = -0.03$ and $\rho_0 = 1.225$ kg/m^3. The aircraft undergoes an $n = 1.5$ manoeuvre at 150 m/s EAS with zero pitch rate. Determine the aerodynamic derivatives and the trim incidence and elevator angle.

$$\begin{bmatrix} Z_\alpha = -2.07 \times 10^6 & Z_\eta = -1.55 \times 10^5 & Z_0 = -5.23 \times 10^4 \text{ N/rad} \\ M_\alpha = -3.89 \times 10^5 & M_\eta = -1.08 \times 10^6 & M_0 = 3.30 \times 10^4 \text{ Nm/rad} \quad \alpha = 2.56^\circ \quad \eta = 0.83^\circ \end{bmatrix}$$

7. Write a MATLAB program to generate the aerodynamic derivatives for a given symmetric flight condition and then to solve the *rigid* aircraft equilibrium manoeuvre for the trim incidence/elevator angle (see companion website). Include the code in a load factor or air speed loop to allow these parameters to be varied. Check the results for the example in this chapter. Then explore the effect of changing air speed, centre of mass position, downwash term and zero lift wing pitching moment coefficient (maintain static stability).

8. Write a MATLAB program to generate the aerodynamic derivatives for a given flight condition and mode type, and then to solve, the *flexible* aircraft equilibrium manoeuvre for the trim incidence/elevator angle (see companion website). Include the code in a load factor or air speed loop to allow these parameters to be varied. Check the results using the example in the chapter. Then explore the effect of changing air speed, centre of mass position, downwash term and zero lift pitching moment coefficient (maintain static stability). Also, generate a new flexible mode for the same aircraft example which is the combination of two of the other modes and investigate the revised behaviour. Finally, develop a data set from another idealized aircraft with a different mass distribution.

13

Dynamic Manoeuvres

In Chapter 12, the balanced aircraft loads, trimmed condition and deformation in a steady symmetric *equilibrium manoeuvre* were considered for both a rigid and a simple flexible aircraft in heave and pitch. Such a 'bookcase' manoeuvre is one in which the aircraft undergoes a steady acceleration normal to the flight path and a steady pitch rate (e.g. steady pull-out or banked turn) and where, by the application of d'Alembert's principle, the aircraft may be considered to be in static equilibrium under aerodynamic, thrust and inertia loads. A balanced/trimmed condition was sought and a model based on inertial axes fixed in space was used, with displacements and angles being the unknown response parameters.

However, a *dynamic manoeuvre* (Lomax, 1996; Howe, 2004) involves applying some form of control input such that a transient (time-varying) response of the aircraft is generated. To obtain accurate responses and loads, a 'rational' calculation will require a more realistic dynamic model and manoeuvre scenario to be used. Certain manoeuvres, e.g. 'unchecked' or 'checked' pitch manoeuvres and also failure cases (see Chapter 23), require a fully rational treatment. The unchecked manoeuvre involves the pilot pulling the stick straight back whereas the checked manoeuvre involves three-quarters of a cycle of harmonic control input; these loading conditions are potential design cases for the rear fuselage and horizontal tailplane. Rolling and yawing manoeuvres may be examined using bookcase or rational calculations (see later in this chapter). Engine failure cases are best treated rationally. The certification requirements for dynamic manoeuvres may be seen in CS-25 and Lomax (1996) and will be discussed further in Chapter 23.

There are two possible approaches to the rational modelling of dynamic manoeuvres: (i) using a linear small displacement model based on inertial axes (rather as will be used later in Chapter 14 for gusts and turbulence) or (ii) using a non-linear flight mechanics model based on body fixed axes. The latter equations allow large angle manoeuvres and are used to assess stability, control and handling qualities etc., to assist in design of the Flight Control System (FCS) and to yield loads in general manoeuvres.

In this chapter, some basic concepts related to the large angle non-linear flight mechanics equations of motion of the aircraft, when undergoing dynamic manoeuvres, will be introduced. The focus will be on ideas relevant to the response and hence internal loads (namely bending moment, torque and shear) in dynamic manoeuvres, and not particularly on performance, stability, control or handling issues. Other texts, e.g. Hancock (1995), Etkin and Reid (1996),

Introduction to Aircraft Aeroelasticity and Loads, Second Edition. Jan R. Wright and Jonathan E. Cooper.
© 2015 John Wiley & Sons, Ltd. Published 2015 by John Wiley & Sons, Ltd.

Schmidt (1998), Cook (1997), Russell (2003), Stengel (2004) and ESDU Data Sheet 67003, should be consulted for a more thorough treatment.

In classical books on flight mechanics, the full 6DoF non-linear equations of motion are usually derived in 3D from first principles with the equations being linearized for some applications. The derivation may be performed using approaches in scalars (Cook, 1997; etc.) or vectors (Etkin and Reid, 1996; Russell, 2003; etc.). In order to minimize the mathematics involved and to allow basic flexible effects to be introduced later in this chapter, the approach taken here will be to carry out the analysis for a symmetric aircraft undergoing 2D longitudinal motion; fore-and-aft motion will be neglected so as to reduce the number of equations further to two, a considerable simplification. An approach based on scalars will be used in order to provide a little more physical appreciation and to avoid an excursion into vectors for only a limited part of the book; however, it must be emphasized that a vector approach is simpler and more elegant for the 3D case.

In regard to the aerodynamic forces and moments for the flight mechanics model, these are defined in terms of stability and control derivatives derived with respect to body fixed (usually stability or wind) axes; this is in contrast to all other chapters. Elsewhere, inertial axes are used for static aeroelastic, flutter, manoeuvre and gust/turbulence cases, with only small excursions (perturbations) from the datum being involved.

The aircraft will initially be considered as rigid but the inclusion of flexible effects (i.e. elastic deformation and dynamic response of the structure) may be important for dynamic manoeuvres. For the model based on inertial axes, adding in flexible modes is relatively straightforward since the entire model is linear. However, the inclusion of flexible modes in the flight mechanics model is not straightforward and is a subject of ongoing research; thus as simple an approach as possible will be adopted here to highlight some of the key issues; the instantaneous centre of mass position in the aircraft is not fixed and the moments of inertia are not constant with time. Performing this analysis will require the introduction of the so-called 'mean axes reference frame' that 'floats' with the aircraft in such a way that its origin always remains at the instantaneous centre of mass. Such a choice of axes allows the inertia coupling terms to be eliminated. The model is also used for assessing the impact of flexibility (i.e. aeroelastic effects) upon the dynamic stability modes (Babister, 1980; Waszak and Schmidt, 1988; Schmidt, 1998; Stengel, 2004), and therefore upon aircraft handling qualities and control effectiveness.

It should be pointed out that in many cases the FCS is designed to filter out dynamic motion of the flexible modes in a dynamic manoeuvre; typically, the control rates for a commercial aircraft are sufficiently slow to not excite dynamics significantly. However, the situation is different for an oscillatory failure of the FCS where dynamic effects do matter.

An alternative to adding flexible modes to the flight mechanics equations is to adjust the rigid body aerodynamic derivatives for static deformation effects, making use of the FE model and zero frequency parameter aerodynamics; aerodynamic derivatives such as the lift curve slope may well be changed significantly due to static deformation involving the flexible modes. Therefore introducing quasi-static flexible corrections to the rigid body aerodynamics may often suffice and eases model adjustment based on results from flight test manoeuvres.

Later in this chapter, the linearized flight mechanics model will be applied to the problems of a rigid or flexible aircraft undergoing a dynamic manoeuvre in heave/pitch; these symmetric cases allow some basic concepts to be seen using a small order mathematical model. The response, dynamic stability modes and control effectiveness will all be calculated. Quasi-steady aerodynamics for the rigid aircraft will be employed in determining the aerodynamic derivatives and these will be defined with respect to wind axes (i.e. body fixed) and not for the inertial axes employed in Chapter 12.

Also, brief consideration will be given to the lateral equilibrium manoeuvre cases of a steady roll rate and steady sideslip, together with analysis of the abrupt application of the aileron or rudder, where the maximum angular acceleration was determined by balancing the applied aerodynamic moment by an inertia moment. These equilibrium manoeuvres are sometimes known as 'bookcases' (see Chapter 20), often somewhat artificial and unrealizable in practice; they may lead to overestimates of the internal loads, but are particularly useful early in the design cycle. However, a more 'rational' approach is often required by the certification specifications and may be used to obtain more realistic loads for some of the bookcases.

It should be noted that in this chapter, the focus will be on determining the aircraft response. The calculation of internal loads will be covered in Chapter 16 since there are many common factors in calculating loads for dynamic, ground and equilibrium manoeuvres, as well as for gusts and turbulence.

The chapter concludes with a brief introduction to the aircraft Flight Control System (FCS), included here because its main function is in connection with the flight mechanics behaviour of the aircraft.

13.1   Aircraft Axes

There are two main types of axes systems in use for aircraft, namely earth and body fixed axes (Cook, 1997).

13.1.1   Earth Fixed Axes

Earth fixed axes ($Ox_E y_E z_E$) are located in the plane of the earth and classically have the x_E axis pointing north and the y_E axis pointing east; the axes assume a 'flat' earth. However, the axes may for convenience be located on the earth surface directly below the aircraft with the x_E axis pointing along the aircraft direction of flight in the initial trimmed state prior to any manoeuvre. Earth axes may also be located with their origin coinciding with the initial origin of the body fixed axes, but thereafter remaining fixed during any manoeuvre. Provided the effect of the rotation of the earth may be neglected, earth axes may be treated as 'inertial' so that accelerations of the aircraft relative to these axes are absolute. A different approach is required where space travel is being considered since the earth's rotation then becomes significant.

The disadvantage of earth fixed axes, when considering the dynamic response of an aircraft in a manoeuvre where significant motion is experienced, is that the moments of inertia of the aircraft will alter as the aircraft orientation to the earth changes. For this reason, body fixed axes are normally used in dynamic manoeuvres, whereas inertial axes are satisfactory for static aeroelasticity, equilibrium manoeuvres, ground manoeuvres, gusts and flutter, where only small perturbations are considered.

13.1.2   Body Fixed Axes

Body fixed axes ($Oxyz$) are located in the aircraft, with their origin usually at the centre of mass (CoM), and translate/rotate with it; the axes are a right-handed system as shown in Figure 13.1. An advantage of using body axes fixed in the aircraft is that the moments of inertia of the

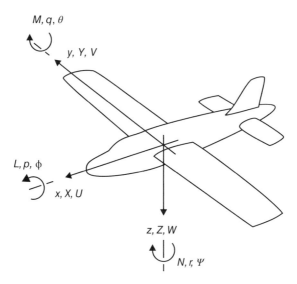

Figure 13.1 Notation for body fixed axes.

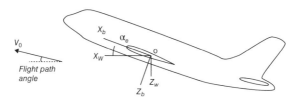

Figure 13.2 Generalized and aerodynamic/wind body axes for a 2D symmetric flight.

aircraft with respect to these axes remain constant during general 3D motion. There are two common types of body axes, as illustrated in Figure 13.2 for 2D.

1. A *generalized body* axis system $(Ox_by_bz_b)$ aligned with the horizontal fuselage datum. For body axes aligned with the principal axes of inertia of the aircraft, by definition the product moments of inertia are zero. However, for non-principal or aerodynamic axes these product terms will exist.
2. A special type of body axis (Hancock, 1995; Cook, 1997), where the *x*-axis is aligned with the velocity vector component in the plane of symmetry and *fixed for a particular flight condition*, is known as an *aerodynamic* or *wind* or *stability* body axis $(Ox_wy_wz_w)$. Such axes have the advantage that the relevant aerodynamic stability derivatives are less difficult to derive.

Generalized and aerodynamic (wind) body axes differ in orientation by the steady (or equilibrium) body incidence value α_e and are used in different situations. Thus it is important to apply the necessary transformations to ensure overall consistency of any model used, e.g. derivatives might be calculated for aerodynamic body axes and then transformed to generalized body axes for response calculation. If calculations are performed using equations for aerodynamic body

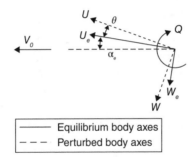

Figure 13.3 Steady and perturbed axes systems in 2D symmetric flight.

axes, then the moments of inertia must be transformed through the steady incidence α_e. The transformations are shown in, for example, Cook (1997).

13.2 Motion Variables

For body fixed axes, the motion is represented by small or large perturbation variables referred to these moving axes. The aircraft may be considered as being initially in a trimmed (or equilibrium) state (subscript e) with total velocity V_0, having components U_e, V_e and W_e along the body axis directions; the trimmed pitch attitude (to the horizontal) is θ_e, incidence is α_e and sideslip β_e. There are no rates of rotation in the initial trimmed state.

When the aircraft is then disturbed from the trimmed state, there are net forces (X axial, Y lateral and Z normal) and moments (L rolling, M pitching and N yawing) together with centre of mass response velocities (U axial, V lateral and W normal) and rates of rotation (P roll, Q pitch and R yaw), all shown in Figure 13.1. Rotations follow the right-hand rule. The forces and moments acting on the aircraft will be due to aerodynamic, control, gravitational and propulsion effects. Positive control rotations introduce positive moments.

The change in orientation of the aircraft due to the disturbance, relative to the equilibrium axis set, is defined by the angles ϕ (roll), θ (pitch) and ψ (yaw). The sign convention for these quantities is also shown in Figure 13.1. If the orientation of the aircraft is referred to earth fixed axes, then the absolute angles defining its position are known as Φ (bank), Θ (inclination) and ψ (azimuth or heading). The steady and perturbed velocities/angles are shown in Figure 13.3 for 2D symmetric flight.

13.3 Axes Transformations

Once the aircraft has started to undergo a manoeuvre such that it experiences a change of position and orientation with respect to the earth axes, it is important to be able to transform the velocities and rates between the earth and body axes systems. Solution of the equations of motion in the body fixed axes system will yield velocities and rates, relative to the body axes, that must then be transformed to earth axes and finally integrated to obtain changes of position and orientation relative to the earth axes.

The three angles defining the relative angular positions of two sets of axes are known as the Euler angles (e.g. see Cook, 1997), in this case ϕ, θ and ψ. Euler angles may be used to transform between body axes and earth axes or between perturbed axes and equilibrium axes.

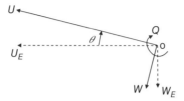

Figure 13.4 Velocity diagram in 2D for body/earth axes transformation.

13.3.1 Transformation in 2D

To demonstrate the basic transformations between body and datum (say, earth) axes using a simple physical approach, consider the behaviour of the aircraft in heave and pitch only. Figure 13.4 shows a velocity diagram with both sets of velocity components. The coincidence of origins does not influence the velocity, rate and angle transformations. By inspection of the figure, it may be seen that the velocities in the datum earth axes directions are given by the summation of the body axes velocity components, namely

$$U_E = U\cos\theta + W\sin\theta \quad W_E = -U\sin\theta + W\cos\theta \qquad (13.1)$$

or in matrix form

$$\begin{Bmatrix} U_E \\ W_E \end{Bmatrix} = \begin{bmatrix} \cos\theta & \sin\theta \\ -\sin\theta & \cos\theta \end{bmatrix} \begin{Bmatrix} U \\ W \end{Bmatrix} = \mathbf{A}_L \begin{Bmatrix} U \\ W \end{Bmatrix} \qquad (13.2)$$

This linear transformation matrix $\mathbf{A}_L$ is in fact the inverse of the matrix of direction cosines in 2D. Also, by inspection it may be seen that the angular velocity Q is identical in both axes systems, so the angular rate transformation is

$$\{\dot\theta\} = \{Q\} = \mathbf{A}_R\{Q\} \qquad (13.3)$$

where the rotational transformation matrix for this simple case is $\mathbf{A}_R = [1]$.

Thus a solution of the equations of motion in body axes will yield U, W and Q, with the above transformations allowing the velocities U_E, W_E and rate of rotation $\dot\theta$ relative to earth axes to be calculated. By integrating these latter velocities and rates, the absolute position and orientation of the aircraft in space may be found.

13.3.2 Transformation in 3D

When the equivalent transformation relationships in 3D are sought, the problem is much more difficult as three angles, not one, need to be considered. The transformation between body and earth axes requires two intermediate axes systems through which transformations must be made (Cook, 1997). The order of axis rotations required to generate the transformations is important, namely ψ, then θ and finally ϕ.

It may be shown that the linear and angular transformations are governed by two 3×3 transformation matrices, each involving the Euler angles. Because the focus in this book is on the basic ideas, seen to a large extent in 2D, the expressions will not be quoted in full; however, they are of the form

$$\begin{Bmatrix} U_E \\ V_E \\ W_E \end{Bmatrix} = \mathbf{A_L} \begin{Bmatrix} U \\ V \\ W \end{Bmatrix} \quad \text{and} \quad \begin{Bmatrix} \dot{\phi} \\ \dot{\theta} \\ \dot{\psi} \end{Bmatrix} = \mathbf{A_R} \begin{Bmatrix} P \\ Q \\ R \end{Bmatrix} \tag{13.4}$$

The complete expressions may be found elsewhere (Cook, 1997). It should be noted that for a linearized analysis with small perturbations, where the angles may be treated as small, the matrices $\mathbf{A_L}$ and $\mathbf{A_R}$ are considerably simplified (see later).

13.4   Velocity and Acceleration Components for Moving Axes in 2D

To determine the flight mechanics equations of motion based on an axes reference frame moving with the aircraft, the kinematic behaviour of the aircraft motion in the moving axes directions is required. The aim is to determine the absolute velocity and acceleration components (i.e. aligned with moving axes directions) of a point in 2D. Then d'Alembert's principle will be used to determine the equations of motion for a rigid symmetric aircraft in 2D.

13.4.1   Position Coordinates for Fixed and Moving Axes Frames in 2D

Consider the 2D axes systems shown in Figure 13.5; here OXZ is an inertial frame (such as earth axes) and oxz is a moving frame. It should be noted that the axes arrangement is chosen in order to match the classical flight mechanics sign convention, e.g. z downwards, x forwards, positive pitch nose up, etc. The absolute position of the origin o is given by the coordinates (X_o, Z_o) and the position of an arbitrary point P is given by coordinates (X_P, Z_P) with respect to the fixed inertial axes and (x, z) with respect to the moving axes.

By studying the geometry of the system in Figure 13.5, the position coordinates of P with respect to OXZ (*fixed axes*) *directions* may be written as

$$X_P = X_o + (x\cos\theta + z\sin\theta) \quad Z_P = Z_o + (-x\sin\theta + z\cos\theta) \tag{13.5}$$

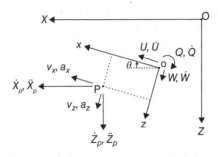

Figure 13.5   Inertial/earth fixed and moving axes frames of reference in 2D.

13.4.2   Differentiation with Respect to Time

When considering systems with both fixed and moving axes systems, differentiation with respect to time may apply for inertial axes (sometimes referred to using d/dt) or for moving axes (sometimes referred to using $\partial/\partial t$ or $\delta/\delta t$); in this chapter, the overdot is used to indicate differentiation with respect to time, with the context showing which axes are intended (e.g. $\dot{X} \equiv \mathrm{d}X/\mathrm{d}t$ and $\dot{x} \equiv \partial x/\partial t$).

13.4.3   Velocity Components for Fixed and Moving Axes in 2D

Now, differentiating the expressions (13.5) with respect to time yields the absolute velocity components of P, again in the OXZ (inertial axes) directions, so

$$\dot{X}_P = \dot{X}_o + (-x\sin\theta + z\cos\theta)\dot{\theta} + (\dot{x}\cos\theta + \dot{z}\sin\theta)$$
$$\dot{Z}_P = \dot{Z}_o + (-x\cos\theta - z\sin\theta)\dot{\theta} + (-\dot{x}\sin\theta + \dot{z}\cos\theta)$$

(13.6)

To understand this differentiation it should be recognized that differentiating a function f of θ with respect to t requires the chain rule; e.g. $f = \theta$

$$\frac{\partial f}{\partial t} = \frac{\partial f}{\partial \theta}\frac{\partial \theta}{\partial t} \qquad \frac{\partial}{\partial t}(\sin\theta) = \frac{\partial}{\partial \theta}(\sin\theta)\frac{\partial \theta}{\partial t} = \cos\theta\,\dot{\theta}$$

(13.7)

Then, to determine the velocity components of point P in the oxz directions, the velocity components of the origin o in the OXZ (inertial axes) directions, namely $\dot{X}_o, \dot{Z}_o$, need to be expressed in terms of the velocity components in the oxz (moving axes) directions. This is essentially the 2D transformation described in Section 13.3.1 and means that

$$\dot{X}_o = U\cos\theta + W\sin\theta \qquad \dot{Z}_o = -U\sin\theta + W\cos\theta$$

(13.8)

Considering the geometry of the velocity components shown in Figure 13.5, the absolute velocity components (v_x, v_z) of P in the oxz (moving axes) directions will be given by

$$v_x = \dot{X}_P\cos\theta - \dot{Z}_P\sin\theta \qquad v_z = \dot{X}_P\sin\theta + \dot{Z}_P\cos\theta$$

(13.9)

If the expressions in Equation (13.6) are substituted into Equation (13.9), and the result simplified using Equation (13.8) and the pitch rate relationship $Q = \dot{\theta}$, then it can be shown that the absolute velocity components of P in the oxz directions are given by the expressions

$$v_x = U + zQ + \dot{x} \qquad v_z = W - xQ + \dot{z}$$

(13.10)

The absolute velocity of point P is therefore equal to the absolute velocity of the moving axes origin o, plus a term involving Q to allow for the rotation of the moving axes and a further term to allow for motion of P relative to o (which only applies for a flexible body). The two 'overdot' terms are velocities with respect to the moving axes origin.

13.4.4   Acceleration Components for Fixed and Moving Axes in 2D

To determine the equivalent relations for the absolute accelerations is more difficult but involves a similar process. Firstly, the expressions in Equation (13.6) are differentiated with respect to time and then the absolute acceleration components of P in the *OXZ* (*inertial axes*) *directions* are

$$\ddot{X}_P = \ddot{X}_o - \dot{x}s\dot{\theta} - xc\dot{\theta}^2 - \dot{x}s\dot{\theta} + \dot{z}c\dot{\theta} - zs\dot{\theta}^2 + zc\ddot{\theta} + \ddot{x}c - \dot{x}s\dot{\theta} + \ddot{z}s + \dot{z}c\dot{\theta}$$

$$\ddot{Z}_P = \ddot{Z}_o - \dot{x}c\dot{\theta} + xs\dot{\theta}^2 - xc\ddot{\theta} - \dot{z}s\dot{\theta} - zc\dot{\theta}^2 - zs\ddot{\theta} - \ddot{x}s - \dot{x}c\dot{\theta} + \ddot{z}c - \dot{z}s\dot{\theta} \tag{13.11}$$

where, in order to simplify the expressions, the shorthand notation $s = \sin\theta$, $c = \cos\theta$ has been used.

The acceleration components of the origin o, namely $\ddot{X}_o, \ddot{Z}_o$, are now expressed in terms of the velocity and acceleration components in the *oxz* directions by differentiating Equation (13.8), so

$$\ddot{X}_o = -U s\dot{\theta} + \dot{U}c + Wc\dot{\theta} + \dot{W}s \qquad \ddot{Z}_o = -Uc\dot{\theta} - \dot{U}s - Ws\dot{\theta} + \dot{W}c. \tag{13.12}$$

The absolute acceleration components (a_x, a_z) of P in the *oxz* (*moving axes*) *directions* are given by a similar expression to that for the velocity in Equation (13.9), namely

$$a_x = \ddot{X}_P \cos\theta - \ddot{Z}_P \sin\theta \qquad a_z = \ddot{X}_P \sin\theta + \ddot{Z}_P \cos\theta \tag{13.13}$$

Now, when Equations (13.11) and (13.12) are substituted into Equation (13.13) and the results simplified, it may be shown that the *acceleration components of P in the oxz (moving axes) directions* seen in Figure 13.5 are given by

$$a_x = \dot{U} + WQ - xQ^2 + z\dot{Q} + \ddot{x} + 2\dot{z}Q \qquad a_z = \dot{W} - UQ - zQ^2 - x\dot{Q} + \ddot{z} - 2\dot{x}Q \tag{13.14}$$

The six terms in each of these expressions have the following meaning (in order from left to right):

a. acceleration of the origin of the moving axes with respect to inertial axes,
b. term to account for the effect of rotation of axes on the acceleration of the origin,
c. centripetal acceleration directed towards the axis of rotation,
d. changes in the tangential acceleration component due to changes in the rate of rotation,
e. acceleration of P relative to the moving frame, and
f. Coriolis acceleration term due to the velocity of P relative to the rotating frame.

The last two terms are zero for a *rigid* aircraft but non-zero for a *flexible* aircraft. In order to distinguish between a rigid and a flexible aircraft, the position of the point P relative to the moving axes origin o may be written as the summation of the position coordinates (x_r, z_r) for the undeformed/rigid aircraft (r) and the displacement components of the elastic deformation (x_e, z_e) of the structure (e), so

$$x = x_r + x_e \qquad z = z_r + z_e \tag{13.15}$$

Because the aircraft is initially assumed to be *rigid*, then from Equation (13.15), $x = x_r$, $z = z_r$ and all the 'overdot' terms in the acceleration component expressions will be zero (i.e. $\dot{x} = \dot{z} = \ddot{x} = \ddot{z} = 0$). Thus, for the rigid aircraft, the acceleration components may be written as

$$a_x = \dot{U} + WQ - x_r Q^2 + z_r \dot{Q} \qquad a_z = \dot{W} - UQ - z_r Q^2 - x_r \dot{Q} \tag{13.16}$$

The treatment of the *flexible* aircraft, where the 'overdot' terms are non-zero, will be considered later in this chapter where the flexible deformations are expressed in terms of free–free normal modes.

13.5 Flight Mechanics Equations of Motion for a Rigid Symmetric Aircraft in 2D

Now, consider using the above acceleration expressions for the symmetric rigid aircraft in longitudinal motion in 2D (i.e. all yaw, sideslip and roll terms are zero). The axes systems are inertial (or earth) axes OXZ and moving body fixed axes oxz, as shown in Figure 13.6. The origin o for the moving axes is taken for convenience to be at the centre of mass, as it allows elimination of several terms in the equations. Some general point P is located at coordinate (x, z) within the aircraft.

13.5.1 Non-linear Equations for Longitudinal Motion

To determine the equations of motion for the longitudinal behaviour of the symmetric aircraft, consider an elemental mass dm at point P and its absolute (inertial) acceleration components in the oxz directions. The elemental masses are assumed to have no rotational inertia. The d'Alembert inertia force on the mass will have components $-dm\, a_x$ and $-dm\, a_z$, aligned with the body axes as shown in Figure 13.7.

If the applied force components in the x and z directions are X and Z, then the body will be in equilibrium, so the applied forces will balance the summation (or integration) of the inertia forces for all the mass elements in the entire body (see Chapter 5) and

$$\int a_x dm = X \qquad \int a_z dm = Z \tag{13.17}$$

where the summations apply to the entire volume of the aircraft. The moment equation may also be obtained using the inertia forces in Figure 13.7 since, for equilibrium, the applied moment must balance the total moment of the inertia forces; therefore

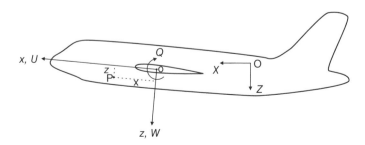

Figure 13.6 Rigid symmetric aircraft in longitudinal motion.

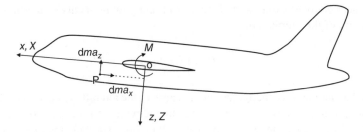

Figure 13.7   Balance of applied and inertia forces for a rigid aircraft.

$$\int (a_x z_r - a_z x_r)\ \mathrm{d}m = M \tag{13.18}$$

where the moment arms are the position coordinates of point P in the rigid aircraft.

Now, if the acceleration components from Equation (13.16) are substituted into Equations (13.17) and (13.18), a number of summation terms are present. Note that the summation terms given by $\int x_r \mathrm{d}m$ and $\int z_r \mathrm{d}m$ are both zero because the origin of the moving axes system was taken at the rigid aircraft centre of mass, and there is no first moment of mass about axes through this point. Also, $m = \int \mathrm{d}m$ is the total aircraft mass and

$$I_y = \int \left(x_r^2 + z_r^2 \right)\ \mathrm{d}m \tag{13.19}$$

is the moment of inertia of the rigid aircraft about the oy axis, i.e. in pitch.

The non-linear coupled equations for longitudinal motion of the symmetric aircraft using body fixed axes may then be shown to be

$$m(\dot{U} + QW) = X \quad m(\dot{W} - QU) = Z \quad I_y \dot{Q} = M \tag{13.20}$$

13.5.2   Non-linear Equations for Combined Longitudinal/Lateral Motion in 3D

If the acceleration components were obtained in 3D, either using an extension of the 2D scalar approach in Section 13.4 or using vectors, a similar analysis could be performed for the symmetric rigid aircraft undergoing combined longitudinal/lateral motion in 3D. The corresponding non-linear equations of motion would be given by (Cook, 1997)

$$
\begin{aligned}
m(\dot{U} - RV + QW) &= X \\
m(\dot{V} - PW + RU) &= Y \\
m(\dot{W} - QU + PV) &= Z \\
I_x \dot{P} - (I_y - I_z)QR - I_{xz}(PQ + \dot{R}) &= L \\
I_y \dot{Q} + (I_x - I_z)PR + I_{xz}(P^2 - R^2) &= M \\
I_z \dot{R} - (I_x - I_y)PQ + I_{xz}(QR - \dot{P}) &= N
\end{aligned}
\tag{13.21}
$$

These non-linear coupled expressions include roll and yaw moments of inertia and also the product moment of the inertia term

$$I_{xz} = \int x_r z_r \mathrm{d}m \tag{13.22}$$

All other product moments of inertia are zero because of the aircraft symmetry and I_{xz} would be zero by definition if the body axes were to be aligned with the principal axes of the aircraft.

These six equations cover general coupled aircraft motion. The equations are non-linear, thus allowing for large variations in response, such as might occur post-stall. The equations require a numerical, not analytical, solution and the forcing terms need to be defined appropriately, i.e. as non-linear functions of incidence, etc. Also, the full Euler angle transformations need to be employed to convert velocities and rates to earth axes.

13.5.3 Linearized Equations of Motion in 3D

Often, the equations of motion are linearized by analytical means for ease of solution, especially for stability and control studies where the behaviour of the aircraft under small perturbations from the trimmed or equilibrium state (denoted by subscript e, not to be confused with its alternative use for elastic deformation) is considered so as to yield transfer functions and dynamic stability modes. Initially, for convenience, the datum flight condition is taken as being steady, straight, horizontal flight without sideslip ($V_e = 0$) and without bank/yaw angles or rates ($\phi = \psi = P_e = Q_e = R_e = 0$).

The total velocity components of the centre of mass in the disturbed flight response are the sum of the equilibrium (trimmed) components and the transient (small) perturbation components (u, v, w), so

$$U = U_e + u \qquad V = V_e + v = v \qquad W = W_e + w \tag{13.23}$$

The equivalent result for the rates of rotation is

$$P = P_e + p = p \qquad Q = Q_e + q = q \qquad R = R_e + r = r \tag{13.24}$$

where the angular disturbance velocities (p, q, r) are also small. All the terms in Equations (13.23) and (13.24) are referred to the moving axes oxz.

Substituting the velocity and rate expressions in Equations (13.23) and (13.24) into Equations (13.21), noting that the trimmed velocities are steady and constant, and neglecting small quantities such as products of rates (e.g. pr, p^2 etc.) yields a set of linearized equations. These may be divided into equations for *longitudinal* motion,

$$m(\dot{u} + W_e q) = X \qquad m(\dot{w} - U_e q) = Z \qquad I_y \dot{q} = M \tag{13.25}$$

which are very similar to Equations (13.20), and for *lateral* motion,

$$m(\dot{v} - W_e p + U_e r) = Y \qquad I_x \dot{p} - I_{xz} \dot{r} = L \qquad I_z \dot{r} - I_{xz} \dot{p} = N \tag{13.26}$$

which are considerably simplified compared to Equation (13.21). The longitudinal and lateral motions are often treated separately for a symmetric aircraft.

Because the angles are small, the full non-linear Euler transformations are not needed. Using the transformation relationships in Equations (13.4), setting $\cos A = 1$ and $\sin A = A$ with all angles assumed small, and neglecting coupling and second-order terms, yields the following linear and rotational transformations

$$U_E = U_e + W_e\theta \qquad W_E = -U_e\theta + W_e + w \qquad \dot{\theta} = q \qquad (13.27)$$

for *longitudinal* motion, and

$$\dot{\phi} = p \qquad \dot{\psi} = r \qquad V_E = \psi U_e + v - \phi W_e \qquad (13.28)$$

for *lateral* motion. These expressions allow transformation of the body axes responses into earth axes.

It should be emphasized that these small disturbance equations cannot be applied to conditions where large perturbation flight manoeuvres are expected (e.g. near stall) since the model is only valid in the region of the initial trimmed condition. However, in many cases, the assumption is adequate. In particular, the linearized equations are used to examine stability and control issues.

13.6   Representation of Disturbing Forces and Moments

So far, the applied/disturbing forces and moments have simply been represented by X, Y, Z, L, M, N. The Z force, for example, may be written as a summation of the different contributions such that

$$Z = Z_a + Z_g + Z_c + Z_p + Z_d, \qquad (13.29)$$

where the subscripts a, g, c, p and d refer to aerodynamic, gravitational, control, propulsive (power) and disturbance (atmospheric gust) terms respectively. These terms must not be confused with the derivatives employed in Chapter 12 .

For the full non-linear analysis, some of the aerodynamic forces and moments are non-linear functions of, for example, incidence and this non-linearity would need to be represented in the model, having been obtained from Computational Fluid Dynamics (CFD) and/or wind tunnel tests; this is clearly important for loads purposes but a study of non-linear aerodynamics is beyond the scope of this book. However, for the linearized analysis often used in stability and control studies, the aerodynamic forces and moments corresponding to small perturbations may be considered. Ignoring the disturbance terms and considering small perturbations, these force and moment components may be written as the summation of *datum* and *perturbation* terms. In this section, a brief explanation will be given as to how some of these terms are obtained; the focus will be on longitudinal motion only and giving limited examples, as other texts cover this ground extensively (e.g. Cook, 1997).

13.6.1   Aerodynamic Terms

The aerodynamic force, for example, may be represented by the summation of datum (equilibrium) and perturbation terms, so

$$Z_a = Z_{a_e} + \Delta Z_a \qquad (13.30)$$

This perturbation for a typical aerodynamic term is normally written using a first-order Taylor expansion involving all the velocities and rates of change of velocity so, for example, in 2D again

$$Z_a = Z_{a_e} + \Delta Z_a = Z_{a_e} + u\frac{\partial Z_a}{\partial u} + w\frac{\partial Z_a}{\partial w} + q\frac{\partial Z_a}{\partial q} + \dot{u}\frac{\partial Z_a}{\partial \dot{u}} + \dot{w}\frac{\partial Z_a}{\partial \dot{w}} + \dot{q}\frac{\partial Z_a}{\partial \dot{q}} \qquad (13.31)$$

In compact notation, this perturbation expression may be written as

$$\Delta Z_a = uZ_u + wZ_w + qZ_q + \dot{u}Z_{\dot{u}} + \dot{w}Z_{\dot{w}} + \dot{q}Z_{\dot{q}} \qquad (13.32)$$

where terms such as Z_w (aerodynamic downwards force per downward velocity derivative) are known as *aerodynamic stability derivatives* (Bryan, 1911; Babister, 1980; Hancock, 1995; Cook, 1997; Russell, 2003; ESDU Data Sheets). Such terms define the perturbation about the equilibrium/datum state for this linearized model so, for example, Z_q is the aerodynamic downwards force per pitch rate derivative. Some derivative terms may be negligible. Similar terms apply for the control forces, for example $\Delta Z_c = (\partial Z_c / \partial \eta)\eta = Z_\eta\eta$ for the elevator input, where Z_η is the aerodynamic downwards force per elevator angle derivative. Aerodynamic derivatives are often refined by or obtained using wind tunnel results.

Note again that the conventional use of the 'over-o' on the dimensional derivative symbol Z_w^o (Cook, 1997) is not employed here since non-dimensional derivatives are not considered and so there is no need to differentiate between them in this way.

Under some circumstances, unsteady/oscillatory as opposed to quasi-steady (constant) aerodynamic derivatives may need to be used (see Chapters 9 and 10) and then there is a reduced frequency dependency, and so the solution approaches would be different (see also Chapter 14 on gusts/turbulence). Derivatives are also used elsewhere in the book when inertial axes are employed (see Chapters 12, 14 and 15), but are generally applied to the total forces and moments acting, not perturbation or incremental quantities.

Aerodynamic derivatives are calculated by considering the forces and moments generated in a perturbed flight condition, most conveniently in terms of wind axes, with a transformation to other types of body axes applied afterwards if required. Examples of calculating derivatives are shown in Appendix E, but other references can be consulted for a fuller treatment. It should again be emphasized that the derivatives in this analysis are calculated for wind axes, in comparison to the total derivatives introduced in Chapter 12, where inertial axes are used. Appendix B tabulates the two sets of longitudinal derivatives for the different axes; most are clearly the same while some are different.

13.6.2   Propulsion (or Power) Term

A convenient way of describing the effect of propulsion is to define thrust derivatives e.g.

$$Z_p = Z_T\tau \qquad (13.33)$$

where Z_T is the normal force due to thrust and τ is the thrust perturbation, controlled by the throttle setting.

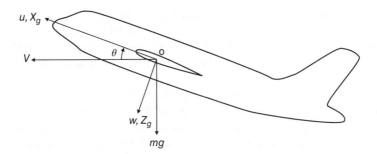

Figure 13.8 Perturbed body axes with gravitational force acting.

13.6.3 Gravitational Term

The aircraft weight resolved along aircraft body axes introduces steady (datum/equilibrium) force components is

$$X_{g_e} = 0 \quad Z_{g_e} = mg \tag{13.34}$$

where the datum inclination of the flight path θ_e has been assumed to be zero for simplicity.

During the small perturbation about the equilibrium condition, the attitude perturbation is θ and Figure 13.8 shows the perturbed body axes. It may be seen that the gravitational force components along the body axes directions in the perturbed state are

$$X_g = -mg\sin\theta = -mg\theta \quad Z_g = mg\cos\theta = mg \tag{13.35}$$

with the simplifications allowed because the attitude perturbation θ is small. Therefore the perturbations in gravitational force about the trim condition are

$$\Delta X_g = -mg\theta \quad \Delta Z_g = 0 \tag{13.36}$$

The expressions are more complex for non-zero initial attitude and may be seen, for example, in (Cook, 1997).

13.7 Modelling the Flexible Aircraft

So far, the emphasis in this chapter has been upon obtaining the basic flight mechanics equations for a *rigid* aircraft. Such an assumption may often suffice for military aircraft where the natural frequencies for the flexible modes (typically > 5 Hz) are significantly greater than the aircraft rigid body frequencies (e.g. short period and Dutch roll motions). The rigid body dynamic and structural dynamic models may then be considered to be uncoupled and the rigid body model used independently. Loads associated with the aircraft response following a control input may then be sufficiently accurately estimated using a purely *rigid* aircraft model.

However, some large commercial aircraft may be very flexible (natural frequencies can be as low as 1 Hz) and structural dynamic effects may not always be negligible. It may be especially important to consider flexibility in the case of oscillatory failures of the FCS or where potential interactions between the pilot and aircraft may occur. Therefore, determining the response to control inputs (and for that matter stability, handling and FCS design) may require an integrated rigid body/flexible aircraft model with flexible modes included. Alternatively, a rigid body model could be used with aerodynamics corrected for flexible effects.

The analysis of a flexible aircraft brings a particular challenge in that, as the aircraft vibrates, the position of the instantaneous centre of mass and the moment of inertia values will both change with time. The choice of a suitable origin and orientation for the axes system becomes important in order to keep the equations as simple as possible and, in particular, to reduce any inertia coupling between the rigid and flexible degrees of freedom. Then, couplings will only occur due to aerodynamic effects. The development of an integrated flight mechanics model for the deformable aircraft is considered in Milne (1961), Babister (1980), Waszak and Schmidt (1988), Schmidt (1998) and Stengel (2004).

In this section, the basic idea of adding flexible modes into the flight mechanics equations will be introduced for a symmetric aircraft in pitch and heave; only a single free–free flexible mode is used in the interests of keeping to a small number of DoF. However, the essential principles that emerge will equally well apply to a full 3D aircraft with multiple flexible modes. Note that care should be taken when comparing different flight mechanics references as there are a number of different notations employed.

13.7.1 Mean Axes Reference Frame

So far, the rigid aircraft in longitudinal motion has a body axis system oxz fixed at the centre of mass, which itself is at a fixed point within the structure. The axes system moves with velocities U, W and rotates with a pitch rate Q. However, when the aircraft is flexible, then in order to minimize any coupling between the rigid body and the flexible mode equations, a 'mean axes reference frame' may be employed (Milne, 1964; Babister, 1980; Waszak and Schmidt, 1988). Such axes move or 'float' in phase with the body motion but they are not attached to a fixed point in the aircraft. The mean axes are defined such that the linear and angular momentum values of the whole aircraft, due to flexible deformation, are zero at every instant of time. Also, the origin of the mean axes system lies at the *instantaneous* centre of mass. In most references, the mean axes conditions and the analysis are expressed in vector form, but here the basic ideas will be explained in scalars.

For the 2D symmetric flexible aircraft, the position of point P defined by rigid (r) and elastic/flexible deformation (e) components (see Equation (13.15)) is shown in Figure 13.9 while momentum components for an elemental mass are shown in Figure 13.10. Hence the linear and angular momentum conditions in scalar form are

$$\int \dot{x}_e \; \mathrm{d}m = 0 \quad \int \dot{z}_e \; \mathrm{d}m = 0 \quad \int \left[(z_r + z_e)\dot{x}_e - (x_r + x_e)\dot{z}_e \right] \mathrm{d}m = 0 \qquad (13.37)$$

If the structural deformation is small, or if the deformation and its rate of change are collinear (i.e. have negligible vector cross-product), then the angular momentum condition may be simplified to

$$\int (z_r \dot{x}_e - x_r \dot{z}_e) \; \mathrm{d}m = 0 \qquad (13.38)$$

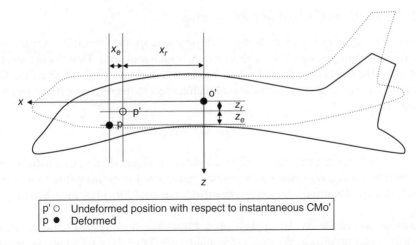

p' ○   Undeformed position with respect to instantaneous CMo'
p ●   Deformed

Figure 13.9   Flexible aircraft in longitudinal motion – rigid and flexible displacement components.

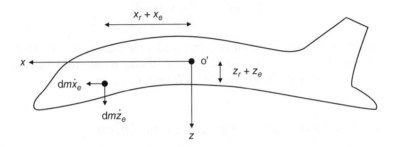

Figure 13.10   Flexible aircraft in longitudinal motion – momentum components.

In addition to these conditions, the axes origin is located at the instantaneous centre of mass, defined in 2D by

$$\int (x_r + x_e)\, dm = 0 \qquad \int (z_r + z_e)\, dm = 0 \tag{13.39}$$

Now the instantaneous moment of inertia in pitch for the flexible aircraft is given by

$$I_y = \int \left[(x_r + x_e)^2 + (z_r + z_e)^2 \right] dm \tag{13.40}$$

and it may be seen that this value will vary in time as the aircraft deforms. However, in the analysis, the moments of inertia are assumed to be constant with time (i.e. variations due to flexible deformation are small).

Thus the position and orientation of these axes have been chosen such that the reference frame motions are inertially decoupled from the flexible deformations; the kinetic energy expression is then considerably simplified, having separate rigid and flexible terms.

13.7.2 Definition of Flexible Deformation

To simplify the analysis somewhat for the 2D flexible aircraft, the aircraft deformation will be considered as occurring in only the z direction (e.g. bending/torsion). Thus the deformation of the structure in the x direction (x_e) will be zero and so will its derivatives with time. The flexible deformation of the aircraft structure measured with respect to the reference frame will then be expressed as

$$z_e(x,y,t) = \kappa_e(x,y)q_e(t) \tag{13.41}$$

where κ_e is a whole aircraft free–free flexible normal mode shape defining modal displacements in the z direction and $q_e(t)$ is the corresponding generalized coordinate (not to be confused with pitch rate q). The mode is clearly symmetric, but could involve wing bending and/or twisting.

If the mean axes conditions are applied when a free–free mode is used to define the deformation, then it may be shown (Waszak and Schmidt, 1988) that the modes must be orthogonal to the rigid body translational and rotational modes; then for 2D motion

$$\int \kappa_e \, dm = 0 \qquad \int \kappa_e x_r \, dm = 0 \tag{13.42}$$

These classical conditions quoted for free–free modes are a consequence of the mean axes assumption. It may also be noted that another way of defining mean axes is that the net inertia force and moment associated with the flexible deformation are zero.

13.7.3 Accelerations in 2D including Flexible Effects

The analysis earlier in Section 13.4 led to the absolute velocity and acceleration components of point P in the oxz axes directions shown in Equations (13.10) and (13.14). Now, when flexible deformation is present only in the z direction ($x_e = \dot{x}_e = \ddot{x}_e = 0$) then the x and z values given in Equation (13.15) may be substituted and the resulting acceleration components are

$$a_x = \dot{U} + WQ - (x_r)Q^2 + (z_r + z_e)\dot{Q} + 2\dot{z}_e Q$$
$$a_z = \dot{W} - UQ - (z_r + z_e)Q^2 - (x_r)\dot{Q} + \ddot{z}_e \tag{13.43}$$

13.7.4 Equations of Motion including Flexible Effects – Motion of Axes

The equations of motion for the flexible aircraft consist of equations governing the motion of the undeformed aircraft (i.e. body fixed axes system) and also the flexible mode response. To determine the first set of equations, consider as before the inertia forces associated with an infinitesimal small mass dm at point P. The equations of equilibrium are similar to those used earlier in Section 13.5.1 except that the instantaneous position of point P (including flexible deformation) is used in the moment equation; the resulting equations are

$$\int a_x dm = X \qquad \int a_z dm = Z \qquad \int [a_x(z_r + z_e) - a_z(x_r + x_e)] \, dm = M \tag{13.44}$$

When the acceleration components are substituted from Equation (13.43) into Equation (13.44) and z_e is replaced by the modal expression in Equation (13.41), the resulting expression initially looks rather complex. However, it may be shown that the equations of motion for the mean axes reference system moving with the body are simply

$$m(\dot{U} + QW) = X \quad m(\dot{W} - QU) = Z \quad I_y \dot{Q} = M \tag{13.45}$$

These are of course the rigid body equations found earlier. Note that there are no couplings to the flexible mode generalized coordinate; they have all been eliminated by using mean axes, free–free modes and the simplifying assumptions introduced earlier. Thus the motion of the reference axes system is independent of the structural deformation until couplings are introduced by aerodynamic effects.

Note that the same results would have been obtained if Lagrange's equations had been employed, but it should be noted that the linear and angular velocities of the centre of mass would need to have been defined with respect to an inertial reference frame. Such an approach is taken in Waszak and Schmidt (1988).

13.7.5   Equations of Motion including Flexible Effects – Modal Motion

There are several ways of determining the equation in q_e, the generalized coordinate for the free–free flexible mode, e.g. Lagrange's equations or the Principle of Virtual Displacements (Davies, 1982). However, for simplicity, Lagrange's equations will be employed.

The *kinetic energy* of the 2D flexible aircraft is given by

$$T = \frac{1}{2} m(U^2 + W^2) + \frac{1}{2} I_y Q^2 + T_s \tag{13.46}$$

where the inertia coupling terms have been omitted, the moment of inertia is constant and the final term in the expression, T_s, is the kinetic energy associated with the flexible mode deformation in the z direction, namely

$$T_s = \frac{1}{2} \int \dot{z}_e^2 \, dm = \frac{1}{2} \int \kappa_e^2 \, dm \, \dot{q}_e^2 = \frac{1}{2} m_e \, \dot{q}_e^2 \tag{13.47}$$

where m_e is the modal mass. The *elastic potential (or strain) energy* term for the flexible aircraft may be expressed in terms of the bending and torsion stiffness distributions over the aircraft, but more conveniently here as

$$U = \frac{1}{2} k_e q_e^2 \tag{13.48}$$

where k_e is the modal stiffness. Then, applying Lagrange's equation for the flexible mode generalized coordinate, adding in a modal damping term c_e and an external modal generalized force Q_{ext} yields the modal equation

$$m_e \ddot{q}_e + c_e \dot{q}_e + k_e q_e = Q_{\text{ext}} \tag{13.49}$$

There are no couplings to Equations (13.45) governing the motion of the mean axes set. It may be seen that this is the classical second-order equation for modal space. If more flexible modes were included, then the orthogonality of the flexible modes would need to be exploited and the final result could be expressed in matrix form.

13.7.6 Full Flight Mechanics Equations with Flexible Modes

Now the equations for the underlying flight mechanics behaviour may be combined together with the equation for the single flexible mode. Consider the linearized version of Equation (13.45), as quoted in Equation (13.25), and omit the variation in the fore-and-aft direction to keep the overall model to 3DoF. Then, adding the modal equation in Equation (13.49), the resulting partitioned equations are

$$
\begin{bmatrix} m & 0 & 0 \\ 0 & I_y & 0 \\ 0 & 0 & m_e \end{bmatrix} \begin{Bmatrix} \dot{w} \\ \dot{q} \\ \ddot{q}_e \end{Bmatrix} + \begin{bmatrix} 0 & -mU_e & 0 \\ 0 & 0 & 0 \\ 0 & 0 & c_e \end{bmatrix} \begin{Bmatrix} w \\ q \\ \dot{q}_e \end{Bmatrix} + \begin{bmatrix} 0 & 0 & 0 \\ 0 & 0 & 0 \\ 0 & 0 & k_e \end{bmatrix} \begin{Bmatrix} \int w \\ \int q \\ q_e \end{Bmatrix} = \begin{Bmatrix} Z \\ M \\ Q_{ext} \end{Bmatrix}
\tag{13.50}
$$

These flight mechanics and modal equations are uncoupled dynamically through the use of mean axes though, of course, a given control input could excite a response in both sets of equations. However, once aerodynamic terms are introduced, coupling will occur through the aerodynamic derivatives but generation of such terms is far from straightforward. Equivalent equations could be written for the non-linear flight mechanics case and for lateral motion.

13.8 Solution of Flight Mechanics Equations for the Rigid Aircraft

Finally, returning to the rigid aircraft and once aerodynamic terms are included, the combined flight mechanics equations (linear or non-linear) could be solved for a variety of control inputs and the responses used to calculate internal loads. Also, the Flight Control System (FCS) could be added and this will introduce further couplings.

13.8.1 Solving the Longitudinal Non-linear Equations of Motion

For response determination, the full set of equations is solved in time, so allowing any non-linear aerodynamic or FCS effects to be included. The resulting accelerations may then be used to obtain inertia force distributions, and hence internal loads, using the force summation approach described later in Chapter 16.

The equations may be expressed in a block diagram form (cf. analogue computer) and solved using a simulation software package such as SIMULINK. The first part of the simulation solves the differential equations for U, W and Q for the 2D rigid aircraft. The second part of the simulation converts velocities to earth axes using the transformation matrices $\mathbf{A}_L$ and $\mathbf{A}_R$ followed by integration to determine the position and orientation of the aircraft. The input forces and moments are calculated according to the gravitational, aerodynamic, control and propulsive actions. MATLAB and SIMULINK programs to solve the dynamic manoeuvre for a rigid aircraft in heave and pitch are included in the companion website.

13.8.2   Dynamic Stability Modes

A classical analysis using the rigid aircraft linearized flight mechanics equations involves examining the dynamic stability of the aircraft in longitudinal or lateral motion. The equations are written in Laplacian form, relevant transfer functions are generated and the characteristic polynomial (given by the denominator of the transfer function) is solved to yield the roots of the constituent motions (Cook, 1997; also see Chapter 6). The motion of the aircraft following a disturbance from the trim condition involves a combination of so-called 'dynamic stability modes' (Cook, 1997). These modes are different to the free-free rigid body modes (see Chapter 3 and Appendix A) since the dynamic stability modes include the effects of the aerodynamic couplings between the rigid body motions. Some dynamic stability modes are oscillatory whereas others are non-oscillatory, governed respectively by complex and real roots in the characteristic equation. The frequencies of the dynamic stability modes are generally well below any flexible mode frequencies except for very large commercial aircraft.

For *longitudinal* motion, there are two dynamic stability modes, both of which are oscillatory. The very low frequency (and lightly damped) *phugoid* mode involves an oscillation in forward speed, coupled with pitch and height variation; this mode requires the fore-and-aft motion of the aircraft to be included. The higher frequency (and more heavily damped) *short period* mode is dominated by a pitching oscillation, with the forward speed remaining essentially constant. Each mode will have positive damping for a commercial aircraft; however, the FCS will be used to create the desired stability and handling characteristics of the aircraft by modifying the dynamic stability modes.

For *lateral* motion, the characteristic equation has two real roots and one pair of complex roots, so there are three dynamic stability modes whose stability has to be examined. The *Dutch roll* mode involves a damped oscillation in yaw, coupled primarily to roll but also to sideslip. The *spiral* mode is non-oscillatory and is a combination of roll, yaw and sideslip motions. The *roll subsidence* mode is a non-oscillatory rolling motion, largely decoupled from the spiral and dutch roll modes. It should be noted that the non-oscillatory modes involve an exponentially decaying motion following a disturbance.

When the flexible aircraft is considered, the basic flight mechanics behaviour will alter since the dynamic stability modes will be modified by elastic contributions and the stability may be compromised. There will be an overall impact on the FCS design to ensure that the handling qualities of the flexible aircraft remain satisfactory and also the flight simulator model will need to be modified to include the influence of flexible mode deformations.

13.9   Dynamic Manoeuvre – Rigid Aircraft in Longitudinal Motion

In this section, the dynamic behaviour of a rigid aircraft responding in heave and pitch to an elevator (or pitch control) input will be considered. Such a model allows the rational unchecked and checked pitch manoeuvres to be simulated. The aircraft representation is shown in Figure 13.11, where the notation used is the same as in Chapter 12. For simplicity, the thrust and drag are assumed to be in line, so that they do not contribute to the pitching moment equation. Later in the chapter, the flexible aircraft case will be considered.

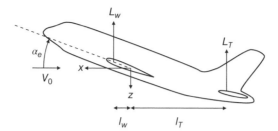

Figure 13.11 Rigid aircraft in dynamic heave/pitch manoeuvre under elevator application.

13.9.1 Flight Mechanics Equations of Motion – Rigid Aircraft in Pitch

The linearized flight mechanics equations of motion for the longitudinal case (symmetric rigid aircraft) were shown earlier in Section 13.5.3. Now, for this example, it is assumed that there are no gravitational terms and no variation in fore-and-aft motion, so u is zero and only two equations are required; this reduced-order model will not therefore represent the long term motion governed by the phugoid mode, but it will approximate the short period motion over moderate timescales. Thus the equations governing the heave/pitch motion are

$$m(\dot{w} - U_e q) = Z \qquad I_y \dot{q} = M \tag{13.51}$$

The normal force Z and pitching moment M may be expressed in terms of aerodynamic stability derivatives as

$$Z = Z_{\dot{w}}\dot{w} + Z_{\dot{q}}\dot{q} + Z_w w + Z_q q + Z_\eta \eta$$

$$M = M_{\dot{w}}\dot{w} + M_{\dot{q}}\dot{q} + M_w w + M_q q + M_\eta \eta \tag{13.52}$$

Note that these derivatives may be expressed in body or wind axes, and a transformation between the axes may be necessary.

13.9.2 Aerodynamic Stability Derivatives in Heave/Pitch

The stability derivatives in Equation (13.52) are related to perturbations about axes fixed in the aircraft (usually wind axes) for dynamic manoeuvres, as opposed to the derivatives used in the equilibrium manoeuvres (see Chapter 12) where small angles and displacements about inertial axes (fixed in space) were considered. However, because the perturbations employed in defining stability derivatives are small, and not all stability derivatives involve perturbations of the aircraft velocity, many of the values turn out to be the same as those obtained for the inertial axes (see the comparison table in Appendix B). It is arguable that using wind axes is preferable to generalized body axes since the aerodynamic derivatives are simpler in form; note that for this simple case of longitudinal motion, the pitch moment of inertia will be the same for both axes sets.

The Z_w derivative in Equations (13.52) for the perturbed normal velocity w is derived in Appendix E as an example of how derivatives involving aircraft velocity perturbations may be determined. However, it may be seen in Appendix B that this 'w stability derivative' differs by a factor of V_0 from the 'α inertial derivative' Z_α because of the difference in the equation variable ($\alpha = w/V_0$) and also the drag coefficient C_D is present due to the velocity perturbations for the stability derivative. The M_w derivative is obtained in a similar way.

The stability derivatives Z_q, M_q, Z_η, M_η for perturbed q and η are found to be the same as those derived in Chapter 12 for inertia axes because there are no perturbations of the flight velocities involved. The aerodynamic stability derivatives for heave and pitch acceleration depend upon the downwash lag effect and aerodynamic inertia terms that are often neglected (Cook, 1997); these derivatives will be ignored here, so

$$Z_{\dot{w}} = Z_{\dot{q}} = M_{\dot{w}} = M_{\dot{q}} = 0 \tag{13.53}$$

Thus, in summary, the non-zero derivatives for wind axes used in Equations (13.52) are given by

$$Z_w = -\frac{1}{2}\rho V_0(S_W a_W + S_T a_T(1-k_\varepsilon) + S_W C_D) \quad Z_q = -\frac{1}{2}\rho V_0 S_T a_T l_T \quad \text{(tailplane effect only)}$$

$$M_w = \frac{1}{2}\rho V_0[S_W a_W l_W - S_T a_T l_T(1-k_\varepsilon)] \qquad Z_\eta = -\frac{1}{2}\rho V_0^2 S_T a_E \tag{13.54}$$

$$M_q = -\frac{1}{2}\rho V_0 S_T a_T l_T^2 \quad \text{(tailplane effect only)} \quad M_\eta = -\frac{1}{2}\rho V_0^2 S_T a_E l_T$$

where the notation in Cook (1997) has been converted to that used in this book. Note that no zero incidence terms are present in the equations for wind axes fixed in the aircraft.

If the equations were to be expressed in generalized body axes (aligned with principal axes), then these wind axes derivatives would need to be transformed through the trim incidence angle α_e, noting that both forces/moments and linear/angular velocities need to be transformed; the methodology is described in (Cook, 1997).

13.9.3   Solution of the Flight Mechanics Equations – Rigid Aircraft

The flight mechanics equations of motion for the rigid aircraft may then be written in matrix form showing terms due to inertia and aerodynamics forces as

$$\begin{bmatrix} m & 0 \\ 0 & I_y \end{bmatrix} \begin{Bmatrix} \dot{w} \\ \dot{q} \end{Bmatrix} + \begin{bmatrix} 0 & -mU_e \\ 0 & 0 \end{bmatrix} \begin{Bmatrix} w \\ q \end{Bmatrix} - \begin{bmatrix} Z_w & Z_q \\ M_w & M_q \end{bmatrix} \begin{Bmatrix} w \\ q \end{Bmatrix} = \begin{Bmatrix} Z_\eta \\ M_\eta \end{Bmatrix} \eta \tag{13.55}$$

with the derivatives defined in Equation (13.54). This equation may be solved to determine the transient response, relative to the wind axes, corresponding to a particular time varying elevator input. The velocity w is difficult to interpret physically so it is arguably easier to consider the variation of incidence using $\alpha = w/V_0$ for small perturbations. A further useful response parameter is the incremental normal acceleration at an arbitrary position x forward of the centre of mass, namely $a_z = \dot{w} - qU_e - x\dot{q}$.

13.9.4   Pitch Rate per Elevator Transfer Function

The transfer function relating pitch rate to elevator angle may be written by transforming Equation (13.55) to the Laplace domain (see Chapter 6) such that

$$\begin{bmatrix} sm-Z_\mathrm{w} & -mU_\mathrm{e}-Z_\mathrm{q} \\ -M_\mathrm{w} & sI_\mathrm{y}-M_\mathrm{q} \end{bmatrix} \left\{ \begin{array}{c} w(s) \\ q(s) \end{array} \right\} = \left\{ \begin{array}{c} Z_\eta \\ M_\eta \end{array} \right\} \eta(s) \tag{13.56}$$

This matrix equation may be solved for the pitch rate per elevator angle, so that

$$\left(\frac{q}{\eta}(s) \right)_{\mathrm{Rigid}} = \frac{smM_\eta + M_\mathrm{w}Z_\eta - M_\eta Z_\mathrm{w}}{D(s)} \tag{13.57}$$

Here the denominator polynomial $D(s)$ is the determinant of the square matrix in Equation (13.56), namely

$$D(s) = s^2 \left[I_\mathrm{y}m \right] + s \left[-I_\mathrm{y}Z_\mathrm{w} - mM_\mathrm{q} \right] + \left[Z_\mathrm{w}M_\mathrm{q} - \left(mU_\mathrm{e}+Z_\mathrm{q} \right)M_\mathrm{w} \right] \tag{13.58}$$

and this will define the characteristic (quadratic) equation of the system. The steady state pitch rate per elevator deflection is given by the transfer function at zero frequency ($s = 0$), namely

$$\left(\frac{q}{\eta} \right)_{\mathrm{Rigid}} = \frac{M_\mathrm{w}Z_\eta - M_\eta Z_\mathrm{w}}{Z_\mathrm{w}M_\mathrm{q} - \left(mU_\mathrm{e}+Z_\mathrm{q} \right)M_\mathrm{w}} \tag{13.59}$$

This expression may be compared later to the flexible aircraft result and the elevator effectiveness then determined.

When the elevator is deflected trailing edge upwards (actually negative) for a rigid aircraft then there is a reduced tailplane lift, the aircraft pitches nose up (positive), generates additional incidence and hence sufficient additional wing lift so as to override the tailplane lift reduction and allow the aircraft to heave upwards; this will be seen in the example for a step elevator input in Section 13.9.8.

13.9.5   Short Period Motion

The roots of the quadratic Equation (13.58) define the characteristic short period motion of the rigid aircraft. The roots are normally a complex conjugate pair in the form

$$-\zeta_\mathrm{S}\omega_\mathrm{S} \pm i\omega_\mathrm{S}\sqrt{1-\zeta_\mathrm{S}^2} \tag{13.60}$$

where ω_S is the undamped frequency of the short period motion (typically 1–10 rad/s) and ζ_S is its damping ratio (stabilizing for a commercial aircraft, typically 20–50% critical). The mode involves a combination of pitch and heave motion. The short period motion may be significantly affected when the FCS is included in the model.

13.9.6   Phugoid Motion

If the longer term motion were to be studied, then the equation governing u would need to be included, as would the pitch rate/pitch angle equation together with gravitational and thrust terms (Cook, 1997). The characteristic equation would then become a quartic polynomial, giving two sets of complex roots for the phugoid and short period motions. The phugoid mode is a lightly damped, low frequency oscillation in speed u that couples with pitch angle and height but has a relatively constant incidence. The Lanchester approximation to the phugoid mode (Cook, 1997) yields a simple value for the phugoid frequency as $\omega_P = \sqrt{2}\frac{g}{V_0}$ (a very low value, e.g. 0.09 rad/s at 150 m/s TAS).

When the transfer function for the reduced model in Equation (13.57) is used to determine the response to a step elevator input, a steady pitch rate results; however, in practice the long term pitch rate will tend to zero if the full longitudinal equations are used, since the gravitational and drag terms will come into play unless the thrust is adjusted (Cook, 1997). However, an on/off or limited sinusoidal elevator input would lead to a more short term motion and the reduced model is considered to be satisfactory to illustrate such a case.

13.9.7 Conversion to Earth Axes Motion

Once the flight mechanics equations for the pitch rate $q(t)$ and heave $w(t)$ with respect to axes moving with the aircraft have been solved, the motion with respect to earth axes needs to be found. To determine the velocities and position in earth fixed axes, the Euler transformation in Section 13.5.3 must be employed, namely

$$U_E = U_e + W_e\theta \qquad W_E = -U_e\theta + W_e + w \qquad \dot{\theta} = q \tag{13.61}$$

This equation has been linearized for small perturbations.

Firstly, the pitch rate $\dot{\theta}$ can be determined simply from q and the pitch angle θ may then be found by integrating $\dot{\theta}$ using a suitable initial condition so as to yield the aircraft pitch motion in space. The other two equations may then be employed to find the component velocities in earth axes, followed by integration to find the aircraft position (X_E, Z_E) and hence the flight path during the manoeuvre. Note that prior to the dynamic manoeuvre caused by application of the elevator, the aircraft is assumed to be in straight and level trimmed or equilibrium (subscript e) flight at velocity U_e $(\theta_e = W_e = 0)$ so the transformation is simplified.

Clearly, the aircraft motion may be seen from the change in position in earth axes and/or by considering the perturbations in the flight path angle γ given by $\gamma = \theta - \alpha \cong \theta - w/V_0$ (Cook, 1997). The steady flight path angle γ_e (note different usage of the torsion mode shape symbol) at the commencement of the manoeuvre will be zero since the aircraft is in level flight $(\theta_e = 0)$ and wind axes are used $(\alpha_e = 0)$.

13.9.8 Example: Rigid Aircraft in Heave/Pitch

Consider an aircraft with the following data (essentially the same as that in Chapter 12): $m = 10\,000$ kg, $I_y = 144\,000$ kg m^2, $S_W = 30$ m^2, $S_T = 7.5$ m^2, $c = 2.0$ m, $l_W = 0.3$ m, $c = 0.6$ m, $l_T = 3.5$ m, $c = 7$ m, $a_W = 4.5$/rad, $a_T = 3.2$/rad, $a_E = 1.5$/rad, $k_\varepsilon = 0.35$, $\alpha_0 = -0.03$ rad, $C_D = 0.1$ and $C_{M_{0W}} = -0.03$. These data are for the rigid aircraft, with further parameters defined later for the flexible aircraft. Note that the data correspond to a static margin of 0.2 m (10% chord). The aim of the example is to determine the short period mode properties and the aircraft response relative to wind and earth axes for the aircraft in straight and level flight at 175 m/s EAS $(U_e = V_0,\ W_e = \theta_e = 0)$ when subjected to various elevator (or pitch) control inputs.

MATLAB and SIMULINK programmes to solve the dynamic manoeuvre for this rigid aircraft in heave/pitch are included in the companion website.

Firstly, it may be shown by calculation that the short period mode has a natural frequency of 2.25 rad/s (0.36 Hz) and damping of 56% at 175 m/s EAS. The frequency varies linearly with velocity but the damping ratio is unaffected. Ideally, the damping would be somewhat lower and this could be achieved by moving the centre of mass further forward. However, it was desired to leave the aircraft centre of mass behind the wing centre of mass in this simple example.

Secondly, when a step elevator input of $-1°$ (i.e. upwards) is applied to the aircraft, then the pitch rate q, incidence α, normal acceleration a_z at the centre of mass, pitch angle θ and flight path angle γ responses are as shown in Figure 13.12. The step elevator input leads to a steady nose up incidence, nose up pitch rate and upwards normal acceleration (positive down); the corresponding pitch and flight path angles increase more or less linearly with time, with the flight path angle lagging the pitch angle by about 0.6 s. Examination of the aircraft position in earth axes (X_E, Z_E) shows the aircraft climbing steadily through about 60 m after 4 s. It was pointed out earlier that using the full equations, including forward speed variation and drag effects, etc., would lead to a different long term response due to the phugoid influence.

Thirdly, a single cycle of a sinusoidal elevator input of frequency 0.25 Hz, duration 4 s and amplitude $-1°$ is applied (similar to the checked pitch manoeuvre). The response to this input is shown in Figure 13.13(a). The aircraft pitches nose up, then nose down and settles back at zero pitch rate, normal acceleration and pitch angle; the maximum incremental acceleration is 1.09 g. The flight path angle also returns to zero. This result, together with the aircraft motion in earth axes shown in Figure 13.13(b), indicates that the aircraft has gained about 25 m in height

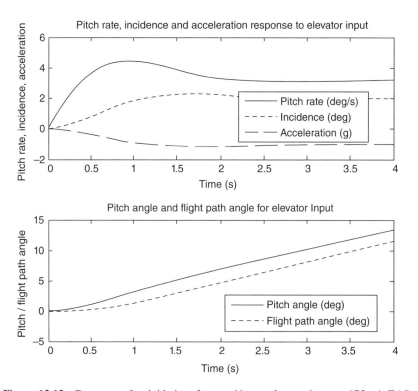

Figure 13.12 Response of a rigid aircraft to a $-1°$ step elevator input at 175 m/s EAS.

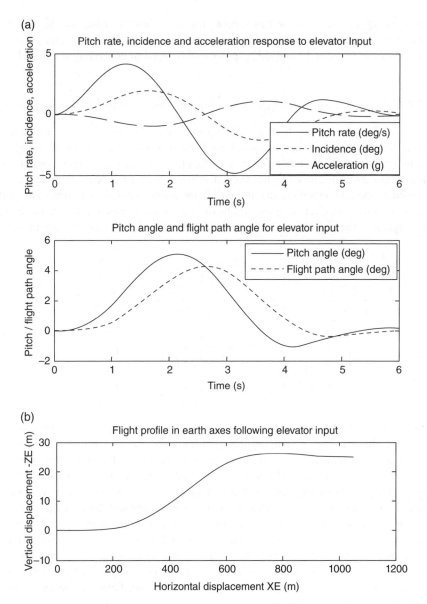

Figure 13.13 Response of a rigid aircraft to a single cycle of sinusoidal elevator input (0.25 Hz/4 s/–1°): (a) responses and (b) aircraft motion in earth axes.

and reached a new level flight condition. Note that different frequencies and numbers of input cycles will yield different responses; e.g. multiple cycles of input will increase the response.

13.10   Dynamic Manoeuvre – Flexible Aircraft Heave/Pitch

In Chapter 12, the equilibrium manoeuvre calculation was shown for a simple flexible aircraft in heave and pitch. The motions involved were rigid body heave and pitch, together with a symmetric free–free mode (see Appendix C) with potential wing bending, wing twist or fuselage bending deformations. In this section, the same flexible mode will be added to the flight mechanics equations, in order to investigate the flexible aircraft in a dynamic manoeuvre. Some results from Chapter 12 and this chapter will be used in generating the model.

13.10.1 Flight Mechanics Equations of Motion – Flexible Aircraft in Pitch

In Section 13.7, it was shown how a flexible mode could be introduced into the equations of motion for the flight mechanics model when mean axes were employed, together with certain assumptions (e.g. small deformations). Then the heave and pitch equations, neglecting fore-and-aft motion as discussed earlier and extending them to include the flexible mode, were written with no inertia coupling terms. If the right-hand side aerodynamic forces are written in aerodynamic stability derivative terms and flexible derivatives added, rather as in Chapter 12 but for wind axes, the equations of motion for the flexible aircraft may be simplified to

$$
\begin{aligned}
&\left\{\begin{bmatrix} m & 0 & 0 \\ 0 & I_y & 0 \\ 0 & 0 & m_e \end{bmatrix} \begin{Bmatrix} \dot{w} \\ \dot{q} \\ \ddot{q}_e \end{Bmatrix} + \left\{ \begin{bmatrix} 0 & -mU_e & 0 \\ 0 & 0 & 0 \\ 0 & 0 & c_e \end{bmatrix} - \begin{bmatrix} Z_w & Z_q & Z_{\dot{e}} \\ M_w & M_q & M_{\dot{e}} \\ Q_w & Q_q & Q_{\dot{e}} \end{bmatrix} \right\} \begin{Bmatrix} w \\ q \\ \dot{q}_e \end{Bmatrix} \right. \\[2mm]
&\left. + \left\{ -\begin{bmatrix} 0 & 0 & Z_e \\ 0 & 0 & M_e \\ 0 & 0 & Q_e \end{bmatrix} + \begin{bmatrix} 0 & 0 & 0 \\ 0 & 0 & 0 \\ 0 & 0 & k_e \end{bmatrix} \right\} \begin{Bmatrix} \int w \\ \int q \\ q_e \end{Bmatrix} = \begin{Bmatrix} Z_\eta \\ M_\eta \\ Q_\eta \end{Bmatrix} \eta \right.
\end{aligned}
\tag{13.62}
$$

This equation may be used to determine the response to any elevator input. Here the matrix partitions separate what are essentially the rigid body and flexible terms; the integral terms are somewhat artificial as they are multiplied by zero, but are included to allow a convenient matrix form of the equations. While the rigid body and flexible equations are actually uncoupled wind-off, the aerodynamic terms introduce couplings between them. If more modes were added, the scalar modal mass, damping and stiffness terms would be replaced by diagonal matrices and additional aerodynamic coupling terms between the modes would be included. It should be noted that the acceleration derivatives have been neglected and that flexible derivatives for both flexible deformation and rate of deformation have been included.

Note that the incremental normal acceleration at position x forward of the centre of mass is $a_z = \dot{w} - qU_e - x\dot{q} + \ddot{z}_s$, where $\ddot{z}_s = \kappa_{ex}\ddot{q}_e$ is the flexible mode contribution to the acceleration and κ_{ex} is the mode shape at position x.

13.10.2 Aerodynamic Derivatives for Flexible Aircraft

In considering the derivatives that need to be calculated, there is no need to obtain again the derivatives that appeared in the rigid aircraft equation of motion (13.55) and were tabulated earlier in Equation (13.54), namely $Z_w, Z_q, Z_\eta, M_w, M_q, M_\eta$. Also, the flexible pitch rate and elevator derivatives Q_q, Q_η and the flexible deformation derivatives Z_e, M_e, Q_e corresponding to perturbations in elastic mode deformation may be shown to be the same as those quoted in Chapter 12 for inertial axes, as seen in Appendix B. This means that only the derivative Q_w and the derivatives $Z_{\dot{e}}, M_{\dot{e}}, Q_{\dot{e}}$ related to perturbations in the rate of elastic deformation need to be obtained here.

To obtain Q_w, a modal force derivative, the incremental work term for the perturbation forces generated by perturbed motion w acting through an incremental flexible deformation may be considered. The perturbation lift forces (upwards) acting on the wing strip and complete tailplane due to perturbation w alone are related to the incidence $\alpha = w/V_0$, so

$$dL_W = \frac{1}{2}\rho V_0^2 c a_W \left(\frac{w}{V_0}\right) dy \quad \Delta L_T = \frac{1}{2}\rho V_0^2 S_T a_T \left(\frac{w}{V_0}\right) \tag{13.63}$$

because of the effective incidence produced by the perturbation in heave motion. The corresponding incremental displacements due to incremental flexible mode deformation at the wing and tailplane aerodynamic centres are $[\kappa_e(y) - l_A \gamma_e(y)]\delta q_e$ and $\kappa_{eT}\delta q_e$ (both downwards) respectively, where functions κ_e, γ_e define the symmetric flexible mode shape (see Appendix C). Thus the incremental work done terms may be determined, differentiated with respect to δq_e, and the result inspected to yield the derivative for perturbed motion

$$Q_w = \frac{\partial Q}{\partial w} = \frac{1}{2}\rho V_0 [-S_W a_W J_2 - S_T a_T (1 - k_\varepsilon)\kappa_{eT}] \tag{13.64}$$

where $J_2 = (1/s)\displaystyle\int_{y=0}^{s} (\kappa_e - l_A \gamma_e) dy$ and again this derivative is similar to Q_α used for the equilibrium manoeuvres.

Now it is assumed that the flexible rate derivatives $Z_{\dot{e}}, M_{\dot{e}}, Q_{\dot{e}}$, rather like the pitch rate derivatives, will be dominated by tailplane effects so, for simplicity, terms involving the wing will be neglected. The perturbation tailplane lift force and pitching moment contributions due to perturbed flexible rate effects will be related to the flexible derivatives $Z_{\dot{e}}$ and $M_{\dot{e}}$ in the heave and moment equations,

$$\Delta L_T = \frac{1}{2}\rho V_0^2 S_T a_T \left(\frac{\kappa_{eT}\dot{q}_e}{V_0}\right) = -Z_{\dot{e}}\dot{q}_e \quad \Delta M_T = -\frac{1}{2}\rho V_0^2 S_T a_T \left(\frac{\kappa_{eT}\dot{q}_e}{V_0}\right) l_T = M_{\dot{e}}\dot{q}_e \tag{13.65}$$

where the vertical velocity at the tailplane due to the flexible rate dictates the effective incidence. The unsteady pitch rate terms associated with the tailplane modal slope γ_{eT} are ignored. Thus, by inspection, the elastic rate derivatives are

$$Z_{\dot{e}} = -\frac{1}{2}\rho V_0 S_T a_T \kappa_{eT} \quad M_{\dot{e}} = -\frac{1}{2}\rho V_0 S_T a_T l_T \kappa_{eT} \tag{13.66}$$

Finally, the flexible rate derivative in the generalized equation $Q_{\dot{e}}$ requires the incremental work done due to the tailplane lift force in Equation (13.65) acting through the incremental displacement in the mode at the tailplane, so

$$\delta W_{\dot{e}} = -\frac{1}{2}\rho V_0^2 S_T a_T \left(\frac{\kappa_{eT}\dot{q}_e}{V_0}\right)(\kappa_{eT}\delta q_e) \tag{13.67}$$

Then the relevant generalized force in the modal equation is

$$Q_{\dot{e}} = -\frac{1}{2}\rho V_0 S_T a_T \kappa_{eT}^2 \tag{13.68}$$

Thus all the derivatives are available so the flight mechanics Equation (13.62) for the flexible aircraft may now be solved and the transformations to earth axes carried out as before.

13.10.3 Pitch Rate per Elevator Transfer Function

The transfer function relating pitch rate to elevator angle for the flexible aircraft may be written by transforming Equation (13.62) to the Laplace domain

$$\begin{bmatrix} sm - Z_w & -mU_e - Z_q & -sZ_{\dot{e}} - Z_e \\ -M_w & sI_y - M_q & -sM_{\dot{e}} - M_e \\ -Q_w & -Q_q & s^2 m_e + sc_e + k_e - sQ_{\dot{e}} - Q_e \end{bmatrix} \begin{Bmatrix} w(s) \\ q(s) \\ q_e(s) \end{Bmatrix} = \begin{Bmatrix} Z_\eta \\ M_\eta \\ Q_\eta \end{Bmatrix} \eta(s) \tag{13.69}$$

These matrix equations may be solved for pitch rate per elevator angle but the expression will be rather complicated so only its general form will be shown, namely

$$\left(\frac{q}{\eta(s)}\right)_{\text{Flexible}} = \frac{N(s)}{D(s)} \tag{13.70}$$

Here the denominator polynomial $D(s)$ is the determinant of the square matrix in Equation (13.69) and defines the characteristic quartic equation of the system and its roots (see later), while the numerator polynomial $N(s)$ is a cubic expression.

The steady-state pitch rate per elevator deflection is given by the above transfer function evaluated at zero frequency, namely

$$\left(\frac{q}{\eta}\right)_{\text{Flexible}} = \frac{N(0)}{D(0)} \tag{13.71}$$

13.10.4 Elevator Effectiveness

In a similar manner to that used in Chapter 8, the ratio of the flexible to rigid values of the steady-state pitch rate per elevator angle provides a measure of elevator effectiveness, namely

$$\mathfrak{I}_{\text{Elevator}} = \frac{(q/\eta)_{\text{Flexible}}}{(q/\eta)_{\text{Rigid}}} \tag{13.72}$$

where the constituent terms are given in Equations (13.71) and (13.59). The action of an elevator deflection on the pitch rate for the rigid aircraft was explained in Section 13.9.4. To assist in understanding the effect of flexibility on elevator effectiveness, the three different mode cases considered earlier in Chapter 12 will be discussed.

13.10.4.1 Fuselage Bending Mode

Consider the aircraft example where the flexible mode involves fuselage bending but the wing is rigid in bending and twist. The downwards tailplane lift due to elevator deflection (trailing edge upwards) will cause bending of the fuselage (in the 'hogging' sense) such that the tailplane incidence will increase, so tending to offset the intended elevator action and reducing its effectiveness.

13.10.4.2 Wing Bending Mode

Consider the aircraft example where the flexible mode involves wing bending only, with no fuselage deformation. When the elevator is deflected, the aircraft will pitch nose up, the incidence will increase and additional wing lift will be developed, so causing the wing to bend upwards. In the steady case, this would not be expected to make any difference to the effectiveness. However, as the wing bends upwards, there will be a small downwards fuselage heave component and a small nose down pitch (seen in the mode shape in Appendix C); then the flexible deformation is expected to introduce a small additional down force on the tailplane, therefore actually increasing the effectiveness slightly.

13.10.4.3 Wing Torsion Mode

Consider the aircraft example where the flexible mode involves wing twist only, with no fuselage deformation. When the elevator is deflected, the aircraft will pitch nose up, the wing incidence will increase and additional wing lift will be developed. However, this lift will cause a nose up twist of the wing and therefore produce further lift. In this case, the effectiveness would increase as the velocity increases.

Therefore, as the velocity increases for a given flexible mode frequency, there will be a loss or increase of control effectiveness depending upon the mode shape. This behaviour will be examined later via an example.

13.10.5 Short Period/Flexible Modes

The roots of the quartic characteristic equation define how the short period motion of the aircraft is affected by flexible deformation. The roots are usually two complex conjugate pairs, namely

$$-\zeta_S \omega_S \pm i\omega_S \sqrt{1-\zeta_S^2} \quad \text{and} \quad -\zeta_E \omega_E \pm i\omega_E \sqrt{1-\zeta_E^2} \qquad (13.73)$$

where ω_S, ω_E are the undamped frequencies and ζ_S, ζ_E are the damping ratios of the constituent motions. The 'S' mode will be a short period type heave/pitch motion modified by the flexible mode while the 'E' mode will be the flexible (or elastic) mode changed by the presence of the rigid body motion. The distinction is rather 'fuzzy' as both roots are likely to end up with short period and flexible mode components; the frequency and mode shape of the flexible mode will dictate how much coupling there will be. Note that it is possible that the aircraft could become unstable in what would in effect be a rigid body/flexible mode coupled flutter; a lot would depend upon the aerodynamics of course. Also, under some circumstances, two real roots could replace the complex roots for the 'S' mode and potentially lead to divergence if one real root were to become positive.

If the phugoid motion were to be included, the additional longitudinal equation would need to be added and a further pair of complex roots obtained from a 6th order characteristic polynomial. It is possible that aircraft flexibility would modify the phugoid behaviour.

13.10.6   Example: Flexible Aircraft in Heave/Pitch

This example will consider the same data as used for the rigid aircraft example in Section 13.9.8 but additional parameters need to be specified to cater for introduction of idealized flexible effects as used in Chapter 12, namely fuselage mass terms $m_F = 1500$ kg, $m_C = 4000$ kg, $m_T = 1500$ kg, wing mass/inertia terms $m_W = 2\mu_W s = 3000$ kg, $I_W = 2\chi_W s = 1330$ kg m^2, pitch moment of inertia $I_y = 144\ 000$ kg m^2, and dimensions $s = 7.5$ m, $l_A = 0.25$ m, $l_E = 0.25$ m, $l_{WM} = 0.1$ m and $l_F = 6.8$ m. The corresponding modal mass and mode shape parameters for (a) fuselage bending, (b) wing bending and (c) wing twist modes being dominant are shown in Appendix C. The effect of air speed and flexible mode natural frequency on the pitch performance (i.e. elevator effectiveness) and stability will be examined using the approaches described earlier.

13.10.6.1   Fuselage Bending Mode

Consider the fuselage bending case where the wing is rigid. Firstly, several natural frequencies, namely 1.5, 3, 4.5 and 6 Hz, were selected and the elevator effectiveness calculated for a range of velocities; the result is shown in Figure 13.14. As would be expected, the lower the natural frequency, the more flexible the fuselage and the more the effectiveness reduces. The trends in

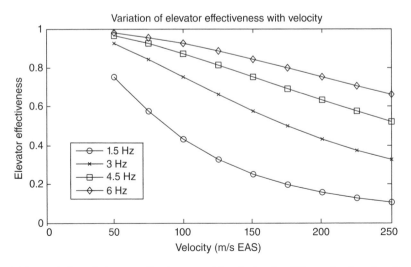

Figure 13.14   Variation of elevator effectiveness with velocity for different fuselage bending natural frequencies.

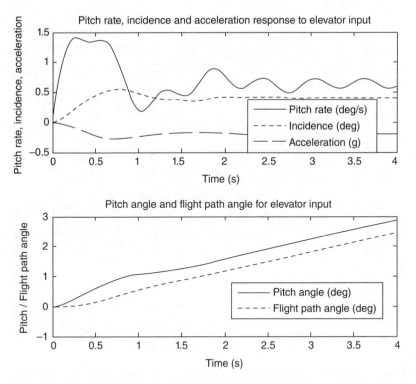

Figure 13.15 Response of a flexible aircraft (1.5 Hz/2% fuselage bending mode) to a 1° step elevator input at 175 m/s EAS.

the figure indicate that effectiveness could become small but that elevator reversal will not occur for this simple aircraft model.

Secondly, the flexible aircraft response to a step elevator input of −1° is calculated at 175 m/s EAS and compared to the rigid aircraft result in Figure 13.12. A low natural frequency of 1.5 Hz with 2% damping is chosen in order to show a noticeable effect on the response, though fuselage modes are higher in frequency for this size of aircraft. The response is presented in Figure 13.15 and the low frequency decaying oscillation due to the flexible mode is apparent, especially on the pitch rate; the response levels are considerably lower (~1/4) than for the rigid aircraft due to the loss of effectiveness of around 80%.

Thirdly, the flexible aircraft stability is examined for the 1.5 Hz fuselage bending mode with 2% damping. The variation of frequency and damping of the two modes ('S' and 'E') with velocity is shown in Figure 13.16. As the velocity increases so does the 'S' mode natural frequency, much more rapidly than for the rigid aircraft (where the frequency at 250 m/s EAS would be 0.51 Hz), and it becomes closer to the 'E' mode frequency. Eventually at 227 m/s EAS the damping crosses zero and in effect a coupled rigid body/flexible mode flutter would occur. For more realistic frequencies of the fuselage bending mode, flutter would not occur at any reasonable velocity.

13.10.6.2 Wing Bending Mode

Now consider the wing bending mode case where the fuselage is rigid but actually has a small heave and pitch component in the mode shape (see Appendix C). Firstly, several natural

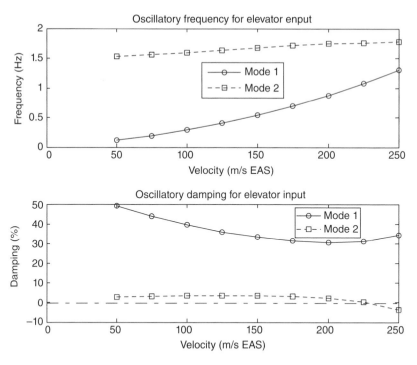

Figure 13.16 Variation of natural frequency and damping with velocity for modes 1 ('S') and 2 ('E') for the flexible aircraft (fuselage bending mode 1.5 Hz/2%).

frequencies, namely 1.5, 3, 4.5 and 6 Hz, are selected and the effectiveness calculated for a range of velocities; the results show, as anticipated from the discussion in Section 13.10.4, that there is only a very small change in effectiveness due to flexibility, namely an increase of about 3% for the lowest natural frequency.

Secondly, the flexible aircraft response to a step elevator input of –1° is calculated at 175 m/s EAS to be compared to the rigid aircraft result in Figure 13.12. A low natural frequency of 1.5 Hz with 2% damping is chosen in order to show whether there is any significant effect on the response; again this value is artificially low. The response to the step input is not presented since there is only a slight difference to the rigid aircraft response (~2%); the damping for the short period mode has changed from 56 to 59% with the addition of flexibility.

Thirdly, the flexible aircraft stability is examined for the 1.5 Hz wing bending mode with 2% damping. The 'S' mode frequency increases with velocity (at about the same rate as for the rigid aircraft) but there is very little change in the 'E' mode frequency. The damping changes are also small, but at 281 m/s EAS the 'E' mode damping crosses zero and a 'soft' flutter occurs (see Chapter 10).

13.10.6.3 Wing Torsion Mode

Now consider the wing torsion case where the fuselage is rigid but the mode shape actually has a small heave and pitch component (see Appendix C). In this case, when the natural frequency was low, the aircraft quickly became unstable. Therefore a higher value of 9 Hz was selected for

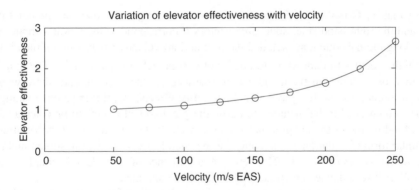

Figure 13.17   Variation of elevator effectiveness with velocity for 9 Hz wing torsion natural frequency.

examination. In practice, wing torsion modes are at significantly higher frequencies than wing bending modes. Firstly, the effectiveness was calculated for a range of velocities for the 9 Hz mode and the result is shown in Figure 13.17. As anticipated from the discussion in Section 13.10.4, there is a significant increase in elevator effectiveness with velocity.

Secondly, the flexible aircraft response to a step elevator input of $-1°$ is calculated at 175 m/s EAS for a natural frequency of 9 Hz with 2% damping. The response is similar to the rigid aircraft response, but somewhat larger because of the increased effectiveness. However, there is no obvious oscillatory behaviour around 9 Hz.

Finally, the flexible aircraft stability is examined for the 9 Hz wing torsion mode with 2% damping. The 'E' mode frequency reduces slowly with little damping variation but the 'S' mode damping increases until at 260 m/s the mode is no longer oscillatory but the motion is represented by two negative (i.e. stable) real roots.

13.11   General Form of Longitudinal Equations

Equations (13.62) for the linearized flight mechanics representation of longitudinal motion, together with flexible modes, may also be written in the more general form with air speed terms shown explicitly, namely

$$
\begin{bmatrix} \mathbf{A}_r & 0 \\ 0 & \mathbf{A}_e \end{bmatrix} \begin{Bmatrix} \dot{u} \\ \ddot{q}_e \end{Bmatrix} + \left\{ \rho V_0 \begin{bmatrix} \mathbf{B}_{rr} & \mathbf{B}_{re} \\ \mathbf{B}_{er} & \mathbf{B}_{ee} \end{bmatrix} + \begin{bmatrix} \mathbf{D}_r & 0 \\ 0 & \mathbf{D}_e \end{bmatrix} \right\} \begin{Bmatrix} u \\ \dot{q}_e \end{Bmatrix}
$$
$$
+ \left\{ \rho V_0^2 \begin{bmatrix} 0 & \mathbf{C}_{re} \\ 0 & \mathbf{C}_{ee} \end{bmatrix} + \begin{bmatrix} 0 & 0 \\ 0 & \mathbf{E}_e \end{bmatrix} \right\} \begin{Bmatrix} \int u \\ q_e \end{Bmatrix} = \rho V_0^2 \begin{bmatrix} \mathbf{F}_r \\ \mathbf{F}_e \end{bmatrix} \eta
$$

(13.74)

Note that the unknown vector for the flexible modes is the vector q_e of modal/generalized coordinates. However, the unknowns for the rigid aircraft in the flight mechanics equations are based on velocity and rate terms so, for example, in the longitudinal case where the fore-and-aft velocity perturbation is ignored, then $u = \{w \ q\}^T$. Also, the diagonal inertia matrix $\mathbf{A}_r$ comprises aircraft mass and moments of inertia, $\mathbf{A}_e$ is the diagonal modal mass matrix, $\mathbf{B}$ is the aerodynamic damping matrix, $\mathbf{D}_r$ is a matrix containing the rigid body velocity coupling terms associated with using body fixed axes (e.g. $-mU_eq$), $\mathbf{D}_e$ is the diagonal modal damping matrix, $\mathbf{C}$ is the aerodynamic stiffness matrix, $\mathbf{E}_e$ is the diagonal modal stiffness matrix and the right-hand side input force vectors $\boldsymbol{F}$ correspond to the elevator input terms.

If the longitudinal equations included the fore-and-aft velocity perturbation u then it would appear in the vector of unknown rigid body motions, namely $u = \{u \quad w \quad q\}^T$. The drag terms would then be included in the aerodynamic matrices for the rigid aircraft, there would be additional forcing vectors for gravitational and thrust effects, and the equations relating rates to angles would need to be included to allow the gravitational effects to be calculated. The model would represent the phugoid behaviour as well as short period and flexible effects.

A similar form of equation would apply for the lateral case with aileron, spoiler and rudder inputs. Similar issues of loss of effectiveness and influence of rigid/flexible couplings on the Dutch roll, roll subsidence and spiral modes could be examined.

Finally, the frequency and damping of the rigid/flexible 'S' and 'E' type modes can be found from the eigenvalue solution of Equation (13.74) (akin to the flutter solution) and the control effectiveness obtained from the steady-state solution of the rigid and flexible aircraft equations. In practice, the flexible/rigid body couplings are not likely to have as severe an effect as shown in the above examples because the natural frequencies were unrealistically low for an aircraft of the size considered and the models employed were very crude. Also, the Flight Control System (FCS) needs to be incorporated into the model and might be expected to have a considerable stabilizing influence on the aircraft dynamics.

13.12 Dynamic Manoeuvre for Rigid Aircraft in Lateral Motion

13.12.1 Fully Coupled Equations

In this section, the dynamic behaviour of a rigid symmetric aircraft in lateral motion will be considered. Three DoF will be required to model the roll, yaw and sideslip characteristics; clearly the addition of flexibility will involve at least four DoF. Because one aim of the book was to keep the number of simultaneous equations to three where possible, only the rigid aircraft will be considered in lateral motion; the treatment of the flexible aircraft follows along similar lines to the longitudinal case considered earlier in the chapter.

The coupled linearized equations for lateral motion with small perturbations are given by

$$m(\dot{v} - W_e p + U_e r) = Y \qquad I_x \dot{p} - I_{xz} \dot{r} = L \qquad I_z \dot{r} - I_{xz} \dot{p} = N \qquad (13.75)$$

The side force, rolling moment and yawing moment may be written in terms of lateral aero-dynamic stability and control derivatives as follows

$$Y = Y_v v + Y_p p + Y_r r + Y_\xi \xi + Y_\delta \delta$$
$$L = L_v v + L_p p + L_r r + L_\xi \xi + L_\delta \delta \qquad (13.76)$$
$$N = N_v v + N_p p + N_r r + N_\xi \xi + N_\delta \delta$$

where ξ and δ are the aileron and rudder angle respectively. The gravitational force components in the small perturbation equations (Cook, 1997) are given by

$$Y_g = mg\phi \cos\theta_e + mg\psi \sin\theta_e \qquad (13.77)$$

where ϕ and ψ are the roll and yaw angles. If the aircraft is in level flight prior to the manoeuvre then $W_e = \theta_e = 0$ where θ_e is the pitch attitude to the horizontal. Therefore the lateral equations of motion may be written

$$
\begin{bmatrix} m & 0 & 0 \\ 0 & I_x & -I_{xz} \\ 0 & -I_{xz} & I_z \end{bmatrix} \begin{Bmatrix} \dot{v} \\ \dot{p} \\ \dot{r} \end{Bmatrix} + \begin{bmatrix} -Y_v & -Y_p & -(Y_r - mU_e) \\ -L_v & -L_p & -L_r \\ -N_v & -N_p & -N_r \end{bmatrix} \begin{Bmatrix} v \\ p \\ r \end{Bmatrix} + \begin{Bmatrix} mg \\ 0 \\ 0 \end{Bmatrix} \phi = \begin{bmatrix} Y_\xi & Y_\delta \\ L_\xi & L_\delta \\ N_\xi & N_\delta \end{bmatrix} \begin{Bmatrix} \xi \\ \delta \end{Bmatrix}
$$

$$(13.78)$$

These equations may be solved to obtain the coupled sideslip, roll and yaw response of the linear rigid aircraft to particular aileron and rudder control inputs; engine thrust inputs could also be added and engine failure cases considered. Considering the non-linear equations with non-linear aerodynamic data added where appropriate would of course yield the large angle responses. Both linear and non-linear equations yield rational solutions.

Transforming Equation (13.78) into Laplace form and determining the relevant transfer functions will yield the dynamic stability modes, namely Dutch roll, spiral and roll subsidence.

Comparing results from a rigid and flexible aircraft would allow the control effectiveness and the impact of flexibility on the dynamic stability modes to be determined, in a similar way to that shown in Section 13.10.

13.12.2 Uncoupled Equation in Roll

As an example of the transient roll response of the aircraft following an aileron input, neglect the couplings with yaw and sideslip motions and note that Equation (13.77) then becomes

$$I_x \dot{p} - L_p p = L_\xi \xi \qquad (13.79)$$

For a step aileron angle ξ_0, the response of this first order system may be shown to be

$$p(t) = -\frac{L_\xi}{L_p} \left[1 - \exp\left(\frac{L_p}{I_x} t\right) \right] \xi_0 \qquad (13.80)$$

The aircraft behaves like a simple lag with a decaying exponent.

Consider a rigid aircraft with the following data: $I_x = 56\ 000$ kg m^2, $s = 7.5$ m, $S_W = 30$ m^2, $a_W = 4.5$/rad, $a_C = 1.5$/rad and $V_0 = 150$ m/s EAS (the drag effect on the roll rate derivative will be neglected). The roll rate (deg/s) and roll angle (deg) responses to a step aileron input of 2 deg are shown in Figure 13.18(a). A steady-state roll rate of 20 deg/s is achieved after about 2 s and the roll angle is steadily increasing. Figure 13.18(b) shows that a steady 20° bank angle is achieved when an aileron angle of 2° is applied for only 1 s and then returned to the neutral position (on/off aileron input). Albeit only employing a simple model, these calculations are rational.

13.13 Bookcase Manoeuvres for Rigid Aircraft in Lateral Motion

In Section 13.12 the behaviour of a fully coupled sideslip, roll and yaw model was considered to permit the determination of response to control inputs and of dynamic stability modes. Such a response model would be suitable for a rational analysis. In this section, simplified bookcase

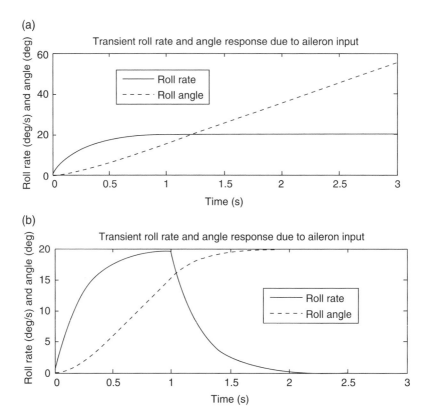

Figure 13.18   Roll rate and roll angle variation for (a) step and (b) on/off aileron input – rigid aircraft.

analyses employing asymmetric equilibrium manoeuvres in roll and yaw will be outlined; such analyses are suitable for use early in the design process, allowing initial sizing to take place (Lomax, 1996).

Some of these equilibrium manoeuvres are steady and involve a balance of aerodynamic moments from different sources; others involve abrupt application of a control with an inertia couple balancing the aerodynamic control moment at that instant, so leading to a conservative (i.e. over) estimate of loads. These rolling and yawing manoeuvres will be considered briefly in this section. Rolling manoeuvres are often important for the outer wing design whereas yawing manoeuvres contribute to rear fuselage and fin design. It is important to recognize that the presence of the Flight Control System (FCS) means that many bookcase loads results are found later to be overestimates.

Note that the treatment of symmetric and asymmetric equilibrium manoeuvres is further considered in Chapter 23. The longitudinal case is the main focus of the book. Some background to static aeroelastic effects for the lateral case may be found in ESDU Data Sheets 01010 and 03011.

In order to justify the way in which the load cases are treated, Equation (13.78) for lateral motion of the aircraft will be used; it is assumed that the aircraft is symmetric and that perturbations about straight and level flight are small. The applied forces and moments are expanded in derivative form. The sideslip angle β is given by the aircraft sideways velocity divided by the total air speed, so $\beta = v/V$ and hence the term $\dot{v} = V\dot{\beta}$ is actually a lateral acceleration term.

13.13.1 Roll Bookcase Analyses

It will be shown later in Chapter 23 that the basic certification requirements for roll involve a steady roll rate and a maximum roll acceleration case. Loads from both these manoeuvres must be superimposed upon an initial steady load factor.

13.13.1.1 Steady Roll Rate

Consider firstly the case of a *steady roll rate* following a defined aileron deflection, in which case all the acceleration terms are zero. The roll angle is assumed to remain small so that the grav-itational term in the y direction may be ignored. The heading is assumed to remain unchanged so the yaw angle and therefore yaw rate are zero. Using these values in Equation (13.78), noting that the roll rate $p = \dot{\phi}$ and rearranging yields

$$
\begin{bmatrix} Y_v & Y_\delta & Y_{\dot\phi} \\ L_v & L_\delta & L_{\dot\phi} \\ N_v & N_\delta & N_{\dot\phi} \end{bmatrix} \begin{Bmatrix} v \\ \delta \\ \dot\phi \end{Bmatrix} = - \begin{Bmatrix} Y_\xi \\ L_\xi \\ N_\xi \end{Bmatrix} \xi
\tag{13.81}
$$

This equation indicates that a rudder angle and a resulting sideslip angle are required to create a balanced condition of pure steady roll rate.

It would be possible to treat the complete solution of Equation (13.81) as a load case with loads developed on all axes. However, for simplicity, cross-couplings between roll and yaw may be neglected so only a roll equation is involved (Lomax, 1996) and the aircraft is consid-ered to be balanced in roll, with only the relevant lifting surface bending loads considered. It may be seen from Equation (13.81) that for the rolling manoeuvre, with coupling terms omitted, the equation reduces to

$$
L_{\dot\phi}\dot\phi = -L_\xi \xi
\tag{13.82}
$$

Thus the resulting steady roll rate is given by

$$
\dot\phi = -\frac{L_\xi}{L_{\dot\phi}}\xi = -\left[\frac{3a_C}{2(a_W + C_D)s}\right]V_o\xi
\tag{13.83}
$$

where the rolling moment derivatives due to aileron and due to roll rate have been substituted from Appendix E. The distribution of these aerodynamic loads over the aircraft will be known so the internal loads may be calculated (i.e. bending moments on each lifting surface).

13.13.1.2 Maximum Roll Acceleration

For the *maximum roll acceleration* case, the bookcase methodology is to react the applied control moment by an inertia couple at the instant of application so that no velocity or steady response values have yet developed; thus both roll and yaw rates will be zero. Also, in this case the acceleration terms will be non-zero, there will be no applied rudder angle and the sideslip angle will be zero. The only relevant derivative terms will be those related to aileron angle so the equations of motion for this condition, starting from a level flight state, may be written as

$$
\begin{bmatrix} m & 0 & 0 \\ 0 & I_x & -I_{xz} \\ 0 & -I_{xz} & I_z \end{bmatrix} \begin{Bmatrix} \dot{v} \\ \ddot{\phi} \\ \ddot{\psi} \end{Bmatrix} = \begin{Bmatrix} Y_\xi \\ L_\xi \\ N_\xi \end{Bmatrix} \xi
\tag{13.84}
$$

This equation would yield the accelerations at the instant of aileron application, prior to any response developing. It may be seen that the yaw and roll accelerations are coupled if the product moment of inertia term is non-zero. Once again (Lomax, 1996), if the coupling terms are neglected then the roll behaviour is given by

$$
I_x\ddot{\phi} = L_\xi\xi \quad \text{or} \quad \ddot{\phi} = \frac{L_\xi}{I_x}\xi = \left[\frac{S_W s a_C \rho}{4I_x}\right] V_o^2 \xi
\tag{13.85}
$$

Hence the instantaneous roll acceleration may be determined and the load distribution due to aileron deployment will be balanced by the inertia loading associated with the roll inertia couple $I_x\ddot{\phi}$; the internal loads (i.e. related to bending of the lifting surfaces) may then be found. The way in which this maximum acceleration is determined will lead to an overestimate of loads (conservative) and a rational approach would produce a more realistic set of loads. In practice, flexible effects will be included, either by adding flexible modes or by determining flexible corrections to the rigid aircraft derivatives.

13.13.2   Yaw Bookcase Analyses

A somewhat similar approach may be taken for the yawing manoeuvre where (a) a rudder input is suddenly applied, (b) a maximum sideslip overswing is reached, followed by (c) a steady sideslip condition and then (d) the rudder is returned to neutral. In effect, steps (a) and (d) involve abrupt rudder application and step (c) requires determination of the steady sideslip condition. The dynamic overswing step (b) (see Chapter 1) may be treated by application of an overswing factor (to allow for the resulting increase in load). Only steps (a) and (c) will be considered in this section, but this manoeuvre will also be considered later in Chapter 23.

13.13.2.1   Abrupt Application of Rudder

The *abrupt application of rudder* will be balanced by inertia effects and the condition examined prior to any response developing so the yaw and roll angles/rates will be zero. Starting from a steady level flight state ($\beta = \phi = \xi = 0$), the result may be shown to be the same as in Equation (13.84) but with the rudder replacing the aileron. Therefore

$$
\begin{bmatrix} m & 0 & 0 \\ 0 & I_x & -I_{xz} \\ 0 & -I_{xz} & I_z \end{bmatrix} \begin{Bmatrix} \dot{v} \\ \ddot{\phi} \\ \ddot{\psi} \end{Bmatrix} = \begin{Bmatrix} Y_\delta \\ L_\delta \\ N_\delta \end{Bmatrix} \delta
\tag{13.86}
$$

If, however, the rudder had been returned abruptly to zero when the aircraft was in a steady sideslip condition (see below), then additional right hand side forcing terms due to the corresponding non-zero steady sideslip, aileron and roll angles would have been included.

The equations could be solved for roll and yaw accelerations, rate of sideslip and loads for all axes considered. Alternatively, for simplicity, the coupling terms could be neglected and the instantaneous yaw acceleration given by

$$I_z \ddot{\psi} = N_\delta \delta \tag{13.87}$$

Then, only the balance of loads in yaw would be considered, rather as for the roll case earlier. In Lomax (1996) reference is made to yawing being a 'flat' manoeuvre, presumably indicating that couplings should be neglected wherever possible so that no roll occurred.

13.13.2.2 Steady Sideslip

The *steady sideslip* condition (with zero heading) may be determined by setting all acceleration terms, roll rate and yaw rate to zero. However, roll angle (assumed small) and aileron terms must be included in the steady sideslip. Thus the equations are given by

$$\begin{bmatrix} Y_v & Y_\xi & Y_\phi \\ L_v & L_\xi & 0 \\ N_v & N_\xi & 0 \end{bmatrix} \begin{Bmatrix} v \\ \xi \\ \phi \end{Bmatrix} = - \begin{Bmatrix} Y_\delta \\ L_\delta \\ N_\delta \end{Bmatrix} \delta \tag{13.88}$$

noting that $Y_\phi = mg$ and $L_\phi = N_\phi = 0$. The solution of these equations (Lomax, 1996) will yield the steady sideslip and the corresponding aileron and roll angles in this balanced case; hence the relevant balanced aerodynamic loads may be determined. If the roll angle had been omitted then the aircraft would have been imbalanced in roll. If, in addition, the aileron angle had been set to zero, then different steady sideslip angles would have been obtained from the side force and yawing moment equations, so one or other would have been imbalanced. Therefore the steady sideslip strictly requires all three equations to be solved and the corresponding aerodynamic loads determined. However, it may be that only the yaw loads would be examined, as these will be dominant for lateral bending of the rear fuselage and tail.

The full yaw manoeuvre case involves four steps, the above two plus the dynamic overswing and return to neutral rudder; this will be discussed further in Chapter 23. A similar bookcase could be carried out for the 'engine out' condition, where the yawing moment due to engine out would be balanced by the rudder application (Lomax, 1996).

13.14 Flight Control System (FCS)

The Flight Control System (FCS) impacts upon manoeuvres, gusts, flutter and static aeroelastic calculations, but its primary function relates to aircraft handling qualities and the flight mechanics model. The FCS is a complex, high integrity system that is essential for safe operation across the full flight envelope and across the full range of expected environmental conditions. It receives commands from the pilot control devices and feedbacks from various aircraft motion and air-data sensors, which (usually) a digital computer translates into commands for control surface deflections; these are signaled to the control surface actuators to enforce control of the aircraft in such a way that the pilot's demands are accomplished while satisfying various other objectives.

Besides this manual mode, the FCS is also used as the inner loop for the autopilot control system, which in principle converts numerical piloting demands (e.g. altitude, glide slope, speed, heading), entered by the pilot through the Flight Control Unit, into FCS piloting instructions. In turn, the autopilot is addressed by the flight management and guidance system which mechanizes the longer term planning of the flight.

The FCS has a number of objectives, namely:

- enabling the controlled aircraft to have the desired stability characteristics and handling qualities uniformly across the whole flight envelope, which includes take-off, landing, cruise, configuration changes and loading conditions (e.g. mass, centre of mass);
- avoiding pilot-induced oscillations (PIOs), also known as aircraft-pilot-coupling (APC, to avoid blaming the pilot);
- providing the pilot with 'carefree handling' such that the aircraft does not exceed aerodynamic, structural or control limits – this addresses more or less usual pilot input sequences (e.g. full stick back does not exceed the load factor limit of 2.5 g for the clean configuration and full roll input does not exceed a maximum allowable bank angle of 67° etc.), while it should not be assumed that any complex sequence of inputs is harmless;
- reducing vertical manoeuvre loads through the use of the Manoeuvre Load Alleviation (MLA) system, e.g. wing root bending moment reduction can be achieved by inboard shifting of the lift distribution (through symmetric outer aileron and spoiler deflection with simultaneous increase of angle of attack, to achieve the commanded load factor, needing the elevator to compensate for the pitch up moment induced by outer wing controls);
- reducing gust and turbulence response (so improving crew and passenger ride quality) and loads (as for the MLA) through the use of a Gust Load Alleviation (GLA) system (gust disturbance rejection achieved using wing controls with induced pitching moment balanced by the elevator);
- providing vibration damping (e.g. lateral engine motion, rear cabin fishtailing, cockpit vibration etc.) to avoid fatigue and discomfort.

The FCS involves the use of a range of sensors to measure the aircraft state, namely air data sensors (e.g. altitude, incidence angle, sideslip angle, air speed, Mach number, dynamic pressure), rate gyros (for body rates and Euler angles) and accelerometers (load factor measurement). The FCS also employs a number of actuators, primarily driving the aerodynamic control surfaces and engine throttle. The aircraft response is fed back, often via multiple feedback loops and control laws, to generate suitable actuation inputs that produce the required behaviour of the aircraft in its closed loop configuration. The aircraft flight condition information, together with mass and centre of mass data, is used to allow the control laws to be 'scheduled' in such a way that they are varied as necessary throughout the flight envelope to produce a homogeneous aircraft response to the pilot's input (e.g. that a stick pull action results in a rapid buildup of 2.5 g, largely compensating for aircraft speed and mass effects). The replacement of analogue by digital control systems has allowed much more versatile and powerful control systems to be developed.

Historically, mechanical linkages have been used to link the pilot stick/pedals to the control actuators, but many modern aircraft are 'fly-by-wire', so electronic signals are used to command the hydraulic or electric actuators, thus avoiding heavy mechanical linkages, although mechanical or electrical back-up is still included for critical controls (e.g. elevator, rudder, aileron).

One important issue for the control of a flexible aircraft is that the sensors will measure not only the rigid body motions but also sense the effect from deformations of flexible modes having frequencies in the sensor bandwidth. These flexible mode contributions will then be processed via the control loops and lead to changes in the actuator demands; the control inputs could then possibly increase the flexible mode responses and lead to an unstable system. Such interaction between the control system and structure is known as structural coupling or aeroservoelasticity (Pratt, 2000; see Chapter 11). To avoid unfavourable effects it is normal to introduce so-called 'notch' filters in order to suppress flexible mode contributions. The design of such filtering becomes more difficult as the flexible modes become closer in frequency to the rigid body modes, a growing problem for large commercial aircraft.

The inclusion of feedback control loops in the aircraft dynamic model, consisting of both rigid body and flexible modes, increases the model complexity significantly. The presence of the FCS in manoeuvres and gust encounters are not considered mathematically in this book, but only some general comments will be made for each case. However, in Chapter 11 it was shown how a simple feedback loop could alter the flutter and response characteristics of a simple binary wing model; also, the way in which the frequency dependency of notch filters could be included in the stability calculations was explained. The same principles would apply for the whole free–free flexible aircraft under manoeuvre or gust conditions. The inclusion of the FCS in loads and aeroelastics calculations is also considered briefly in Chapter 21.

The aircraft structural design benefits from the FCS (which in general reduces the loads on the structure) because the proper functioning of the FCS is supervised with a monitoring system that detects relevant failure cases and even tries to attenuate their impact on the aircraft loads and handling by initiating reconfiguration actions. The dynamic behaviour induced by system failure and any corrective actions undertaken by the FCS must be considered for the structural design as this may offset any benefits.

There are many features of the FCS that are beyond the scope of this book to cover in any detail; however, the reader is directed to Pratt (2000) for a comprehensive treatment of the FCS, particularly from an industrial viewpoint.

13.15   Representation of the Flight Control System (FCS)

For dynamic rational manoeuvres simulated in the time domain, the relevant features of the non-linear FCS, coupled to the aircraft dynamics, should be represented, including a Manoeuvre Load Alleviation (MLA) system if appropriate (see Chapter 21). Bookcase manoeuvres would only require FCS limits to be imposed. Any transfer function or other studies involving frequency domain analysis would utilize a linearized FCS.

13.16   Examples

1. Obtain expressions for the short period natural frequency and damping ratio for the *rigid* aircraft in terms of the derivatives. Also, determine an expression for the transfer functions relating incidence ($\alpha = w/V_0$) and normal acceleration to the elevator angle.

2. Write a MATLAB/SIMULINK program (see the companion website) to determine (a) the aerodynamic derivatives in wind axes for a given flight condition of the *rigid* aircraft, (b) the static margin, (c) the short period frequency and damping, (d) the response of the aircraft (pitch rate/incidence, etc.) to an elevator input and (e) the aircraft motion in earth axes. Check the results using the example in the chapter. Then explore the effect of changing the centre of mass position, tailplane area, downwash term, etc.

3. For the rigid aircraft used in the text, flying at 175 m/s EAS, determine the pitch rate, normal acceleration and response at the centre of mass relative to earth axes for an oscillatory elevator input of $+1°$ at a range of frequencies. What frequency leads to the largest acceleration response in the steady state and how is this related to the undamped or damped short period frequency? Note that this problem may be solved using a simulation or using the transfer function expression evaluated at different frequencies. [Short period mode undamped 0.36 Hz, 56%; damped 0.30 Hz – maximum acceleration 1.06 g close to damped frequency]

4. Repeat the exercise in Example 2 for the *flexible* aircraft with a fuselage bending mode. Check the results using the example in the chapter. Then explore the effect of changing the centre of mass position; the fuselage mass distribution and hence modal information will need to be altered (see Appendix C). Also, generate a new flexible mode for the same aircraft example, which is the combination of two of the other modes, and investigate the revised behaviour. Finally, develop a data set from another idealized aircraft with a different mass distribution.

5. For the aircraft used in the text, flying at 175 m/s EAS and having a fuselage bending mode with a 2 Hz natural frequency (2% damping), determine the undamped natural frequency, damping ratio and undamped natural frequency for the 'S' and 'E' modes. Then determine the pitch rate and normal acceleration for an oscillatory input of $+1°$ at these frequencies. What frequency leads to the largest acceleration response in the steady state? Note that this problem may be solved using a simulation or using the transfer function expression evaluated at different frequencies. [0.586 Hz, 35.1%; 2.198 Hz, 3.55%; 0.50 g close to 'S' damped frequency and 0.28 g at 'E' frequency – less than for a rigid aircraft]

6. Determine the cubic characteristic equation for the flexible aircraft and solve it (using MATLAB function *roots*) for Example 5 at a range of velocities and hence determine the coupled rigid/flexible flutter condition. [303 m/s]

7. Evaluate the elevator effectiveness for Example 5. [0.31]

8. Determine an expression for the initial roll acceleration when a step aileron input is applied to a simple *rigid* aircraft in roll.

9. Consider the rigid aircraft with the following data: $I_x = 40\ 000$ kg m^2, $s = 6.25$ m, $S_W = 20$ m^2, $a_W = 4.5$/rad, $a_C = 2$/rad and $V_0 = 150$ m/s EAS. Calculate the maximum roll rate and roll angle responses when the aileron is suddenly deployed at an angle of $2°$ for 1 s and then returned to zero. This example may be solved using a MATLAB/SIMULINK program (see companion website) or using the superposition of two steps. Note the effect of varying the input pulse length. [29.8 deg/s and 32°]

10. Derive the expressions for the derivatives L_e, Q_p for the *flexible* wing antisymmetric torsion mode.

11. Repeat Example 9 for the flexible aircraft with $I_W = 850 \text{ kg m}^2$, $c = 1.6 \text{ m}$, $l_A = 0.2 \text{ m}$ and modal damping $\zeta_e = 4\%$ by extending the program in Example 9 to cover flexible effects (see also in companion website). The natural frequency for the flexible antisymmetric wing torsion mode is 6 Hz. Note the effect of varying the input pulse length.
    [6.2 deg/s and 5.8°]

12. Use the characteristic equation and the roll rate/aileron gain transfer functions to determine the variation of aileron effectiveness with velocity, and reversal and oscillatory divergence speeds for Example 10 with a natural frequency of 8 Hz and damping of 2%.
    [222 and 248 m/s]

14

Gust and Turbulence Encounters

It is a well known but unfortunate feature of air travel that aircraft regularly encounter atmospheric turbulence (or 'rough air') of varying degrees of severity. Turbulence may be considered as movement of the air through which the aircraft passes. Any component of the velocity of the air (so-called 'gust velocity') that is normal to the flight path, as illustrated in Figure 14.1 for the vertical gust case, will change the effective incidence of the aerodynamic surfaces, so causing changes in the lift forces and hence a dynamic response of the aircraft involving flexible deformation; gust inputs are also considered when acting across or along the flight path.

The response will involve both the rigid body and flexible modes, may give rise to passenger and crew discomfort, and will introduce internal loads that need to be considered for aircraft safety. Thus it is important for the safe design of the aircraft to calculate the response and internal loads generated under the conditions defined by the Airworthiness Authorities, and to evaluate the effect on the fatigue life. Gust and turbulence loads may be significant for all the components of the aircraft.

In this chapter, gusts and turbulence will be considered in two idealized forms, namely discrete gust and continuous turbulence. The process of determining the response of an aircraft to a discrete gust or continuous turbulence input will be shown using a progression of models of differing complexity and using an analysis based in the time or frequency domain as appropriate. The main features of the response analysis are introduced and illustrated one at a time, so that the range of issues involved may be seen as clearly as possible. Note that the focus will be on determining the aircraft response, with the calculation of loads covered later in Chapter 16.

The way in which gusts and turbulence have been treated for design has changed significantly over the years, particularly because of the difficulty of defining the external load input and of performing the calculations. A historical perspective is given in Fung (1969), Hoblit (1988), Flomenhoft (1994), Fuller (1995) and Bisplinghof *et al.* (1996), and, of course, the latest requirements are given in the current certification specifications (CS-25 and FAR-25). Many of the loads calculations required are explained in Hoblit (1988), Howe (2004) and Lomax (1996). The ESDU series also includes an item on gusts (ESDU Data Sheet 04024). This book will not attempt to catalogue the changes in requirements and philosophy, but rather point out the key issues and methods involved by using earlier rigid and flexible aircraft dynamic models. The approach used in industry and the certification requirements (in particular CS-25) are outlined later in Chapter 23.

Introduction to Aircraft Aeroelasticity and Loads, Second Edition. Jan R. Wright and Jonathan E. Cooper.
© 2015 John Wiley & Sons, Ltd. Published 2015 by John Wiley & Sons, Ltd.

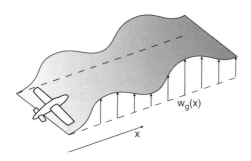

Figure 14.1   Aircraft encountering turbulence.

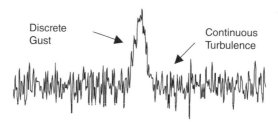

Figure 14.2   Continuous and discrete turbulence.

It should be noted that the flight mechanics equations of motion will not be used here, since the aircraft encountering a gust is normally in a steady flight condition and the excursions from this datum are relatively small. Thus, an inertial axis system may be employed and rigid body characteristics (e.g. rigid body physical or generalized coordinates) may be combined with flexible modes in the flexible aircraft analysis. The equations of motion are similar to those in Chapter 12 but now with aerodynamic rate derivatives, aircraft acceleration effects and gust aerodynamics present. The similarity of the equations of motion to previous manoeuvre cases means that it is once again convenient to write the equations in terms of aerodynamic derivatives defined with respect to inertial (and not body fixed) axes where displacements and rotations are the unknowns. The derivatives employed here are tabulated in Appendix B.

14.1   Gusts and Turbulence

Atmospheric turbulence, although a complicated phenomenon, is normally considered for design purposes in one of two idealized categories, namely:

a. *discrete gusts*, where the gust velocity varies in a deterministic manner, usually in the form of a '1-cosine' shape (i.e. there is an idealized discrete 'event' that the aircraft encounters), and
b. *continuous turbulence*, where the gust velocity is assumed to vary in a random manner.

The difference between the two components of turbulence may be seen in Figure 14.2. The discrete gust response is solved in the time domain whereas the continuous turbulence response is usually determined in the frequency domain via a Power Spectral Density (PSD) method.

Gusts and turbulence may be vertical, lateral or at any orientation to the flight path, but vertical and lateral cases are normally treated separately. Thus, for a symmetric aircraft, a vertical gust will give rise to heave (or plunge) and pitch motions whereas a lateral gust will cause sideslip, yaw and roll motions. Note that it is normally assumed that there is no variation of the gust velocity across the aircraft span (or with the aircraft height in the case of lateral gusts). In this chapter, only the symmetric vertical gust case will be considered. Chapter 23 includes a brief discussion on lateral and round-the-clock gusts and some detail on the magnitude of the gust and turbulence inputs used in certification.

14.2 Gust Response in the Time Domain

Initially, the 'sharp-edged' and '1-cosine' discrete gusts will be considered, and the so-called 'gust penetration effect' introduced; the sharp-edged gust is included to aid understanding though it is no longer relevant to the gust clearance of large commercial aircraft. The time domain solution for a rigid aircraft entering a sharp-edged gust and responding in heave only will firstly be considered by ignoring any pitching effects from separate wing and tailplane aerodynamics. It will also be assumed that the aerodynamic surfaces enter the sharp-edged gust instantaneously and develop a lift force proportional to the effective incidence angles associated with the response and gust, i.e. quasi-steady aerodynamics. Then, the effect of employing *response-dependent* and *gust-dependent* unsteady aerodynamics, the latter allowing for a non-instantaneous lift build-up due to the penetration of the gust by the aerofoil, will be shown using the Wagner and Küssner functions respectively (see Chapter 9).

Later on, the response of a rigid aircraft, with pitching effects included, to a more general gust input (i.e. '1-cosine' gust) will be considered by including separate tailplane aerodynamics and the penetration delay between the wing and tailplane. Finally, the gust response of a simple flexible aircraft will be considered. This exercise will allow the general form of the equations of motion to be developed.

14.2.1 Definition of Discrete Gusts

14.2.1.1 'Sharp-edged' Gust

The early work on gust response considered the gust input to be in the form of a *sharp-edged* or *step gust* (Fung, 1969; Fuller, 1995), where the aircraft entered instantaneously into a uniform gust velocity field (see Figure 14.3a), defined spatially by

$$w_g(x_g) = \begin{cases} 0 & x_g < 0 \\ w_{g0} & x_g \geq 0 \end{cases} \tag{14.1}$$

where w_{g0} is the constant gust velocity and where x_g is the position of the aircraft in the spatial description of the gust relative to a convenient fixed origin. However, as the prediction methodologies developed, this input was considered to be too unrealistic and others were adopted. (It should be noted that the symbol w_g is a classical notation for gust analysis and should not be confused with the use of w as an aircraft velocity for the flight mechanics model in Chapter 13, or for downwash in Chapter 18 – all these usages are classical in different areas.)

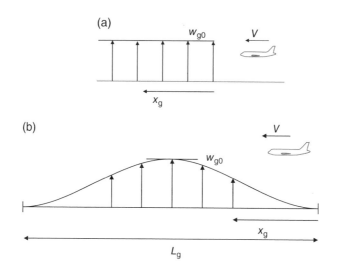

Figure 14.3   (a) 'Sharp-edged' discrete gust (b) '1-cosine' discrete gust.

14.2.1.2   '1-cosine' Gust

Discrete gusts are currently represented for design purposes by a so-called '*1-cosine' gust*, where the variation in the velocity of the air normal to the path of the aircraft is as shown in Figure 14.3b.

The expression governing the *spatial* behaviour of the '1-cosine' gust is

$$w_g(x_g) = \frac{w_{g0}}{2}\left(1 - \cos\frac{2\pi x_g}{L_g}\right) \qquad 0 \le x_g \le L_g \tag{14.2}$$

where w_{g0} is the value of the peak, or design, gust velocity and L_g is the gust length (or twice the so-called 'gust gradient' H). The design gust velocity w_{g0} is taken to vary with gust length, altitude and speed, discussed in Chapter 23 (CS-25; Hoblit, 1988).

To calculate the aircraft response in the time domain, the gust velocity expression needs to be transformed from a spatial into a temporal function. Consider the aircraft encountering a gust whose vertical velocity varies spatially as shown in Figure 14.3b, where the aircraft wing enters the gust at time $t = 0$ and position $x_g = 0$. Any tailplane would enter the gust shortly after the wing, the so-called 'gust penetration effect' (see later). If the aircraft has constant velocity V then the gust velocity w_g experienced by the aircraft may be thought of as a function of time since the amount by which the aircraft wing has penetrated the gust will be given by $x_g = Vt$, where in this case V needs to be TAS (as does the gust velocity). For the spatial variation in gust velocity quoted above for the '1-cosine' gust, then the corresponding *temporal* variation of gust velocity, as seen by the aircraft, will be given by

$$w_g(t) = \frac{w_{g0}}{2}\left(1 - \cos\frac{2\pi V}{L_g}t\right) \tag{14.3}$$

Gust response calculations may be performed in the time domain, using a solution method introduced in Chapter 1, for a series of different gust lengths to seek the so-called 'discrete tuned gust' that causes the maximum (and minimum) value of each internal load (e.g. wing root shear force, torque and bending moment). Non-linear effects due to the FCS may be included readily. A similar approach will be used for the taxiing case in Chapter 15 where a pair of (1-cosine) bumps on the runway is considered.

14.3 Time Domain Gust Response – Rigid Aircraft in Heave

Initially, the time domain solution for a rigid aircraft entering a 'sharp-edged' gust and responding in heave only will be considered in this section. Both quasi-steady and unsteady aerodynamic assumptions will be examined and the effect on the heave response discussed. The approach will be extended to a more general gust shape. The arguments presented are partially responsible for the development of the gust alleviation factor, used historically in the load factor approach. Nevertheless, unsteady aerodynamic effects are also important when considering current analysis methodologies.

14.3.1 Gust Response of Rigid Aircraft in Heave using Quasi-Steady Aerodynamics

To simplify the analysis as much as possible, consider that the whole aircraft (wing plus tailplane) encounters a sharp-edged gust instantaneously (Fung, 1969) and assume that the aircraft will heave (i.e. move up or down) without pitching; thus penetration effects are ignored. The term 'plunge' is sometimes used in place of 'heave'. The aircraft is in a trimmed level flight condition before encountering the gust (with lift equal to weight), so the forces and motions shown below are relative to the initial trimmed state, with the symbol Δ indicating incremental values.

Now, consider the aircraft of mass m as being rigid, moving forwards with velocity V and downwards with heave velocity $\dot{z}_C$ at any instant of time (i.e. relative to the initial trimmed state). Because the air is moving upwards with the instantaneous gust velocity w_g, it may be seen from the relative velocity between the aircraft and air in Figure 14.4 that there is an effective instantaneous increase in incidence $\Delta\alpha_g$, given by

$$\Delta\alpha_g \approx \frac{w_g + \dot{z}_C}{V} \tag{14.4}$$

where small angles are assumed $\left(w_g \text{ and } \dot{z}_C \ll V\right)$. The aircraft heave velocity must be absolute, hence the need to use TAS rather than EAS for the aircraft and gust velocities. Also, the change in the total velocity vector is neglected, so the total velocity in the gust $V_g \approx V$. If the lift is assumed to be developed instantaneously (essentially the quasi-steady assumption), then the incremental whole aircraft lift due to the gust velocity is

$$\Delta L = \frac{1}{2}\rho V^2 S_W a \left(\frac{w_g + \dot{z}_C}{V}\right) = \frac{1}{2}\rho V S_W a \left(w_g + \dot{z}_C\right) \tag{14.5}$$

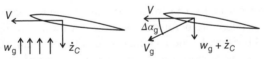

Figure 14.4 Effective incidence due to gust velocity and aircraft response.

where ρ is the air density, S_W is the wing area and $a = dC_L/d\alpha$ is the lift curve slope for the whole aircraft (using the wing area as the reference). It may thus be seen from Equations (14.4) and (14.5) that the lift developed due to gust velocity is proportional to V (not V^2) and $(w_g + \dot{z}_C)$. Since pitching effects are neglected, then, using Newton's second law, the heave equation of motion of the aircraft is given by

$$m\ddot{z}_C = -\Delta L = -\frac{1}{2}\rho V S_W a (w_g + \dot{z}_C)$$

$$m\ddot{z}_C + \frac{1}{2}\rho V S_W a \dot{z}_C = -\frac{1}{2}\rho V S_W a w_g \tag{14.6}$$

or

$$m\ddot{z}_C + \Delta L_z = -\Delta L_g \tag{14.7}$$

where the subscripts z and g indicate that the incremental lift terms are due to the response and gust effects respectively. This form of the equation is included to allow the important distinction between response-dependent and gust-dependent aerodynamic terms to be recognized. The equation could also be written in derivative form, and this will be done later for the rigid (heave/pitch) and flexible aircraft cases.

Dividing Equation (14.7) through by m and defining $\eta_g = \rho V S_W a/(2m)$ (using the subscript g to distinguish from the symbol used for elevator angle) leads to the differential equation of motion in the form

$$\ddot{z}_C + \eta_g \dot{z}_C = -\eta_g w_g \tag{14.8}$$

Given that $w_g(t)$ is a step function of magnitude w_{g0}, the so-called 'sharp-edged' gust, then the displacement and acceleration responses following zero initial conditions $z_C = \dot{z}_C = 0$ at $t = 0$ may be shown to be

$$z_C = \frac{1}{\eta_g} w_{g0}(1 - e^{-\eta_g t}) - w_{g0}t \quad \text{and} \quad \ddot{z}_C = -\eta_g w_{g0}e^{-\eta_g t} \tag{14.9}$$

Consider an aircraft with the following data: $m = 10\,000$ kg, $S_W = 30$ m^2, $S_T = 7.5$ m^2, $c = 2.0$ m, $a_W = 4.5$/rad, $a_T = 3.2$/rad and $k_\varepsilon = 0.35$ so the whole aircraft lift curve slope is $a = 5.02$/rad. Responses to a step gust of $w_{g0} = 6.25$ m/s TAS are shown in Figure 14.5 for $V = 187.5$ m/s TAS and $\rho = 0.784$ kg/m^3 (14 000 ft). The heave displacement z_C shows the aircraft climbing steadily after an initial transient while the acceleration $\ddot{z}_C$ decays from the initial maximum value at $t = 0$. The maximum total load factor n (see Chapter 12) occurs at $t = 0$ and is given by

$$n = 1 + \frac{\Delta L}{W} = 1 + \Delta n = 1 - \frac{\ddot{z}_{C\,\text{max}}}{g} = 1 + \frac{\rho V w_{g0} a}{2W/S_W} \tag{14.10}$$

where Δn is the increment of load factor and the minus sign appears because both heave displacement and acceleration are defined as positive downwards whilst the gust velocity is positive upwards. In this example, the incremental load factor is $\Delta n = 0.705$ and $\eta_g = 1.107$. Note that the load factor is proportional to gust velocity, aircraft velocity, lift-curve slope and the inverse of the wing loading.

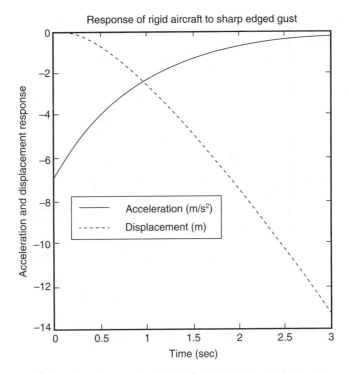

Figure 14.5 Response of a rigid aircraft to a sharp-edged gust.

There are significant shortcomings in the above simplified analysis that lead to an incorrect estimation of loads, namely:

- without the inclusion of a separate tailplane, penetration effects introducing a pitching motion, which in turn modify the effective incidence of both wing and tailplane at any instant of time, are not included;
- the lift force is not developed instantaneously because the wing enters the gust progressively and the changes in incidence do not have an instant effect (i.e. unsteady aerodynamics);
- the 'sharp-edged' gust is not realistic since in practice the gust velocity takes a finite time to build up; and
- a rigid model doesn't exhibit the aircraft flexible modes that modify the response.

To account for the instantaneous lift assumption and for replacing the sharp-edged gust by a more 'real' discrete gust, a 'gust alleviation factor' K_g (0.7–0.8) can be included in Equation (14.10) to produce more realistic loads.

14.3.2   Gust Envelope

Historically, Equation (14.10) has been written with K_g included and in terms of EAS (equivalent air speed), so

$$n = 1 + \frac{\rho_0 w_{g0\mathrm{EAS}}\, a K_g}{2W/S_\mathrm{W}} V_{\mathrm{EAS}} \qquad (14.11)$$

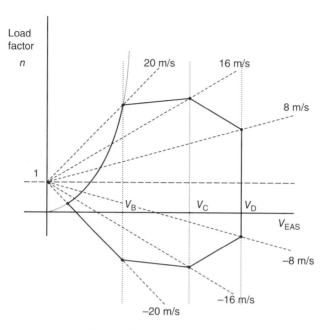

Figure 14.6   Gust envelope.

where ρ_0 is the air density at sea level (see Chapter 4). This is known as Pratt's equation and was used to define the load factors for design at different flight conditions. One way of presenting the flight load factors graphically is by generating the 'gust envelope' shown in Figure 14.6. This envelope is constructed by using Equation (14.11) to draw lines of load factor against velocity V_{EAS} for different sharp-edged gust velocities w_{g0EAS}. The points that define the gust envelope occur at the value of the appropriate gust load factor expression for the specified flight velocity (i.e. V_B = speed for maximum gust intensity, V_C = cruise speed and V_D = design dive speed). The gust velocities at these three flight speeds were defined as 8, 16 and 20 m/s (or 25, 50 and 66 ft/s), with the gust values reducing at higher altitudes.

Design calculations were then carried out at all the corner (and some intermediate) points on the gust envelope for different flight conditions (i.e. altitudes, centres of mass, weight etc.) and the worst case responses and internal loads obtained, in a somewhat similar way to the manoeuvre envelope in Chapter 12.

However, this approach is no longer used for large commercial aircraft (it does not appear in CS-25) and gust loads for aircraft are obtained using more accurate approaches, as explained later. In particular, the aircraft model used for gust calculations will include heave and pitch motion together with flexible modes, will approximate the penetration effects, will take account of unsteady aerodynamic effects and will use both discrete tuned '1-cosine' gust and continuous turbulence representations.

14.3.3   Unsteady Aerodynamic Effects in the Time Domain

So far, it has been assumed that the aerodynamic forces develop instantly following the change in effective incidence caused by the gust and response (i.e. quasi-steady assumption). However, the time taken for the wing to penetrate the gust and experience the change in effective

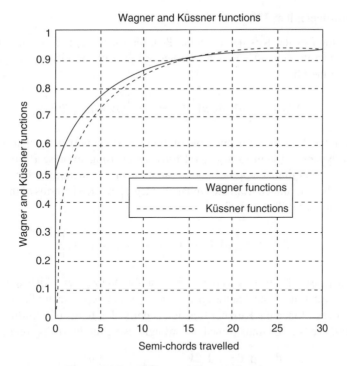

Figure 14.7   Wagner and Küssner functions.

incidence due to the gust velocity will mean that the *gust-dependent aerodynamic forces* will not be developed instantaneously, but there will be a finite build-up to the steady value. This effect is represented classically by the *Küssner* function $\Psi(\tau)$ (Bisplinghoff *et al.*, 1996), introduced in Chapter 9; it is the ratio between the transient lift and the final steady-state lift on an aerofoil section penetrating a sharp-edged gust. Figure 14.7 shows the Küssner function, where the non-dimensional (or dimensionless) time quantity

$$\tau = \frac{2V}{c}t \qquad (14.12)$$

may be interpreted as the distance travelled by the aerofoil, measured in semi-chords.

Also, the *response-dependent aerodynamic forces* will take time to develop following any change in incidence due to the heave (or plunge) velocity response and so will differ from the quasi-steady behaviour. The *Wagner* function $\Phi(\tau)$, also described in Chapter 9 and shown in Figure 14.7, indicates the change in lift following a step change in incidence.

14.3.4   Gust Response of Rigid Aircraft in Heave using Unsteady Aerodynamics

In this section, the earlier analysis in Section 14.3.1 for the rigid aircraft in heave will be extended to include unsteady aerodynamic effects and a more general input will be allowed.

14.3.4.1 Gust-dependent Lift

From the definition of the Küssner function $\Psi(t)$, the time variation of the *gust-dependent* lift produced when the aircraft enters a sharp-edged gust of velocity w_{g0} is given in terms of dimensional time t by

$$\Delta L_g(t) = \frac{1}{2}\rho V^2 S_W a \left(\frac{w_{g0}}{V}\right)\Psi(t) = \frac{1}{2}\rho V S_W a w_{g0}\Psi(t) \tag{14.13}$$

In Chapter 1 the idea of using the convolution (or Duhamel) integral to determine the response of a dynamic system to a general force input using Impulse Response Functions was introduced. In a similar approach, the variation in lift increment for a *general gust shape* $w_g(t)$ may be determined from the Duhamel integral, written in terms of the step response (Fung (1969); Hoblit (1988)), namely

$$\Delta L_g(t) = \frac{1}{2}\rho V S_W a \left\{\int_{t_0=0}^{t}\Psi(t-t_0)\dot{w}_g(t_0)\mathrm{d}t_0\right\} \tag{14.14}$$

Here $w_g(t_0)$ is the gust velocity at t_0, an integration variable and $\dot{w}_g = \mathrm{d}w_g/\mathrm{d}t$ is the rate of change of gust velocity with time. Note that the term $\dot{w}_g(t_0)\mathrm{d}t_0$ is the step gust amplitude w_g used in the convolution process. Since the Kussner function is defined in terms of non-dimensional time, the lift equation needs to be transformed accordingly using the following relationships

$$\frac{\mathrm{d}}{\mathrm{d}t} = \frac{\mathrm{d}}{\mathrm{d}\tau}\frac{\mathrm{d}\tau}{\mathrm{d}t} = \frac{\mathrm{d}}{\mathrm{d}\tau}\frac{2V}{c} \quad\text{and}\quad \mathrm{d}t = \frac{c}{2V}\mathrm{d}\tau. \tag{14.15}$$

Thus the non-dimensional lift is

$$\Delta L_g(\tau) = \frac{1}{2}\rho V S_W a\left\{\int_{\tau_0=0}^{\tau}\Psi(\tau-\tau_0)w_g'(\tau_0)\mathrm{d}\tau_0\right\} = \frac{1}{2}\rho V S_W a\left\{w_g'{}^*\Psi\right\} \tag{14.16}$$

where the dash $'$ indicates differentiation with respect to non-dimensional time and the symbol * indicates convolution (not complex conjugate in this case).

14.3.4.2 Response-dependent Lift

Equivalent expressions for the *response-dependent* lift, involving the Wagner function $\Phi(t)$, are

$$\Delta L_z(t) = \frac{1}{2}\rho V S_W a\left\{\int_{t_0=0}^{t}\Phi(t-t_0)\ddot{z}_C(t_0)\mathrm{d}t_0\right\}$$

$$\Delta L_z(\tau) = \frac{1}{2}\rho V S_W a\frac{2V}{c}\left\{\int_{\tau_0=0}^{\tau}\Phi(\tau-\tau_0)z_C''(\tau_0)\mathrm{d}\tau_0\right\} = \frac{1}{2}\rho V S_W a\frac{2V}{c}\left\{z_C''{}^*\Phi\right\} \tag{14.17}$$

14.3.4.3 Equation of Motion

The aircraft model with quasi-steady aerodynamics may now be revised to account for unsteady aerodynamics. For convenience, non-dimensional time will be retained so

$$m\left(\frac{2V}{c}\right)^2 z_C'' + \Delta L_z(\tau) = -\Delta L_g(\tau) \tag{14.18}$$

where both lift functions are defined in Equations (14.16) and (14.17) and are the result of convolutions. Thus the equation of motion is

$$m\left(\frac{2V}{c}\right)^2 z''_C + \frac{1}{2}\rho V S_W a \frac{2V}{c}\{z''_C * \Phi\} = -\frac{1}{2}\rho V S_W a \{w'_g * \Psi\} \tag{14.19}$$

Now, defining a mass parameter as

$$\mu_g = \frac{2m}{\rho a S_W c} \tag{14.20}$$

then the differential equation (14.19) may be rewritten as

$$z''_C + \frac{1}{2\mu_g}\{z''_C * \Phi\} = -\frac{1}{4\mu_g}\frac{c}{V}\{w'_g * \Psi\} \tag{14.21}$$

and clearly μ_g is a key parameter.

For a deterministic gust such as the '1-cosine', the rate of change of gust velocity may be determined and the right hand side of Equation (14.21) evaluated. The aircraft heave response may then be calculated by solving the resulting differential equation numerically; alternatively, the equation may be solved via the frequency domain by including the Theodorsen and Sears functions in the relevant acceleration per gust velocity FRF (see Chapter 1).

14.3.4.4 Gust Alleviation Factor

Such an approach may be used to estimate the gust alleviation factor K_g mentioned earlier; the response to a '1-cosine' gust of gradient distance of 25 semi-chords was used to calculate the maximum load factor $\Delta n_{1-\text{cosine}_{US}}$ for a range of values of μ_g with unsteady aerodynamics included. The results were compared to the load factor $\Delta n_{\text{Sharp}-\text{edged}_{QS}}$ found earlier for the sharp-edged gust of the same amplitude but only using quasi-steady aerodynamics (Hoblit, 1988). The gust alleviation factor K_g was then found from the relationship

$$\Delta n_{1-\text{cosine}_{US}} = K_g \Delta n_{\text{Sharp}-\text{edged}_{QS}} \tag{14.22}$$

An expression for K_g in terms of the aircraft mass parameter μ_g was obtained from a fit to the computational results obtained with different mass parameters. This gust alleviation factor was in effect an attempt to represent unsteady aerodynamic and real gust effects in terms of the mass parameter before the ability to handle more advanced calculations became available.

As an example, the acceleration response to a '1-cosine' gust of 25 semi-chords length is shown in Figure 14.8 for $\mu_g = 84.7$ with both quasi-steady and unsteady aerodynamics; the gust alleviation factor found is $K_g = 0.77$. Note that the unsteady aerodynamic effects are seen to reduce the severity of the gust response when compared to quasi-steady aerodynamics.

The analysis shown in this section applies to the rigid aircraft in heave only. When the rigid aircraft in heave and pitch is considered, then a similar approach would be used, except that the Wagner and Küssner functions would need to be applied to both the wing and tailplane lift forces, recognizing that the wing and tailplane chords, and corresponding non-dimensional

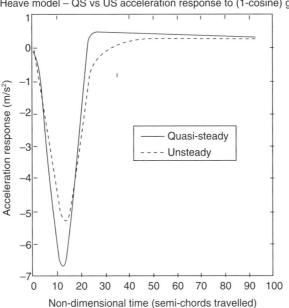

Figure 14.8   Response of heave-only model to a '1-cosine' gust of 25 semi-chords wavelength with quasi-steady and unsteady aerodynamics.

time definitions, will be different. For a swept wing, each strip contribution could be treated individually. Penetration effects would also need to be included.

14.4   Time Domain Gust Response – Rigid Aircraft in Heave/Pitch

So far, it has been assumed that the tailplane experienced the same incidence change as the wing and at the same time, so ignoring the penetration delay. The effects of including a separate tailplane and gust penetration will now be considered.

14.4.1   Gust Penetration Effect

It was noted earlier that the wing and tailplane encounter features of the gust at different times, as shown in Figure 14.9, and this delay needs to be represented in the gust response calculations. If $l \, (= l_W + l_T)$ is the distance between the wing and tailplane aerodynamic centres, then the wing experiences the gust velocity $w_g(t)$ while the tailplane experiences $w_g(t - l/V)$. Clearly, this assumes that the spatial variation of the gust velocity remains the same in the time taken for the aircraft to pass. Downwash effects on the tailplane are ignored at this point. However, even this argument is somewhat simplistic and only applies for lifting surfaces without sweep, since when wing sweep is present then different parts of the wing will encounter a particular feature of the gust at different times; the wing could then be split into sections, each encountering the gust at different times. Alternatively, the penetration effects can be allowed for when using a three-dimensional panel method (see later in Chapters 18 and 19).

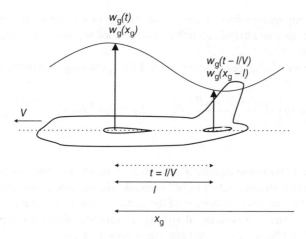

Figure 14.9   Gust penetration effect.

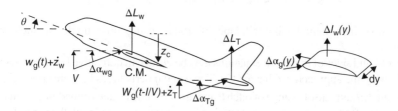

Figure 14.10   Rigid aircraft with heave/pitch motions showing incremental quantities.

14.4.2   *Equations of Motion – Rigid Aircraft including Tailplane Effect*

Consider the rigid aircraft shown in Figure 14.10 where the heave motion z_C (positive downwards) and pitch motion θ (positive nose up) are both referred to the centre of mass. The incremental lift forces ΔL_W, ΔL_T are considered, since the aircraft is assumed to be in a trimmed state prior to entering the gust and the gust loads will then be added to the steady flight loads. The lift forces are shown acting at the aerodynamic centres of the wing and tailplane. For simplicity, the quasi-steady aerodynamic representation will be employed, though unsteady aerodynamic effects are normally included (see earlier). Downwash effects will initially be excluded here to simplify the expressions, but will be added later when the derivative formulation is used.

It may be seen that the instantaneous pitch angle and heave rate of the aircraft influence the effective incidence of the wing and tailplane. From Figure 14.10, the increments in wing and tailplane incidence are given by

$$\Delta\alpha_{Wg} = \frac{w_g(t) + \dot{z}_W}{V} + \theta \qquad \Delta\alpha_{Tg} = \frac{w_g(t - l/V) + \dot{z}_T}{V} + \theta \tag{14.23}$$

where the heave velocities at the aerodynamic centres are now different for the wing and tailplane due to the pitch rate effects, namely

$$\dot{z}_W = \dot{z}_C - l_W\dot{\theta}, \qquad \dot{z}_T = \dot{z}_C + l_T\dot{\theta} \tag{14.24}$$

The unsteady aerodynamic effect of the rate of change of pitch $\dot{\theta} = q$ (i.e. pitch damping) is neglected here for the gust analysis though it was included for the wing in the earlier treatment of flutter in Chapter 10.

Thus the incremental changes in wing and tail lift (excluding downwash) are given by

$$\Delta L_W = \frac{1}{2}\rho V^2 S_W a_W \left[\frac{w_g(t) + \dot{z}_C - l_W \dot{\theta}}{V} + \theta\right] \quad \Delta L_T = \frac{1}{2}\rho V^2 S_T a_T \left[\frac{w_g(t - l/V) + \dot{z}_C + l_T \dot{\theta}}{V} + \theta\right]$$

$$(14.25)$$

It is now possible to write the equations of motion in heave and pitch to allow the response (relative to the initial trimmed state) to be obtained. The net downward force on the aircraft will be equal to the (mass × acceleration of the centre of mass) and the net nose up pitching moment about the centre of mass is equal to the (pitch moment of inertia I_y × pitch acceleration) using the generalized Newton's second law (see Chapter 5), so

$$m\ddot{z}_C = -\Delta L_W - \Delta L_T \quad I_y \ddot{\theta} = \Delta L_W l_W - \Delta L_T l_T \quad (14.26)$$

where positive incremental wing/tailplane lift forces cause nose up/down pitch responses respectively.

The right-hand sides of these two equations may be written in derivative form (based on inertial axes, following the approach in Chapter 12). The α and q derivatives here are appropriate for the change of pitch θ and rate of change of pitch $\dot{\theta}$ and are the same as determined in Chapter 12, with pitch rate terms for the wing being neglected. Also, terms associated with the zero lift condition are not required for this incremental case. However, derivatives for the heave velocity $\dot{z}_C$ appear for the first time in the gust analysis and these are essentially very similar to the w derivatives used for wind axes (see Chapter 13). Values for these derivatives are given in Appendix B.

Thus, building on earlier results, the equations of motion may be written compactly in matrix form as follows:

$$\begin{bmatrix} m & 0 \\ 0 & I_y \end{bmatrix} \begin{Bmatrix} \ddot{z}_C \\ \ddot{\theta} \end{Bmatrix} + \begin{bmatrix} -Z_{\dot{z}} & -Z_q \\ -M_{\dot{z}} & -M_q \end{bmatrix} \begin{Bmatrix} \dot{z}_C \\ \dot{\theta} \end{Bmatrix} + \begin{bmatrix} 0 & -Z_\alpha \\ 0 & -M_\alpha \end{bmatrix} \begin{Bmatrix} z_C \\ \theta \end{Bmatrix}$$

$$= \begin{Bmatrix} Z_{gW} \\ M_{gW} \end{Bmatrix} w_g(t) + \begin{Bmatrix} Z_{gT} \\ M_{gT} \end{Bmatrix} w_g\left(t - \frac{l}{V}\right)$$

$$(14.27)$$

where it may be seen that gust-related derivatives have now been introduced, namely

$$Z_{gW} = -\frac{1}{2}\rho V S_W a_W \qquad Z_{gT} = -\frac{1}{2}\rho V S_T a_T(1 - k_\varepsilon)$$

$$(14.28)$$

$$M_{gW} = \frac{1}{2}\rho V S_W a_W l_W \qquad M_{gT} = -\frac{1}{2}\rho V S_T a_T l_T(1 - k_\varepsilon)$$

There is only a dependency upon V (and not V^2) because the incidence due to the gust is w_g/V. Also, downwash terms involving k_ε (see Chapter 12) have been included because they influence the tailplane aerodynamics and therefore the effect of the gust at the tail. It should be noted of course that only the first term on the right-hand side of Equation (14.27) would be included for the initial part of the solution, i.e. before the tailplane enters the gust; this is indicated by the use of square brackets where, if the contents are negative, the term is ignored (i.e. the Heaviside function). The gust tail derivatives are now assumed to be affected by downwash effects and the other derivatives in Appendix B will also be used with downwash effects.

Solution of the equations of motion will yield the heave and pitch responses; the acceleration at the tailplane, for example, will be given by $\ddot{z}_T = \ddot{z}_C + l_T\ddot{\theta}$. The response will be more complicated than for the simple heave model considered earlier. For example, in a sharp-edged gust the aircraft would initially pitch nose up and heave upwards as the wing lift increased over the very short time prior to the tailplane entering the gust, but then the aircraft would pitch nose down due to the tailplane lift, so reducing the wing lift. The aircraft would tend to oscillate in pitch in its short period mode, but eventually settle down into a steady condition.

Even though this more advanced model of the aircraft allows for the tailplane effect, the penetration delay and a more complex gust shape, lift forces are still allowed to develop instantaneously and the aircraft is still rigid.

14.4.3   Example: Gust Response in the Time Domain for a Rigid Aircraft with Tailplane Effects

Consider an aircraft with the following data: $m = 10\,000$ kg, $S_W = 30$ m^2, $S_T = 7.5$ m^2, $c = 2.0$ m, $a_W = 4.5$/rad, $a_T = 3.2$/rad, $k_\varepsilon = 0.35$, $l_W = 0.6$ m and $l_T = 7$ m. These data are for the rigid aircraft, with further parameters being defined later for the flexible aircraft. The aircraft flies at 150 m/s EAS at 14 000 ft ($\sqrt{\sigma} \approx 0.8$), so the velocity is 187.5 m/s TAS. Consider the aircraft entering (a) a '1-cosine' gust of variable length or (b) a sharp-edged gust, both with maximum velocity 5 m/s EAS (or 6.25 m/s TAS). The responses will be evaluated firstly using the heave/pitch model and then using the heave only model discussed earlier (with quasi-steady aerodynamics). MATLAB and SIMULINK programmes to solve the time domain gust response for this rigid aircraft in heave and pitch are presented in the companion website.

14.4.3.1   '1-cosine' Gust

Firstly, the centre of mass (CoM) accelerations for '1-cosine' gusts of various lengths are determined. The variation of minimum and maximum values of acceleration for a whole range of gust lengths is shown in Figure 14.11; the maximum acceleration is around +0.51 g for a 400 m gust length (i.e. 200 m gust gradient) and the minimum of –0.65 g occurs for a 40 m gust length. In Figure 14.12, the variations of pitch rate, pitch angle and nose/CoM/tailplane accelerations

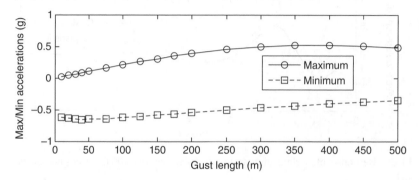

Figure 14.11   Minimum/maximum acceleration response of a rigid aircraft to variable length '1-cosine' gust – heave/pitch model.

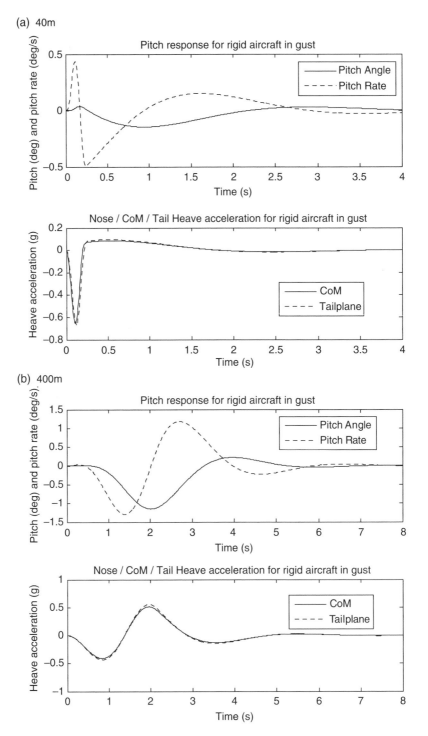

Figure 14.12 Response of a rigid aircraft to (a) 40 m and (b) 400 m '1-cosine' gusts – heave/pitch model.

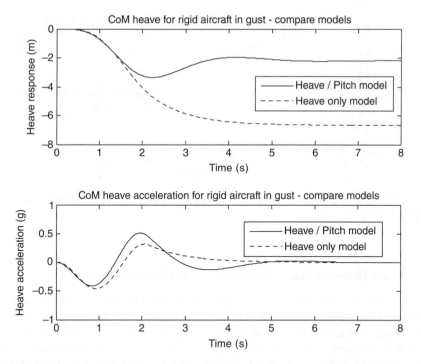

Figure 14.13 Response of a rigid aircraft to the 400 m '1-cosine' gust – comparison of heave only and heave/pitch models.

with time for the 'tuned' 40 m and 400 m gust lengths are presented (note the different time axes). The aircraft is seen to pitch nose up very slightly, pitch nose down as the tailplane enters the gust and the tailplane lift increases, and then pitch nose up again; the aircraft finishes with zero attitude and pitch rate, having climbed to a slightly higher altitude. The CoM acceleration first peaks at a negative value (i.e. upwards), as the aircraft initially encounters the gust, and then peaks at a positive value (i.e. downwards) as the nose down pitch takes effect; the relative magnitude of the two peaks depends upon the gust length. What is interesting is that there is barely any difference between the accelerations at the CoM and tailplane.

Secondly, in Figure 14.13 the results for the heave/pitch model are compared to the heave only model for the 400 m gust length. The heave/pitch model shows the aircraft to have gained less height than the heave only model because the nose down pitch offsets the initial incidence change. The accelerations for the two models are again surprisingly similar, although the heave only model underestimates the peaks somewhat because of the absent pitch effect. It should be noted that the 400 m wavelength is longer than would need to be considered under the certification requirements (see Chapter 23).

14.4.3.2 Sharp-edged Gust

Some equivalent results for the sharp-edged gust response are presented in Figure 14.14; the overshoot in the acceleration for the heave/pitch model when compared to the heave only model should be particularly noted. The aircraft ends up climbing steadily with no CoM acceleration; the heave only model leads to about twice the rate of climb compared to the heave/pitch model. These results confirm the discussions for the '1-cosine' gust above and show some differences between the two models.

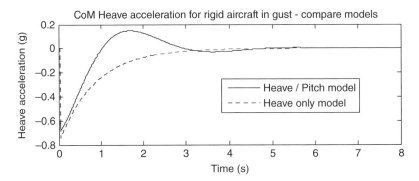

Figure 14.14 Response of a rigid aircraft to the sharp-edged gust – heave only model compared to the heave/pitch model.

14.5 Time Domain Gust Response – Flexible Aircraft

So far, only the rigid aircraft has been considered, but in practice the flexible modes of the aircraft will make a difference to the loads developed, especially if any significant frequency content in the gust time history coincides with one or more of the natural frequencies of the aircraft. In this section, in order to keep the mathematics at a manageable level, the flexible aircraft considered earlier in Chapter 12 will be used, i.e. rigid body heave and pitch modes together with a single flexible mode. The equations will refer to inertial axes.

The flexible mode was described in Chapter 12 and Appendix C. However, as a reminder, it permits free–free flexible modes to be considered with fuselage bending, wing bending or wing twist as the dominant components. The mode shape for the general case involves the wing bending and twist shapes defined by the functions $\kappa_e(y)$, $\gamma_e(y)$ respectively and the fuselage deformations defined by κ_{eF}, κ_{e0}, κ_{eC}, κ_{eT} at the front fuselage, wing mass axis, aircraft centre of mass/centre fuselage and tailplane respectively. The tailplane slope in the mode is γ_{eT}.

14.5.1 Equations of Motion – Flexible Aircraft

In order to save effort and space, the equation of motion for this gust case will simply be written by inspection. Remembering the gust equations for the rigid aircraft and how the flexible mode terms appeared in the equations for the equilibrium and dynamic manoeuvres in Chapters 12 and 13, the equations will take the form

$$
\begin{bmatrix} m & 0 & 0 \\ 0 & I_y & 0 \\ 0 & 0 & m_e \end{bmatrix} \begin{Bmatrix} \ddot{z}_C \\ \ddot{\theta} \\ \ddot{q}_e \end{Bmatrix} + \begin{bmatrix} -Z_{\dot{z}} & -Z_q & -Z_{\dot{e}} \\ -M_{\dot{z}} & -M_q & -M_{\dot{e}} \\ -Q_{\dot{z}} & -Q_q & c_e - Q_{\dot{e}} \end{bmatrix} \begin{Bmatrix} \dot{z}_C \\ \dot{\theta} \\ \dot{q}_e \end{Bmatrix}
$$

$$
+ \begin{bmatrix} 0 & -Z_\alpha & -Z_e \\ 0 & -M_\alpha & -M_e \\ 0 & -Q_\alpha & k_e - Q_e \end{bmatrix} \begin{Bmatrix} z_C \\ \theta \\ q_e \end{Bmatrix} = \begin{Bmatrix} Z_{gW} \\ M_{gW} \\ Q_{gW} \end{Bmatrix} w_g(t) + \begin{Bmatrix} Z_{gT} \\ M_{gT} \\ Q_{gT} \end{Bmatrix} w_g \left[t - \frac{l}{V} \right]
$$

(14.29)

The equations are in much the same form as those for the heave/pitch rigid aircraft model in Section 14.4 above, but now including aerodynamic terms associated with the flexible mode and also flexible mode mass and elastic stiffness terms. The general form of the time domain equations will be considered later in Section 14.6.

In this case, there are only two new derivatives that need to be determined, namely those corresponding to the flexible mode generalized force per gust velocity for the wing and tailplane, namely Q_{gW}, Q_{gT}. These derivatives require evaluation of the incremental work done for the wing and tailplane lift forces (due to gust velocity) moving through incremental flexible deformations of the mode. The incremental lift force on the wing strip dy due to wing gust velocity is

$$\Delta l_{gW}(y) = \frac{1}{2}\rho V^2 c dy a_W \frac{w_g(t)}{V} \tag{14.30}$$

Also, the effective incremental lift on the tailplane due to tailplane gust velocity (including downwash effects) is

$$\Delta L_{gT} = \frac{1}{2}\rho V^2 S_T a_T (1 - k_\varepsilon) \left[\frac{w_g(t - l/V)}{V} \right] \tag{14.31}$$

The corresponding incremental displacements at the wing strip and tailplane aerodynamic centres due to flexible mode deformation will be $(\kappa_e - l_A \gamma_e)\delta q_e$ and $\kappa_{eT}\delta q_e$ (both downwards) respectively. Thus, writing the incremental work done by the lift forces due to the gust moving through the flexible mode displacements, the flexible mode generalized forces due to gust velocity may be obtained and it may be shown that the corresponding flexible mode derivatives are

$$Q_{gW} = -\frac{1}{2}\rho V S_W a_W J_2 \quad Q_{gT} = -\frac{1}{2}\rho V S_T a_T \kappa_{eT}(1 - k_\varepsilon) \tag{14.32}$$

Here $J_2 = (1/s)\int_{y=0}^{s}(\kappa_e - l_A \gamma_e)\,dy$ as before. All the derivatives in Equation (14.29) are shown in Appendix B.

The equations of motion in Equation (14.29) will allow the generalized responses to a gust input to be determined and the result may be transformed into the physical response (e.g. acceleration) of the aircraft at any location, depending upon the mode deformation there. Thus, for example at the tailplane, the downwards acceleration is

$$\ddot{z}_T = \ddot{z}_C + l_T\ddot{\theta} + \kappa_{eT}\ddot{q}_e \tag{14.33}$$

The responses may also be used to generate the internal loads as described later in Chapter 16.

14.5.2   Example: Gust Response in the Time Domain for a Flexible Aircraft

This example uses the same data as in the earlier rigid aircraft example in Section 14.4.2 but additional parameters are specified to cater for idealized flexible effects, namely fuselage mass terms (see Chapter 12 and Appendix C) $m_T = 1500\,\text{kg}$, wing mass / inertia terms $m_W = 2\mu_W s = 3000\,\text{kg}$, $I_W = 2\chi_W s = 1330\,\text{kg m}^2$, aircraft pitch moment of inertia $I_y = 144\,000\,\text{kg m}^2$, and dimensions $s = 7.5\,m$, $l_A = 0.25\,m$, $l_E = 0.25$ m, $l_{WM} = 0.1$ m and $l_F = 6.8$ m. The modal mass and mode shape parameters for modes dominated by (a) fuselage bending, (b) wing bending and (c) wing twist are shown in Appendix C. Only the vertical '1-cosine' gust will be considered for the flexible aircraft. Again, it should be noted that the natural frequencies are chosen to be artificially low to highlight flexible effects.

Consider firstly the dominant *fuselage bending* mode with a natural frequency of 2 Hz and damping of 4% for the same flight condition as the rigid aircraft. In Figure 14.15 the pitch rate and CoM accelerations are shown for '1-cosine' gusts of 50, 100, 150 and 200 m lengths. The variation of minimum and maximum values of acceleration with gust length was obtained; for the flexible aircraft, the maximum acceleration is around +0.59 g for a 225 m gust length, with a minimum of −0.67 g for a 50 m gust length. Clearly, the tuned gust length and accelerations have altered when flexible effects are included. The oscillatory modal response is superimposed upon the short period type motion, with the flexible contribution depending on the gust wavelength.

In Figure 14.16, the variations with time of pitch rate, pitch angle and nose/CoM/tailplane accelerations for the 'tuned' 50 m gust length are presented. The presence of the flexible mode may be seen on the pitch rate and particularly on the tailplane and nose accelerations; some of the values are greater than for the rigid aircraft. In any practical case, it is quite possible that a particular internal load may be greater for the flexible case.

Secondly, the dominant *wing bending* case with a flexible mode of frequency 3 Hz and damping 4% was considered. The acceleration responses to gusts of 20, 40, 60 and 80 m lengths are shown in Figure 14.17. The gust length corresponding to the maximum (negative)

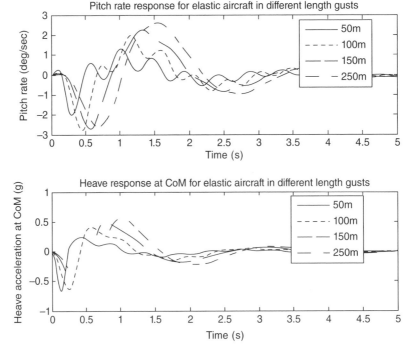

Figure 14.15   Response of a flexible aircraft to a '1-cosine' gusts of various/lengths – heave/pitch mode with fuselage bending mode (2 Hz/4%).

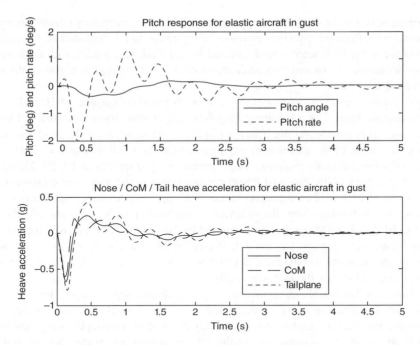

Figure 14.16   Response of a flexible aircraft to a tuned 50 m '1-cosine' gust – heave/pitch model with fuselage bending mode (2 Hz/4%).

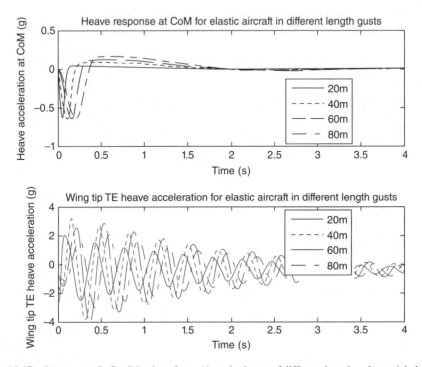

Figure 14.17   Response of a flexible aircraft to a '1-cosine' gust of different lengths – heave/pitch model with wing bending mode (3 Hz/4%).

acceleration at the CoM (–0.65 g) was approximately 40 m; the oscillatory response is not very significant on the fuselage, with the behaviour being similar to the rigid aircraft. However, the motion of the wing tip is seen to be dominated by the flexible response, with the maximum wing tip deflection of 108 mm and acceleration of 3.5 g occurring for a gust in the region of a 40–60 m length.

Finally, the dominant *wing torsion* case with a flexible mode of frequency 9 Hz and damping of 4% was considered. For a relatively long wavelength gust, there is no real evidence of flexible mode oscillations and the behaviour is much the same as for the rigid aircraft. For shorter gusts nearer to the wavelength required to tune the flexible response, there is clear evidence of a flexible mode response. The responses to gust lengths of 10, 20, 35 and 50 m are shown in Figure 14.18. The gust length corresponding to the maximum (negative) acceleration at the CoM (–0.72 g) was approximately 35 m; the oscillatory response is not very significant on the fuselage, with the behaviour being similar to the rigid aircraft. However, the motion of the wing tip is seen to be dominated by the flexible response, with the maximum wing tip twist of 0.5° and acceleration of 2.88 g being for a gust in the region of 20 m length. Obviously, the unsteady aerodynamic attenuation effect will be considerable at such a high frequency and will reduce the response levels.

The impact of the flexible mode on gust response clearly depends upon the mode shape, natural frequency and gust wavelength as well as on the response position on the aircraft. Sweep will also affect the gust response, not only because of the coupled wing bending and torsion in the modes but also because of the delayed penetration of the wing tip compared to the root.

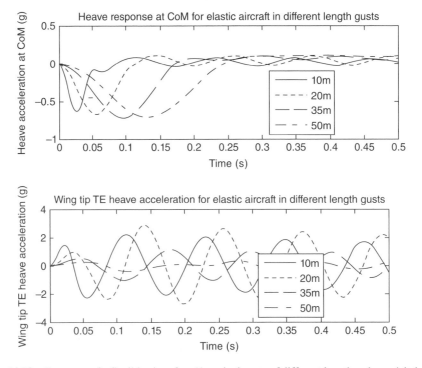

Figure 14.18   Response of a flexible aircraft to '1-cosine' gusts of different lengths – heave/pitch model with wing torsion mode (9 Hz/4%).

14.6   General Form of Equations in the Time Domain

So far, the examples chosen have deliberately been aimed at introducing different features of gust response calculations in turn. However, in practice all these features will be included together, e.g. realistic gust histories, rigid body plus flexible modes, gust penetration effects, unsteady aerodynamics, etc. In this section, a more general expression will be introduced; if the models developed earlier are examined and common features sought, it may be seen that the time domain equations of motion can be written in the general form

$$\mathbf{A}\ddot{\mathbf{q}} + \rho V \mathbf{B}\dot{\mathbf{q}}^* \Phi + \mathbf{E}\mathbf{q} + \rho V^2 \mathbf{C}\mathbf{q}^* \Phi = \rho V \mathbf{R}_{\mathrm{gW}} \dot{w}_g(t)^* \Psi + \rho V \mathbf{R}_{\mathrm{gT}} \dot{w}_g \left[t - \frac{l}{V} \right]^* \Psi \qquad (14.34)$$

where q are generalized coordinates (now both rigid body and flexible modes), $\mathbf{A}$ is an inertia matrix, $\mathbf{B}$ and $\mathbf{C}$ are aerodynamic *response*-dependent matrices, $\mathbf{E}$ is a structural stiffness matrix (with zero elements corresponding to the rigid body terms) and $\mathbf{R}_{\mathrm{gW}}$, $\mathbf{R}_{\mathrm{gT}}$ are *gust*-dependent aerodynamic vectors for the wing and tailplane; all these aerodynamic terms are related to the quasi-steady aerodynamic derivatives. The convolutions with the Wagner and Küssner functions are included to account for unsteady aerodynamic effects. A structural damping matrix $\mathbf{D}$ may also be introduced if desired.

To examine the response for any general gust time history, for example the '1-cosine' gust, these simultaneous differential equations (14.34) would be solved in the time domain. Non-linear effects (e.g. FCS) could also be introduced. The equations shown would apply for strip theory type calculations, but if a three-dimensional panel method was used for the unsteady aerodynamics, then both the response and gust dependent aerodynamic terms would be a function of reduced frequency and the results could be expressed in a time domain state space representation using the Rational Function Approximation approach (see Chapter 19).

14.7   Turbulence Response in the Frequency Domain

In the first part of this chapter, the response to discrete gusts was seen to be calculated in the *time domain*, normally by numerical integration of the equations of motion. However, the treatment of random continuous turbulence usually requires calculations to be performed in the *frequency domain* using a Power Spectral Density (PSD) approach as the computations would otherwise be too lengthy; this approach is possible because the continuous turbulence input can be defined via its PSD.

The relationship between the PSD of response and excitation for a system was introduced in Chapter 1 where it was shown that the response PSD may be calculated from the excitation PSD and the relevant Frequency Response Function (FRF). This process will yield statistical information about the response amplitude, but nothing about its phase. For a continuous turbulence excitation, the relevant FRF is that between the response and a unit, harmonically oscillating, one-dimensional gust field (or harmonic gust) determined over the frequency range of interest. The harmonic gust is essentially a deterministic gust whose response can be obtained from the earlier equations transformed into the frequency domain.

Because frequency domain analysis using the PSD is a linear approach, it is important to recognize that any non-linearity of the aircraft (e.g. via the FCS) will have to be linearized in some way, leading to an approximation in the predicted response. An alternative approach, that permits non-linear effects to be included, is to generate a random time signal with the required spectral characteristics to represent the turbulence and then to calculate the response directly in the time domain; this methodology will be considered briefly in Chapter 23.

In the remainder of this chapter, the basic ideas underpinning analysis of the response to turbulence are illustrated for both quasi-steady and unsteady response-dependent and gust-dependent aerodynamics. The methodology will initially be based on the heave only model and then extended to the heave/pitch rigid and flexible aircraft cases. This exercise will also allow the equations of motion to be developed in a general form.

14.7.1 Definition of Continuous Turbulence

Continuous turbulence is represented by a random variation in the velocity of the air normal to (or sometimes along) the flight path of the aircraft, where the random variable is assumed to have a Gaussian distribution with zero mean and is represented by a turbulence (or gust) PSD $\Phi_{gg}(\Omega)$ with units of $(m/s)^2/(rad/m)$. The subscript 'g' is used here in place of classical forced excitation symbol 'f' (see Chapter 1) to highlight that the turbulence excitation may be thought of as a combination of harmonic gusts.

A commonly used turbulence (or gust) spectrum that matches extensive experimentally observed data is that according to von Karman (Fung, 1969; Hoblit, 1988; CS-25), namely

$$\Phi_{gg}(\Omega) = \sigma_g^2 \frac{L}{\pi} \frac{1 + (8/3)(1.339\Omega L)^2}{\left[1 + (1.339\Omega L)^2\right]^{11/6}} \tag{14.35}$$

where $\Omega = \omega/V$ is the scaled frequency (rad/m), σ_g is the root-mean-square turbulence (or gust) velocity intensity (m/s TAS) and V is the air speed (m/s TAS). L is the characteristic scale wavelength of the turbulence (typically 762 m, but usually quoted in ft, namely 2500 ft) that dictates the variation of the PSD with frequency. Calculations are usually carried out for a root-mean-square turbulence (or gust) intensity of 1 m/s and then results factored for the design turbulence velocities specified in CS-25 at the particular flight condition considered. It should be noted that the area under the PSD $\Phi_{gg}(\Omega)$ is equal to the mean square of the turbulence σ_g^2 and so the root-mean-square value σ_g is the square root of this area (see Chapter 1). For continuous turbulence analysis, the PSD is defined as a one-sided spectrum so only a positive frequency axis is relevant.

The variation of the turbulence (or gust) PSD with scaled frequency for a root-mean-square turbulence intensity of 1 m/s and a scale length of 2500 ft is shown in Figure 14.19. Note that use of the scaled frequency means that the graph is independent of aircraft velocity. The area under this curve is $1(m/s)^2$ which is the mean square value of the turbulence intensity, so confirming that the root-mean-square value is 1 m/s.

The PSD may also be expressed in terms of frequency (Hz) instead of rad/m by replacing L/π in Equation (14.35) with $2L/V$; the PSD units would then be $(m/s)^2/Hz$ and the frequency axis would be in Hz. The example given later in this chapter will use these dimensional units. Care needs to be taken, when determining the root-mean-square value from the area under the PSD, that the correct combination of units is used.

14.7.2 Definition of a Harmonic Gust Velocity Component

Continuous turbulence, as a random phenomenon, may be thought of as a combination of harmonic gust velocity components. The frequency domain approach requires a harmonic gust

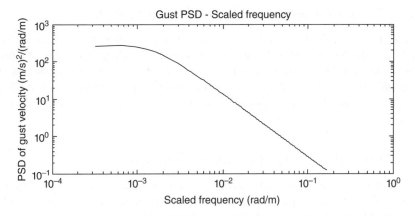

Figure 14.19 Von Karman turbulence (or gust) PSD as a function of scaled frequency.

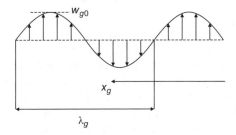

Figure 14.20 Harmonic gust component.

excitation component to be defined in order to perform the PSD analysis and process each frequency value separately. Consider a harmonic gust component of amplitude w_{g0} and wavelength λ_g, defined spatially by the expression

$$w_g(x_g) = w_{g0} \sin \frac{2\pi x_g}{\lambda_g} \tag{14.36}$$

as shown in Figure 14.20. Since $x_g = Vt$ then the spatial variation of the harmonic gust velocity at the wing may be transformed into a temporal variation at frequency $\omega = 2\pi V/\lambda_g$ rad/s, namely

$$w_g(t) = w_{g0} \sin \frac{2\pi V}{\lambda_g} t = w_{g0} \sin \omega t = w_{g0} \exp(i\omega t). \tag{14.37}$$

Here a complex algebra approach has been used, as introduced in Chapter 1. In order to include the penetration delay for the tailplane, the tailplane gust velocity must also be written in complex algebra form but using a phase lag to account for the delay; therefore at the tailplane

$$w_g\left(t - \frac{l}{V}\right) = w_{g0} \sin \frac{2\pi V}{\lambda_g}\left(t - \frac{l}{V}\right) = w_{g0} \sin \omega \left(t - \frac{l}{V}\right) = w_{g0} \exp\left(i\omega \left(t - \frac{l}{V}\right)\right). \tag{14.38}$$

The response for each harmonic gust component may now be obtained, as illustrated in Chapters 1 and 2.

14.7.3   FRFs for Response per Harmonic Gust Velocity

In order to determine the PSD of a response, knowing the turbulence (or gust) PSD from section 14.7.1, the FRF between the response and the harmonic gust velocity at a particular frequency is needed. For a particular response term (r), the FRF per harmonic gust velocity is $H_{rg}(\omega)$ and this is determined using a complex algebra approach as shown in Chapters 1 and 2. It should be noted that the approach to the analysis of a 'response' may also be extended to internal loads (see Chapter 16).

14.7.4   PSD of Response to Continuous Turbulence

The von Karman turbulence (or gust) PSD $\Phi_{gg}(\omega)$ was defined in Equation (14.35). Thus the PSD $\Phi_{rr}(\omega)$ of a particular response parameter (r) may be found by the approach introduced in Chapter 1 (and Chapter 2 for multiple degrees of freedom), so

$$\Phi_{rr}(\omega) = \left| H_{rg}(\omega) \right|^2 \Phi_{gg}(\omega) \tag{14.39}$$

where this expression only applies for a single response parameter. It is important that care be taken with units e.g. if the turbulence (or gust) PSD has units of $(m/s)^2/Hz$, then the acceleration per gust velocity FRF has units of $g/(m/s)$ and the acceleration response PSD will have units of $(g)^2/Hz$. Therefore, Equation (14.39) shows that a simple frequency by frequency multiplication of the modulus squared value of the relevant FRF and the turbulence (or gust) PSD will yield the response PSD. The root-mean-square value of the response σ_r may then be found from the square root of the area under the response PSD curve such that

$$\sigma_r = \sqrt{\int_0^{\omega_{max}} \Phi_{rr}(\omega) \, d\omega} \tag{14.40}$$

where ω_{max} is the maximum frequency for which the one-sided PSD data are generated. Alternatively, the scaled frequency form of the PSDs may be used. Examples of calculating PSDs and root-mean-square values for responses such as acceleration will be given later in this chapter, but the transformation from response PSDs to internal load PSDs is considered later in Chapter 16.

In addition to the root-mean-square of the response, a further statistical parameter employed is given by

$$\bar{A} = \frac{\sigma_r}{\sigma_g} \tag{14.41}$$

which is the ratio of the root-mean-square of response to the root-mean-square turbulence (or gust) velocity (in TAS); it will be used later when loads are considered.

14.8   Frequency Domain Turbulence Response – Rigid Aircraft in Heave

In this section, the determination of FRFs of response per harmonic gust velocity will be considered for the rigid aircraft in heave with quasi-steady and unsteady aerodynamics. The method for determining the FRF of response per harmonic gust velocity is illustrated and examples shown.

14.8.1　FRF for Rigid Aircraft Response in Heave per Harmonic Gust Velocity – Quasi-Steady Aerodynamics

The approach to obtaining an FRF between response and harmonic gust velocity will be first illustrated using the simple rigid aircraft in heave with quasi-steady aerodynamics, as considered earlier in section 14.3.1. The methodology based in the frequency domain is introduced in Chapter 1. Because the quasi-steady assumption requires the lift forces to depend on the instantaneous gust velocity, then the equation of motion for the sharp-edged gust may be revised in terms of a harmonic gust input as

$$m\ddot{z}_\mathrm{C} + \Delta L_z(t) = -\Delta L_\mathrm{g}(t) \tag{14.42}$$

or

$$m\ddot{z}_\mathrm{C} + \frac{1}{2}\rho V S_\mathrm{W} a \dot{z}_\mathrm{C} = -\frac{1}{2}\rho V S_\mathrm{W} a w_{\mathrm{g}0} \mathrm{e}^{i\omega t} \tag{14.43}$$

where ΔL_z, ΔL_g are the response-dependent and gust-dependent lift forces. It is now assumed that the steady-state response to the harmonic gust component defined in Equation (14.37) is given by

$$z_\mathrm{C}(t) = \tilde{z}_\mathrm{C} \mathrm{e}^{i\omega t} \tag{14.44}$$

where ~ indicates a complex quantity (i.e. including phase information), as introduced in Chapter 1. Also, the lower case is used for frequency domain amplitude to avoid possible confusion with other symbols (in particular the aerodynamic derivatives). The steady-state response-dependent and gust-dependent lift forces are given by

$$\Delta L_z(t) = \Delta \tilde{L}_z \mathrm{e}^{i\omega t} \qquad \Delta L_\mathrm{g}(t) = \Delta \tilde{L}_\mathrm{g} \mathrm{e}^{i\omega t} \tag{14.45}$$

where $\Delta \tilde{L}_z$, $\Delta \tilde{L}_\mathrm{g}$ are the complex amplitudes of the relevant lift forces. Then substituting for $z_\mathrm{C}(t)$, $\Delta L_z(t)$ and $\Delta L_\mathrm{g}(t)$ in Equation (14.43), cancelling out the $\mathrm{e}^{i\omega t}$ term and simplifying yields the frequency domain expression

$$\left(-\omega^2 m + i\omega \frac{1}{2}\rho V S_\mathrm{W} a \right) \tilde{z}_\mathrm{C} = -\Delta \tilde{L}_\mathrm{g} = -\frac{1}{2}\rho V S_\mathrm{W} a w_{\mathrm{g}0} \tag{14.46}$$

where it may be noted that the quasi-steady gust-dependent lift force amplitude is in-phase with the gust velocity component. Hence, rearranging Equation (14.46) yields

$$H_{z\mathrm{g}}(\omega) = \frac{\tilde{z}_\mathrm{C}}{w_{\mathrm{g}0}} = \frac{-\dfrac{1}{2}\rho V S_\mathrm{W} a}{-\omega^2 m + i\omega \dfrac{1}{2}\rho V S_\mathrm{W} a}, \tag{14.47}$$

where $H_{z\mathrm{g}}(\omega)$ is the Frequency Response Function (FRF) at frequency ω between the (downwards) heave response and the (upwards) harmonic gust velocity component. This process effectively allows the FRF to be determined over a suitable range of frequencies; only amplitude data are required as use of the modulus squared FRF eliminates phase. The equivalent acceleration FRF result would be obtained by multiplying by $-\omega^2$.

14.8.2   Unsteady Aerodynamic Effects in the Frequency Domain

The analysis in Section 14.8.1 obtained the response of a simple rigid aircraft to a harmonic turbulence component using *quasi-steady* aerodynamic effects. In this section, the representation of *unsteady* aerodynamic effects will be explained.

To include the *gust-dependent* unsteady aerodynamics, consider the aircraft wing 'immersed' in a harmonic gust as shown in Figure 14.20. It may be shown (Fung, 1969) that the attenuation and phase lag associated with a wing moving through a harmonic gust field can be allowed for by employing the Sears function $\phi(k)$, a complex function of the reduced frequency $k = \omega c/(2V)$, introduced in Chapter 9. The Sears function is actually very similar to the Theodorsen function at small reduced frequencies. The effect of using unsteady aerodynamics on the *gust-dependent* lift in the frequency domain is to multiply the lift term shown in Equation (14.46) by the Sears function, so

$$\tilde{\Delta}L_{\mathrm{Wg}} = \frac{1}{2}\rho VS_{\mathrm{W}} a w_{\mathrm{g0}}\, \phi(k) = \frac{1}{2}\rho VS_{\mathrm{W}} a w_{\mathrm{g0}}\, \phi\!\left(\frac{\omega c}{2V}\right) \tag{14.48}$$

This is because convolution in the time domain is equivalent to multiplication in the frequency domain (see Chapter 1). The Sears function attenuates the lift force and introduces a phase lag, the effects becoming more significant as the frequency increases.

In a similar way, the Theodorsen function $C(k)$ may be introduced to account for unsteady aerodynamic effects on the *response-dependent* lift in the frequency domain, so

$$\tilde{\Delta}L_z = i\omega \frac{1}{2}\rho VS_{\mathrm{W}} a \tilde{z}_{\mathrm{C}}\, C\!\left(\frac{\omega c}{2V}\right) \tag{14.49}$$

where again multiplication has replaced convolution.

The Theodorsen and Sears functions would only be used for large commercial aircraft if strip theory were to be employed (see Chapter 21). More commonly, unsteady aerodynamic computations based on three-dimensional panel methods (e.g. the Doublet Lattice method; see Chapters 18 and 19) are employed and in effect the Theodorsen and Sears effects are automatically embedded in the complex Aerodynamic Influence Coefficient (AIC) matrices. These functions are covered in this book to assist in understanding the effect of unsteady aerodynamics, because strip theory type approaches are still used sometimes and also for historical reasons.

14.8.3   FRF for Rigid Aircraft Response in Heave per Harmonic Gust Velocity – Unsteady Aerodynamics

The FRF between response and harmonic gust velocity at frequency ω was obtained in Section 14.8.1 using quasi-steady aerodynamics. Because the unsteady response-dependent and gust-dependent aerodynamic effects may be introduced in the frequency domain by a multiplication process, as shown in the previous section, the corresponding FRF may be written by inspection of Equations (14.43), (14.48) and (14.49) as

$$H_{zg}(\omega) = \frac{\ddot{z}_C}{w_{g0}} = \frac{-\frac{1}{2}\rho V S_W a}{-\omega^2 m + i\omega \frac{1}{2}\rho V S_W a\, C(\omega c/(2V))}\phi\left(\frac{\omega c}{2V}\right) \qquad (14.50)$$

Thus the aerodynamic attenuation and phase shift are directly applied to the FRF, rather like a filter.

14.8.4 Example: Turbulence Response in the Frequency Domain for a Rigid Aircraft in Heave with Quasi-Steady Aerodynamics

In Section 14.8.1, the determination of FRFs between response and harmonic gust velocity was considered. Given the turbulence (or gust) PSD, the response PSD can be evaluated using the analysis outlined in Section 14.7. In this section, an example showing the FRF and PSD for the acceleration response will be shown for the rigid aircraft in heave with quasi-steady aerodynamics.

For the examples here and in the remainder of this chapter, the von Karman turbulence (or gust) PSD is employed with a scale length of 2500 ft and a reference root-mean-square turbulence intensity of 1 m/s; results for design turbulence velocities may be obtained later by simple scaling. The aircraft is flying at 150 m/s EAS at 14 000 ft ($\sqrt{\sigma}\approx0.8$). The turbulence (or gust) PSD is shown in Figure 14.21 plotted against frequency in Hz, so the units are in (m/s)^2/Hz. Note that the graphical results would look different if presented in terms of scaled frequency (see Figure 14.19) but the final root-mean-square values of response should of course be the same. The graphs are mostly presented in log–log format, allowing the low frequency behaviour to be seen more clearly.

For the rigid aircraft considered earlier, using the heave only model with quasi-steady aerodynamics, the modulus squared FRF relating CoM acceleration to gust velocity, namely $|H_{zg}(\omega)|^2$, is shown as a function of frequency (Hz) in Figure 14.22. The acceleration response PSD (g^2/Hz) is then found by multiplying the squared FRF by the turbulence (or gust) PSD; the result is also shown in Figure 14.22. The effect of the roll-off in the turbulence (or gust)

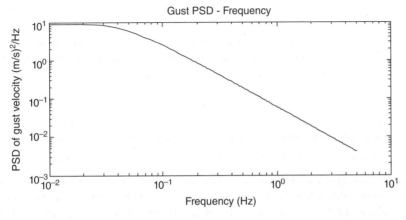

Figure 14.21   Von Karman turbulence (or gust) PSD as a function of frequency.

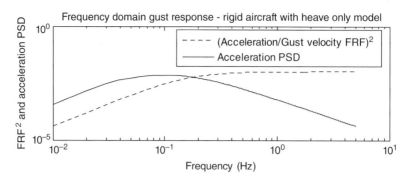

Figure 14.22 Frequency domain gust results for a rigid aircraft heave only model with quasi-steady aerodynamics: (CoM acceleration/gust velocity FRF)^2 and CoM acceleration PSD.

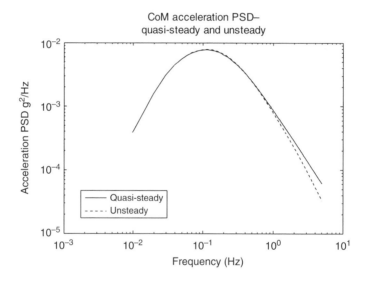

Figure 14.23 PSD for heave acceleration response – quasi-steady and unsteady aerodynamics.

PSD may be seen by comparing the two curves. The root-mean-square value of the CoM acceleration is 0.056 g, found from the square root of the area under the acceleration PSD curve. The peak of the acceleration PSD occurs at around 0.1 Hz.

14.8.5 Example: Turbulence Response in the Frequency Domain for a Rigid Aircraft in Heave with Unsteady Aerodynamics

The example in Section 14.8.4 is now repeated to allow the comparison of the response PSD for both quasi-steady and unsteady aerodynamics. The aircraft heave response acceleration PSD is plotted in Figure 14.23. The effect on the higher frequency response (such as for flexible modes) will be important.

14.9   Frequency Domain Turbulence Response – Rigid Aircraft in Heave/Pitch

So far, the frequency domain analysis has been restricted to the rigid aircraft in heave alone. However, in practice the aircraft will have a separate tailplane and will experience pitch due to the penetration effect, so the desired FRFs would need to be extracted from the heave and pitch equations. Only quasi-steady aerodynamics will be considered here though unsteady effects could be incorporated.

14.9.1   FRF for Rigid Aircraft Response in Heave/Pitch per Harmonic Gust Velocity

In order to illustrate the idea, consider the earlier example of the rigid aircraft in heave/pitch in Section 14.4. The key time domain equation is (14.27) and if a harmonic gust input is considered, then the steady-state heave and pitch responses can be written in exponential form as

$$z_C = \tilde{z}_C e^{i\omega t} \qquad \theta = \tilde{\theta} e^{i\omega t} \tag{14.51}$$

If the time domain equation is transformed into the frequency domain by using the complex algebra approach employed in Chapters 1 and 2, then using the expressions for the harmonic gust velocity at the wing and tailplane in Equations (14.37) and (14.38), it may be shown that

$$
\left\{ -\omega^2 \begin{bmatrix} m & 0 \\ 0 & I_y \end{bmatrix} + i\omega \begin{bmatrix} -Z_{\dot{z}} & -Z_q \\ -M_{\dot{z}} & -M_q \end{bmatrix} + \begin{bmatrix} 0 & -Z_\alpha \\ 0 & -M_\alpha \end{bmatrix} \right\} \left\{ \begin{matrix} \tilde{z}_C \\ \tilde{\theta} \end{matrix} \right\}
$$
$$
= \left\{ \begin{matrix} Z_{gW} \\ M_{gW} \end{matrix} \right\} w_{g0} + \left\{ \begin{matrix} Z_{gT} \\ M_{gT} \end{matrix} \right\} w_{g0} e^{-i\frac{\omega l}{V}} \tag{14.52}
$$

where the delay due to the penetration effect is accounted for using the phase lag term for the tailplane. Then, solving these simultaneous equations for the harmonic response amplitudes in terms of the system dynamics and the harmonic gust velocity amplitude leads to the expression

$$
\tilde{q} = \left\{ \begin{matrix} \tilde{z}_C \\ \tilde{\theta} \end{matrix} \right\} = \left[-\omega^2 \begin{bmatrix} m & 0 \\ 0 & I_y \end{bmatrix} + i\omega \begin{bmatrix} -Z_{\dot{z}} & -Z_q \\ -M_{\dot{z}} & -M_q \end{bmatrix} + \begin{bmatrix} 0 & -Z_\alpha \\ 0 & -M_\alpha \end{bmatrix} \right]^{-1}
$$
$$
\left\{ \left\{ \begin{matrix} Z_{gW} \\ M_{gW} \end{matrix} \right\} + \left\{ \begin{matrix} Z_{gT} \\ M_{gT} \end{matrix} \right\} \exp\left(-i\frac{\omega l}{V} \right) \right\} w_{g0} = \boldsymbol{H}_{qg} w_{g0} \tag{14.53}
$$

where $\boldsymbol{H}_{qg}$ is the vector of FRFs between the heave and pitch responses and the harmonic gust velocity. Then, using Equation (14.33) as an example of obtaining the physical response from the generalized responses above, the relationship between the tailplane harmonic displacement response and the harmonic gust velocity may be found and converted to acceleration by multiplying by $-\omega^2$, so

$$\tilde{z}_{T\,Acc} = -\omega^2 \{ 1 \quad l_T \} \boldsymbol{H}_{qg} w_{g0} = H_{T\,Acc\,g} w_{g0} \tag{14.54}$$

This acceleration FRF $H_{T\,Accg}$ will show a peak for the short period mode. An example is shown in Section 14.9.2. MATLAB code for this frequency domain calculation is included in the companion website. Unsteady aerodynamics effects could be allowed for simply by multiplying the relevant derivatives by the Theodorsen and Sears functions.

14.9.2   Example: Turbulence Response in the Frequency Domain for a Rigid Aircraft in Heave/Pitch

Here, the rigid aircraft heave/pitch model from the time domain study is used. The FRF modulus squared for CoM acceleration per harmonic gust velocity is shown as a function of frequency (Hz) in Figure 14.24. The corresponding acceleration response PSD (g^2/Hz) is also shown in Figure 14.24. The root-mean-square value of the CoM acceleration is 0.051 g, close to that of 0.056 g for the heave only model. The acceleration PSDs for the CoM and tailplane are similar for the rigid aircraft, with the tailplane root-mean-square value being 0.053 g. The peak of the acceleration PSD corresponds to the short period mode and occurs around 0.3 Hz for this flight condition.

Finally, the heave/pitch and heave only models are compared in terms of acceleration PSDs in Figure 14.25; although the root-mean-square values are fairly similar, the results look rather different for the logarithmic axes. The major difference between the two results occurs below 0.3 Hz. Note that some sample results are shown in (Hoblit, 1988) for different non-dimensional parameters and it is pointed out that the heave only case displays a 'short period' type peak, but of course this is not a mode.

14.10   Frequency Domain Turbulence Response – Flexible Aircraft

So far, the frequency domain analysis has been restricted to the rigid aircraft in heave/pitch alone. However, in practice the aircraft will be flexible. The flexible aircraft result will simply be quoted, being obtained by inspection of the rigid aircraft results and analysis in earlier chapters. Only quasi-steady aerodynamics will be considered.

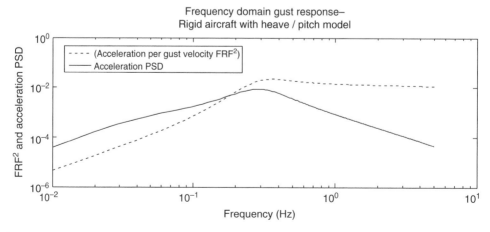

Figure 14.24   Frequency domain gust results for a rigid aircraft heave/pitch model: (CoM acceleration/ gust velocity FRF)^2 and CoM Acceleration PSD.

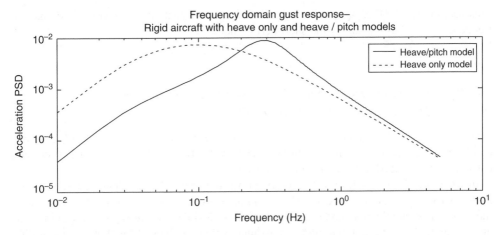

Figure 14.25 Comparison of frequency domain gust results for rigid aircraft heave only and heave/pitch models: CoM acceleration PSD.

14.10.1 FRF for Flexible Aircraft Response in Heave/Pitch per Harmonic Gust Velocity

Consider the earlier example of a flexible aircraft in Section 14.5; the behaviour of the rigid aircraft in heave/pitch may then be seen by inspection. If a harmonic gust input is considered, then the responses can be written as

$$z_C = \tilde{z}_C e^{i\omega t} \qquad \theta = \tilde{\theta} e^{i\omega t} \qquad q_e = \tilde{q}_e e^{i\omega t} \qquad (14.55)$$

where $\tilde{q}_e$ is used instead of $\tilde{Q}_e$ to avoid confusion with the flexible derivative Q_e. If the time domain equation (14.29) is transformed into the frequency domain by using the complex algebra approach employed in Chapters 1 and 2, then using the expressions for the harmonic gust velocity at the wing and tailplane in Equations (14.37) and (14.38), it may be shown that the final expression equivalent to Equation (14.53) is

$$\tilde{q} = \left\{ \begin{matrix} \tilde{z}_C \\ \tilde{\theta} \\ \tilde{q}_e \end{matrix} \right\} = \left[-\omega^2 \begin{bmatrix} m & 0 & 0 \\ 0 & I_y & 0 \\ 0 & 0 & m_e \end{bmatrix} + i\omega \begin{bmatrix} -Z_{\dot{z}} & -Z_q & -Z_{\dot{e}} \\ -M_{\dot{z}} & -M_q & -M_{\dot{e}} \\ -Q_{\dot{z}} & -Q_q & c_e - Q_{\dot{e}} \end{bmatrix} + \begin{bmatrix} 0 & -Z_\alpha & -Z_e \\ 0 & -M_\alpha & -M_e \\ 0 & -Q_\alpha & k_e - Q_e \end{bmatrix} \right]^{-1}$$

$$\left\{ \left\{ \begin{matrix} Z_{gW} \\ M_{gW} \\ Q_{gW} \end{matrix} \right\} + \left\{ \begin{matrix} Z_{gT} \\ M_{gT} \\ Q_{gT} \end{matrix} \right\} \exp\left(-i\frac{\omega l}{V} \right) \right\} w_{g0} = \boldsymbol{H}_{qg} w_{g0} \qquad (14.56)$$

where $\boldsymbol{H}_{qg}$ is the vector of FRFs between the heave, pitch and elastic mode responses and the harmonic gust velocity. Then, using Equation (14.33) as an example of obtaining the physical response from the generalized responses above, the relationship between the tailplane harmonic acceleration and the harmonic gust velocity will be

$$\tilde{z}_{T\,Acc} = -\omega^2 \{ 1 \quad l_T \quad \kappa_{eT} \} \boldsymbol{H}_{qg} w_{g0} = H_{T\,Acc\,g} w_{g0} \qquad (14.57)$$

This acceleration FRF $H_{T\,Acc\,g}$ will now show a resonant peak at the flexible mode natural frequency as well as a peak for the short period mode.

14.10.2 Example: Turbulence Response in the Frequency Domain for a Flexible Aircraft

Consider the flexible aircraft used earlier and include a 2 Hz/4% *fuselage bending* mode, combined with heave and pitch motions. The MATLAB code for this frequency domain calculation is included in the companion website.

In Figure 14.26, the CoM, tailplane and wing tip trailing edge (TE) acceleration PSDs (g^2/Hz) are shown as a function of frequency (Hz). Peaks are seen corresponding to what are essentially the short period and flexible modes, and the increased flexible mode response at the tailplane is also evident. The corresponding root-mean-square value for the CoM is 0.063 g (about 20% larger than the rigid aircraft value at the CoM) and for the tailplane is 0.113 g (80% larger than at the CoM – a good reason for sitting nearer to the centre of the aircraft for comfort). The wing tip trailing edge results are almost identical to those for the CoM because for this mode there is no significant wing bending or twist.

Clearly, modes involving any significant wing bending or twist would be apparent on the wing response PSDs. For example, for a *wing bending* mode of 3 Hz frequency and 4% damping, the response PSDs are shown in Figure 14.27; the wing tip response is much larger than the fuselage response and the root-mean-square values for the fuselage and wing tip were 0.06 g and 0.31 g respectively.

Finally, the results for a *wing torsion* mode of 9 Hz frequency and 4% damping are shown in Figure 14.28 where the wing tip root-mean-square value is 0.17 g, noting the different frequency axis used.

It should be noted that the PSD and hence root-mean-square value for load parameters such as the wing root bending moment may also be determined from the responses using the so-called auxiliary equation (see Chapter 16).

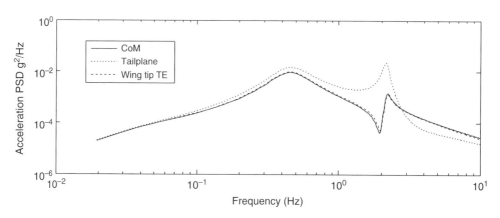

Figure 14.26 Frequency domain gust results for a flexible aircraft with the fuselage bending mode at 2 Hz/4% showing the CoM, tailplane and wing tip TE acceleration PSDs.

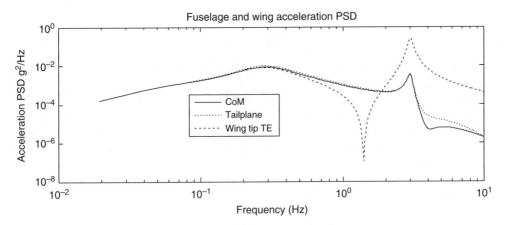

Figure 14.27   Frequency domain gust results for a flexible aircraft with the wing bending mode at 3 Hz/ 4% showing the CoM, tailplane and wing tip TE acceleration PSDs.

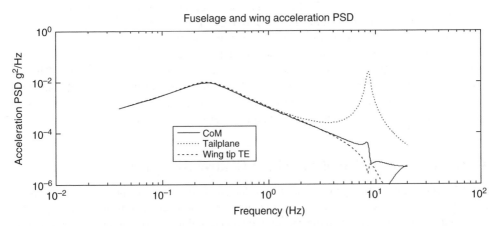

Figure 14.28   Frequency domain gust results for a flexible aircraft with the wing torsion mode at 9 Hz/4% showing the CoM, tailplane and wing tip TE acceleration PSDs.

14.11   General Form of Equations in the Frequency Domain

In Sections 14.7 to 14.10, the frequency domain approach was illustrated. When the resulting equations are examined their general form may be seen as shown for the time domain in Section 14.6. Thus, when a harmonic gust velocity is introduced, the generalized response vector is assumed to be

$$q(t) = \tilde{q}e^{i\omega t} \tag{14.58}$$

where the frequency domain amplitude is again left as lower case. Then the response may be written, including flexible effects (see Equation (14.56)) but omitting unsteady aerodynamic effects, as

$$\tilde{q} = \left[-\omega^2 \mathbf{A} + i\omega\rho V\mathbf{B} + \left(\mathbf{E} + \rho V^2 \mathbf{C}\right)\right]^{-1} \left\{\rho V\mathbf{R}_W + \rho V\mathbf{R}_T \exp\left(-\frac{i\omega l}{V}\right)\right\} w_{g0} \tag{14.59}$$

or

$$\tilde{q} = H_{qg}(\omega) w_{g0} \tag{14.60}$$

where H_{qg} is the vector of FRFs relating multiple generalized responses to the harmonic gust velocity at the relevant flight condition. Note that the Theodorsen and Sears functions are omitted in this equation for clarity, but they would multiply the left-hand and right-hand side aerodynamic matrices if used. Employing three-dimensional panel aerodynamics instead of strip theory would mean that the explicit presence of the Theodorsen and Sears functions would be omitted and the unsteady effects would be catered for via the Aerodynamic Influence Coefficient matrix (see Chapters 18 and 19).

The generalized responses may then be transformed to the desired physical accelerations $\tilde{z}_{Acc}$ at different positions on the aircraft; e.g.

$$\tilde{z}_{Acc} = -\omega^2 T_{zq}\tilde{q} = -\omega^2 T_{zq}H_{qg}w_{g0} = H_{zAccg}w_{g0} \tag{14.61}$$

where the transformation matrix T_{zq} includes values of the rigid body displacements and flexible modal displacements at the desired spatial positions. This equation may therefore be compared to Equation (14.57) for the tailplane acceleration using $T_{zq} = \{1 \quad l_T \quad \kappa_{eT}\}$. Finally, Equation (14.61) will enable the PSD for each different physical response in the vector to be determined term by term, rather as in Equation (14.39).

Later in Chapter 16, the transformation to load quantities and processing of loads and responses from time and frequency domain analyses will be considered briefly. There is considerable statistical post-processing involved in some aspects of the analysis of continuous turbulence response and Hoblit (1988) covers this thoroughly; mention is made of current certification clearance philosophy for loads due to turbulence in Chapter 23.

14.12 Representation of the Flight Control System (FCS)

The FCS (Pratt, 2000) often has an important effect upon the aircraft response and loads caused by a gust or turbulence encounter. Of particular interest is the effectiveness of any Gust Load Alleviation (GLA) system upon the resulting response and load levels; such a system uses the normal acceleration to deploy the ailerons and spoilers and so aims to counteract the effects of gusts and turbulence (see Chapter 21). Chapter 11 showed a simple model to demonstrate load alleviation. For the time domain discrete gust simulations, it is possible to incorporate the relevant non-linear features of the FCS directly into the aircraft dynamic model. However, for turbulence calculations in the frequency domain, a linearized FCS model must be employed. When the effect of non-linearity on turbulence response is to be considered, then the non-linear FCS behaviour may be incorporated into a stochastic time domain turbulence representation (see Chapter 24).

14.13 Examples

Note that some of the examples in Chapters 1 and 2 might be helpful to solve the following.

1. A rigid aircraft has the following characteristics: $m = 10000$ kg, $S_W = 40$ m^2, $a = 0.09$/deg. The aircraft is flying at 100 m/s at sea level when it meets an upgust of 10 m/s. Find the incremental load factor due to the gust.
   [1.28]

2. Draw a gust envelope for the aircraft with the following characteristics: $m = 10\,200$ kg, $S_W = 50$ m^2, maximum lift coefficient 1.6, $a = 5$/rad, $V_C = 110$ m/s EAS, $V_D = 140$ m/s EAS and $\rho_0 = 1.225$ kg/m^3. Gust velocities are as follows: 20 m/s maximum, 16 m/s at cruise speed and 8 m/s at dive speed. What is the maximum load factor? Ignore the gust alleviation factor. [3.69 at cruise speed]

3. An aircraft has the following characteristics: $m = 20000$ kg, $S_W = 80$ m^2, $a_W = 5$/rad, $S_T = 15$ m^2, $a_T = 4$/rad (including downwash). The aircraft is flying at 120 m/s at sea level when it enters a sharp-edged upgust of 10 m/s. Assuming the whole aircraft enters the gust instantly and ignoring unsteady aerodynamic effects, determine the incremental wing and tailplane lift forces and hence the incremental load factor. [294 kN, 44 kN and 1.72]

4. Write a MATLAB/SIMULINK program to solve the problem of a rigid aircraft with heave only motion flying through a '1-cosine' gust, check the results given in the chapter for 40 and 400 m gust lengths, and determine the tuned condition that yields the maximum load factor; use quasi-steady aerodynamics.

5. Extend the program in Example 4 to generate the aerodynamic derivatives for a given flight condition and set up the equations of motion and time solution for (a) the heave/pitch model and (b) the flexible model. Repeat the calculations in Example 4. Explore the effect on the solution of changes in aircraft parameters.

6. The data in the table below corresponds to the frequency (Hz), gust PSD (m/s)^2/Hz for a 1 m/s root-mean-square turbulence velocity (von Karman spectrum) and (CoM acceleration per gust velocity FRF)^2 in $[(m/s^2)/(m/s)]^2$; a flexible mode is present. Determine the acceleration PSD values, plot the results on linear scales and obtain a rough estimate of the root-mean-square of the acceleration. [0.57 m/s^2]

Frequency	Gust PSD	FRF^2	Frequency	Gust PSD	FRF^2
0	8.128	0	1.6	0.027	0.86
0.2	0.853	0.05	1.8	0.023	0.64
0.4	0.275	2.05	2.0	0.019	1.07
0.6	0.140	3.00	2.2	0.016	3.56
0.8	0.087	2.01	2.4	0.014	4.01
1.0	0.060	1.59	2.6	0.012	3.46
1.2	0.044	1.33	2.8	0.011	3.07
1.4	0.034	1.10	3.0	0.010	2.82

7. Consider the rigid aircraft heave model used in the examples of this chapter. Obtain the FRF relating heave acceleration per harmonic gust velocity and hence the acceleration PSD for turbulence modelled by the von Karman spectrum with a scale length of 2500 ft and gust velocity of 1 m/s root-mean-square. Obtain the root-mean-square value of the acceleration. Now introduce the Theodorsen and Sears functions into the FRF and obtain the revised plots and root-mean-square values.

8. Extend the program in Example 6 for quasi-steady aerodynamics to cater for (a) the heave/ pitch model and (b) the flexible mode model. Check the results presented in this chapter. Explore the effect of changing aircraft and modal parameters on the resulting shapes of the acceleration response PSDs.

15

Ground Manoeuvres

The behaviour of an aircraft when in contact with the ground is not straightforward, mainly because of the complexity of the landing gears used to absorb energy on landing and to allow ground manoeuvrability. Dynamic loads are developed in the landing gear, and therefore in the airframe, when taxiing, turning, taking off, landing and braking (Lomax, 1996; Howe, 2004); all these load cases may be important for sizing of some aircraft or landing gear components.

The certification requirements for ground loads are shown in CS-25 and FAR-25 for large commercial aircraft and some of the calculations are discussed in Lomax (1996) and Howe (2004), though some of the requirements have been revisited since their publication because larger aircraft with more than two main landing gears have been designed. The ground loads certification is considered under the headings of (a) landing and (b) ground handling (taxi, take-off and landing roll, braked roll, turning, towing, pivoting, tie-down, jacking and shimmy). The calculations outlined in CS-25 may be defined as being 'rational' or 'bookcase'. Rational calculations employ a model that seeks to represent more accurately the real physics and dynamics of the system whereas bookcase calculations tend to be more artificial and usually require ground reactions to be balanced by inertia forces (and sometimes inertia moments).

Because of the complexity of the landing gear and some of these ground operations, the treatment in this chapter will be kept fairly simple in order to provide a basic understanding of some of the key issues involved. The landing gear will be introduced, followed by a relatively simple treatment of the rational taxiing and landing cases, bookcases for braking and turning, and simple modelling of wheel spin-up/spring-back and shimmy. Other load cases and approaches used by industry to meet the requirements are discussed in Chapter 24. Note that the calculation of internal loads in the airframe will be considered later in Chapter 16.

15.1 Landing Gear

The landing gears for modern aircraft are extremely complex, being required to provide energy absorption on landing and manoeuvrability on the ground, as well as needing to be retracted and stowed in flight through one of a variety of kinematic arrangements (Currey, 1988; Niu, 1988; Howe, 2004). It is most common for large commercial aircraft to have a nose/main gear layout; typical main and nose landing gears are shown in Figure 15.1.

Introduction to Aircraft Aeroelasticity and Loads, Second Edition. Jan R. Wright and Jonathan E. Cooper.
© 2015 John Wiley & Sons, Ltd. Published 2015 by John Wiley & Sons, Ltd.

Figure 15.1 Typical nose and main landing gears for a commercial aircraft (reproduced by permission of Messier-Bugatti-Dowty Ltd).

Detailed non-linear mathematical models of the landing gears are required, and loads in each component have to be determined over the whole range of ground manoeuvres. Treatment of landing gear design and construction may be found in Currey (1988) and Niu (1988), though the process of developing a detailed dynamic model for calculation of dynamic responses and loads is not included in these references. In this book, some basic concepts of response calculations will be introduced by treating the gear as a linear or non-linear 'black box'.

15.1.1   Oleo-pneumatic Shock Absorber

The main landing gear component relevant to energy absorption and carrying static loads is the shock absorber. Oleo-pneumatic shock absorbers, of which the gas spring is an integral part, are the most common form of landing gears used on medium to large aircraft, primarily because of their improved efficiency, weight and reliability. The basic idea is shown schematically in Figure 15.2. During compression, oil is forced through the compression orifices, dissipating energy and compressing the gas (typically nitrogen) as the shock absorber is loaded. Then the rebound of the aircraft is controlled by the pressured gas forcing oil back through the recoil orifices. The recoil of a landing gear shock strut can be classified as one of two types: controlled recoil or free recoil. Controlled recoil is defined by the continuous ground contact of the landing gear tyre as the gear recoils. This behaviour is a result of the performance of the aircraft controlling the rate of extension; the gear recoil speed is greater than the aircraft lift off speed. Free recoil occurs when the gear fully extends without any ground contact. In these cases the recoil damping performance of the shock strut controls the rate of extension.

The shock absorber provides a 'gas spring' to support the aircraft weight, to prevent 'bottoming' and to provide ride comfort. The *stiffness* characteristics of the gas spring are non-linear, controlled by the gas law governing the compression of the gas and assuming that the oil is incompressible for a simple model. The ideal gas law (Duncan *et al.*, 1962) is

$$PV^\gamma = C \tag{15.1}$$

where P is the absolute pressure (gauge + atmospheric), V is the volume, C is a constant and γ is the polytropic constant. For *static* loading conditions where there is a steady rate of compression

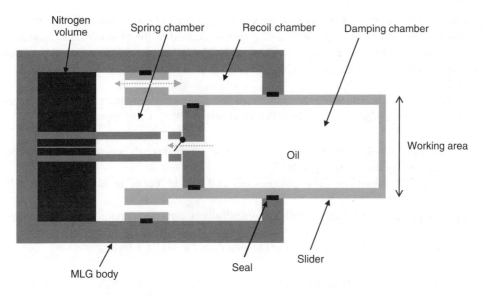

Figure 15.2   Oleo-pneumatic shock absorber arrangement.

(e.g. taxiing, loading the aircraft with payload), the process may be viewed as isothermal (constant temperature) and a value of $\gamma = 1$ may be used. However, for *dynamic* loading conditions where there is a rapid rate of compression (e.g. landing), the process is assumed to be adiabatic (no heat transfer) and a value of $\gamma = 1.3$–1.4 may be used.

When the gas law is applied to the system to relate the two conditions where the piston is fully extended (subscript ∞) and then partially compressed by an amount z, it may be shown that

$$P(V_\infty - Az)^\gamma = P_\infty V_\infty^\gamma \tag{15.2}$$

where A is the piston area. Hence, the absolute pressure/displacement relationship is given by

$$\frac{P}{P_\infty} = \left(1 - \frac{z}{z_\infty}\right)^{-\gamma} \tag{15.3}$$

where $z_\infty = V_\infty / A$ is the fully extended to fully 'bottomed' distance (about 10% greater than the stroke z_S).

In order to obtain an estimate of the stroke required for the shock absorber, the energy balance on landing should be considered (Currey, 1988; Howe, 2004). Now, it is normally assumed in the critical part of the landing process for commercial aircraft that lift remains equal to the weight. Then, neglecting the tyre deformation (and therefore the associated energy dissipation), the aircraft kinetic energy must be absorbed by the work done in the shock absorber, so

$$\frac{1}{2}mW_e^2 = \eta_S F_{LG_{max}} z_S = \eta_S n_{LG} W z_S \qquad z_S = \frac{W_e^2}{2\eta_S n_{LG}} g \tag{15.4}$$

where m is the part of the aircraft mass supported by each main gear shock absorber (static load $W = mg$), W_e is the vertical landing (or sink/descent) velocity, $F_{LG_{max}}$ is the maximum landing gear force, η_S is the shock absorber efficiency (taking account of losses, typically 0.8) and z_S is the stroke. The load factor n_{LG} for a landing gear is defined as the ratio of (static + dynamic reaction load) to (static load); typical values of 2–2.5 for commercial aircraft will occur on landing (Lomax, 1996). As an example, if $W_e = 3$ m/s, $W = 50$ kN, $\eta_S = 0.8$, $n_{LG} = 2.0$ then $z_S = 0.287$ m.

In order to estimate values of other parameters for use in later simulations involving the gas spring, the following process may be followed (Currey, 1988), using typical values:

- determine area $A = 0.005$ m^2 by assuming that the static pressure P_S is approximately 100 bar when supporting the aircraft weight;
- choose P_C (C subscript = fully compressed) and P_∞ to be of order $3P_S$ and $0.25P_S$ respectively;
- apply the gas law ($\gamma = 1$) to relate fully extended/fully compressed conditions and so yield the volume ratio $V_\infty/V_C = 12$; and
- recognize that $V_\infty = V_C + Az_S = 12\ Az_S/11$, then $z_\infty = 12\ z_S/11 = 1.091\ z_S$.

The stiffness curve that may be derived from Equation (15.3) for the static and dynamic cases is shown in Figure 15.3a in terms of the stiffness force (normalised per static load) against compression (normalized by stroke); note that the atmospheric pressure must be accounted for when the pressure relationship is used in a simulation to generate forces.

The shock absorber also provides friction and a high level of *damping* to help absorb the energy of a landing impact and avoid excessive recoil. The damping is controlled by oil flow through the compression and recoil orifices. Using Bernoulli's theorem and the orifice discharge characteristics (Duncan et al., 1962), it may be seen that the damping is proportional to velocity squared, i.e. non-linear. The oleo damping force F_D is thus given by

$$F_D = D_C \dot{z}^2 \qquad \dot{z} \geq 0 \qquad F_D = -D_R \dot{z}^2 \quad \dot{z} < 0 \qquad (15.5)$$

where D_C, D_R are the compression and recoil damping coefficients and z is the shock absorber compression (or closure); note that D_R is of the order of 15–20 times greater than D_C so as to minimize recoil. The damping force (normalized by static load) is plotted against velocity (normalized by descent velocity W_e) in Figure 15.3b for damping coefficients of $D_C = 16$ and $D_R = 120$ kN s^2/m^2.

15.1.2 Wheel and Tyre Assembly

A landing gear has two major dynamic elements in series, namely the shock absorber and the wheel/tyre assembly. The dynamic behaviour of this assembly is complex and has an impact upon landing, taxiing, braking, turning and shimmy. It is important for the accurate estimation of the landing gear internal loads, and the loads at the airframe attachment points, that the overall system is modelled adequately. A very simple representation that includes tyre characteristics is the 2DoF model shown in Figure 15.4 where the unsprung mass consists of the sliding tube, axle and wheel/tyre/brake assembly.

The behaviour of tyres under static and dynamic vertical and lateral loading is complex (Pacejka, 2005) and a detailed treatment is beyond the scope of this book. The model depends upon material properties, pressure, temperature, tyre wall flexibility and ground conditions.

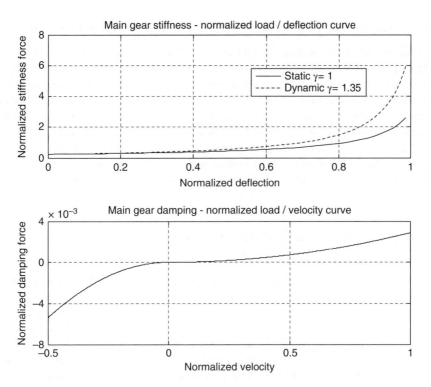

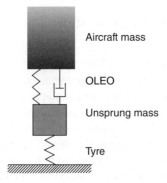

Figure 15.3 Oleo-pneumatic shock absorber – (a) normalized compression and (b) normalized damping curves.

Figure 15.4 Simplified landing gear/tyre model.

A linear, undamped tyre model with linear stiffness proportional to tyre pressure may be used for simple calculations involving vertical motion of the aircraft, namely landing, taxiing and braking; damping may also be considered. The friction force due to the tyre-to-ground contact under combined rolling and skidding conditions will be covered later.

The lateral model, particularly important for turning and shimmy, is significantly non-linear and involves coupling between the yaw, roll and sideslip motions.

15.1.3   Determinate and Statically Indeterminate Landing Gear Layouts

When there are only three landing gears on an aircraft (single aisle), the static loads in each one may be determined by equilibrium considerations alone since the problem is statically determinate. However, for some large commercial aircraft (typically twin aisle), there are more than three landing gears (e.g. Airbus A340, Boeing 747); in such cases the problem of finding the static loads and shock absorber compression is statically indeterminate. This means that the elastic deformation of the aircraft and the flexibility of each landing gear (including non-linear effects) need to be considered when determining the loads carried by the gear units. A non-linear static balance calculation on the ground will yield these loads. Also, any variation in elevation of the ground profile across the runway needs to be considered.

15.1.4   Determinate and Statically Indeterminate Landing Gear Attachments

A further issue involving landing gears concerns the attachment of the landing gears to the aircraft structure. A statically determinate arrangement allows the loads transferred from the gear to the airframe to be determined by equilibrium considerations alone; consequently the detailed airframe structure around the attachment does not need to be modelled and a simple transformation of loads from the ground to the airframe may be used. On the other hand, if the attachment is statically indeterminate (i.e. redundant) then the distribution of loads into the airframe will depend upon the airframe stiffness at the gear attachments, as well as the gear stiffness, and so the FE model of the airframe needs to be included in the landing gear analysis (see also Chapter 24).

15.2   Taxi, Take-Off and Landing Roll

Taxiing loosely describes the entire phase of ground movement prior to final take off and following landing (Lomax, 1996; Howe, 2004). Clearly the aircraft will need to brake and turn, but these are considered as separate operations, with taxiing simply being concerned with straight line motion on the ground. If the runway and taxiways were completely smooth, then taxiing would not be a problem. However, in practice runways and taxiways are uneven to some extent, i.e. the elevation along the length of the runway does not vary linearly. As a consequence, the aircraft responds dynamically during taxiing, with rigid body and flexible modes being excited and dynamic loads generated that need consideration for design. Also, the comfort of the crew and passengers is affected. An aircraft that experienced significant dynamic responses during taxiing was the Concorde, since the fuselage was slender and flexible, with the pilot located a long way forward of the nose landing gear when compared to more conventional commercial aircraft; also, the high take-off and landing speeds meant that longer wavelengths present in the runway profile became important.

   For simplicity, in forming the equations in this section the landing gear will initially be assumed to be a simple linear spring/damper, resulting in a set of linear equations. Later, the true non-linear nature of the gear will be introduced and it will be seen that the aircraft and landing gear equations are arguably better written separately and coupled via the gear forces. The dynamic calculations described here for taxiing are rational and broadly similar to the industrial practice used when meeting certification requirements (CS-25; Lomax, 1996).

15.2.1   Runway Profile

To calculate the aircraft dynamic response during taxiing, the runway profile over which the aircraft is operating must be specified. Note that every runway in the world has a different profile. However, it is not possible to consider them all in the design process so only a representative non-smooth runway is considered, namely San Francisco 28R prior to resurfacing (Lomax, 1996; see also Chapter 24). The elevation profile $h(x_r)$ of a runway is defined relative to a flat datum as shown in Figure 15.5, where the distance along the runway x_r is measured with respect to a convenient origin. For convenience, the profile is taken as downwards positive, to be consistent with downwards positive displacements on the aircraft. There is no variation of profile across the runway. The analysis has some similarities to that for the simple vertical gust analysis considered earlier in Chapter 14.

Sometimes the response to one or more 'dips' or 'bumps' on an otherwise flat runway may be required, e.g. in meeting the discrete load condition in the certification requirements; the term 'dip' will be used to describe a negative 'bump'. For example, a '1-cosine' dip of depth Δh_r and length L_r as shown in Figure 15.6 (not to be confused with the lateral stability derivative given the same symbol) would have a profile

$$h(x_r) = \frac{\Delta h_r}{2}\left(1 - \cos\frac{2\pi x_r}{L_r}\right) \tag{15.6}$$

The aircraft nose and main gears would pass over the 'dip' at different times, so causing both heave and pitch motions of the aircraft (similar to the gust penetration effect in Chapter 14). By considering 'dips' of different lengths, then the critical length that maximizes the response could be determined. Alternatively, a pair of 'dips' could be optimally spaced so as to be 'tuned' to the rigid body heave and/or pitch mode frequency of the aircraft on its landing gears (or indeed to a flexible mode frequency); thus a worst case pair of dips could be envisaged.

When the equations are set up, both $h(t)$ and $\dot{h}(t)$ (i.e. the runway profile and rate of change with time of the profile, as seen by each gear) are required; thus a temporal definition must be obtained from the spatial profile. The value of $h(t)$ is equal to the value of $h(x_r)$ at the current location of the aircraft during the taxiing process. The time rate of change of the profile $\dot{h}(t)$ depends upon the local runway slope and the aircraft velocity. Let the aircraft forward velocity

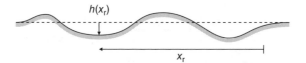

Figure 15.5   Definition of a runway profile.

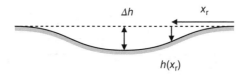

Figure 15.6   '1-Cosine' runway 'dip' profile.

be V (TAS) at the instant of time considered. Then, using the chain rule, the rate of change of profile with time is given by

$$\dot{h} = \frac{\mathrm{d}h}{\mathrm{d}t} = \frac{\mathrm{d}h}{\mathrm{d}x_\mathrm{r}} \frac{\mathrm{d}x_\mathrm{r}}{\mathrm{d}t} = V \frac{\mathrm{d}h}{\mathrm{d}x_\mathrm{r}} \qquad (15.7)$$

where the runway slope may be estimated from the profile values near to the point of interest. The treatment in this chapter only covers the effect of the runway profile upon the vertical gear forces; in a more complete model, the effect of a bump or dip upon the wheel drag force would also be included.

15.2.2   Rigid Aircraft Taxiing

Consider a rigid aircraft supported on linear spring/damper gears as shown in Figure 15.7 (see also Chapter 2). The aircraft response is represented by the heave z_C (at the centre of mass - subscript C) and the pitch θ, relative to any horizontal datum; zero motion corresponds to the aircraft at rest in its static equilibrium position on the datum runway. Thus responses are actually calculated *relative to* this datum state and so are incremental. The aircraft has mass m, pitch moment of inertia about the centre of mass I_y, nose (subscript N) and main (subscript M) gear stiffnesses K_N, K_M, viscous damping coefficients C_N, C_M and distances l_N, l_M from the centre of mass; clearly the main gears on both sides of the aircraft have been combined into one unit for simplicity.

Consider the aircraft position at any instant, being defined with the centre of mass at distance x_r from the runway origin, the nose gear at x_N and the main gear at x_M. The profile values at the gear positions are then

$$h_\mathrm{N} = h(x_\mathrm{N}) = h(x_\mathrm{r} + l_\mathrm{N}) \qquad h_\mathrm{M} = h(x_\mathrm{M}) = h(x_\mathrm{r} - l_\mathrm{M}) \qquad (15.8)$$

Note that the use of the symbol x here is different to that used for one of the aircraft axes.

The energy dissipation and work done functions depend upon expressions for compression and the rate of compression of the landing gear springs and dampers, and these values will

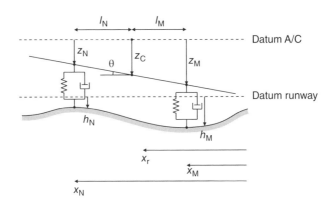

Figure 15.7   Rigid aircraft with a linear landing gear during taxiing.

depend on the runway profile and rate of change of the profile. Thus, the compression and rate of compression of the nose and main gears are

$$\Delta_N = z_N - h_N = z_C - l_N\theta - h_N \qquad \dot{\Delta}_N = \dot{z}_C - l_N\dot{\theta} - \dot{h}_N$$
$$\Delta_M = z_M - h_M = z_C + l_M\theta - h_M \qquad \dot{\Delta}_M = \dot{z}_C + l_M\dot{\theta} - \dot{h}_M \tag{15.9}$$

The kinetic energy, elastic potential energy and dissipation function are then given by

$$T = \frac{1}{2}\, m\dot{z}_C^2 + \frac{1}{2}\, I_y\dot{\theta}^2 \qquad U = \frac{1}{2}\, K_N\Delta_N^2 + \frac{1}{2}\, K_M\Delta_M^2 \qquad \Im = \frac{1}{2}\, C_N\dot{\Delta}_N^2 + \frac{1}{2}\, C_M\dot{\Delta}_M^2 \tag{15.10}$$

Then, applying Lagrange's equations with generalized coordinates (z_C, θ), it may be shown that the equations of motion for the aircraft taxiing on the uneven runway are

$$
\begin{bmatrix} m & 0 \\ 0 & I_y \end{bmatrix} \begin{Bmatrix} \ddot{z}_C \\ \ddot{\theta} \end{Bmatrix} + \begin{bmatrix} C_N + C_M & -l_N C_N + l_M C_M \\ -l_N C_N + l_M C_M & l_N^2 C_N + l_M^2 C_M \end{bmatrix} \begin{Bmatrix} \dot{z}_C \\ \dot{\theta} \end{Bmatrix}
$$
$$
+ \begin{bmatrix} K_N + K_M & -l_N K_N + l_M K_M \\ -l_N K_N + l_M K_M & l_N^2 K_N + l_M^2 K_M \end{bmatrix} \begin{Bmatrix} z_C \\ \theta \end{Bmatrix} = \begin{bmatrix} C_N & C_M \\ -l_N C_N & +l_M C_M \end{bmatrix} \begin{Bmatrix} \dot{h}_N \\ \dot{h}_M \end{Bmatrix}
$$
$$
+ \begin{bmatrix} K_N & K_M \\ -l_N K_N & l_M K_M \end{bmatrix} \begin{Bmatrix} h_N \\ h_M \end{Bmatrix} \tag{15.11}
$$

It may be seen clearly that the left-hand side of this equation is the same as that in Chapter 2; however, right-hand side excitation terms are now present due to the variation in the elevation of the runway profile. Aerodynamic effects have been ignored in this simple model, but steady effects would normally be included to allow the nominal gear compressions at the relevant speed to be determined. The taxiing response of the rigid aircraft may be found by solving these differential equations in time, as discussed in Chapters 1 and 2.

15.2.3   Example of Rigid Aircraft Taxiing

A rigid aircraft is considered to pass over a '1-cosine' dip and the resulting heave displacement and acceleration at the nose and main landing gear positions will be examined. The aircraft parameters are $m = 10\,000$ kg, $I_y = 144\,000$ kgm², $l_N = 6.8$ m, $l_M = 0.75$ m, $C_N = 3200$ N s/m, $C_M = 19\,200$ N s/m, $K_N = 80\,000$ N/m and $K_M = 240\,000$ N/m. The fuselage mass distribution and component moments of inertia are the same as used for the example in Chapter 12 and Appendix C, but these data only become relevant when the flexible mode is considered later in this chapter. The natural frequencies and damping ratios for the rigid body modes of the aircraft on its linear landing gear are approximately 0.70 Hz/15.1% and 1.00 Hz/16.1%. MATLAB and SIMULINK programs for a rigid aircraft taxiing over a '1-cosine' dip are presented in the companion website.

Consider a case where the taxiing velocity is 30 m/s and the depth of the dip is 30 mm. The simulation is carried out over 10 s and the nose gear encounters the dip at the start of the simulation. Care needs to be taken when setting up the simulation to ensure that the nose and main gears encounter the dip with the appropriate time lag. When a range of different

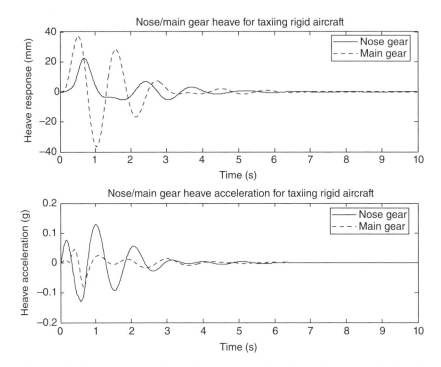

Figure 15.8 Response to taxiing at 30 m/s over a 15 m/30 mm dip for a rigid aircraft.

dip lengths were examined, the largest acceleration response was found to occur for a dip around 25 m long; the approach is similar to that adopted for the gust in Chapter 14, but the detailed results are not presented here. However, because the flexible aircraft example given later has a maximum acceleration response for a 15 m dip, only this dip length will be considered for the rigid and flexible aircraft; this corresponds to an effective 0.5 s duration pulse input. The displacement and acceleration responses for the rigid aircraft are shown in Figure 15.8, with maximum values of approximately 36.4 mm (at the main gear) and 0.129 g (at the nose gear); the motion is predominantly pitching.

15.2.4 Flexible Aircraft Taxiing

In practice, the taxiing case is always considered with flexible effects included for commercial aircraft. Also, because the landing gear is a complex non-linear dynamic system in its own right, the aircraft and landing gear equations may best be formed separately, with intercomponent forces (see Chapter 12) and kinematic constraints employed to link the two sets of equations, as shown in Figure 15.9. In this section, the idea will be illustrated together with the inclusion of a flexible mode, rather as was done in Chapters 12 and 14, except that here the aerodynamic forces are ignored.

The sensible option is again to take the deformation of the aircraft at rest on a smooth runway as the datum. The response during taxiing is then *relative to* the initial deformation and will include flexible contributions. The deformation of the flexible aircraft will be treated via the summation of rigid body and flexible free–free normal modes, such as would emerge from a Finite Element analysis of a free–free aircraft. Such a model is suitable for taxiing and gust

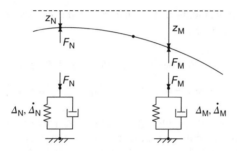

Figure 15.9 Flexible aircraft with linear landing gear during taxiing.

response calculations since the basic dynamics of the airframe (without landing gear) are linear and only relatively small motions away from the datum path occur. The model uses inertial axes with unknown displacements and rotations, so there is therefore no need to consider axes moving with the aircraft, as was the case for dynamic flight manoeuvres in Chapter 13.

15.2.4.1 Flexible Airframe Equations

Consider the aircraft shown in Figure 15.9, which is rather similar to the earlier rigid aircraft but now with elastic deformation present and the landing gears and aircraft treated as separate systems. Let the deformation of the aircraft be expressed as the summation of three free–free normal modes for the aircraft without landing gear; the first two modes will be rigid body modes (i.e. heave, pitch) and the other a flexible mode (e.g. fuselage bending). Whole aircraft modes are considered in Chapter 3 and in Appendices A and C. Thus a modal approach including the rigid body behaviour has replaced the previous direct use of heave and pitch coordinates.

The aircraft (downwards) displacement $z(x, y)$ at any position on the airframe and relative to the equilibrium state will then be given by

$$z(x,y) = \kappa_h(x,y)q_h + \kappa_p(x,y)q_p + \kappa_e(x,y)q_e \qquad (15.12)$$

where q_h, q_p, q_e are the modal coordinates used to define the displacements and κ_h, κ_p, κ_e are the heave, pitch and flexible mode shapes. In Appendix A, it is seen that the modal mass values are $m_h = m$, $m_p = I_y$ if the rigid body mode shapes are normalized such that $\kappa_h = 1$, $\kappa_p = -x$. Equation (15.12) then simplifies to

$$z(x,y) = q_h - xq_p + \kappa_e(x,y)q_e \qquad (15.13)$$

The kinetic energy may be expressed in terms of all the modal quantities, whereas the elastic potential energy for the airframe only exists for the flexible mode, so

$$T = \frac{1}{2} m_h \dot{q}_h^2 + \frac{1}{2} m_p \dot{q}_p^2 + \frac{1}{2} m_e \dot{q}_e^2 \quad U = \frac{1}{2} k_e q_e^2 \qquad (15.14)$$

where m_h, m_p, m_e are the modal masses and k_e is the modal stiffness for the flexible mode. The flexible mode quantities depend upon the mode shape and mode normalization (see Chapter 12 and Appendix C).

The (downwards) displacements of the airframe at the nose and main gear positions are given, using equation (15.13), by

$$z_N = q_h - l_N q_p + \kappa_e(l_N, 0)q_e = q_h - l_N q_p + \kappa_{eN} q_e$$
$$z_M = q_h + l_M q_p + \kappa_e(-l_M, \pm d_M)q_e = q_h + l_M q_p + \kappa_{eM} q_e \quad (15.15)$$

where $\pm d_M$ are the y positions of the main gear and κ_{eN}, κ_{eM} are the values of the flexible mode shape at the nose and main landing gear positions. Also, the intercomponent forces (compressive positive) between the landing gears and the airframe will be F_N and F_M (refer to Chapter 5). Then, by determining the incremental work done by these forces moving through incremental nose and main gear displacements, the modal forces due to the presence of the landing gear may be determined, so

$$\delta W = -F_N \delta z_N - F_M \delta z_M$$

or   (15.16)

$$\delta W = -F_N \left(\delta q_h - l_N \delta q_p + \kappa_{eN} \delta q_e \right) - F_M \left(\delta q_h + l_M \delta q_p + \kappa_{eM} \delta q_e \right)$$

Thus, applying Lagrange's equations, the 3DoF equations of motion for the flexible aircraft (see Chapter 3) are found to be

$$\begin{bmatrix} m & 0 & 0 \\ 0 & I_y & 0 \\ 0 & 0 & m_e \end{bmatrix} \begin{Bmatrix} \ddot{q}_h \\ \ddot{q}_p \\ \ddot{q}_e \end{Bmatrix} + \begin{bmatrix} 0 & 0 & 0 \\ 0 & 0 & 0 \\ 0 & 0 & k_e \end{bmatrix} \begin{Bmatrix} q_h \\ q_p \\ q_e \end{Bmatrix} = - \begin{Bmatrix} 1 \\ -l_N \\ \kappa_{eN} \end{Bmatrix} F_N - \begin{Bmatrix} 1 \\ -l_M \\ \kappa_{eM} \end{Bmatrix} F_M \quad (15.17)$$

It may be seen that the matrix partitions separate the rigid body and flexible modes. To add more flexible modes would simply mean adding further diagonal modal mass and stiffness terms together with additional mode shape values at the landing gear positions. Damping would also normally be introduced for the flexible mode, and landing gear damping could be included once the two sets of equations were coupled. These equations are in the general modal form

$$\mathbf{M}_q \ddot{q} + \mathbf{K}_q q = -\kappa_N F_N - \kappa_M F_M \quad (15.18)$$

where, for example, $\kappa_M = \{ 1 \quad l_M \quad \kappa_{eM} \}^T$ and by inspection of Equation (15.15) it may be seen that $z_M = \kappa_M^T q$. The outcome of this analysis is that Equations (15.17) and (15.18) relate the airframe response to the intercomponent forces applied at the landing gear positions.

In this analysis for the flexible aircraft, the response has been expressed in the classical way as the combination of rigid body heave/pitch modes and free-free flexible modes, as illustrated in Figure 15.10a below; the two landing gear positions are shown by dots. In order to establish an accurate response at the landing gear positions for this approach then a considerable number of flexible modes might need to be superimposed.

However, in practice the so-called Craig-Bampton modes (Craig and Kurdila, 2006) might be employed (for both rational taxiing and landing calculations) as illustrated in Figure 15.10b. Here the rigid body (so-called 'constraint' modes) and flexible modes are defined with respect to the landing gear positions as shown and thus fewer flexible modes need to be superimposed to obtain accurate displacements at the landing gears.

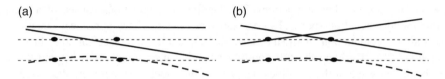

Figure 15.10 Rigid body plus flexible modes: (a) classical and (b) Craig-Bampton approaches.

15.2.4.2 Landing Gear Equations – Linear

The landing gear equations relate intercomponent forces to the gear compressions, and thus to the aircraft response and runway profile. The compression and rate of compression vectors for the nose and main gears are

$$\begin{Bmatrix} \Delta_N \\ \Delta_M \end{Bmatrix} = \begin{Bmatrix} z_N \\ z_M \end{Bmatrix} - \begin{Bmatrix} h_N \\ h_M \end{Bmatrix} \qquad \begin{Bmatrix} \dot{\Delta}_N \\ \dot{\Delta}_M \end{Bmatrix} = \begin{Bmatrix} \dot{z}_N \\ \dot{z}_M \end{Bmatrix} - \begin{Bmatrix} \dot{h}_N \\ \dot{h}_M \end{Bmatrix} \tag{15.19}$$

The relationship between the gear forces and compressions for a simple *linear* landing gear is

$$\begin{Bmatrix} F_N \\ F_M \end{Bmatrix} = \begin{bmatrix} C_N & 0 \\ 0 & C_M \end{bmatrix} \begin{Bmatrix} \dot{\Delta}_N \\ \dot{\Delta}_M \end{Bmatrix} + \begin{bmatrix} K_N & 0 \\ 0 & K_M \end{bmatrix} \begin{Bmatrix} \Delta_N \\ \Delta_M \end{Bmatrix} \tag{15.20}$$

The aircraft taxiing response is obtained by simultaneous solution of Equations (15.17) and (15.20), with the kinematic constraints between the airframe and landing gears given by Equations (15.19) and the relationship between the physical and modal coordinates given by Equation (15.15). Solution in the time domain yields the incremental dynamic response of the flexible aircraft as a combination of rigid body and flexible mode motions.

The modal responses and accelerations can then be transformed into physical displacements (and accelerations) anywhere on the airframe, e.g. at the pilot station, the vertical acceleration is given by

$$\ddot{z}_{Pilot} = \ddot{q}_h - l_{Pilot}\ddot{q}_p + \kappa_{e\,Pilot}\ddot{q}_e \tag{15.21}$$

where $\kappa_{e\,Pilot}$ is the flexible mode shape value at the pilot position, a distance l_{Pilot} ahead of the centre of mass. A mode will only be excited during taxiing if there is a finite value of the mode shape at one or both of the landing gear positions, i.e. both the gears are not located at node points. Internal airframe loads may also be determined as described later in Chapter 16. Landing gear internal loads may be found knowing the leg forces and compressions, together with the gear dynamic model.

15.2.4.3 Landing Gear Equations – Non-linear

Since the landing gear is actually *non-linear*, then the landing gear forces will be a non-linear function of the compressions and rates of compression, and it should be noted that the sets of dynamic equations for each gear will be uncoupled, so

$$F_N = f_{NL}\left(\Delta_N, \dot{\Delta}_N\right) \qquad F_M = g_{NL}\left(\Delta_M, \dot{\Delta}_M\right) \tag{15.22}$$

where $f_{\mathrm{NL}}, g_{\mathrm{NL}}$ are non-linear functions. The landing gear equations consist of a complete dynamic model, including shock absorber and tyre effects. These non-linear equations can be solved in conjunction with the airframe equations in place of the linear gear Equation (15.20). (Deriving the full non-linear landing gear equations is beyond the scope of this book, but some of the non-linear features of the shock absorber were discussed earlier in Section 15.1.1 and will be considered later for landing.)

15.2.5 Example of Flexible Aircraft Taxiing

A flexible aircraft will now be considered to pass over a '1-cosine' dip and the resulting heave displacement and acceleration at the nose and main gear positions is examined. The aircraft parameters are the same as those used for the rigid aircraft taxiing example, but with a fuselage bending mode added, having a natural frequency of 2 Hz and damping of 2% critical; this artificially low frequency (for this size of aircraft) has again been selected to highlight dynamic effects. The fuselage mass idealization and the mode shape are the same as used in Chapter 12 and Appendix C. The mode is the idealized case of fuselage bending with no wing twist or bending. The mode shape values at the nose and main gear locations are $\kappa_{\mathrm{eN}} = -2.382$ and $\kappa_{\mathrm{eM}} = 1$ respectively; these values are obtained by assuming the nose gear to be located at the front fuselage mass position and the main gear on the rigid centre fuselage section (see Appendix C). The modal mass is $m_{\mathrm{e}} = 23\,340\,\mathrm{kg}$ and the modal stiffness is chosen to generate the chosen 2 Hz natural frequency.

The taxiing velocity is 30 m/s, the depth of the dip is 30 mm and its length is 15 m, as before. The displacement and acceleration responses at the landing gear mounting points are shown in Figure 15.11, with maximum values of approximately 43.2 mm and 0.300 g respectively at the

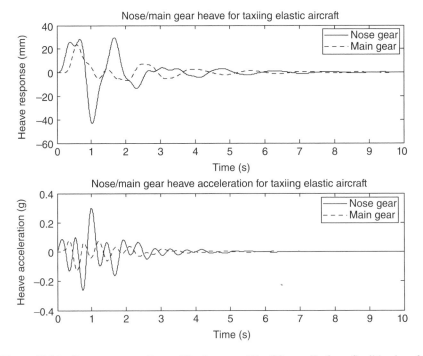

Figure 15.11   Response to taxiing at 30 m/s over a 15 m/30 mm dip for a flexible aircraft.

nose gear; the motion caused by the dip now shows a significant response at the natural frequency of the flexible mode, especially in the acceleration, as would be expected. Clearly, inclusion of the flexible mode has increased the dynamic response and will change the internal loads.

15.3   Landing

Landing is a critical load case for both the airframe and landing gear (Lomax, 1996; Howe, 2004) since a significant amount of energy has to be dissipated. The main case considered is for the aircraft landing on all the *main* gears simultaneously; the nose gear is then still airborne since the aircraft has a positive pitch attitude at the moment of landing prior to final nose down rotation and the nose wheel impacting the ground. The basic landing case (CS-25 and FAR-25) involves a descent velocity of 3 m/s (or 10 ft/s) at the design landing weight and with the trimmed attitude/air speed for the 1 g condition being varied (see also Chapter 24). It will be assumed during the landing of a commercial aircraft that lift remains equal to the weight and that the aircraft is trimmed in attitude on approach by a suitable choice of elevator angle and thrust settings (see Chapter 12). The lift drops off when lift dumpers/spoilers are activated and the aircraft rotates nose down until the nose wheel comes into contact; the dumping of lift is timed to transfer the weight on to the gear and to avoid the aircraft losing contact with the runway during the landing.

Most of the landing calculations specified in CS-25 are rational, but some bookcases are also carried out (see Chapter 24). In this section, some simple landing cases will be considered in a rational manner, considering a non-linear shock absorber with and without the tyre. The effect of the flexible aircraft will be considered briefly. Note that no wheel spin-up and consequent leg spring-back effects are included in these models but these concepts will be considered later. Some reference to bookcase calculations will also be made.

15.3.1   Rigid Aircraft Landing – Non-linear Shock Absorber but No Tyre

Consider the aircraft of mass m (ignoring gear mass) with the main gear shock absorbers represented by the non-linear model introduced earlier. If the shock absorber were linear, then a considerable rebound could occur. However, in reality, the landing gear is highly non-linear and capable of absorbing the landing impact with minimal rebound.

A single DoF non-linear model is considered as shown in Figure 15.12, where any fore-and-aft offset of the main gear from the aircraft centre of mass, and consequent pitch motion,

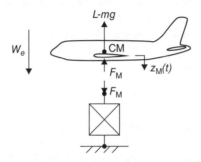

Figure 15.12   Rigid aircraft landing with non-linear landing gear.

has been ignored. The equations for the aircraft and landing gear as the gear comes into contact with the ground are then

$$m\ddot{z}_M = mg - F_M - L = 0 \quad F_M = g_{NL}(\dot{z}_M, z_M) \tag{15.23}$$

or

$$m\ddot{z}_M + g_{NL}(\dot{z}_M, z_M) = mg - L = 0 \tag{15.24}$$

Here z_M is the shock absorber compression (equal to the aircraft motion measured from the position with the leg uncompressed) and g_{NL} is the non-linear function described earlier in Section 15.1.1, being a combination of the gas spring and orifice damping effects. The weight mg has been included as a steady force because the final solution (once lift has reduced to zero) must show a steady leg deformation equal to the sag of the aircraft on its landing gear. However, with lift L present during the initial landing impact, sag will not occur since lift offsets the weight; the weight will only transfer onto the gear once the lift is 'dumped'. Equation (15.24) would be solved with the initial conditions of $z_M(0) = 0$, $\dot{z}_M(0) = W_e$.

A numerical example of using a non-linear shock absorber without a tyre model is now considered. The parameters for a single leg shock absorber are those introduced earlier, namely: $D_C = 16$ kN s^2/m^2, $D_R = 120$ kN s^2/m^2, $A = 0.005$ m^2, $z_S = 0.287$ m, $V_\infty = 0.0013$ m^3, $z_\infty = 0.313$ m, $P_\infty = 25$ bar (2500 kN/m^2) and $\gamma = 1.35$. The aircraft weight per leg is 50 000 N (i.e. mass of 5000 kg per leg descending) and the descent velocity is 3 m/s. The simulation was carried out using so-called 'look-up' tables in SIMULINK to cater for the non-linear damping and stiffness features of the shock absorber; a 'look-up' table contains arrays of the force and displacement/velocity, determined using the formulae presented earlier, and SIMULINK interpolates between the values during the solution.

The displacement, velocity and deceleration shown in Figure 15.13 show minimal rebound but the deceleration shows peaks at $3.18\,g$ and $1.86\,g$ for the damping and stiffness actions respectively; the initial nonzero deceleration value occurs because the damper acts immediately when subject to the 3 m/s initial velocity, but in practice the presence of the tyre dynamics means that the deceleration actually starts from near zero (see the next section). The maximum compression is 83% of the stroke. The normalized ground load shown in Figure 15.14 is equal to the aircraft deceleration multiplied by the mass and divided by the weight; the high load peak at the moment of impact is unrealistic and requires inclusion of the tyre to reduce it.

15.3.2 Rigid Aircraft Landing – Non-linear Shock Absorber with Tyre

In the previous section example, it was stated that the instantaneous load build-up on the landing gear was unrealistic and occurred because there was no tyre model. A model of the unsprung mass m_T, supported by a linear tyre stiffness k_T, is now included; the small tyre damping forces are neglected in this simple landing model. The equations of motion for the model shown in Figure 15.4 are

$$m\ddot{z}_M + g_{NL}(\dot{z}_{SA}, z_{SA}) = mg - L = 0 \quad m_T\ddot{z}_T - g_{NL}(\dot{z}_{SA}, z_{SA}) + k_T z_T = 0 \tag{15.25}$$

where $z_{SA} = z_M - z_T$ is the compression of the shock absorber. Note that the initial velocity for both masses must be equal to the descent velocity W_e. The normalized ground load may be calculated from the relative tyre to ground motion. A MATLAB/SIMULINK model for this system is shown in the companion website, where the use of look-up tables may be seen.

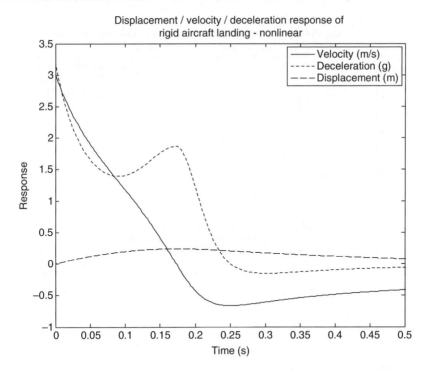

Figure 15.13   Displacement, velocity and deceleration responses for landing of a rigid aircraft with a non-linear shock absorber without tyre.

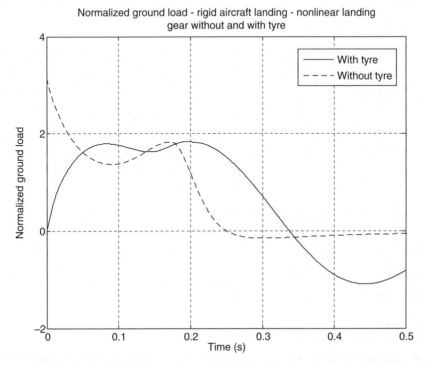

Figure 15.14   Normalized ground load – landing of a rigid aircraft with a non-linear shock absorber with and without tyre.

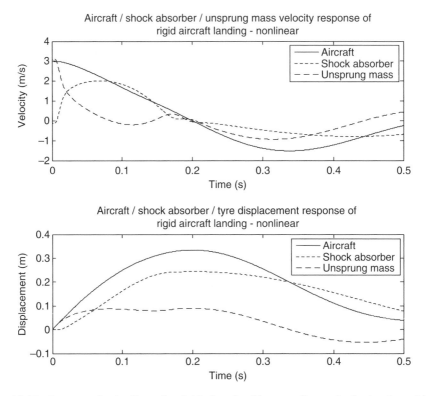

Figure 15.15 Response for landing of a rigid aircraft with a non-linear shock absorber with tyre: (a) velocity and (b) displacement.

Consider the earlier numerical example in Section 15.3.1, but now include an unsprung mass $m_T = 100\,\text{kg}$ and tyre stiffness $k_T = 1000\,\text{kN/m}$ (equivalent to tyre deformation of approximately 50 mm under the aircraft weight). The displacements and velocities of the aircraft and unsprung mass, and the relative shock absorber motions, are shown in Figure 15.15. It may be seen that the total displacement of the aircraft is now a combination of both the shock absorber and tyre deformations. It may be seen that the tyre will rebound and lose contact with the runway after 0.34 s; in practice the lift will be dumped so that the aircraft will not usually rebound for a normal landing; obviously the ground load result shown beyond 0.34 s is meaningless since the tyre load drops to zero when contact is lost.

The ground load normalized to the aircraft weight is shown in Figure 15.14, overlaid on the results without a tyre. The important part of the landing occurs within the first 0.3 s where the loads peak. It may be seen that adding this simple tyre model allows the load to rise from zero, with normalized ground load peaks of 1.80 and 1.84. The inclusion of the tyre avoids the acceleration peak on initial impact. These results are encouragingly similar to those experienced in practice during a drop test. The values of the first and second load peaks depend upon the nonlinear damping and stiffness in compression respectively.

15.3.3 Flexible Aircraft Landing

When the aircraft is flexible, the situation is more complicated since at the moment of impact the aircraft is in a trimmed state, albeit descending at constant velocity, and elastic deformations of the aircraft will exist because the steady gravitational and aerodynamic forces are differently distributed over the airframe. In Chapter 12, the trim of a flexible aircraft was discussed. It is therefore sensible to consider the response of the aircraft during landing as being *relative to* the deformation of the aircraft on impact with the ground (i.e. incremental quantities). The incremental landing loads are then be added to the steady 1 g flight loads present on approach.

The equations for landing of a flexible aircraft on a linear or non-linear main landing gear may be expressed as a variant of the taxiing Equations (15.18) with only the main gear included in the key phase of the landing such that

$$\mathbf{M}_q\ddot{q} + \mathbf{K}_q q = -\kappa_\mathrm{M} F_\mathrm{M}$$

$$F_\mathrm{M} = g_\mathrm{NL}(\dot{z}_\mathrm{M}, z_\mathrm{M}) \quad z_\mathrm{M} = \kappa_\mathrm{M}^\mathrm{T} q,$$

(15.26)

and where vector q now defines the aircraft deformation (heave/pitch/flexible mode) relative to the pre-landing state. The steady aerodynamic and gravitational terms have cancelled out in this incremental calculation. Alternatively, the large angle non-linear flight mechanics equations for the flexible aircraft developed in Chapter 13 may be employed. Also, high incidence aerodynamic terms may be added as required. Note that the equations will need to be modified to cater for the subsequent nose wheel impact if desired; in practice, the nose gear impact is influenced by pilot control actions, any FCS contribution, aerodynamics etc.

The initial conditions that need to be set are not immediately obvious. At the moment of impact, the aircraft will not have deformed elastically relative to its pre-landing state so the initial displacement condition is $q(0) = \mathbf{0}$. Also, since only the heave mode has any rate of change at the moment of impact due to the descent velocity, with the pitch and flexible mode generalized coordinates having a zero rate of change, the initial velocity vector may be shown to be $\dot{q}(0) = \{W_\mathrm{e} \ 0 \ 0\}^\mathrm{T}$. Equations (15.26) may then be solved simultaneously in the time domain to determine the aircraft behaviour during the landing. The response and loads will be affected by the flexible mode contribution, being greatest at the aircraft extremities.

15.3.4   Friction Forces at the Tyre-to-runway Interface

So far, only the vertical characteristics of the aircraft landing process have been considered. However, in practice the rational treatment of landing needs to consider the fore-and-aft behaviour introduced via friction forces acting at the tyre-to-runway interface.

The coefficient of friction between the tyre and the runway behaves in a complex manner and is a non-linear function of the so-called slip ratio SR (ESDU Data Sheets 71025 and 71026) defined by

$$\mathrm{SR} = 1 - \frac{r\dot{\theta}}{V}$$

(15.27)

where r is the effective wheel radius, $\dot{\theta}$ is the angular velocity of the wheel and V is the forward speed of the wheel axle relative to the ground (which may vary with time if the main gear leg bends during landing). The slip ratio is a measure of the relative slip velocity between the tyre and the runway surface. An approximation to the variation of the coefficient of friction μ against slip ratio may be seen in Figure 15.16. When the slip ratio is zero, the wheel is rolling freely so a low rolling coefficient of friction will apply (typically ~ 0.02); however, when the slip ratio is unity, the wheel is locked without rolling, so a skidding coefficient of friction will

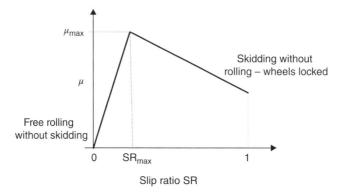

Figure 15.16   Variation of the friction coefficient with slip ratio (redrawn from ESDU Data Sheet 71025, Figure 3, with modifications).

apply (typically ~ 0.25 at higher forward speeds and increasing rapidly towards ~ 0.75 as the forward speed reduces). At an intermediate slip condition, typically around a slip ratio of $SR_{max} \sim 0.05$–0.3, the coefficient of friction reaches a maximum value of $\mu_{max} \sim 0.8$; at this condition, the tyre is able to grip more effectively than when skidding. These values are quoted for a dry concrete runway and are strongly affected by the condition of the tyre and runway.

This behaviour of the coefficient of friction is relevant for both wheel 'spin-up' on landing and also for braking. For braking, when a braking torque is applied, the operating point on Figure 15.16 moves from left to right whereas in the wheel 'spin-up' case on landing, the operating point moves from right to left.

15.3.5   'Spin-up' and 'Spring-back' Conditions

Assuming that the wheels are non-rotating just prior to landing, then the wheels need to 'spin-up' (Lomax, 1996; Howe, 2004) from zero rotation speed to an angular speed consistent with rolling at the landing speed of the aircraft (prior to the braking phase); this happens very quickly and is a dynamic process that introduces fore-and-aft loads into the landing gear at the tyre-to-runway interface. The angular acceleration of the wheel causes the slip ratio to reduce from unity (i.e. skidding without rolling) to zero (i.e. rolling without skidding). Thus the friction force due to tyre-to-runway contact will be governed by the curve in Figure 15.16, and these non-linear characteristics will need to be included. As the wheels 'spin-up', the force generated at the tyre-to-runway interface will increase to a maximum value until an optimal slip ratio (SR_{max}) is reached; beyond this point the contact force will reduce until there is no slip. The effect of the varying fore-and-aft force introduced as the wheels 'spin-up' will be to cause the landing gear leg to bend backwards and forwards, the so-called 'spring-back' phenomenon.

Clearly the 'spin-up' and 'spring-back' process will be combined with the vertical landing dynamics explained earlier. Indeed there will be some coupling of the two since the vertical ground reaction will vary with time and influence the friction forces.

As a simple model to illustrate the 'spin-up' part of the whole landing process, consider forces acting on a representative aircraft landing gear leg and corresponding set of wheels as shown in Figure 15.17; the diagram shows the friction force F acting at the tyre-to-runway interface, the normal reaction N at the ground and intercomponent loads A, B between the wheels and the remainder of the aircraft at the axle bearings. The wheels experience a forward

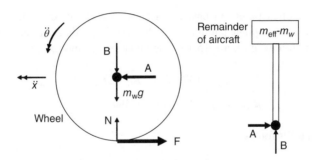

Figure 15.17   Wheel forces during 'spin-up'.

acceleration $\ddot{x}$ and angular acceleration $\ddot{\theta}$ as shown (for the positive sign convention); if the leg is assumed to be *rigid in fore-and-aft bending*, then the aircraft and wheels are assumed to experience the same acceleration. The wheels have mass m_W, moment of inertia I_W and effective tyre radius r (somewhat less than the overall wheel radius due to the tyre deformation); the effective mass of the aircraft per supporting gear is m_{eff}.

The equations of motion for the wheels in the 'spin-up' case are given by

$$\text{Aircraft} \quad -F = m_{eff}\ddot{x}$$
$$\text{Wheels} \quad Fr = I_W\ddot{\theta} \quad F = \mu N \quad A - F = m_W\ddot{x} \quad B + m_Wg = N \tag{15.28}$$

The vertical reaction N at the ground will vary due to the vertical landing dynamics as seen earlier in Section 15.3.2. A simulation may be set up using SIMULINK in which the current value of the friction force F may be used together with the dynamics of the system to calculate $\ddot{\theta}$ and $\ddot{x}$; these values may be integrated to estimate the velocities $\dot{\theta}$ and $\dot{x}$, which may then be used to obtain the corresponding slip ratio SR. The slip ratio will then yield the coefficient of friction μ according to Figure 15.16 and hence a revised value for the friction force may be obtained using the current vertical reaction. Note that the variation with time of the aircraft speed will have to be accounted for when calculating the slip ratio in Equation (15.27) if the forward speed is not constant. The equations governing wheel 'spin-up' are therefore non-linear because of the tyre-to-runway and shock absorber characteristics.

However, in practice the main landing gear is not rigid but *flexible in fore-and-aft bending*, and so the frictional force due to the wheel 'spin-up' will cause the leg to deform rearwards in bending and then 'spring-back' in the forwards direction (Lomax, 1996; Howe, 2004); the gear leg will therefore respond in its bending modes of vibration. Thus, due to this oscillation of the flexible gear, there will be relative motion between the wheels and the aircraft structure, and this will need to be accounted for in the calculation of slip ratio at any instant. A suitable dynamic model for the aircraft plus flexible landing gear leg would need to be developed so that it may be coupled to the wheel dynamic equations. A further complication is that the leg modes will be a function of the vertical closure of the shock absorber at any instant of time and so some form of interpolation between modal properties and leg length will be required.

15.3.6   Bookcase Landing Calculations

Sometimes, in the early design phase of the aircraft, a non-linear dynamic model for landing is not yet available. In such a case, a bookcase analysis may be used for initial loads estimation with load factors applied based on past experience to account for dynamic effects. As an example, consider a three point landing with the free body diagram showing load factors, reaction and inertia forces in Figure 15.18. The equations of motion are

$$R_N + 2R_M = nW - L \qquad F_N + 2F_M = n_x W$$
$$R_N(a+b) = nWa - La + n_x Wh \tag{15.29}$$

and the friction and lift forces are

$$F_N = \mu R_N \qquad F_M = \mu R_M \qquad L = \chi W \tag{15.30}$$

where χ defines the proportion of weight reacted by the lift and μ is the coefficient of friction at the ground. Solving these equations allows the nose and main wheel reactions to be defined via equivalent load factors, namely

$$R_N = (n-\chi)\frac{a+\mu h}{a+b}W = n_{Nequiv}W \qquad 2R_M = (n-\chi)\frac{b-\mu h}{a+b}W = n_{Mequiv}W \tag{15.31}$$

Similar calculations may be carried out for one and two point landings, where inertia moments need to be included for equilibrium.

Further into the design process, where rational landing calculations have been carried out, the critical vertical landing gear load from level landings provides a reference load used in the certification process to define other bookcases such as lateral drift and single point landings, as well as drag load and side load cases. As an example, consider the side load bookcase shown in Figure 15.19. The equations of motion are

$$T = 1.4R_M h = I_x\ddot{\phi} \qquad n_y = \frac{1.4R_M}{W} \qquad nW - L = (n-\chi)W = 2R_M \tag{15.32}$$

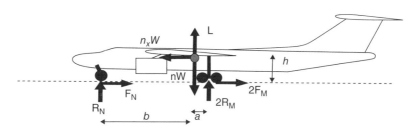

Figure 15.18   Three point landing bookcase.

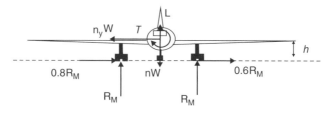

Figure 15.19   Side load bookcase.

where R_M is 50% of the maximum vertical load on each leg in a level rational landing, n_y is the lateral load factor, T is the roll inertia torque, I_x is the moment of inertia in roll and $\ddot{\phi}$ is the roll acceleration. The inertia loads may be determined from the roll acceleration and internal loads obtained.

15.4   Braking

Braking from a high speed landing (or aborted take-off) is a critical design case (Lomax, 1996; Howe, 2004), though braking usually occurs in conjunction with reverse thrust. Part of the forward kinetic energy of the aircraft is converted into heat energy in the main gear braking system. The forces exerted on each wheel due to the application of disc brakes lead to a torque about the axle, and hence to a braking force applied at the tyre-to-runway interface, so decelerating the aircraft. The limiting braking that can occur without skidding depends upon the coefficient of friction available between the tyre and the runway, and typically the maximum possible coefficient of friction is 0.8. However, the value depends upon air speed, tyre tread, tyre pressure, runway surface, antiskid system and any torque limitations (ESDU Data Sheets 71025 and 71026); experimental data may be employed in the braking model. Both bookcase and rational calculations for braking will be considered briefly (see also Chapter 24).

15.4.1   Bookcase Braked Roll

Application of the brakes will cause changes in the normal reactions at each gear e.g. main wheel braking will cause a nose down pitch and hence lead to an increase in the nose gear reaction load. The basic *steady* braked roll case in the certification requirements involves the aircraft with a total braking force F acting at the main gears, corresponding to a coefficient of friction of 0.8 (so $F = 0.8R_M$), and balanced by a fore-and-aft load factor n_X (i.e. inertia forces n_XW); the vertical loads are for a load factor of 1.0 at the design ramp weight (i.e. the maximum weight for ground handling).

Two steady braking cases are considered: (a) with the nose gear in contact and (b) with the nose gear not in contact (and pitching effects balanced by a pitch inertia moment). For the first case, the forces are shown in Figure 15.20; a similar example where overall loads are calculated is shown in Chapter 5. In each case, the airframe internal loads could then be determined with the inertia forces distributed over the airframe according to the mass distribution (see Chapters 5 and 16 on loads).

The *dynamic* braked roll bookcase specifies an enhanced nose gear reaction, 'in the absence of a more rational calculation'; this phrase is often quoted in CS-25 in order to give the

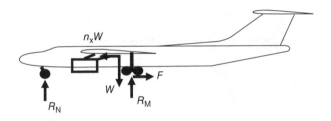

Figure 15.20   Bookcase loads for a steady braked roll.

manufacturer the opportunity to justify reduced loads by using a more representative calculation or test results.

15.4.2   Rational Braked Roll

If a rational braked roll calculation is carried out for the steady or dynamic condition, then detailed dynamic models are required to study the loads on the aircraft during braking. The braking force may be applied directly using a suitable coefficient of friction adjusted for the anti-skid system in use and justified by test. Alternatively, the braking torque may be considered to be applied with a suitable rise time and a tyre-to-runway contact model. The calculation is rather complex.

15.5   Turning

It is also important for an aircraft to be able to steer in a circular path on the ground without turning over or generating excessive loads (Lomax, 1996; Howe, 2004). A sketch of the basic turning kinematics is shown in Figure 15.21 (Currey, 1988). Usually, the nose wheel is steerable and the main wheels follow, with a small degree of tyre 'scrub' (i.e. rotation about a vertical axis if they are not being steered). Lateral tyre forces are developed to generate the required centripetal acceleration V^2/r in the turn of radius r.

The turning bookcase in CS-25 (see also Chapter 24) is for a 0.5 g turn where the side forces on each gear are equal to 0.5 times the vertical reaction per gear for the static case; the side forces are balanced by lateral inertia forces and the vertical forces are balanced by the weight. The idea is shown in Figure 15.22 where the subscripts 'o' and 'i' on the main gear reactions refer to the outside and inside of the turn respectively. The equations for the unknown reactions R_{Mi}, R_{Mo}, R_N, and hence turning forces, may be derived by equilibrium considerations (zero net vertical force, zero rolling moment and zero pitching moment). Internal loads in the airframe may be found by introducing distributed inertia loads. Obviously, far more rational and therefore complex calculations may be carried out if desired, especially since the 0.5 g turn is quite a severe manoeuvre.

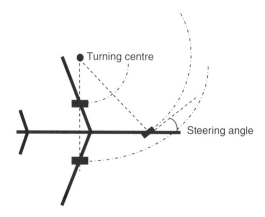

Figure 15.21   Simple turning kinematics.

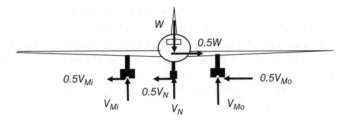

Figure 15.22   Bookcase loads for a 0.5 g turn.

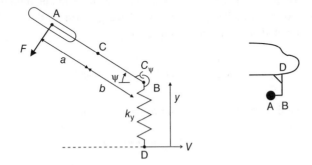

Figure 15.23   Simplified shimmy model.

15.6   Shimmy

Shimmy (Den Hartog, 1984; Pacejka, 2007) occurs when a wheel (or set of wheels) oscillates at often large amplitudes about its caster axis (i.e. the vertical axis about which the wheel system rotates), with a frequency typically between 10 and 30 Hz. It is in essence a self-excited insta-bility, but one that manifests itself as a limit cycle oscillation; i.e. due to non-linear effects, it reaches a maximum motion. It is most common for nose or main gear cantilevered twin wheel units and does not occur on bogie units. It can occur, for example, when severe vibration occurs at the wheels of a supermarket shopping trolley.

Shimmy can occur due to inadequate torsional stiffness, torsional freeplay, wheel imbalance, etc. It is normally countered by careful design or use of dampers. It is an extremely difficult phenomenon to model as it is non-linear and requires knowledge of complex tyre response characteristics (Pacejka, 2007). The tyre lateral force causes a sideslip angle of the landing gear and in turn the tyre sideslip angle changes, thus modifying the lateral force and potentially lead-ing to an unstable system. Once models are developed, the stability boundary may be deter-mined and if non-linear effects are included then any limit cycle behaviour may be explored.

A highly simplified 2DoF linear model of the shimmy phenomenon for a trailing wheel nose gear can be developed, following on from Den Hartog (1984). Figure 15.23 shows the wheel and suspension arrangement viewed from above and the position of the landing gear on the front fuselage. The gear-to-aircraft attachment point D moves forward at the ground speed V of the aircraft and is unaffected by feedback from the nose gear dynamic motion (i.e. front fuselage flexibility effects are being ignored). Point B is the bottom of the strut and so it may be seen that the gear strut is flexible, bending through a small displacement y against an effective strut bending stiffness k_y. The tyre is assumed to be rigid. The wheel, whose centre is at point A, lies behind point B by a mechanical trailing arm $a + b$ and yaws through the shimmy angle ψ (assumed small), with the yaw motion restricted by a viscous damper c_ψ. Note that the dynamics would alter if a yaw stiffness k_ψ were present and this could be examined as an exercise. The centre of mass of the combined landing gear is at point C. A follower friction force F acts from the ground on the wheel tread to prevent any sideslip motion of the tyre.

The kinetic energy term is

$$T = \frac{1}{2} m(\dot{y} + b\dot{\psi})^2 + \frac{1}{2} I_C \dot{\psi}^2 \tag{15.33}$$

where m is the mass of the combined landing gear and I_C is the moment of inertia of the landing gear about the centre of mass C. The elastic potential energy term for the strut bending and the dissipative term associated with the yaw damper are

$$U = \frac{1}{2} k_y y^2 \quad \text{and} \quad \Im = \frac{1}{2} c_\psi \dot{\psi}^2 \tag{15.34}$$

Finally, the incremental work done due to the friction force is

$$\delta W = -F[\delta y + (a + b)\delta \psi] \tag{15.35}$$

Applying Lagrange's equations leads to

$$m\ddot{y} + mb\ddot{\psi} + k_y y = -F$$
$$mb\ddot{y} + (I_C + mb^2)\ddot{\psi} + c_\psi \dot{\psi} = -F(a + b) \tag{15.36}$$

Now the unknown friction force F may be eliminated to combine the two equations into one kinetic equation (with several possible forms); e.g.

$$ma\ddot{y} + (mab - I_C)\ddot{\psi} - c_\psi \dot{\psi} + k_y(a + b)y = 0 \tag{15.37}$$

A further equation may be obtained from the kinematic relationship that there must be no sideslip of the wheel; i.e. the net velocity of the centre of the wheel must act at the angle ψ to the forward direction of the aircraft, so

$$-\frac{\dot{y} + (a + b)\dot{\psi}}{V} = \tan \psi \approx \psi \tag{15.38}$$

and

$$\dot{y} + (a + b)\dot{\psi} + V\psi = 0 \tag{15.39}$$

Combining Equations (15.37) and (15.39) in matrix form leads to the equation

$$\begin{bmatrix} ma & mab - I_C \\ 0 & 0 \end{bmatrix} \begin{Bmatrix} \ddot{y} \\ \ddot{\psi} \end{Bmatrix} + \begin{bmatrix} 0 & -c_\psi \\ 1 & a + b \end{bmatrix} \begin{Bmatrix} \dot{y} \\ \dot{\psi} \end{Bmatrix} + \begin{bmatrix} k_y(a + b) & 0 \\ 0 & V \end{bmatrix} \begin{Bmatrix} y \\ \psi \end{Bmatrix} = \begin{Bmatrix} 0 \\ 0 \end{Bmatrix} \tag{15.40}$$

giving rise to a nonconservative system. Assuming that the strut displacement and shimmy angle are governed by the exponent exp (λt) as per a flutter solution (see Chapter 10), the roots of the system are given by setting the 2×2 determinant to zero, thus yielding a cubic equation in λ:

$$\left(I_C + ma^2\right)\lambda^3 + \left(maV + c_\psi\right)\lambda^2 + k_y(a+b)^2\lambda + k_y(a+b)V = 0 \qquad (15.41)$$

Applying the Routh–Hurwitz criterion (see Chapter 6) yields the following condition for stability

$$(mab - I_C)V + c_\psi(a+b) > 0 \qquad (15.42)$$

Thus the damping, not present in Den Hartog's analysis, is stabilizing and leads to a finite shimmy speed

$$V_{crit} = \frac{c_\psi(a+b)}{I_C - mab} \qquad (15.43)$$

Above this critical speed, shimmy will occur as an instability involving both y and ψ motions, though of course in practice non-linear effects will then come into play to limit the response amplitude (i.e. a limit cycle oscillation; see Chapter 10). Clearly, based on this model, shimmy will never occur if $I_C \leq mab$, but this condition is unlikely for practical cases (Den Hartog, 1984). Equation (15.43) also shows that it is important to include damping.

The analysis carried out here is very simplistic. For a real aircraft it should be recognized that the flexibility of the fuselage and landing gear and complex non-linear flexible tyre dynamics need to be included.

15.7   Representation of the Flight Control System (FCS)

The extent to which the FCS needs to be represented in ground manoeuvres depends upon which, if any, of the FCS functions are relevant to the particular manoeuvre being considered; e.g. if the rotation and de-rotation phases of landing are to be modelled, then the appropriate control laws will be included. The steering and braking control systems should be modelled as required for rational calculations. Bookcase calculations will require any relevant FCS limits to be considered.

15.8   Examples

Note that some of the examples in Chapters 1, 2 and 5 might be helpful.

1. An aircraft taxiing may be represented as a single degree of freedom system consisting of a mass m supported by a landing gear of linear stiffness k and damping c. The aircraft moves at a velocity V over a harmonically undulating runway surface of wavelength λ and amplitude $\pm h$. Determine the equation of motion for the mass and the amplitude of the steady-state harmonic response as a function of the velocity. Hence find an expression for the most unfavourable value for the velocity of the aircraft.

2. An aircraft of mass 50 000 kg is travelling at a steady speed in high speed taxiing trials. Neglecting aerodynamic forces, the aircraft and its landing gear/tyres can be modelled as a system with a single (vertical) DoF with stiffness 30 MN/m and damping coefficient 1.6 MN s/m. The surface of the runway has a sinusoidal profile with amplitude of 10 mm and a wavelength of 15 m. The wheels always remain in contact with the runway. Find the amplitude of vertical displacement of the aircraft and its maximum vertical acceleration when the forward speed of the aircraft is 30 m/s.
   [12 mm, 0.19 g]

3. Write a MATLAB/SIMULINK program to solve the problem of an aircraft taxiing over (a) a harmonically undulating runway and (b) a '1-cosine' dip. The aircraft on its landing gear is modelled as a single DoF system, rather as in Examples 1 and 2 above. Choosing the parameters in Example 2, use part (a) of the program to check the result in Example 2. Then use part (b) to solve for the variation of maximum displacement response of the aircraft as a function of forward speed for a given 'dip' wavelength.

4. Consider the general form of the two DoF heave/pitch taxiing model used in this chapter. The aircraft moves at a velocity V over a harmonically undulating runway surface of wavelength λ and amplitude $\pm h$. Write the equations governing the steady-state solution in the frequency domain (refer to Chapter 14 for the analogous harmonic gust treatment).

5. Use the taxiing program given in the companion website for the rigid aircraft in pitch and heave to confirm the results given in this chapter for the worst case velocity for a given 'dip'.

6. A shock absorber has a piston area 0.01 m², stroke 0.4 m, pressure when fully extended of 30 bar and fully compressed of 300 bar, polytropic constant 1.35, compression and recoil damping coefficients 18 and 270 kN s²/m² and effective aircraft weight of 100 000 N per leg (i.e. 10 000 kg mass descending per leg). Generate a MATLAB/SIMULINK model for the landing case, using look-up tables for the stiffness and damping characteristics. Determine the displacement, velocity and acceleration response following a 3 m/s landing and find the maximum value of acceleration. Examine the effect of changing the values of the two damping coefficients on the relative magnitude of the two acceleration peaks.
   [1.65 g]

7. Repeat the solution of Example 6 by adding a tyre representation of an unsprung mass of 200 kg and tyre stiffness 2000 kN/m.
   [1.41 g and 1.52 g]

8. An aircraft of mass 10 000 kg has two main legs with four wheels on each leg. The rolling radius of each wheel is 0.25 m, the wheel mass is 30 kg and the radius of gyration in rolling is 0.2 m. A braking torque of 1250 N m is applied to each wheel. Determine the aircraft deceleration and the horizontal load acting on each leg. What is the ratio of braking force to normal reaction (i.e. effective friction value?
   [3.94 m/s², 19 200 N, 0.39]

9. An aircraft in Figure 15.24 undergoes a steady braked roll where the braking force at the main wheels is 0.8 times the main wheel reaction. Obtain expressions for the fore and aft

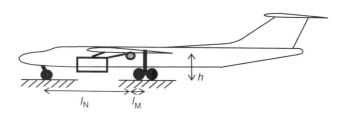

Figure 15.24

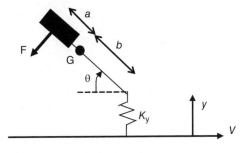

Figure 15.25

load factor n_x (if the vertical load factor is unity) for (a) nose gear in contact and (b) nose gear not in contact during braking. Assume that $h = 2l_M$ and $l_N = 6l_M$.
[24/43; 0.8]

10. Determine the critical condition for shimmy to occur for the undercarriage system shown in Figure 15.25 (Den Hartog, 1984). Determine the critical shimmy condition for the addition of (a) a translational damper C_y and (b) a rotational stiffness K_θ.

16

Aircraft Internal Loads

So far, in Chapters 12 to 15, the different manoeuvre and gust/turbulence inputs have been considered from the point of view of calculating the aircraft response in the rigid body and flexible mode generalized coordinates, and therefore potentially at any physical location on the structure. In this chapter, the way in which these responses are used to obtain the internal loads (i.e. moment, axial, shear and torque – 'MAST' – loads) at reference cross-sections in the airframe will be considered; the method of summation of forces is employed. The focus will be on the internal loads present in the main slender structural components of the wing and fuselage.

Some important concepts of loads were introduced in Chapter 5, and it will be assumed that the reader has a grasp of these issues when reading this present chapter. In particular, the use of d'Alembert's principle to allow an accelerating member to be treated as being in effective static equilibrium by adding distributed inertia forces was considered. This approach is powerful in that it allows standard static methods of determining internal loads to be applied to an accelerating slender member subject to a loading which varies with time and is non-uniformly distributed spatially. Indeed, the airworthiness regulations (CA-25 and FAR-25) include direct reference to balancing the applied loads by inertia forces and couples.

In Chapter 5, the slender members were considered firstly as continuous, and then represented in a discretized manner. This is because, in practice, the determination of internal loads requires the structure and loading to be idealized into discrete finite width strips and a summation will then replace the integral in the loads analysis. The analysis of the structural behaviour using a discretized representation and the relevant strip theory aerodynamics is introduced in Chapters 17 and 4 respectively. Further consideration of generating a suitable aerodynamic model for a discretized aircraft and coupling it to the structure will be considered in Chapters 18 and 19.

In this chapter, the basic concepts introduced in Chapter 5 are applied to more representative aircraft cases, including scenarios where the inertia and aerodynamic loadings are both non-uniformly distributed and time varying, such as would occur in dynamic manoeuvres and gust/turbulence encounters. Lifting surfaces (e.g. wings) are considered firstly as continuous, in order to be consistent with the Rayleigh–Ritz treatment used extensively earlier in the book, and then as discretized. The effect of discrete external load inputs, such as engine thrust and landing gear reactions, is explained. The loads in the fuselage will only be considered on a discretized basis because the fuselage geometry and mass distribution are not ideal for treating as continuous.

The determination of internal loads using load summation will cover the equilibrium, dynamic and ground manoeuvres, as well as discrete gusts. However, the special case of turbulence will be

---

Introduction to Aircraft Aeroelasticity and Loads, Second Edition. Jan R. Wright and Jonathan E. Cooper.
© 2015 John Wiley & Sons, Ltd. Published 2015 by John Wiley & Sons, Ltd.

introduced where a spectral treatment is required to generate values of the load Power Spectral Density (PSD).

A brief explanation of load sorting, to obtain critical cases for dimensioning, and the importance of loads envelopes is included. Also, the process of using these internal loads to obtain loads and stresses acting on internal structural elements will be mentioned; however, it is a complex procedure and beyond the scope of this book. Some explanation is provided of areas where terminology employed for loads can be confusing. Practical issues are further discussed in Chapters 20 and 24.

16.1   Limit and Ultimate Loads

Strength requirements for certification (see Chapter 20) are specified in terms of:

a. *limit loads*, which are the maximum loads to be expected in service and which the structure needs to support without 'detrimental permanent deformation' and
b. *ultimate loads* (limit loads multiplied by a factor of safety, normally 1.5 unless otherwise stated), which the structure must be able to support without failure/rupture.

Loads specified in the certification requirements are almost entirely limit loads. Thus the internal loads calculated for manoeuvres and gust/turbulence encounters are limit loads. The requirements also specify that compliance with the strength and deformation conditions must be demonstrated for each critical loading condition. Further comments on limit and ultimate loads, and on fatigue and damage tolerance, are included in Chapter 20.

16.2   Internal Loads for an Aircraft

In this section, a brief introduction to the origin of internal loads in a wing and fuselage will be presented. The ESDU series includes an item on internal loads (ESDU Data Sheet 94945).

16.2.1   Internal Loads for a Wing

In general, for an aircraft under three-dimensional motion, the external loading is distributed such that the wing is subject to bending moments and shear forces in the vertical and horizontal planes, as well as to torque. A typical Free Body Diagram (FBD) for a section of *wing* is shown in Figure 16.1. Vertical bending and shear occur due to a spanwise imbalance between the lift and inertia distribution along the wing whereas horizontal (fore-and-aft) bending and shear occur due to a spanwise imbalance between the drag and fore-and-aft inertia distributions and the engine thrust (for wing-mounted engines). Torque occurs because of a chordwise imbalance of wing lift and inertia distribution, as well as of drag, fore-and-aft inertia and thrust. Landing gear loads also influence the internal loads on the inboard wing.

It should be noted that there is more weight, and hence inertia loading, in the fuselage than in the wings whereas the converse is true for the aerodynamic loading. Therefore on the wings the inertia loads associated with the wing structure (and fuel) and any wing-mounted engines will counteract the aerodynamic loads and so 'relieve' the bending moments and shear forces caused by the aerodynamic loads, as illustrated in Figure 16.2. This inertia relief is demonstrated in a later example, and is one reason why engines are often placed on the wings.

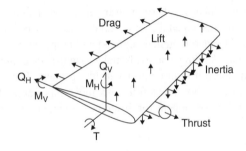

Figure 16.1   External and internal loads on a wing section.

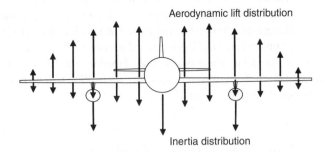

Figure 16.2   Inertia and aerodynamic loading distribution on a wing showing inertia relief.

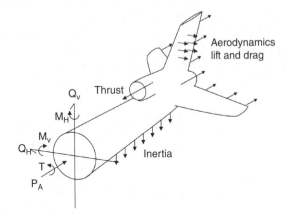

Figure 16.3   External and internal loads on a rear fuselage section.

In the limit of this argument, it is interesting to note that for a flying wing, if lift and weight (and therefore inertia) could be equally distributed spanwise, then there would be no vertical shear force or bending moment.

16.2.2   Internal Loads for a Fuselage

An FBD for a rear *fuselage* section is shown in Figure 16.3. Vertical bending and shear are experienced due to the imbalance between tailplane/rear fuselage lift and inertia forces, while torsion, lateral bending and shear result from an imbalance of lateral aerodynamic loads and inertia forces acting on the fuselage/tail. Also, axial loads will occur due to the imbalance of engine thrust (for rear fuselage mounted engines) and tailplane drag forces (and also cabin pressurization loads). Again, landing gear loads make a significant contribution to the front and centre fuselage when the aircraft is in contact with the ground. The convention adopted for positive internal loads must be chosen carefully and, most importantly, be consistent for loads calculations along any given component for different manoeuvre and gust cases.

16.3 General Internal Loads Expressions – Continuous Wing

In this section, the general case of an aircraft wing experiencing vertical shear/bending/torsion loads under the action of inertia and aerodynamic lift forces during a manoeuvre or gust encounter is considered; note that the treatment of turbulence is different and will be considered later. The approach introduced here could also be used for other lifting surfaces such as the fin and horizontal tailplane. An example will be given of a tapering wing for an aircraft undergoing an equilibrium manoeuvre. In the subsequent sections, application will be made to the aircraft model employed earlier in Chapters 12 to 15.

16.3.1 General Expression for Internal Loads

Consider the unswept wing shown in Figure 16.4 where the aerodynamic lift and inertia forces per unit span are $\lambda_A(\eta, t)$ and $\lambda_I(\eta, t)$ respectively; η defines the spanwise position and μ is the mass per span. Both force distributions are defined as positive upwards; the inertia force acts upwards if acceleration $\ddot{z}$ is defined as positive downwards, as has been the case earlier in the book (apart from in Chapter 5 where acceleration was defined positive upwards for convenience – in the direction of the positive applied force). The FBD for the section of wing created by a 'cut' at the spanwise position y is also shown in Figure 16.4, with internal loads

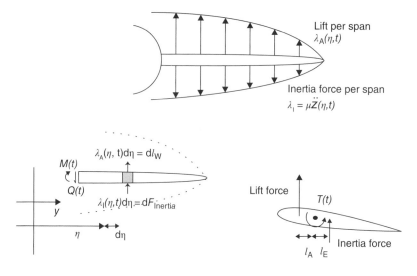

Figure 16.4   Loads on a continuous wing.

introduced to permit equilibrium of the cut section; thus, in essence, the analysis starts at the tip and works towards the root such that the wing root reactions are not required a priori.

Taking moments about the cut for equilibrium, the shear force and bending moment expressions are given by integrating over the cut section, so

$$Q(y,t) = \int_{\eta=y}^{s} [\lambda_A(\eta, t) + \lambda_I(\eta, t)] d\eta \quad M(y,t) = \int_{\eta=y}^{s} [\lambda_A(\eta, t) + \lambda_I(\eta, t)](\eta-y) d\eta \quad (16.1)$$

The shear force and bending moment may therefore be calculated at any spanwise position as a function of time provided the lift distribution is defined as a function of the wing motion (and, where relevant, gust velocity). Since the lift and inertia forces act at different chordwise positions, as shown in Figure 16.4, then a torque is also present and would need to be calculated using a similar integration approach to that applied above for the shear force and bending moment. For an untapered/unswept wing, then the nose up torque about the elastic axis is

$$T(y,t) = \int_{\eta=y}^{s} [l_A \lambda_A(\eta,t) - l_E \lambda_I(\eta,t)] d\eta \quad (16.2)$$

where the lift acts a distance l_A forward of the elastic axis and the inertia force a distance l_E behind it. However, the torque expression would differ for a swept or tapered wing.

16.3.2 *Example: Equilibrium Manoeuvre – Continuous Wing*

An aircraft whose wing planform is shown in Figure 16.5 has a total mass of 50 000 kg and experiences a steady symmetric pull-up manoeuvre with an incremental acceleration of 2 g, so that the load factor n is 3. Each wing with a full fuel load has a mass of 5000 kg. For ease of analysis the wing is unswept, and it may be assumed that all the aircraft lift is carried by the wings and that the wing mass (including fuel) and wing lift are distributed in proportion to the wing chord; thus any tip effects or flexible mode deformation effects are ignored. At any spanwise location, the lift acts at the quarter chord and the mass axis is assumed to be at 40 % chord. Determine the distributions of the wing shear force, bending moment and torque (about 50% chord), assuming the sign convention shown in Figure 16.6. Firstly, treat the problem as continuous; the results will be compared later to those obtained from a discretized version. Take $g = 10 \text{ m/s}^2$ for simplicity.

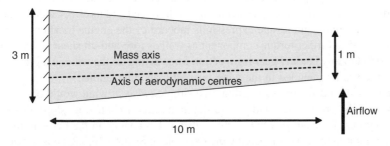

Figure 16.5 Diagram of the tapered wing example.

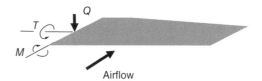

Figure 16.6   Internal load sign convention for the tapered wing example.

The total wing lift in the manoeuvre is $L = nW = 3 \times (50\,000 \times 10)/1000 =$ kN, so the lift per wing is 750 kN. Since the area per wing is 20 m², the wing aerodynamic loading is $750/20 = 37.5$ kN/m². The inertia force per wing (acting downwards and therefore negative in the sign convention) is $nW_{wing} = 3 \times (5000 \times 10)/1000 = 150$ kN so the wing inertia loading is -7.5 kN/m². The variation of chord with distance η from the root is given by $c(\eta) = 3 - 0.2\eta$. The lift and inertia forces per unit span are then given by $\lambda_A = 37.5c$ and $\lambda_I = -7.5c$ kN/m respectively, so they vary with the chord. Then, considering the forces acting on an element $d\eta$ and using the approach taken earlier in Section 16.3.1, the shear force, bending moment and torque (at 50 % chord) at position y outboard of the wing root are given by the following expressions:

$$Q(y) = \int_{\eta=y}^{10} (37.5 - 7.5)c\,d\eta = \int_{\eta=y}^{10} (37.5 - 7.5)(3 - 0.2\eta)\,d\eta = 600 - 90y + 3y^2$$

$$M(y) = \int_{\eta=y}^{10} (37.5 - 7.5)c(\eta - y)\,d\eta = 2500 - 600y + 45y^2 - y^3 \qquad (16.3)$$

$$T(y) = \int_{\eta=y}^{10} \left(37.5c\frac{c}{4} - 7.5c\frac{c}{10}\right)d\eta = 373.75 - 77.625y + 5.175y^2 - 0.115y^3$$

The values of these internal loads at the root and 4 m outboard of the root are $Q(0) = 600$ kN, $Q(4) = 288$ kN, $M(0) = 2500$ kN m, $M(4) = 756$ kN m, $T(0) = 373.75$ kN m and $T(4) = 138.69$ kN m. These are 'exact' values within the approximations made in the problem and will be seen later to compare closely with the results obtained for the discretized representation when plotted along the wing span.

16.4   Effect of Wing-mounted Engines and Landing Gear

If a wing-mounted engine were to be present, then several internal load values are affected. Firstly, a discrete vertical inertia force needs to be introduced and its effect added into the shear force, bending moment and torque expressions inboard of the engine location. Secondly, the engine thrust adds a further torque component as well as fore-and-aft shear and bending. Similarly, the effect of any landing gear leg reaction force acting on the wing in a ground manoeuvre would need to be included in the wing internal load calculations.

The effect of such discrete loads acting on a simple *unswept* wing would be to introduce a step change in the shear force and torque diagrams and a change of slope in the bending moment diagrams; for a *swept* wing, there is also a step change in the bending moment.

Now, returning to the earlier tapered wing example in Section 16.3.2, consider an engine of mass 4000 kg to be mounted on each wing (instead of on the fuselage) a distance 3 m outboard

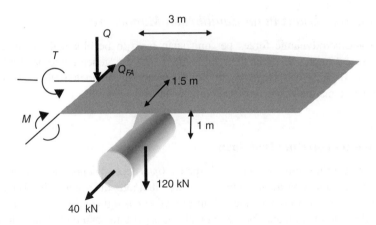

Figure 16.7   Addition of an under-wing engine.

from the root and with its centre of mass being 1.5 m forward of the wing centre line, as shown in Figure 16.7. A thrust of 40 kN acts at a distance 1 m below the wing. Assume that the wing lift in this manoeuvre is unchanged.

The thrust contributes 40 kN m nose up torque and 40 kN fore-and-aft shear force on the inner wing, as well as a fore-and-aft bending moment (120 kN m at the root). The downwards inertia force associated with the engine is $3 \times (4000 \times 10)/1000 = 120$ kN, so contributing $120 \times 1.5 = 180$ kN m nose down torque about the wing centre line. The net nose down torque due to the thrust and inertia load for the engine is $180 - 40 = 140$ kN m. Therefore the vertical root shear force, bending moment and torque calculated earlier will reduce by 120 kN, 360 kN m and 140 kN m respectively. The presence of the wing-mounted engine thus alleviates (or relieves) the wing loading in a manoeuvre as mentioned earlier.

16.5   Internal Loads – Continuous Flexible Wing

In this section, the general expressions for internal loads will be considered for a flexible aircraft undergoing manoeuvres or encountering gusts. The analysis employs a continuous representation of the wing so as to be compatible with the earlier consideration of manoeuvres and gusts in Chapters 12 to 15 where a single flexible mode was included together with heave and pitch motions; the discretized case will be considered later.

16.5.1   Steady and Incremental Loads

At this point it is worth noting that under some circumstances the dynamic model only yields the incremental response, and hence loads, relative to a steady underlying condition; the steady and incremental loads then need to be added together. This is the case for the gust encounter and also for flight manoeuvres; depending upon the way in which the equations are formed, it may also be the case for some of the ground manoeuvres. Obviously, the equilibrium manoeuvre only generates a steady load.

16.5.2   Internal Loads in an Equilibrium Manoeuvre

The inertia and aerodynamic forces per unit span need to be obtained for an equilibrium manoeuvre (see Chapter 12) and then substituted into the general internal load equations derived above. Because it is a steady, and not time-varying, manoeuvre this is a simpler case than the dynamic manoeuvre or gust encounter.

16.5.2.1   Inertia Force per Unit Span

The expression for the inertia force per unit span $\lambda_I(\eta, t)$ depends upon the type of manoeuvre. For an equilibrium pitching manoeuvre at a steady pitch rate, the acceleration is constant with time and the same at every point on the airframe whether it is rigid or flexible; the acceleration is dictated by the load factor n and for a wing with mass per unit span μ the inertia force per unit span is given by

$$\lambda_I(\eta, t) = -\mu n g \tag{16.4}$$

The minus sign occurs because the inertia force is defined as positive upwards for a positive downwards acceleration whereas the aircraft actually accelerates upwards for a positive load factor (this was the case for the earlier tapered wing example). If a manoeuvre involved steady pitch acceleration, then the load factor would vary along the aircraft.

16.5.2.2   Aerodynamic Force per Unit Span

Now consider the aerodynamic lift per span $\lambda_A(\eta, t)$, again acting positive upwards. In this case, for an equilibrium manoeuvre, the lift will be constant with time though it will not normally have a uniform spatial distribution along the wing. Lift will be a function of the aircraft rigid body and flexible mode responses. Again, referring back to the earlier analyses in Chapter 12, the wing lift per unit span in a symmetric equilibrium manoeuvre for a flexible aircraft is given by

$$\lambda_A(\eta) = \frac{1}{2}\rho V^2 c a_W [\alpha - \alpha_0 + \gamma_e(\eta) q_e] \tag{16.5}$$

Thus, the steady lift for this equilibrium manoeuvre is simply a function of incidence, angle for zero lift and wing torsional deformation in the flexible mode; however, these quantities depend upon the trimmed state and therefore on the rear fuselage deformation and elevator angle. Note here that the symbol η used for the integration variable along the wing should not be confused with that used for the elevator angle.

16.5.2.3   Internal Loads in an Equilibrium Manoeuvre

Now that expressions have been obtained for the inertia and aerodynamic forces per unit span in Equations (16.4) and (16.5), these may be substituted into Equation (16.1) to determine the steady internal loads. For simplicity, consider only the wing root bending moment ($y = 0$). For the flexible aircraft used earlier, the result is given by

$$M(0) = \frac{1}{2}\rho V^2 S_{\mathrm{W}} a_{\mathrm{W}} \frac{s}{4}\left[\alpha - \alpha_0 + \gamma_{\mathrm{e}0}\left(1 + \frac{2B}{3}\right)q_{\mathrm{e}}\right] - \mu n g \frac{s^2}{2} = f\{\alpha, \alpha_0, q_{\mathrm{e}}, n\} \qquad (16.6)$$

where B is a constant defining the torsional mode shape (see Appendix C). Note that, for any location other than the wing root, each of the four terms in this expression would also be a function of y, the spanwise coordinate. It may be seen that the wing twist will increase the root bending moment value due to the outboard shift in the centre of pressure; this phenomenon is illustrated by some wing root bending moment results presented later for the example in Section 16.5.4. The other mode types considered earlier, namely wing bending and fuselage bending, will not affect the distribution of internal loads in the wing when compared to the rigid aircraft. The expressions for wing root shear force and torque are similar in appearance to the bending moment. The loads on a swept wing will differ (see Chapter 12).

16.5.3   Internal Loads in a Dynamic Manoeuvre/Gust Encounter

In this section, the inertia and aerodynamic forces per unit span will be determined for a symmetric gust encounter (see Chapter 14) and then substituted into the general expression for internal loads in the continuous wing. In this case the loadings are time varying. For a gust encounter, the incremental internal loads due to the gust need to be added to the internal loads for the initially trimmed steady level flight case ($n = 1$) prior to encountering the gust (see Chapters 14 and 23). The approach is essentially the same for ground manoeuvres but landing gear leg reaction forces need to be included.

For a dynamic manoeuvre where a body fixed axes system is employed (see Chapter 13), the principles are the same, but the different axes system needs to be taken into account. Also, the loads in the initial trimmed condition must be added to the incremental loads experienced in the dynamic manoeuvre and control angle terms would be present in the internal load expressions.

16.5.3.1   Inertia Force per Unit Span

For a general gust encounter case, the acceleration will vary over the airframe and with time; also, it will have both rigid body and flexible mode components. Thus the inertia force per unit span on the wing will be given by

$$\lambda_{\mathrm{I}}(\eta, t) = \mu \ddot{z}_{\mathrm{WM}}(\eta, t) \qquad (16.7)$$

where $\ddot{z}_{\mathrm{WM}}$ is the downwards acceleration along the wing mass axis for the spanwise position under consideration. Referring back to Chapters 12 and 14, the displacement at the wing mass axis is dependent upon heave, pitch and flexible mode motions and is given by

$$z_{\mathrm{WM}}(y) = z_{\mathrm{C}} - l_{\mathrm{WM}}\theta + [\kappa_{\mathrm{e}}(y) + l_{\mathrm{E}}\gamma_{\mathrm{e}}(y)]q_{\mathrm{e}} \qquad (16.8)$$

Therefore the inertia force per unit span is given by

$$\lambda_{\mathrm{I}}(\eta, t) = \mu \ddot{z}_{\mathrm{WM}}(\eta, t) = \mu\left[\ddot{z}_{\mathrm{C}}(t) - l_{\mathrm{WM}}\ddot{\theta}(t) + [\kappa_{\mathrm{e}}(y) + l_{\mathrm{E}}\gamma_{\mathrm{e}}(y)]\ddot{q}_{\mathrm{e}}(t)\right] \qquad (16.9)$$

The inertia contribution to the internal loads is thus a function of the heave, pitch and flexible mode accelerations. It should be noted that if the aircraft motion involves torsional acceleration of the flexible mode, then the resulting inertia pitching moments for the wing will affect any torque generated and will have to be included.

16.5.3.2   Aerodynamic Force per Unit Span

When encountering a gust, the wing lift will be a function of the wing incidence and twist, of the heave, pitch and flexible mode velocities, and also of the gust velocity. Thus, referring back to Chapter 14, and using the same notation, the incremental wing lift per unit span in a gust encounter, for example, is given by

$$\lambda_A(\eta, t) = \frac{1}{2}\rho V c a_W \left\{ w_g + \dot{z}_C - l_W \dot{\theta} + [\kappa_e(\eta) - l_A \gamma_e(\eta)]\dot{q}_e \right\} + \frac{1}{2}\rho V^2 c a_W [\theta + \gamma_e(\eta)q_e] \quad (16.10)$$

where both rate- and incidence-dependent terms are present and the dependency of the response quantities on time has been omitted for clarity.

Note that the unsteady aerodynamics effects have also been omitted but could be included by employing a convolution process (see Chapters 9 and 14). Also, it is important to note that when more advanced aerodynamic panel methods are used (see Chapters 18 and 19), the lift distribution will be expressed as a function of the motion of the entire wing and not just the motion of the element (or strip) under consideration; the aerodynamic forces would also be a function of reduced frequency.

16.5.3.3   Internal Loads in a Gust Encounter

Now that expressions have been obtained for the inertia and aerodynamic forces per unit span in Equations (16.9) and (16.10), these may be substituted into Equation (16.1) to determine, for example, the bending moments on the wing. For simplicity, again consider only the wing root bending moment ($y = 0$) and, for the aircraft model used in the earlier chapter on gusts, the result may be shown to be

$$M(0, t) = \mu \frac{s^2}{2}\left\{ \ddot{z}_C - l_{WM}\ddot{\theta} + \left[\kappa_{e0}\left(1 + \frac{A}{2}\right) + l_E \gamma_{e0}\left(1 + \frac{2B}{3}\right) \right]\ddot{q}_e \right\}$$

$$+ \frac{1}{2}\rho V_0 a_W c \frac{s^2}{2}\left\{ \dot{z}_C - l_W \dot{\theta} + \left[\kappa_{e0}\left(1 + \frac{A}{2}\right) - l_A \gamma_{e0}\left(1 + \frac{2B}{3}\right) \right]\dot{q}_e + w_g \right\} \quad (16.11)$$

$$+ \frac{1}{2}\rho V_0^2 a_W c \frac{s^2}{2}\left[\theta + \gamma_{e0}\left(1 + \frac{2B}{3}\right)q_e \right]$$

where A and B are constants defining the wing bending and torsional mode shapes (see Appendix C). The bending moment, response variables and gust velocity are all functions of time. Thus knowing the gust velocity and the generalized coordinate responses, the root bending moment time history can be calculated. For any other spanwise location, each term would also be a function of y. Similar expressions exist for shear force and torque.

16.5.4 Example: Internal Loads during a '1-Cosine' Gust Encounter

To illustrate some results for time-varying internal loads, consider the same time domain gust response example for the simple flexible aircraft used in Chapter 14. The flight case considered was for 150 m/s EAS at 14 000 ft (4267 m) with the aircraft encountering 5 m/s EAS '1-cosine' gusts of different lengths. In Chapter 14, only the responses were obtained whereas the corresponding wing root bending moment, shear force and torque results will be shown here.

16.5.4.1 Steady Loads experienced Prior to the Gust Encounter

To determine the overall loads in a gust encounter, both the *steady* and *incremental* components of internal load need to be found. The steady loads are those in the equilibrium flight condition that the aircraft is in prior to encountering the gust (see Equation (16.6) for the wing root bending moment) whereas the incremental loads are those generated by the gust (see Equation (16.11)).

For the *rigid* aircraft in steady level flight, the wing root bending moment, shear force and torque values for this example are 120.3 kN m, 32.1 kN and 15.4 kN m respectively. However, for the *flexible* aircraft in steady flight, the internal load values depend upon the mode shapes and natural frequencies. For the modes where fuselage bending or wing bending are dominant, the internal loads are unchanged from the rigid aircraft value since there is no change in the local wing incidence due to the flexible modes if the wing is unswept. However, for the mode where wing torsion is dominant, the bending moment will increase when compared to the rigid aircraft value since the wing twists and the lift shifts outboard; the wing root bending moment values of 124.7, 130.5 and 145.0 kN m correspond to torsional natural frequencies of 9, 6 and 4 Hz respectively. Shear force and torque are also affected by the torsion mode.

16.5.4.2 Incremental Loads in the Gust Encounter

It should be noted that the values presented in the figures below are actually the *incremental* internal loads due to the gust. These loads need to be added to the steady level flight loads (i.e. for an $n = 1$ equilibrium manoeuvre).

Firstly, consider the mode where *fuselage bending* is dominant, with 2 Hz frequency and 4% damping; note that in this mode shape (see Appendix C) there is a finite heave displacement of the wing and so the mode is excited by a gust input. The variation of the maximum and minimum incremental wing root bending moment with gust wavelength yields a maximum value of 68.4 kN m for around an 80 m wavelength and a minimum value of –62.2 kN m for a 225 m wavelength. A comparison of the wing root bending moment for several different wavelengths is shown in Figure 16.8; both the short period and flexible mode contributions may be seen in the bending moment time history. Note that the initial internal load peak tends to 'follow' the gust velocity peak. The loads will not necessarily peak at the same wavelength as the response.

The equivalent result for a mode where *wing bending* is dominant, with 3 Hz frequency and 4% damping, is shown in Figure 16.9; the maximum value of 102.8 kN m occurs for a wavelength of 70 m whereas the minimum value of –98.9 kN m occurs for a 60 m gust. For this mode and low natural frequency, the incremental bending moment is dominated by the oscillatory flexible response.

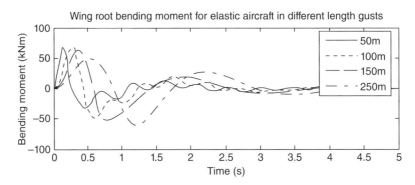

Figure 16.8 Incremental wing root bending moment for a flexible aircraft in a '1-cosine' gust of different lengths – heave/pitch model with a fuselage bending mode (2 Hz/4%).

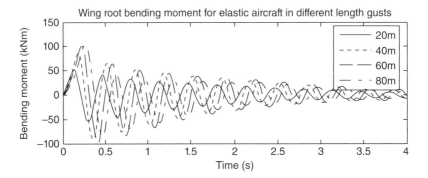

Figure 16.9 Incremental wing root bending moment for a flexible aircraft in a '1-cosine' gust of different lengths – heave/pitch model with a wing bending mode (3 Hz/4%).

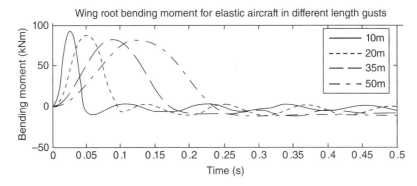

Figure 16.10 Incremental wing root bending moment for a flexible aircraft in a '1-cosine' gust of different lengths – heave/pitch model with a wing torsion mode (4 Hz/4%).

The results for the mode where *wing torsion* is dominant, with 9 Hz frequency and 4% damping, show the peak response occurring for small wavelengths where, as seen from the results in Figure 16.10, the incremental bending moment essentially 'follows' the gust velocity. However, it should be noted that these (and the earlier) calculations did not include the unsteady

Küssner and Wagner effects (see Chapters 9 and 14) and so in reality these initial peaks would be different.

Although the behaviour of each of the three different flexible modes have been shown separately, in practice a significant number of flexible modes would be superimposed and the approach involved would be extended by summing the effects of multiple modes, as indicated in Chapter 14.

16.5.5   Form of Internal Loads for a Continuous Wing Representation

Having established typical expressions for inertia and aerodynamic lift forces per unit span for a continuous wing, then by substituting them into the internal load Equations (16.1) the form of the expression for any internal load on the wing for the flexible aircraft may be determined. Equation (16.11) showed a typical expression for the wing root bending moment in a gust encounter; this may be written in the more general form for the heave, pitch and a single flexible mode:

$$\text{Internal load } (y, t) = \{A_1 \quad A_2 \quad A_3\}\ddot{q} + \{B_1 \quad B_2 \quad B_3\}\dot{q} + \{C_1 \quad C_2 \quad C_3\}q + Dw_g \qquad (16.12)$$

where $q = \{z_C \ \theta \ q_e\}^T$ is the response and $A_j, B_j, C_j \quad (j = 1, 2, 3)$ and D are all functions of y (in general), calculated via the integrals in the internal load Equations (16.1). Note that in some circumstances, terms related to control input would be included, e.g. if spoilers were deployed on the wing. Thus, once the responses $q(t)$, $\dot{q}(t)$, $\ddot{q}(t)$ in a manoeuvre or gust encounter are calculated, the incremental internal loads may be obtained using Equation (16.12) and then added to the steady flight loads to yield the overall loads.

16.6   General Internal Loads Expressions – Discretized Wing

Since the loads on a real aircraft are not expressed analytically, this section considers the case where both the wing structure and the loading need to be discretized. There are different ways of carrying out the discretization, one possibility being outlined here. In this section, the approach used to obtain internal loads for a discretized wing will be considered (see Chapter 5), together with an example allowing the general form of the expression to be seen. The way in which the aircraft may be represented discretely is discussed further in Chapter 21.

16.6.1   Wing Discretization

Consider the wing span to be divided into N strips of equal or unequal width. Assume that the Finite Element model is either defined as a beam, or condensed to be 'beam-like' (see Chapters 17 and 21), with the beam axis defining a reference axis with load stations being defined on this axis at the inner end of each strip, as seen in Figure 16.11 for example. The aerodynamic force per strip will nominally lie at the aerodynamic centre of the strip, when using the two-dimensional strip theory assumption, or by aggregating three-dimensional panel results on to a strip. The mass of each strip is located at the strip centre of mass (i.e. on the mass axis), with any offset of the strip centre of mass from the reference axis being accounted for in the analysis (see Chapter 17).

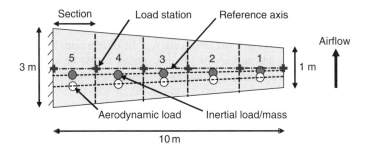

Figure 16.11 Diagram of a discretized tapered wing example.

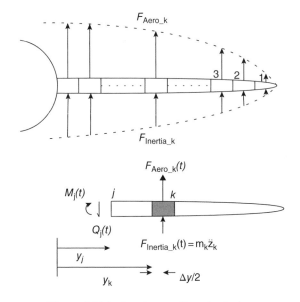

Figure 16.12 Loads on a discretized wing.

16.6.2   General Expression for Internal Loads – Discretized Wing

Consider the discretized wing with N strips shown in Figure 16.12. In this case, the aerodynamic lift and inertia forces on the kth strip of the wing are given by $F_{\text{Aero\_}k}(t)$ and $F_{\text{Inertia\_}k}(t)$ respectively (both positive upwards). Following a similar argument to that used earlier for the continuous wing, the shear force and bending moment at the inboard end of the jth strip are

$$Q_j(t) = \sum_{k=1}^{j}\left[F_{\text{Aero\_}k}(t) + F_{\text{Inertia\_}k}(t)\right]$$

$$M_j(t) = \sum_{k=1}^{j}\left[F_{\text{Aero\_}k}(t) + F_{\text{Inertia\_}k}(t)\right]\left(y_k - y_j + \frac{\Delta y}{2}\right) \quad j = 1, 2, \ldots, N \tag{16.13}$$

Thus the internal loads may be calculated at any spanwise position, defined by the interface between two strips, and as a function of time, provided the lift and inertia distributions are known as a function of the aircraft motion (and, if appropriate, gust velocity). These two internal loads do not depend upon the chordwise position of the loadings. However, a similar expression may be derived for torque, where the chordwise position of the loads is important.

These shear force and bending moment expressions are generally applicable to any unswept wing, although the same principles would apply for a swept wing provided that the strips and internal loads were defined relative to streamwise (i.e. aircraft and not wing) axes. Note that for a swept wing, the torque expression needs to allow for the effect of sweep on the moment arm associated with the lift and inertia forces.

16.6.3   Example: Equilibrium Manoeuvre – Discretized Wing

In order to illustrate this discretization approach further, consider the same wing example as covered in Section 16.3.2 but with the tapered wing shown in Figure 16.5 now divided into five strips of equal width $\Delta y = 2$ m, as seen in Figure 16.11. The mass axis lies along the 40% chord and the aerodynamic centre axis lies along the quarter chord. Assume that the reference axis lies along the mid-chord. Having defined the discretization for the earlier example in the previous section, the internal loads at each strip may now be calculated using the direct approach introduced above in Section 16.6.2. The equivalent expression for wing torque (nose up positive) for this particular example, where the mass axis is at 40% chord, may be shown to be given by

$$T_j(t) = \sum_{k=1}^{j} \left[F_{\text{Aero}\_k}(t) \frac{c_k}{4} + F_{\text{Inertia}\_k}(t) \frac{c_k}{10} \right] \tag{16.14}$$

where torque is referred to the wing reference axis (i.e. mid-chord). Note that this torque expression only applies to the specific chordwise positions for the aerodynamic lift and inertia forces used in this example (i.e. distances $c_k/4$ and $c_k/10$ ahead of the reference axis), but a more general formula could be derived.

The aerodynamic lift and inertia forces per strip of the wing are given in Table 16.1, remembering that the aerodynamic lift and inertia force per unit span (upwards positive) were obtained earlier as $37.5c$ and $-7.5c$ kN/m. The table also shows the internal loads calculated using Equations (16.13) and (16.14). The variation of these three internal loads along the wing is shown in Figure 16.13. At the wing root, the bending moment is 2520 kN m (sag positive)

Table 16.1   Wing properties, external loads and internal loads per strip.

Strip j	Mean strip chord (m) c_j	Lift force (kN) F_{Aero_j}	Inertia force (kN) F_{Inertia_j}	Net upwards force (kN) $(F_{\text{Aero}\_j} + F_{\text{Inertia}\_j})$	Shear force (kN) Q_j	Bending moment (kN m) M_j	Torque (kN m) T_j
0	—	—	—	—	0	0	0
1	1.2	90	−18	72	72	72	24.8
2	1.6	120	−24	96	168	312	69
3	2.0	150	−30	120	288	768	138
4	2.4	180	−36	144	432	1488	237.4
5	2.8	210	−42	168	600	2520	372.6

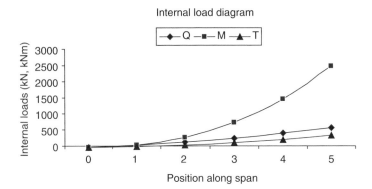

Figure 16.13 Variation of internal loads along a discretized wing for the equilibrium manoeuvre example.

so the wings tend to bend upwards. The wing root torque is 372.6 kN m, with the wing tending to twist nose up. The values in the table agree well with the 'exact' values calculated earlier in Section 16.3.2; the shear forces are identical whereas there are relatively small differences (~1%) in the bending moments and torques, due to the discretization; using more strips would improve the agreement.

The effect of flexibility can be included by adding the relevant modal terms to the aerodynamic and inertia forces per strip. Also, the dynamic manoeuvre or gust encounter would be treated in a similar way, with the inertia and aerodynamic loads evaluated at the required instant of time.

16.6.4   Form of Internal Loads for a Discretized Wing Representation

In Section 16.5.5, the general form for a *continuous* wing internal load at the spanwise position y was illustrated via Equation (16.12). In the general case for any location on the wing, using a *discretized* representation, all the internal loads required at all stations may, for convenience, be expressed in a matrix form. This idea will be illustrated for the shear forces on a wing structure, but the same approach may be taken for other internal loads and aircraft components. The force summation shown for the wing in Equations (16.13) and (16.14) could be written in the following way for the *inertia* contribution (subscript I), namely

$$Q_I = \begin{Bmatrix} Q_1 \\ Q_2 \\ \vdots \\ Q_N \end{Bmatrix}_I = \begin{bmatrix} 1 & 0 & \cdots & 0 \\ 1 & 1 & \cdots & 0 \\ \vdots & \vdots & \cdots & \vdots \\ 1 & 1 & \cdots & 1 \end{bmatrix} \begin{Bmatrix} F_{I1} \\ F_{I2} \\ \vdots \\ F_{IN} \end{Bmatrix} = T_I F_I, \qquad (16.15)$$

where the transformation (T_I) matrix for the shear force allows the contribution of the inertia forces on each strip to be added into the shear forces at the desired positions. The equivalent T matrix for the bending moment includes the moment arm between the load station and the inertia loading point of action; also, the roll inertia moment per strip could be included. If torque loads were being calculated, the pitch inertia moment (and of course any aerodynamic pitching moment) would need to be included. The inertia forces are given by the product of the mass and acceleration for every strip where the mass is lumped, so

$$F_I = \left\{ \begin{array}{c} m_1 \ddot{z}_1 \\ m_2 \ddot{z}_2 \\ \vdots \\ m_N \ddot{z}_N \end{array} \right\} = \begin{bmatrix} m_1 & 0 & \cdots & 0 \\ 0 & m_2 & \cdots & 0 \\ \vdots & \vdots & \cdots & \vdots \\ 0 & 0 & \cdots & m_N \end{bmatrix} \left\{ \begin{array}{c} \ddot{z}_1 \\ \ddot{z}_2 \\ \vdots \\ \ddot{z}_N \end{array} \right\} = \mathbf{M}\ddot{z}, \qquad (16.16)$$

where the vector $\ddot{z}$ defines the physical acceleration at each mass station. This may in turn be related to the generalized coordinates q for the heave, pitch and flexible modes via the modal transformation $z = \mathbf{\Phi}q$, where $\mathbf{\Phi}$ is the modal matrix, so finally

$$Q_I = \mathbf{T}_I \mathbf{M} \mathbf{\Phi} \ddot{q} = \mathbf{A}_M \ddot{q} \qquad (16.17)$$

where $\mathbf{A}_M$ is a modal transformation matrix for the shear force due to inertia effects.

For the *aerodynamic* contribution to the shear force (subscript A), an equivalent expression $Q_A = \mathbf{T}_A \mathbf{F}_A$ applies. These aerodynamic forces may be expressed in terms of the velocities and displacements at the reference positions and the control input η and/or gust velocity w_g (if appropriate), using

$$F_A = \mathbf{B}\,\dot{z} + \mathbf{C}z + R_c \eta + R_g w_g \qquad (16.18)$$

where $\mathbf{B}, \mathbf{C}$ are the aerodynamic matrices defining the rate and incidence dependent terms and R_c, R_g are the vectors defining the control input and gust velocity terms. Unsteady aerodynamic effects are neglected here for simplicity. The aerodynamic contribution to the shear force is then

$$Q_A = \mathbf{T}_A \left\{ \mathbf{B}\mathbf{\Phi}\dot{q} + \mathbf{C}\mathbf{\Phi}q + R_c\eta + R_g w_g \right\} = \mathbf{B}_M \dot{q} + \mathbf{C}_M q + R_{Mc}\eta + R_{Mg} w_g \qquad (16.19)$$

Therefore the *overall* expression for the wing shear forces is given by the supplementary or *auxiliary equation*

$$Q = \mathbf{A}_M \ddot{q} + \mathbf{B}_M \dot{q} + \mathbf{C}_M q + R_{Mc}\eta + R_{Mg} w_g \qquad (16.20)$$

Thus the internal loads may be calculated directly from the generalized responses, control input and gust velocity using an equation that looks somewhat similar to the basic equation of motion of the aircraft. Clearly, the final form of the load expression is somewhat similar to that found for the continuous wing in Equation (16.12). These auxiliary matrices are conveniently formed when the initial aeroelastics and loads models are generated. The same form of expression would apply for bending moments and torques except that the content of the transformation matrices would differ. These internal loads may be supplemented by other 'interesting quantities' such as accelerations, rates, control forces, strains, etc. (see Chapter 24).

Finally, it should be noted that the approach described above for determination of internal loads is often known as the 'force summation' method; it is the most intuitive of methods available. Other approaches sometimes used are the so-called 'mode displacement' or 'mode acceleration' methods (CA-25).

16.7   Internal Loads – Discretized Fuselage

The previous sections considered the internal loads present on a continuous or discretized wing in an equilibrium manoeuvre or gust encounter, but the approach could equally well apply for any other aerodynamic lifting surface such as a tailplane or fin. The case of internal loads in a fuselage is somewhat different, but uses similar principles. In this section, only the discretized fuselage will be considered; the use of an analytical approach would be possible by integrating over the fuselage and adding in the tailplane effects directly, but this is omitted here as it is more 'messy' than the wing case and the wing examples are thought to be adequate to illustrate the ideas for continuous analysis.

16.7.1   Separating Wing and Fuselage Components

The wing and fuselage may be considered crudely as separate components by introducing intercomponent forces and moments (see Chapter 5). For example, Figure 16.14 illustrates the simple arrangement where the wing may be assumed to connect with the fuselage at the front and rear spar positions, where only vertical loads A, B and horizontal load C are present; note that the intercomponent forces will differ if the engines are mounted on the fuselage instead of the wing. These intercomponent loads may be obtained by considering equilibrium of the separated wing and fuselage components. This approach is reasonable for aircraft where the wing is connected above or below the fuselage using a statically determinate connection arrangement at the main and rear spars (as is the case for some small commercial aircraft). However, it is somewhat idealized for aircraft where the wing/fuselage junction is more integrated

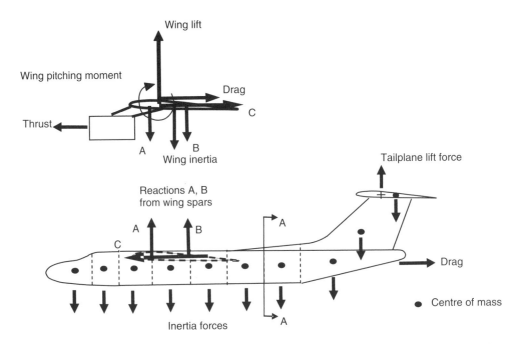

Figure 16.14   Separation of wing and fuselage components on a discretized fuselage/tail unit using intercomponent loads.

and the load paths are less distinct; nevertheless, the net effect of the connection may be represented by discrete intercomponent forces as shown here.

16.7.2   Discretized Fuselage Components

A discretization of the fuselage and tail (sometimes called 'empennage') is also shown in Figure 16.14. These sections of the fuselage, fin and tailplane have associated masses; the interfaces between fuselage sections would usually correspond to frame/bulkhead locations. The fuselage, fin and tailplane are subject to vertical inertia forces for each discrete mass (and to horizontal inertia forces if there is fore-and-aft acceleration). There will also be reactions from the wing spars (essentially intercomponent loads). The tailplane experiences a lift force and the drag can be distributed over the rear fuselage sections. The fuselage/tail unit will be in equilibrium under the forces shown in Figure 16.14. The internal loads may be obtained by considering cuts at each fuselage section, such as when analysing the wing, and working from the tailplane or nose towards the centre section/wing junction.

16.7.3   Example: Equilibrium Manoeuvre – Discretized Fuselage

In Chapter 12, an equilibrium manoeuvre example for a rigid aircraft was shown, involving the thrust and drag being out-of-line so that an iterative process was required to solve for the balanced condition. The case examined was for a load factor of 2.5 at 250 m/s EAS. The tailplane lift in the equilibrium manoeuvre was $L_T = -93.7$ kN (i.e. acting downwards). In the example here, a sample fuselage internal load calculation is undertaken for this flight condition.

To determine internal loads (see Chapter 5), the tailplane and rear fuselage are 'cut' away from the remainder of the aircraft at position AA in Figure 16.14 in order to expose the internal loads at this fuselage location, as shown in Figure 16.15. The subsystem under consideration is

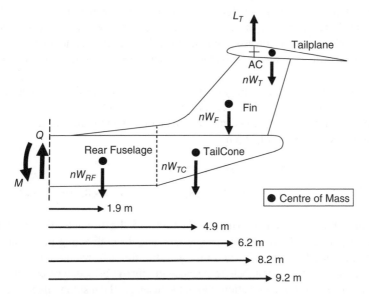

Figure 16.15   Rear fuselage loads in the equilibrium manoeuvre.

Table 16.2 Mass and inertia force per component.

	Tailplane	Fin	Tail cone	Rear fuselage
Mass (kg)	1000	900	400	3200
Inertia force (kN)	24.5	22.1	9.8	78.5

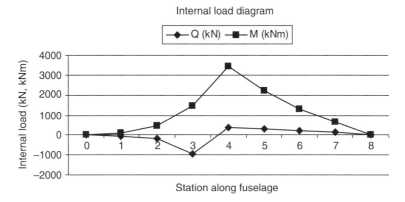

Figure 16.16 Internal load diagram for the fuselage/tail unit example.

divided into sections comprising the tailplane, fin, tail cone and rear fuselage; the tailplane lift force and the vertical inertia forces per component are shown on the diagram. The objective is to determine the shear force and bending moment at position AA in the rear fuselage section, as indicated in the figure; this will be an example of the analysis approach that could also be applied at other sections. There is no fuselage torque because the manoeuvre, and hence the loading, is symmetric. Axial loads would be present if the effects of aerodynamic drag, engine thrust, fore-and-aft acceleration or cabin pressurization loads were to be included. The component masses are tailplane m_T, fin m_F, tail cone m_{TC} and rear fuselage m_{RF}; values are shown in Table 16.2, together with the corresponding inertia forces. Therefore, as an example, for a load factor of 2.5 the inertia force on the fin will be $nW_F = nm_Fg = 2.5 \times (900 \times 9.81)/1000 = 22.1$ kN (downwards).

The rear fuselage/fin/tailplane subsystem shown will be in equilibrium under the action of the inertia forces, the tailplane lift force and the internal loads. Considering the equilibrium relationships of zero net vertical force and zero net moment about the cut, and recognizing that the tailplane lift acts downwards, then

$$Q = 24.5 + 22.1 + 9.8 + 78.5 - (-93.7) = 228.6 \text{ kN}$$
$$M = 24.5 \times 9.2 + 22.1 \times 6.2 + 9.8 \times 4.9 + 78.5 \times 1.9 - (-93.7 \times 8.2) = 1328 \text{ kN m}$$
$$(16.21)$$

Thus, examining the sign convention used in Figure 16.15, the fuselage will bend downwards at the tail (i.e. 'hog'), with the inertia forces and tailplane lift both acting in the same direction. However, even if the tailplane lift acted upwards, the fuselage would still bend downwards during a positive load factor manoeuvre because the inertia effects will be dominant. When the process is repeated for other positions along the fuselage, the resulting internal load diagram is shown in Figure 16.16; values calculated in the above example are at section 6. The assumed wing spar/fuselage attachment points are at locations 3 and 4 in this figure. Also, as expected, the internal loads are zero at each end of the aircraft.

16.7.4   Internal Loads for General Manoeuvres and Gusts

The above example was for an equilibrium manoeuvre. However, if the aircraft was subject to a more general acceleration, such as when undergoing a dynamic flight manoeuvre, encountering a gust or landing, the accelerations could differ from section to section. Internal loads could be calculated at any instant in time by using the appropriate tailplane (and maybe fuselage) lift as well as the accelerations at each mass section and any landing gear reaction. The general form of the internal load equations is similar to those for the wing.

16.8   Internal Loads – Continuous Turbulence Encounter

So far in this chapter, the generation of internal loads has been sought in the *time* domain using the output from simulations of steady/dynamic manoeuvres and discrete gust encounters. However, in the analysis of a continuous turbulence response, it is normally required to express the loads behaviour in the *frequency* domain in terms of spectral incremental loads quantities. In this section, the approach for calculating root-mean-square (RMS) loads from generalized coordinate responses will be outlined. It may be applied to continuous or discretized components but is defined using notation from the latter obtained in Section 16.6.4.

In Chapter 14, the frequency domain Power Spectral Density (PSD) approach was used to obtain the root-mean-square values of response parameters. However, it is now possible, by expressing Equation (16.20) in the frequency domain, to define transfer functions between each internal load, the vector of generalized coordinates and the gust velocity (neglecting any control input for simplicity). Thus, in the frequency domain (referring back to Chapter 14 for the methodology), it may be seen that the vector of shear forces is given by

$$\tilde{Q} = \left[-\omega^2 \mathbf{A}_M + i\omega \mathbf{B}_M + \mathbf{C}_M \right] \tilde{q} + \mathbf{R}_M w_{g0} = \mathbf{H}_{Qq} \tilde{q} + \mathbf{R}_M w_{g0} \qquad (16.22)$$

where $\mathbf{H}_{Qq}$ is the Frequency Response Function (FRF) matrix relating the internal load vector (shear force in this case) to the generalized response vector. However, since it was shown in Chapter 14 that these generalized responses could be related to the gust velocity via $\tilde{q} = H_{qg} w_{g0}$, then

$$\tilde{Q} = \mathbf{H}_{Qq} H_{qg} w_{g0} + \mathbf{R}_M w_{g0} = \left[\mathbf{H}_{Qq} H_{qg} + \mathbf{R}_M \right] w_{g0} = H_{Qg} w_{g0} \qquad (16.23)$$

Thus an FRF vector $\mathbf{H}_{Qg}$ relating the internal loads (shear forces in this case) directly to the gust velocity may be obtained; the dynamics of the linear aircraft are embedded in this expression. This approach allows the continuous turbulence PSD approach to be extended to consideration of internal loads; e.g. the root-mean-square incremental wing root shear force (and therefore relevant stresses) experienced in turbulence may be determined from the relevant internal load PSD.

Once the PSDs for internal loads are found, the statistical implications for fatigue life may be considered; e.g. the limit load may be found from the steady 1 *g* load and the limit turbulence intensity multiplied by the ratio of the root-mean-square load to the root-mean-square of gust velocity (see Chapters 14 and 23). Also, positions of the peaks in the PSD dictate the number of cycles of a particular stress level, which can be used with techniques such as rainflow counting and Miner's rule. This area is a complex process and other texts (Hoblit, 1988; Niu, 1988; Megson, 1999) should be consulted for the detail. Further comments on loads and loads processing are made in Chapter 23.

16.9   Loads Generation and Sorting to yield Critical Cases

Calculations need to be carried out for a large range of different loading actions, flight conditions (i.e. air speed, altitude) and centre of mass/mass cases within the design envelope. The aim is to determine the critical cases that will dictate the dimensioning of each part of the aircraft (see also Chapters 20 and 24). Many sets of internal loads values or time histories are generated from different manoeuvre and gust cases, with steady loads added to incremental results as appropriate. The loads are then used to obtain suitable one-dimensional, two-dimensional and multi-dimensional loads envelopes (explained below); this is part of a sorting process employed in order to determine correlated sets of loads that correspond to critical loadings at each location of interest on the aircraft.

The term 'correlated loads' is used to mean a set of loads that is consistent across the aircraft, e.g. extracted at the same instant of time, thus providing balanced loads for subsequent FE analysis. Note that a special approach is required to extract such correlated loads from turbulence response data in the frequency domain. These sorted loads are made available to the stress office or component loads customers. They may also be converted into equivalent loads acting on the reference axes for the FE model of components (see later in this chapter and Chapter 24).

16.9.1   One-dimensional Load Envelopes

A one-dimensional load envelope shows the maximum and minimum internal loads at each load station along a component such as a wing as part of the loads sorting process, as illustrated in Figure 16.17; the enclosed shaded area contains all the smaller and therefore non-critical values. Such a diagram may be drawn for each loading action (equilibrium, dynamic or ground manoeuvre, gust/turbulence etc.) and the overall worst cases then shown as a composite of all loading actions. It is not possible a priori to determine the relative effects of different load cases since that depends on many factors; however, the different loading actions can often yield quite similar results and are all important somewhere on the aircraft. Such an envelope would be

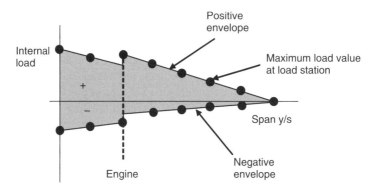

Figure 16.17   One-dimensional load envelope along the wing.

appropriate to use on parts of the aircraft where the design stresses only depend upon one of the 'MAST' quantities, e.g. the effect of the wing bending moment on stresses in parts of the wing cover skins dominated by bending.

16.9.2   Two-dimensional Load Envelopes

In the previous section, the one-dimensional load envelope allowed the worst case for a given internal load (e.g. the bending moment) to be seen at a particular load station. More often, the design stresses in structural elements within modern aerospace structures can only be adequately represented by a more complex combination of two 'MAST' loads.

As an example, for areas where shear stresses dominate, such as spar caps / webs, a step improvement in fidelity can be achieved by considering a two-dimensional envelope of shear and torque since these both contribute to the shear stresses. It is therefore sometimes useful to plot these loads quantities against each other at a particular load station for all the load cases; such a process will help to identify the sets of correlated loads that correspond to the critical design cases for parts of the wing box not dominated by bending.

To illustrate this, consider the approach shown earlier in Section 16.5.4 for obtaining the steady plus incremental loads in response to a '1-cosine' gust. In this section, four different wavelengths of 20, 50, 100 and 200 m will be considered for a 10 m/s EAS '1-cosine' gust at the same flight condition of 150 m/s EAS at 14 000 ft. The incremental wing root shear force and torque for the mode with wing bending dominant are added to the steady internal loads ($n = 1$). When the torque and shear force are plotted, with the locus of the line following time, the result in Figure 16.18 is obtained; the two phases of the load variation involving the

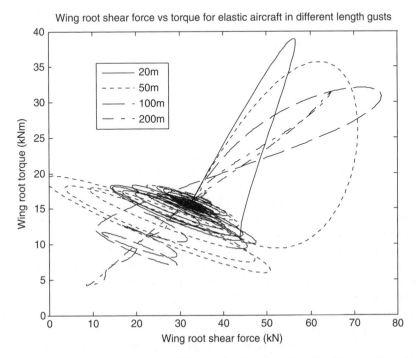

Figure 16.18 Variation of total wing root torque against shear force for a flexible aircraft in a '1-cosine' gust of different lengths – heave/pitch model with a wing bending mode (3 Hz/4%).

response to the initial gust input and the decay of the flexible mode following the input may be seen to give two zones in the plot. The shape of these loci would obviously be different at other load stations, for other loading actions, for other modes and for aircraft with more than one flexible mode.

These two-dimensional plots can be generated using a superposition of the many thousands of loads cases investigated (i.e. equilibrium, dynamic, ground and gust) for different flight conditions and mass/centre of mass values to show a complete picture of the torque/shear force load combinations that the aircraft is expected to experience. A 'two-dimensional load envelope' may be defined to embrace this composite plot and so allow the most severe combinations of the shear force and torque to be investigated. It is important that the load sets defined are correlated, i.e. will give rise to a balanced load set. Such composite two-dimensional plots (sometimes called 'potato' plots) are not always simple in shape but depend upon the aircraft modes involved, load station, etc. Clearly the detailed time responses that lie within this load envelope are unimportant and would not need to be plotted. Similar plots could involve the variation of bending moment and torque and also the variation of the wing bending moments for vertical and fore-and-aft bending.

To improve the fidelity of design case sorting/selection further is clearly possible by more directly representing the more complex dependencies of stress within each structural element on the 'MAST' loads. However, this would considerably expand the number of 'design cases' analyzed. At this point it may begin to be worthwhile reformulating the whole loads solution in terms of structural DoF, again going beyond the scope of this book.

Finally, it should be pointed out that in some cases, more than two internal loads may be involved in dictating the stress in a particular component, e.g. critical cases for the wing front spar may occur due to a combination of engine loads and wing shear, bending moment and torque. It is possible to derive a multi-dimensional transformation matrix between multiple loads quantities and approximate element stresses and this matrix may be used in the loads process to calculate the approximate stresses and then select cases based on the maximum and minimum of these load quantities.

16.10 Aircraft Dimensioning Cases

Consider now the importance of different loadings on the aircraft design (Howe, 2004). Shear tends to be of most significance when combined with torque, because these effects generate shear loading in the wing box cover skins and the spar webs (Donaldson, 1993; Megson, 1999; Sun 2006). Shear and torque are also relevant to the inner wing spar and landing gear attachment design, with ground loads such as landing being of most importance. For the outer wing spar design, the flight loads are usually larger, though dynamic inertia loading on landing can be important in some aircraft. Wing torque is less important than wing bending, with bending loads tending to be most important for the design of spars and of top wing box cover skins (to avoid buckling of the stiffened stringer panels under compression loading in upwards bending); this is the case examined under the ultimate load test (see Chapter 25). Usually the gust/turbulence, dynamic and 2.5 g equilibrium manoeuvre load envelopes are fairly close in magnitude, depending upon wing loading (lift per area). Downwards bending of the wing is less significant, with bottom cover skins tending to be designed by fatigue/damage tolerance issues (the tension loading present due to upwards bending). The presence of a wing-mounted engine introduces a step change in internal loads due to thrust and inertia effects (see the earlier example) and possibly also nacelle aerodynamic effects. Also, the envelope of the vertical load factor (i.e. acceleration) is important, with flight and ground loadings causing different levels of acceleration as the different frequency content of the excitation excites different wing modes. High levels of acceleration (typically 20 g at the wing tip) can introduce large local loads due to system component attachments, fuel inertia loading, etc. Other texts (Niu, 1988; Donaldson, 1993; Megson, 1999) should be sought for further understanding of aircraft loads.

16.11   Stresses derived from Internal Loads – Complex Load Paths

In classical stress analysis for simple slender structures where load paths are well defined, the internal loads found from the external aerodynamic and inertia loads present in various manoeuvres and gust encounters could be used directly to estimate stresses (see Chapter 5). There is therefore no problem of terminology in regard to the use of the terms 'internal' and 'external' loads.

However, in the analysis of complex built-up aircraft structures where load paths are ill-defined, the determination of stresses is far less straightforward. The shear forces, bending moments etc. could be converted into external loads acting, for example, on the reference axis of the component (e.g. the wing) and making up a balanced load case. Alternatively, the internal loads could be converted into external 'nodal' loads distributed as realistically as possible directly on to the load-carrying nodes within the structural FE model; such a distribution of loads could be assisted by use of detailed mass and aerodynamic data.

Then the behaviour of different aircraft components could be analysed using the FE method, with the analysis yielding further 'internal' loads or stresses acting on structural elements such as wing spars and ribs. This issue is also addressed briefly in Chapters 20 and 24; the explanation given here is only one of several approaches that can be used for this complex task.

There is clearly some room for confusion over terminology since what are referred to as 'internal loads' in this book are sometimes called 'external loads' in industry, since they are used to generate equivalent external loads acting on the nodes of the FE model.

16.12   Examples

Note that it would be helpful for the reader to carry out some of the examples in Chapter 5 first to ensure that the basic principles for determining internal loads are well understood.

1. An aircraft has the following data: $m = 48\,000$ kg, $S_W = 160$ m^2, $C_{MOW} = -0.015$, $c = 4$ m, $l_W = 0.4$ m and $l_T = 9$ m. Section AA is 2 m aft of the centre of mass and the section of fuselage/tailplane aft of AA has a mass distribution that can be approximated by lumped masses of 5000, 3000 and 2000 kg at positions that are 2, 4 and 6 m respectively aft of AA. Determine the tailplane lift and the bending moment at AA for a flight case where the air speed is 150 m/s EAS and the load factor $n = 2$.
   [26. kN, 485.1 kN m]
2. For the aircraft in Example 1 above, the wing lift acts at quarter chord, the net wing/engine inertia force acts at 40% chord and the front and rear spars lie at 20% and 60% chord. The wing and engine mass is 16 000 kg. Determine the intercomponent forces acting between the wings and centre fuselage at the front and rear spars.
   [Front 558.2 kN and rear –42.6 kN]
3. An aircraft of mass 30 000 kg has the tapered wing planform shown in Figure 16.19. Each wing has a structural mass of 2000 kg and a fuel mass of 2000 kg, with the fuel located only

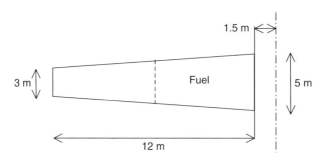

Figure 16.19

in the inboard half of the wing, as shown in the figure. The wing/fuel mass may be assumed to be uniformly distributed over the relevant wing area whereas the lift may be assumed to be uniformly distributed over the wing and centre fuselage area (i.e. ignore the tailplane effects). The aircraft experiences an $n = 3$ manoeuvre. Determine the shear force and bending moment at the wing root assuming that the structure and loading are continuously distributed. With the wing divided into eight strips, determine and sketch the shear force and bending moment distribution. Would these internal loads increase or reduce if the same manoeuvre were performed with wings (and therefore the aircraft) empty of fuel?
[Continuous: wing root shear force 264 kN, bending moment 1606 kN m; Discretized: 274 kN, 1608 kN m]

4. For the aircraft in Example 3 above, consider the engines moved from the rear fuselage to the wing. Each engine mass is 2000 kg and the engine is placed 3.75 m outboard of the fuselage, 2 m forward of the wing centre line and 1 m below it. The engine thrust is 20 kN. Determine and sketch the changes to the internal load diagrams that will result from moving the engines in this way for the full fuel case.
[Wing root shear force 205 kN, bending moment 1387 kN m plus torque 98 kN m]

5. During a heavy landing, at the instant when the vertical force on the landing gear leg reaches a maximum, an aircraft is decelerating in the vertical direction at $2\,g$ so the load factor is $n = 3$. The spanwise dimensions of the aircraft are shown in Figure 16.20 and both the wing structure mass and lift may be assumed to be uniformly distributed across the span (including the section of wing enclosed by the fuselage). The mass of the complete aircraft is 100 000 kg, the wing mass is 15 000 kg and each engine has a mass of 5000 kg. The attitude and speed of the aircraft are such that the wing lift is equal to 90% of the aircraft

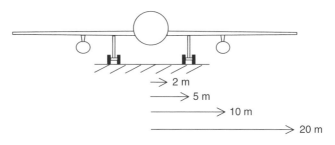

Figure 16.20

weight. Determine the landing gear reaction force and the bending moment experienced by the wing (a) at the section where the landing gear is attached and (b) at the wing root.
[1030 kN, (a) 506 kN m, (b) 3701 kN m]

6. The aircraft, whose rear fuselage and tail are shown in Figure 16.21, experiences a manoeuvre such that, at some instant of time, the aircraft centre of mass accelerates upwards at 10 m/s^2 (so effectively $n = 2$) and simultaneously the aircraft accelerates nose up in pitch at 1 rad/s^2 about the centre of mass. The aircraft may be assumed to be in a horizontal flight attitude at this instant. The rear fuselage mass of 1200 kg is divided equally at stations A, B and C and the fin and tailplane have masses of 200 and 100 kg respectively. The two engines have a total mass of 1500 kg and develop a net thrust of 30 kN. The net tailplane lift in the manoeuvre is 5 kN and the rear fuselage, fin and tailplane drag forces are 1, 0.5 and 0.5 kN respectively. Determine the inertia loads and draw an FBD. Estimate the bending moment, shear force and axial force defined on the centre line at the fuselage section XX which lies m aft of the centre of mass. Note that each section of the structure experiences a different acceleration. Assume for convenience that $g = 10 \text{ m/s}^2$.

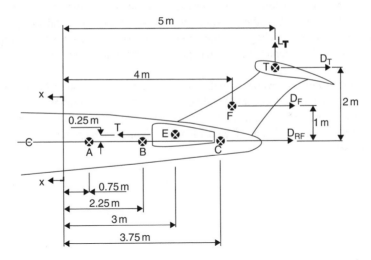

Figure 16.21

[86.4 kN m, 37.5 kN and 28. kN]

7. An aircraft of mass 12 000 kg has the wing planform shown in Figure 16.22, with the mass of each wing (outboard of the fuselage) being 2000 kg and uniformly distributed over the span. The radius of gyration of the aircraft (without wings) in roll is 1 m. The aircraft is flying at 150 m/s EAS when the ailerons are deflected through $5°$. The aileron lift curve slope over the local aileron span is $a_C = 1.5/\text{rad}$ and the aileron lift may be assumed to be distributed uniformly over the aileron span. Determine the rolling moment, the aircraft roll moment of inertia, the rolling acceleration and the incremental wing root bending moment in the manoeuvre at the instant of applying the ailerons.
[101 kN m, 115 300 kg m^2, 0.88 rad/s^2, 8.5 kN m]

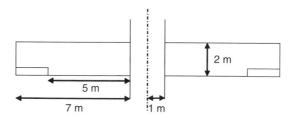

Figure 16.22

8. Obtain the expression for the wing root shear force, equivalent to that for the bending moment presented in Section 16.5.2.3 of this chapter, for the flexible aircraft in an equilibrium manoeuvre.

9. Obtain the expression for the incremental wing root shear force, equivalent to that for the bending moment presented in Section 16.5.3.3 of this chapter, for the flexible aircraft in a gust encounter.

10. Modify the programs developed in the example section of Chapter 12 to include evaluation of the wing root shear force and bending moment in an equilibrium manoeuvre for a rigid and flexible aircraft. Use the programs to determine these internal loads for the aircraft parameters used in that chapter at the corner points of the manoeuvre envelope developed in Example 1 in Chapter 12. Which point on the manoeuvre envelope has the most severe loads?

11. Modify the programs developed in the example section of Chapter 14 to include evaluation of the wing root shear force and bending moment in a '1-cosine' gust for a rigid and flexible aircraft. Use the programs to determine these internal loads for the aircraft parameters used in that chapter and hence the gust length at which each of these loads has a maximum absolute value.

17

Vibration of Continuous Systems – Finite Element Approach

In Chapter 3, the vibration of continuous systems was considered using the Rayleigh–Ritz approach, based on describing the motion with a summation of assumed shapes. This approximate analysis led to differential equations of motion expressed in terms of the unknown coefficients (or generalized coordinates) that multiply each shape. Standard MDoF analysis approaches as described in Chapter 2 could then be employed to solve for natural frequencies, normal modes and the response to various forms of excitation.

However, in this chapter the vibration of continuous systems will be approached using a physical discretization of the system. An early approach to this discretization process employed flexibility influence coefficients (Rao, 1995), but this was superseded by the Finite Element (FE) method (NAFEMS, 1987; Cook *et al.*, 1989; Rao, 1995). The move towards using the Finite Element method has been aided by the availability of many dedicated software packages.

17.1 Introduction to the Finite Element Approach

The FE method involves dividing the structure into a number of so-called 'finite elements' interconnected at discrete points known as 'nodes'. The unknowns in the analysis are the displacements and, where relevant, rotations at the nodes. The deformation within each element is approximated in terms of these unknowns using a polynomial representation. The behavior of the structure is represented by determining element stiffness and mass matrices and assembling global matrices for the structure. The nodal displacements then become the unknowns for which the overall equations of motion are formulated, so that the continuum structure is reduced to a discretized one. Damping can be incorporated into the model, but only as an approximation based on past experience or measured data.

The Finite Element method may be used for static, dynamic and thermal problems. In this book, only linear analysis will be covered, where strains and displacements/rotations are small and material properties are linear elastic. However, advanced analyses, with geometric non-linearities (large strain, large displacement/rotation or boundary conditions) and/or material non-linearity, may also be performed (Prior, 1998).

Introduction to Aircraft Aeroelasticity and Loads, Second Edition. Jan R. Wright and Jonathan E. Cooper.
© 2015 John Wiley & Sons, Ltd. Published 2015 by John Wiley & Sons, Ltd.

The Finite Element approach is widely used in aircraft aeroelastic and loads calculations. For typical commercial aircraft with high aspect ratio wings, there are two main approaches to FE modelling of the thin-walled 'box-like' nature of the major aircraft components:

- use a 'beam-like' model with beam elements of known bending, torsion, axial and shear properties or
- use a full structural model with membrane/bar or shell/beam elements, with the 'box-like' nature of the component retained.

Later, the application of the Finite Element approach to more complex structural representations will be considered briefly. However, in this chapter, only the 'beam-like' representation will be discussed in any detail. An example of a beam modelled with two elements is shown in Figure 17.1.

The FE approach is firstly to determine the dynamic properties of each element in the form of *element* stiffness and mass matrices and then to assemble all the elements to form *global* (or overall structure) mass and stiffness matrices from which modes and responses may be determined. The element assembly process satisfies exact *compatibility* of displacements and rotations between elements (i.e. the nodes common to adjacent elements have the same displacements/rotations); however, *equilibrium* is normally only satisfied in an approximate manner over the entire structure except for very simple problems.

The element stiffness and mass matrices are obtained by an energy approach, assuming a form for the displacement variation *within* the element. This is somewhat similar to the

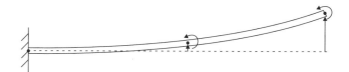

Figure 17.1   Built-in beam represented by two elements.

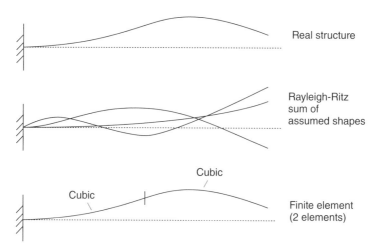

Real structure

Rayleigh-Ritz sum of assumed shapes

Cubic

Cubic

Finite element (2 elements)

Figure 17.2   Complete versus piecewise displacement representation.

Rayleigh–Ritz approach except that in the latter the displacement variation over the *whole* structure is represented by a summation of assumed shapes. Thus the Finite Element methodology is rather like a 'piecewise' Rayleigh–Ritz approach, as illustrated in Figure 17.2, where three assumed shapes are compared with two finite elements, each having an assumed cubic displacement variation. The advantages of the Finite Element method are that more elements may be used in regions where the displacement and stress are expected to vary more rapidly and that more complex problems may be handled.

17.2 Formulation of the Beam Bending Element

17.2.1 Stiffness and Mass Matrices for a Uniform Beam Element

A typical uniform beam element of length L, mass per unit length μ and flexural rigidity EI is shown in Figure 17.3. For simplicity, bending in only one plane, with no shear, axial or torsion deformation, is considered. It should be noted that in this chapter, the bending deformation will again be denoted using the symbol z.

17.2.1.1 Element Shape Functions

The nodal 'displacements' (it is usually implied that some are rotations) shown in Figure 17.3 are denoted by the column vector $d = \{d_1\ d_2\ d_3\ d_4\}^{\mathrm{T}}$. In order to write the elastic potential and kinetic energy terms for the element, the variation of displacement within the element needs to be expressed as a function of the nodal displacements. The variation of the transverse displacement $z(y)$ along the element may be expressed as a cubic polynomial in y, namely

$$z = a_0 + a_1 y + a_2 y^2 + a_3 y^3 \tag{17.1}$$

where $a_0, \dots, a_3$ are unknown coefficients that must be determined such that the polynomial matches the nodal displacements at the ends $y = 0, L$. When the displacements and slopes at each end of the beam are equated to the nodal displacements $d_1, \dots, d_4$ then the following equations are found:

$$y = 0 \quad \text{displacement } d_1 = a_0,$$

$$y = 0 \quad \text{slope } d_2 = a_1,$$

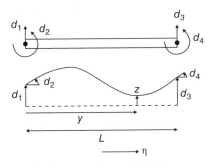

Figure 17.3 Two-node beam bending element.

$$y = L \quad \text{displacement } d_3 = a_0 + a_1 L + a_2 L^2 + a_3 L^3, \tag{17.2}$$

$$y = L \quad \text{slope } d_4 = a_1 + 2a_2 L + 3a_3 L^2.$$

Equations (17.2) may be solved to yield expressions for the polynomial coefficients $a_0, \ldots, a_3$ in terms of the nodal displacements $d_1, \ldots d_4$. The final polynomial then takes on the following form:

$$z = N_1 d_1 + N_2 d_2 + N_3 d_3 + N_4 d_4 = \mathbf{N}^T \mathbf{d} \tag{17.3}$$

where $\mathbf{N}$ is a column vector of the so-called 'shape functions' $N_1, \ldots, N_4$, each being a cubic polynomial in y. For example, it can be shown that the shape functions N_1, N_2 are given by the polynomials

$$N_1 = \frac{1}{4}(1-\eta)^2(2+\eta) \quad N_2 = \frac{L}{8}(1-\eta)^2(1+\eta) \tag{17.4}$$

where $\eta = 2y/L - 1, +1 \geq \eta \geq -1$ is a non-dimensional local coordinate commonly used in FE analysis (see Figure 17.3); the shape functions N_3, N_4 are very similar. The shape functions, shown in Figure 17.4, have distinctive shapes in that N_k is the polynomial corresponding to $d_k = 1$ and $d_j = 0 \ \forall \ j \neq k$.

The fact that the displacement varies as a cubic function of y along the element means that the bending moment (proportional to curvature), and therefore the bending stress, will vary linearly. Thus sufficient elements must be used to allow the exact stress variation along the beam to be represented reasonably well by a piecewise linear approximation. The accuracy of the FE method depends upon the number and type of elements used e.g. a quintic polynomial is used for the higher order three-node beam element.

17.2.1.2 Element Equation of Motion

In the FE approach, forces and moments may only be applied to the element at the nodes, as shown in Figure 17.5; these are termed nodal 'forces' (usually implied that some are moments) $\mathbf{P} = \{P_1 \ P_2 \ P_3 \ P_4\}^T$. The element equation relating nodal forces, displacements and accelerations will then be sought. The element mass and stiffness matrices are known and may be assembled for the structure.

Lagrange's equations are used to determine the equation of motion for the beam element, with nodal displacements $\mathbf{d}$ acting as the coordinates, so ensuring that equilibrium applies

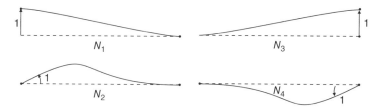

Figure 17.4   Shape functions for a two-node beam element.

Figure 17.5 Nodal forces for a two-node beam element.

on average over the element. The elastic potential and kinetic energy terms are the same as those used in Chapter 3, but with the revised displacement description in Equation (17.3) being used. Thus the elastic potential energy is

$$U = \frac{1}{2}\int_0^L EI\left(\frac{\partial^2 z}{\partial y^2}\right)^2 dy = \frac{1}{2}\int_0^L EI\left(d^T N''\right)\left(N''^T d\right) dy \quad \text{so} \quad U = \frac{1}{2}d^T\left[\int_0^L EI\left(N'' N''^T\right) dy\right] d$$

(17.5)

where the shorthand notation $' = \partial/\partial y$ and $'' = \partial^2/\partial y^2$ is used. The kinetic energy is

$$T = \frac{1}{2}\int_0^L \mu \dot{z}^2 \, dy = \frac{1}{2}\int_0^L \mu\left(\dot{d}^T N\right)\left(N^T \dot{d}\right) dy \quad \text{so} \quad T = \frac{1}{2}\dot{d}^T\left[\int_0^L \mu\left(NN^T\right) dy\right] \dot{d}$$

(17.6)

Here the matrix approach introduced in Chapter 3 has been used. The incremental work done by the applied nodal forces acting through the nodal displacements will be given by

$$\delta W = P_1 \delta d_1 + P_2 \delta d_2 + P_3 \delta d_3 + P_4 \delta d_4 = P^T \delta d$$

(17.7)

When Lagrange's equations are employed, the differential equation of motion for the element is

$$m\ddot{d} + kd = P$$

(17.8)

where $\mathbf{m}$, $\mathbf{k}$ are the 4×4 element mass and stiffness matrices respectively, given by

$$\mathbf{m} = \left[\int_0^L \mu\left(NN^T\right) dy\right] \quad \text{and} \quad \mathbf{k} = \left[\int_0^L EI\left(N'' N''^T\right) dy\right]$$

(17.9)

Introducing the relevant shape function polynomials $N_1, \ldots, N_4$ into Equations (17.9) and performing the matrix multiplications and integrations (including the transformation from y to η), it may be shown that, for a *uniform* beam element, these matrices are given by

$$\mathbf{m} = \frac{\mu L}{420}\begin{bmatrix} 156 & 22L & 54 & -13L \\ 22L & 4L^2 & 13L & -3L^2 \\ 54 & 13L & 156 & -22L \\ -13L & -3L^2 & -22L & 4L^2 \end{bmatrix} \quad \mathbf{k} = \frac{EI}{L^3}\begin{bmatrix} 12 & 6L & -12 & 6L \\ 6L & 4L^2 & -6L & 2L^2 \\ -12 & -6L & 12 & -6L \\ 6L & 2L^2 & -6L & 4L^2 \end{bmatrix}$$

(17.10)

Here it may be seen that both the matrices are symmetric and fully-populated. The matrices quoted are only in this precise form provided that the ordering and sign convention of the element displacements and rotations are preserved.

17.2.1.3   Consistent and Lumped Mass Models

The mass representation used above is known as a '**consistent**' mass model because it is the most accurate, matching the kinetic energy corresponding to the deformation within the element. An alternative mass representation is the more simple '**lumped**' mass model, where the element mass matrix is diagonal. For a two-node beam element, half the mass is 'lumped' at each node. The rotational terms on the diagonal may be zero or else take on some intermediate value to allow for rotary inertia effects (Cook *et al.*, 1989). The lumped mass matrix is

$$\mathbf{m}_{\text{Lumped\_No\_Rotary\_Inertia}} = \frac{\mu L}{24}\text{diag}\begin{bmatrix} 12 & 0 & 12 & 0 \end{bmatrix}$$

or   (17.11)

$$\mathbf{m}_{\text{Lumped\_Rotary\_Inertia}} = \frac{\mu L}{24}\text{diag}\begin{bmatrix} 12 & L^2 & 12 & L^2 \end{bmatrix}$$

The consistent mass matrix is the most rigorous approach to handling *structural* mass i.e. that associated with the load carrying structure. However, the *non-structural* mass present in an aircraft (e.g. fuel, payload, systems etc.) will need to be represented by a lumped mass model. Structural mass may in fact be represented by a consistent or lumped mass model. By employing sufficient elements, the errors involved in using a lumped mass approximation will be considered small.

17.2.1.4   Kinematically Equivalent Nodal Forces

When applied forces are distributed over the structure then, as part of the FE idealization, they need to be replaced by forces acting at the nodes themselves, i.e. so-called 'kinematically equivalent nodal forces', defined on the basis that they do equivalent work to the true distributed forces when the element deforms. For the two-node beam element example, if a uniformly distributed force of q per length is applied over the element, as shown in Figure 17.6, then the distributed load is represented by nodal loads. The vector of 'kinematically equivalent nodal forces' (Cook *et al.*, 1989) for this example is given by

$$P_{\text{KinEq}} = \frac{qL}{2}\left\{ 1 \quad \frac{L}{6} \quad 1 \quad -\frac{L}{6} \right\}^{\text{T}}$$ (17.12)

Similar equivalent nodal forces account for other effects distributed over the element, such as thermal loading, initial strains, etc. When an aircraft component is idealized using FE beam elements, then the inertia and aerodynamic loads will tend to be added at each node and so are not kinematically equivalent; by employing sufficient elements, the approximation is small.

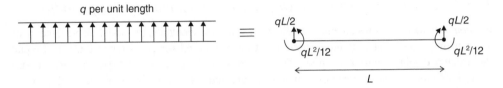

Figure 17.6   Kinematically equivalent nodal forces for a two-node beam element under distributed loading.

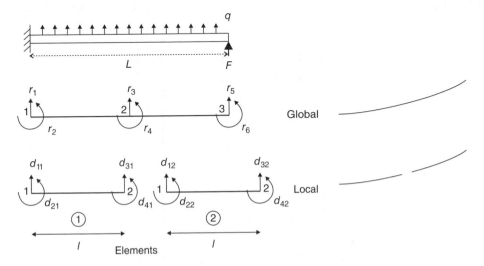

Figure 17.7   Built-in beam assembly with two elements.

17.3   Assembly and Solution for a Beam Structure

Once the element mass and stiffness matrices have been determined, they may be assembled to generate the global mass and stiffness matrices for the structure. This process is automated within any FE package, with the user providing the structure geometry, properties, boundary conditions, and the element type and topology (i.e. how the elements are interconnected).

17.3.1   Element and Structure Notation

For simplicity, the assembly process will be shown longhand using the two-element/three-node built-in beam example in Figure 17.7; the structure (or 'global') displacements at the three nodes are given by $r = \{r_1, r_2, \ldots, r_6\}^{\mathrm{T}}$ and the structural forces by $R = \{R_1, R_2, \ldots, R_6\}^{\mathrm{T}}$. The element (or 'local') nodal displacements are given by d_1, d_2, where the subscripts 1 and 2 refer to the two elements, and the element nodal force sets by P_1, P_2. The structure forces are either *applied* (e.g. applied point loads at nodes, distributed loads represented by kinematically equivalent nodal forces) or *reactive* (supplied via any supports). Note that nodal forces are a combination of external loads and the equal and opposite internal forces acting between the elements.

17.3.2 Element and Structure Displacements – Imposing Compatibility

For *compatibility* of displacements between the assembled elements, the nodal displacements of all elements meeting at a common node must be equal, and for consistency must also equal the structure displacements at that node. Thus, referring to Figure 17.7, the displacements at nodes $\{1\ 2\}$ for elements 1 and 2 must map on to structure nodes $\{1\ 2\}$ and $\{2\ 3\}$ respectively, so

$$\{d_{11}\ d_{21}\ d_{31}\ d_{41}\} = \{r_1\ r_2\ r_3\ r_4\} \quad \text{for element 1}$$
$$\{d_{12}\ d_{22}\ d_{32}\ d_{42}\} = \{r_3\ r_4\ r_5\ r_6\} \quad \text{for element 2}$$

(17.13)

where d_{ij} is the ith nodal displacement for the jth element. Alternatively, in matrix form,

$$d_1 = \begin{Bmatrix} d_{11} \\ d_{21} \\ d_{31} \\ d_{41} \end{Bmatrix} = \begin{bmatrix} 1 & 0 & 0 & 0 & 0 & 0 \\ 0 & 1 & 0 & 0 & 0 & 0 \\ 0 & 0 & 1 & 0 & 0 & 0 \\ 0 & 0 & 0 & 1 & 0 & 0 \end{bmatrix} \begin{Bmatrix} r_1 \\ r_2 \\ r_3 \\ r_4 \\ r_5 \\ r_6 \end{Bmatrix} = \mathbf{\Gamma}_1 r$$

$$d_2 = \begin{Bmatrix} d_{12} \\ d_{22} \\ d_{32} \\ d_{42} \end{Bmatrix} = \begin{bmatrix} 0 & 0 & 1 & 0 & 0 & 0 \\ 0 & 0 & 0 & 1 & 0 & 0 \\ 0 & 0 & 0 & 0 & 1 & 0 \\ 0 & 0 & 0 & 0 & 0 & 1 \end{bmatrix} \begin{Bmatrix} r_1 \\ r_2 \\ r_3 \\ r_4 \\ r_5 \\ r_6 \end{Bmatrix} = \mathbf{\Gamma}_2 r \quad (17.14)$$

The jth element 'maps' on to the structure via $d_j = \mathbf{\Gamma}_j r$, where $\mathbf{\Gamma}_j, j = 1, 2$ is the assembly matrix.

17.3.3 Assembly of the Global Stiffness Matrix – Imposing Equilibrium

The imposition of *equilibrium* at each structure node is essentially a statement that the structure force is 'shared' between the elements at that node and when combined with the element load/displacement relationships will yield the structure load/displacement relationship based on assembled element stiffness matrices. More formally, equilibrium is imposed because the incremental work done by the two equivalent sets of loads moving through the corresponding incremental displacements must be equal, so

$$\delta W = \delta r^{\mathrm{T}} R = \sum_{j=1}^{2} \delta d_j^{\mathrm{T}} P_j \quad (17.15)$$

If the mass terms are ignored at this stage, the relationship between the nodal forces and displacements for the two elements is

$$P_j = \mathbf{k}_j d_j \quad j = 1, 2 \quad (17.16)$$

where $\mathbf{k}_1$, $\mathbf{k}_2$ are the 4×4 element stiffness matrices derived earlier. What is being sought is the global force/displacement relationship for the structure, namely

$$\mathbf{R} = \mathbf{K}_r r \tag{17.17}$$

where $\mathbf{K}_r$ is the 6×6 structure stiffness matrix (prior to boundary conditions being imposed). Now, combining Equations (17.15) to (17.17) with the compatibility relationship in Equation (17.14), and simplifying, yields

$$\delta W = \delta r^{\mathrm{T}} \mathbf{K}_r r = \sum_{j=1}^{2} \delta \left(\mathbf{\Gamma}_j r \right)^{\mathrm{T}} \left(\mathbf{k}_j \mathbf{\Gamma}_j r \right) = \delta r^{\mathrm{T}} \left[\sum_{j=1}^{2} \mathbf{\Gamma}_j^{\mathrm{T}} \mathbf{k}_j \mathbf{\Gamma}_j \right] r \tag{17.18}$$

and so

$$\mathbf{K}_r = \sum_{j=1}^{2} \mathbf{\Gamma}_j^{\mathrm{T}} \mathbf{k}_j \mathbf{\Gamma}_j \tag{17.19}$$

In practice, the complete assembly matrices $\mathbf{\Gamma}_j$ are not stored, and the matrix operation in Equation (17.19) is not actually carried out, because of the considerable number of zero values present in the assembly matrices. Instead, it may be seen that the effect of Equation (17.19) is that the element stiffness matrices are 'added into' the structure stiffness matrix in positions corresponding to the 'mapping' between the structure and element displacements, i.e. the so-called 'element topology' defined by the nonzero values in the assembly matrices.

Consider the two-element example in Figure 17.7 with the length of *each* element being taken as l for convenience. Using the element stiffness matrix derived earlier in Equation (17.10) then Equation (17.18) yields the assembled 6×6 structure stiffness matrix

$$\mathbf{K}_r = \frac{EI}{l^3} \begin{bmatrix} 12 & 6l & -12 & 6l & 0 & 0 \\ 6l & 4l^2 & -6l & 2l^2 & 0 & 0 \\ -12 & -6l & 12+12 & -6l+6l & -12 & 6l \\ 6l & 2l^2 & -6l+6l & 4l^2+4l^2 & -6l & 2l^2 \\ 0 & 0 & -12 & -6l & 12 & -6l \\ 0 & 0 & 6l & 2l^2 & -6l & 4l^2 \end{bmatrix} \tag{17.20}$$

On careful examination, it may be seen that the first element stiffness matrix appears in rows/columns 1–4 and that the second element stiffness matrix is added into rows/columns 3–6; this matrix structure is defined by the assembly matrices. An equivalent result may be obtained for the assembled mass matrix.

17.3.4 Matrix Equation for the Assembled Structure

The mass matrix is included into the assembly process by adding the inertia terms from Equation (17.8) into Equation (17.16). The final equation of motion for the assembled elements in the structure is

$$\mathbf{M}_r \ddot{r} + \mathbf{K}_r r = \mathbf{R} \tag{17.21}$$

where $\mathbf{M}_r$ is the structure mass matrix and R represents all the assembled external applied forces.

17.3.5 Solution Process for the Assembled Structure

Once the structure mass and stiffness matrices have been assembled, the solution can proceed as follows:

a. The 'boundary conditions' need to be defined (corresponding to zero or prescribed nodal displacements). Since the beam is built in at node 1, as shown in Figure 17.7, the boundary conditions are $r_1 = r_2 = 0$. The forces R_1, R_2 are the corresponding reaction forces required to prevent support translation or rotation.
b. The applied loads need to be defined at unconstrained nodes 2 and 3. Loads may be applied directly at the nodes or be distributed over the structure.
c. Once the boundary conditions and applied loads are defined, the solution of the structure equations is best seen by partitioning the equation of motion in Equation (17.21) so as to separate out the equations for reactions and those for unknown responses, namely

$$\begin{bmatrix} \mathbf{M}_{aa} & \mathbf{M}_{ab} \\ \mathbf{M}_{ba} & \mathbf{M}_{bb} \end{bmatrix} \begin{Bmatrix} \ddot{r}_a \\ \ddot{r}_b \end{Bmatrix} + \begin{bmatrix} \mathbf{K}_{aa} & \mathbf{K}_{ab} \\ \mathbf{K}_{ba} & \mathbf{K}_{bb} \end{bmatrix} \begin{Bmatrix} r_a \\ r_b \end{Bmatrix} = \begin{Bmatrix} R_a \\ R_b \end{Bmatrix} \qquad (17.22)$$

where a and b refer to the partitioned quantities. In this equation, r_a are the known (or prescribed) support displacements (i.e. r_1, $r_2 = 0$ in the example), R_a are the corresponding unknown reactions at the supports (or the forces required to impose any prescribed displacements), R_b are the known applied forces and r_b are the corresponding unknown displacements. Once the partition has been defined, the solution can proceed. The second equation in Equation (17.22) may be written

$$\mathbf{M}_{ba}\ddot{r}_a + \mathbf{M}_{bb}\ddot{r}_b + \mathbf{K}_{ba}r_a + \mathbf{K}_{bb}r_b = R_b \qquad (17.23)$$

and assuming that $r_a = 0$ (i.e. fixed support), then Equation (17.23) may be rewritten as

$$\mathbf{M}_{bb}\ddot{r}_b + \mathbf{K}_{bb}r_b = R_b \qquad (17.24)$$

where this is a 4 DoF set of equations in the example. For a static problem where R_b is known, then Equation (17.24) may be solved for the unknown displacements r_b. The first equation in Equation (17.22) will then yield the corresponding support reactions R_a if required. If there is no applied force, then the normal modes of the structure may be determined using classical matrix eigenvalue methods (see Chapter 2).

Consider now the solution of the three-node built-in beam example used above in Figure 17.7. Assume that the beam has length $s = 2l = 10$ m, flexural rigidity $EI = 4 \times 10^6$ Nm^2 and mass per length $\mu = 100$ kg/m and that the two elements have equal length $l = 5$ m. Two examples will be considered, namely a static load and a normal modes analysis. The effect of increasing the number of elements will be considered.

17.3.5.1 Static Loading Analysis: Two Elements

Consider an applied force $F = 1000$ N acting upwards at the tip and a distributed load of $q = 100$ N/m acting upwards over the entire length. Using the earlier result for kinematically equivalent loads, the applied force vector is

$$R^T = \{R_1 \ R_2 \ 0 \ 0 \ F \ 0\} + \frac{ql}{2}\left\{1 \ \frac{l}{6} \ 1 \ -\frac{l}{6} \ 0 \ 0\right\} + \frac{ql}{2}\left\{0 \ 0 \ 1 \ \frac{l}{6} \ 1 \ -\frac{l}{6}\right\} \quad (17.25)$$

Here the first term in the expression is for the point loads, with R_1, R_2 being the unknown reactions, whereas the second and third terms are the kinematically equivalent forces for the two elements; the component of the distributed load at the support will be ignored as it acts at a restrained point.

Once the equations are partitioned and boundary conditions imposed, the final load/displacement equation is

$$\mathbf{K}_{bb}\mathbf{r}_b = \mathbf{R}_b \quad \text{or} \quad \frac{EI}{l^3}\begin{bmatrix} 12+12 & -6l+6l & -12 & 6l \\ -6l+6l & 4l^2+4l^2 & -6l & 2l^2 \\ -12 & -6l & 12 & -6l \\ 6l & 2l^2 & -6l & 4l^2 \end{bmatrix}\begin{Bmatrix} r_3 \\ r_4 \\ r_5 \\ r_6 \end{Bmatrix} = \begin{Bmatrix} ql \\ 0 \\ F+ql/2 \\ -ql^2/12 \end{Bmatrix} \quad (17.26)$$

Using the numerical values for the parameters leads to a value for tip displacement (r_5) of 32.7 mm, agreeing with the exact value. The exact theoretical displacement is a quartic function, matched well by the combination of the two cubic functions.

17.3.5.2 Normal Modes Analysis: Two Elements

The structure mass and stiffness matrices may now yield normal modes for the structure; employing the consistent mass representation in Equation (17.25) gives

$$\frac{\mu l}{420}\begin{bmatrix} 312 & 0 & 54 & -13l \\ 0 & 8l^2 & 13l & -3l^2 \\ 54 & 13l & 156 & -22l \\ -13l & -3l^2 & -22l & 4l^2 \end{bmatrix}\begin{Bmatrix} \ddot{r}_3 \\ \ddot{r}_4 \\ \ddot{r}_5 \\ \ddot{r}_6 \end{Bmatrix} + \frac{EI}{l^3}\begin{bmatrix} 12+12 & -6l+6l & -12 & 6l \\ -6l+6l & 4l^2+4l^2 & -6l & 2l^2 \\ -12 & -6l & 12 & -6l \\ 6l & 2l^2 & -6l & 4l^2 \end{bmatrix}\begin{Bmatrix} r_3 \\ r_4 \\ r_5 \\ r_6 \end{Bmatrix} = 0$$

$$(17.27)$$

Using the numerical parameter values gives the first two natural frequencies of 2.095 and 13.23 Hz for the consistent mass representation; this compares to the exact values of 2.092 and 13.12 Hz, clearly a very good agreement. The equivalent result for the lumped mass representation is that the natural frequencies are (a) 1.808 and 8.61 Hz for the case where rotary inertia effects are included or (b) 1.879 and 9.68 Hz for the case where rotary inertia effects are ignored. Thus the consistent mass approach yields more accurate results than the lumped mass representation, as expected.

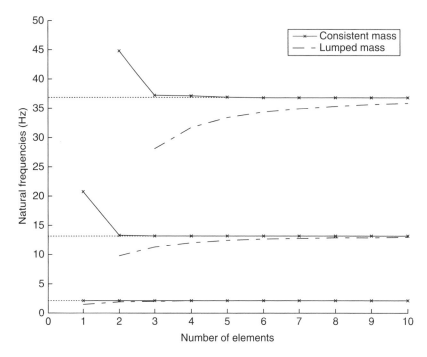

Figure 17.8 Variation of modes 1 to 3 natural frequencies with number of elements for consistent and lumped masses (no rotary inertia effects) (- - - - - exact values).

17.3.5.3   Normal Modes Analysis – Effect of Increasing the Number of Elements

The accuracy of the results for these, and higher, modes may be improved by introducing more elements. It is good practice to increase the number of elements until the results of interest stabilize, as seen in Figure 17.8 for the consistent and lumped mass representations; here the natural frequencies for the first three modes are shown plotted against the number of elements used. The exact values are also shown and clearly the consistent mass results converge far more quickly than those for the lumped mass representation.

17.4   Torsion Element

So far, the focus has been on beam bending. However, in aircraft applications, it is important to model the torsional behavior of slender members and so a brief introduction to torsion elements will be given here. A typical two-node torsion element is shown in Figure 17.9. The twist for a two-node torsion element varies linearly along its length. The two shape functions are therefore linear polynomials and the element stiffness and mass matrices may be evaluated as before; the elastic potential and kinetic energy terms for a member under torsion are the same as used earlier in Chapter 3. It may be shown that the element stiffness and consistent mass matrices are as follows:

$$\mathbf{k} = \frac{GJ}{L}\begin{bmatrix} 1 & -1 \\ -1 & 1 \end{bmatrix} \quad \mathbf{m} = \frac{\chi L}{6}\begin{bmatrix} 2 & 1 \\ 1 & 2 \end{bmatrix} \tag{17.28}$$

Figure 17.9   Two-node torsion element.

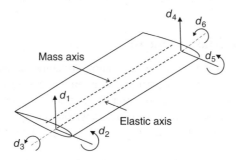

Figure 17.10   Combined bending/torsion element.

where GJ is the torsional rigidity, L is the length and χ is the torsional moment of inertia per unit length.

It may also be shown that the element stiffness and mass matrices for a two-node bar element under axial extension/compression are the same as those for the torsion element except that the axial rigidity EA replaces GJ and the mass per unit length μ replaces χ.

17.5   Combined Bending/Torsion Element

Having obtained the element mass and stiffness matrices for the beam bending and torsion elements based on using independent notations for nodal displacements, the elements may be combined using an integrated set of nodal displacements to obtain a single 2D uniform beam element having 6×6 matrices and able to bend and twist. The nodal displacements are conveniently defined with reference to the elastic axis (i.e. the axis where a bending load causes no twist and a torque causes no bending). The form of the matrices depends upon the numbering system for the nodal displacements. For example, if $\{d_1\ d_2\ d_4\ d_5\}$ corresponds to the bending displacements/rotations and $\{d_3\ d_6\}$ to the twists, as shown in Figure 17.10, this would lead to matrices where the bending and torsion terms are interspersed, as shown in Chapter 19. There is no bending/torsion stiffness coupling in the stiffness matrix and no mass coupling if the mass and elastic axes are coincident.

However, if the mass axis does not coincide with the elastic axis then there will be inertia coupling present, so that inertia forces associated with bending acceleration will cause twist and vice versa. Then, whether the mass is distributed or else lumped and attached to each node via a rigid connection (sometimes the case for an aircraft, as discussed in Chapter 21), a mass matrix will be generated having bending/torsion coupling terms involving the mass offset (i.e. the product of the mass and the offset distance, with an appropriate sign).

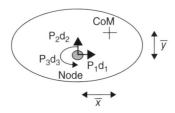

Figure 17.11   Concentrated mass element.

17.6   Concentrated Mass Element

When the FE method is used to analyse continuum structures where the mass is structural and is distributed in a similar way to stiffness (e.g. engine casting), then a consistent mass matrix may be employed. However, for aircraft structures a considerable portion of the mass is non-structural (e.g. associated with fuel, payload and systems) and a lumped mass representation will be required. Thus it is common practice to employ concentrated mass elements.

For simplicity, a 2D concentrated mass element is shown in Figure 17.11; there is an offset $(\bar{x},\bar{y})$ from the node to the centre of mass (CoM). The element has mass M and moment of inertia about the centre of mass I_z. The nodal displacements/rotations are $\{d_1, d_2, d_3\}$. The kinetic energy, based on a small angle assumption, is given by

$$T = \frac{1}{2}M\left\{\left(\dot{d}_1 - \bar{y}\dot{d}_3\right)^2 + \left(\dot{d}_2 + \bar{x}\dot{d}_3\right)^2\right\} + \frac{1}{2}I_z\dot{d}_3^2 \tag{17.29}$$

Thus, employing Lagrange's equations, the element mass matrix is

$$\mathbf{m} = \left[\begin{array}{cc|c} M & 0 & -M\bar{y} \\ 0 & M & M\bar{x} \\ \hline -M\bar{y} & M\bar{x} & I_z \end{array}\right] \tag{17.30}$$

and this includes coupling terms due to the offsets. The 3D element mass matrix is more complex, including product moments of inertia.

17.7   Stiffness Element

In some cases it is helpful to connect two nodes using a stiffness element representing a link of defined stiffness K_S. A 2D element inclined at angle θ to the x axis with four nodal displacements $\{d_1, d_2, d_3, d_4\}$ is shown in Figure 17.12. The elastic potential energy is given by

$$U = \frac{1}{2}K_S\left\{(d_3c + d_4s) - (d_1c + d_2s)\right\}^2 \quad \text{where } s = \sin\theta \; c = \cos\theta \tag{17.31}$$

Again, applying Lagrange's equations yields the 4 x 4 element stiffness matrix

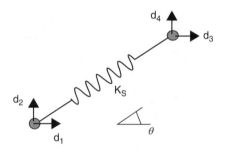

Figure 17.12   Stiffness element.

$$\mathbf{k} = K_S \left[\begin{array}{cc|cc} c^2 & sc & -c^2 & -sc \\ sc & s^2 & -sc & -s^2 \\ \hline -c^2 & -sc & c^2 & sc \\ -sc & -s^2 & sc & s^2 \end{array} \right] \tag{17.32}$$

It is important to recognize that this stiffness element should not be used to represent a rigid link or connection by setting the stiffness K_S to a very large value since this will cause the overall solution to be ill-conditioned. Instead, the concept of Multi-Point Constraints (MPCs) should be employed as discussed in the next section.

17.8   Rigid Body Elements

There are two basic forms of rigid body element that are used in aircraft FE modeling: rigid body element with an *infinite* constraint and rigid body element with an *interpolation* constraint.

17.8.1   Rigid Body Element with an Infinite Constraint

This classical rigid body element with an *infinite* constraint is used to impose Multi-Point Constraints (MPCs). The element allows one or more *dependent* (or slave) nodes to be constrained in translation and/or rotation with respect to only one *independent* (or master) node. Consider the 2D element in Figure 17.13 having two nodes in total and with both displacement and rotation constraints imposed, that is, the connection is completely rigid. Here the independent nodal displacements are $\{d_1, d_2, d_3\}$ and the dependent nodal displacements are $\{d_4, d_5, d_6\}$. The geometric relationships required to impose a rigid (i.e. infinite stiffness) connection between the nodes are as follows

$$d_4 = d_1 - d_3 L \sin\theta \quad d_5 = d_2 + d_3 L \cos\theta \quad d_6 = d_3 \tag{17.33}$$

The MPCs can be implemented by a method such as Lagrange Multipliers and this ensures that the problem is well conditioned.

As a simple illustration, consider the 2D circular frame shown in Figure 17.14a with the eight dependent nodes on the circumference connected to a central independent node using a rigid

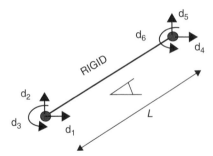

Figure 17.13   Rigid body element with infinite constraint.

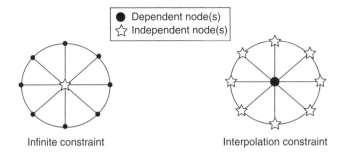

Figure 17.14   Illustration of rigid body elements on circular frame.

body element with a set of infinite constraints. Because the connections have infinite stiffness, the rigid body element has the effect of preventing the frame from deforming under external loading.

17.8.2   *Rigid Body Element with an Interpolation Constraint*

The other, more recently developed, rigid body element with an *interpolation* constraint defines the motion at one *dependent* node as the weighted average of the motion at a set of other *independent* nodal points. The average is essentially derived from a least squares process. Returning to the 2D circular frame example, but this time using a rigid body element with an interpolation constraint as seen in Figure 17.14b, it may be seen that the central node is dependent whereas the outer frame nodes are independent. The effect of this revised model is to permit the frame to deform under external loading whilst the central node moves as the weighted average (or interpolated value) of the frame nodes.

In modeling an aircraft structure, one use of this element is to place the mass of a 3D wing or fuselage section at a dependent central node and connect it to independent nodes on the outer wing box or fuselage structure (see Chapter 21).

17.9   Other Elements

Of course there are a number of other elements employed in modeling aircraft structures but these will not be considered in detail here. Suffice it to simply list those particularly relevant to thin-walled structures:

a. Membrane element – tension/compression/shear;
b. Plate element – bending/twisting; and
c. Shell element – combination of membrane and plate.

17.10   Comments on Modelling

17.10.1   'Beam-like' Representation of Slender Members in Aircraft

When an aircraft component such as a wing or fuselage is to be represented by a 'beam-like' model using the FE method, a three-dimensional beam element, more complete than the two-dimensional bending version considered earlier in this chapter, needs to be employed. The element will have bending in two directions, axial extension and torsion defined by six DoFs per node, plus allowance for shear deformation. The distributions of flexural and tor-sional rigidities (EI and GJ) along the wing have traditionally been estimated from the structural stiffness behaviour of the wing box boom/skin model (Donaldson, 1993; Megson, 1999; Sun, 2006). Note that a simple example of bending/torsion will be considered in Chapter 19.

Clearly, models may be set up for individual 'branches' (called substructures) or for the whole aircraft (see Chapter 3). Lumped masses could be attached to any beam using concentrated mass elements. If the free–free modes of the aircraft are required, then the boundary conditions are fully free for all nodes and six rigid body modes each with zero frequency will be found. Note that since the stiffness matrix is singular for a free–free structure, the eigenvalue problem needs to be formulated using the inverse of the mass matrix.

17.10.2   'Box-like' Representation of Slender Members in Aircraft

So far, the emphasis has been upon analysis using a 'beam-like' representation. However, to perform more accurate analyses, it is important to represent the stiffened thin-walled structure and its complex load paths more comprehensively in the FE model. In order to do this, the structure may be represented by a 'box-like' FE model such as the simple untapered and unswept box shown in Figure 17.15.

Here booms/stringers are represented by axial (or rod/bar) elements that react the axial loads (i.e. tension/compression). Cover skins/spar webs/rib webs are modelled using membrane elements that carry in-plane axial and shear loads. Typical two-node axial and four-node

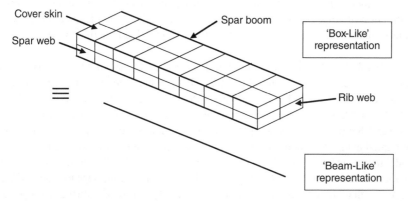

Figure 17.15   Simple 'box-like' representation of a wing box structure.

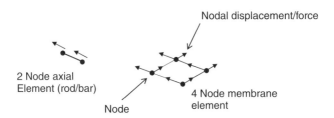

Figure 17.16   'Low order' axial and membrane elements.

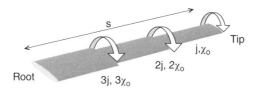

Figure 17.17

membrane elements are shown in Figure 17.16; these elements are known as 'low order' elements, because the assumed variation of displacement within the element is linear and so the variation of stress (NAFEMS, 1987) may be shown to be (approximately) constant. On the other hand, the 'higher order' three-node axial and eight-node membrane elements are more accurate, since the displacement variation within the element is represented as a quadratic function and the stress variation is (approximately) linear. It is also possible to represent the booms/stringers by beam elements and the skins by shell elements, where these elements allow for local bending of the booms and skin.

In practice, the aircraft wing is much more complex than the simple geometry shown here, so the model may have to use significant approximations of local structural features. For example, a stiffened cover panel with holes may be represented by a uniform panel of equivalent thickness; once the load paths (i.e. how the loads are distributed through this panel) are known from a preliminary FE analysis using the relatively coarse type of model discussed above, a local FE or other type of analysis (e.g. a plate buckling analysis or an analysis based on data sheets) may be performed on a more detailed representation of the panel. It should be noted that for a complex aerospace type structure, many parts of the FE model are not suitable for direct extraction of stresses due to the approximations made in the idealization. Instead, nodal forces are obtained and employed to define the external loads acting on the relevant local structure.

Having obtained such a 'box-like' model, it is possible to use an approach known as static 'condensation' (see Chapter 21 and Appendix D) to reduce the size of the FE model and to replace the 'box-like' model by an equivalent beam along a reference axis, as illustrated in Figure 17.15 (see also Chapter 21); this beam can be used in the aeroelasticity and loads analyses (see Chapter 19), being coupled with inertia loads (see Chapter 5) and aerodynamic loads (obtained using strip theory or panel methods; see Chapters 4, 18 and 21). Thus a 'beam-like' model may now be extracted from a more realistic 'box-like' FE model, under the assumption that the condensed DoF behave linearly with respect to the master DoF, and will be much more accurate than if a traditional beam with crudely estimated stiffness properties were employed from scratch.

Note that the mass may be represented by concentrated mass elements attached to the condensed beam. Alternatively, the 'box-like' representation may be retained without condensation and lumped masses attached to local parts of the box model using a rigid body element with an interpolation constraint (see Chapter 21).

17.11   Examples

1. Using the same approach as taken above for developing the element stiffness matrix for a two-node bending element, determine the shape functions and hence the 2×2 stiffness and consistent mass matrices for a two-node torsion element of length L, torsional rigidity GJ and torsional moment of inertia χ per unit length. (An equivalent result may be obtained for a two node bar element, i.e. one under axial loading.)

$$\left[k = \frac{GJ}{L} \begin{bmatrix} 1 & -1 \\ -1 & 1 \end{bmatrix} \quad m = \frac{\chi L}{6} \begin{bmatrix} 2 & 1 \\ 1 & 2 \end{bmatrix} \right]$$

2. Consider a clamped–clamped member with length $4\,l$, flexural rigidity EI and mass per length μ. The fundamental natural frequency is to be obtained and since the corresponding mode shape will be symmetric, only half of the member needs to be modeled, in this case using two 2-node bending elements. Determine the 3×3 overall stiffness and consistent mass matrices, recognizing the zero slope boundary condition at the line of symmetry. Then determine the fundamental natural frequency and mode shape using MATLAB (the symbols may be ignored in the calculations and added in later).

$$\left[1.400 \sqrt{EI/\mu l^4} \quad \text{cf. exact } 1.398 \right]$$

3. The element stiffness and consistent mass matrices for the two-node torsion element shown in Example 1 above may be used in this question. In order to estimate the first torsion natural frequency of a tapered unswept wing, the wing is modelled by three equal length uniform torsion elements of different dimensions, as shown in Figure 17.13. The wing is clamped at the root. The relevant parameters for each section are shown on the diagram, i.e. the torsion constants for the three sections are $3\,J$, $2\,J$, J (root to tip) and the equivalent torsional moments of inertia per unit length are $3\chi_0$, $2\chi_0$, χ_0. Determine the 3×3 stiffness and consistent mass matrices for the overall system after boundary conditions have been applied. Use MATLAB to determine the natural frequency estimates (again ignoring symbols and adding in later) and the mode shape. If a tank with moment of inertia I_t in torsion were to be added to the wing tip, explain how the analysis would change.

$$[\mathbf{K} = GJ/s\,[15 \quad -6 \quad 0; \quad -6 \quad 9 \quad -3; \quad 0 \quad -3 \quad 3]$$

$$\mathbf{M} = \chi_0 s/18\,[10 \quad 2 \quad 0; \quad 2 \quad 6 \quad 1; \quad 0 \quad 1 \quad 2], \quad 2.094\sqrt{GJ_0/\chi_0 s}$$

[cf. 2.121 using Rayleigh–Ritz with linear assumed shape and piecewise integration]

4. A wing/tip store combination may be idealized as a uniform member, built-in at one end, with an offset tip store (e.g. fuel tank). The wing has a mass per length of 75 kg/m, moment of inertia in twist per length of 25 kg m^2/m, span 6 m, flexural rigidity 2×10^6 N m^2 and

torsional rigidity $5 \times 10^5 \, \text{N m}^2$. The tip store has a mass of 100 kg and moment of inertia in pitch of 25 kg m^2 about its centre of mass, which itself is offset by 0.5 m forward of the wing centre line. It may be assumed that the elastic and mass axes coincide at mid-chord so there are no couplings between the bending and torsion behavior for the basic wing. Using a single element comprising both bending and torsion behavior (this 6×6 matrix is a simple combination of the 4×4 bending and 2×2 torsion matrices), obtain the 3×3 stiffness and mass matrices for the built-in member. Then add suitable mass terms to the mass matrix in order to account for the store inertia. One approach for doing this is to consider the inertia forces and moments acting on the store, then apply them to the FE model as right-hand side forces and express the force vector as a matrix multiplied by an acceleration vector; rearranging the equations then yields a mass matrix augmented with the store effect. Use MATLAB to obtain the first two natural frequencies and sketch the mode shapes. Compare the results with those from the Rayleigh–Ritz assumed modes of Example 7 in Chapter 3.
[1.81 and 5.00 Hz; cf. 2.18 and 5.178 Hz]

18

Potential Flow Aerodynamics

The two-dimensional strip theory aerodynamics that has been used so far in this book, for convenience, implies a number of major assumptions about the aerodynamic load distribution (e.g. it neglects tip effects) and is only moderately accurate for low speed, high aspect ratio and unswept wings. Of particular importance is the assumption that the aerodynamic forces acting on one chordwise strip have no effect on other chordwise strips. In order to perform a more accurate aeroelastic analysis, aerodynamic theories have been developed that are able to define more accurate pressure distributions over the entire wing. The so-called three-dimensional *panel* methods were developed to model the interaction between the aerodynamic forces on different parts of the lifting surfaces (wings, fin and horizontal tail surfaces) more accurately. It will be shown in Chapter 19 how it is possible to tightly couple the panel method aerodynamics with a Finite Element (FE) model; consequently, panel methods are the primary aerodynamic tool used by industry for aeroelastic analysis. However, it should be noted that panel methods cannot be used to give accurate lift distributions in the transonic flight regime and corrections based upon wind tunnel tests are often employed. Also, only the induced drag is able to be estimated. Consequently, there is an increasing use of higher fidelity Computational Fluid Dynamics (CFD) methods, using Euler or Reynolds Averaged Navier–Stokes (RANS) aerodynamic methods coupled with FE models to determine the time response to some initial displacement in the transonic region.

In this chapter, some aspects of inviscid flow analysis will be introduced that enable the velocity at any part of the flow, around a wing for instance, to be defined. Knowing that the flow component normal to the surface of an aerofoil must be zero, the entire flow and resulting pressure distribution can be determined. Following the introduction of the concept of vorticity, the analysis of the flow around two-dimensional thin aerofoils is introduced. Through the use of the Biot–Savart law, the extension to three-dimensional wings is discussed, leading to the panel method analysis of the three-dimensional steady and unsteady cases. The reader is directed to *Low Speed Aerodynamics* by Katz and Plotkin (2001), which provides a complete overview of panel method aerodynamics.

18.1   Components of Inviscid, Incompressible Flow Analysis

The classical two-dimensional inviscid flow theory provides a basis for analyzing flow motion. Of particular interest are the resulting velocities at a point in the flow which is described by the streamlines (see Chapter 4). The underlying stream function and/or velocity potential of the

Introduction to Aircraft Aeroelasticity and Loads, Second Edition. Jan R. Wright and Jonathan E. Cooper.
© 2015 John Wiley & Sons, Ltd. Published 2015 by John Wiley & Sons, Ltd.

flow for Cartesian and polar coordinates need to be determined first, and then the velocities at any point can be calculated.

The *stream function* $\Psi(x, z)$ is constant for each streamline and the velocities in the horizontal x, vertical z, radial q_r and tangential q_θ directions are given as

$$u = \frac{\partial \psi}{\partial z} \quad w = -\frac{\partial \psi}{\partial x} \quad q_r = \frac{1}{r}\frac{\partial \psi}{\partial \theta} \quad q_\theta = -\frac{\partial \psi}{\partial r} \tag{18.1}$$

The *velocity potential* $\phi(x, z)$ is defined as the amount of fluid flowing between two points on a streamline (Houghton and Carpenter, 2001); the equivalent expressions for the velocity components are

$$u = \frac{\partial \phi}{\partial x} \quad w = \frac{\partial \phi}{\partial z} \quad q_r = \frac{\partial \phi}{\partial r} \quad q_\theta = \frac{1}{r}\frac{\partial \phi}{\partial \theta} \tag{18.2}$$

From inspection of Equations (18.1) and (18.2) it is possible to determine the stream function and velocity potential for a range of different flow components, as shown in the following examples.

18.1.1 Uniform Flow

For the flow of velocity V parallel to the x axis, as shown in Figure 18.1, using the Cartesian definitions gives

$$u = V \quad w = 0 \quad \Rightarrow \quad \psi = Vz \quad \text{and} \quad \phi = Vx \tag{18.3}$$

With the flow along the z axis the expressions change to

$$u = 0, \quad w = V \quad \Rightarrow \quad \psi = -Vx \quad \text{and} \quad \phi = Vz \tag{18.4}$$

and for a flow inclined at $\tan^{-1}(V/U)$ then

$$\psi = V(z - x) \quad \text{and} \quad \phi = V(x + z). \tag{18.5}$$

18.1.2 Point Source and Point Sink

For the source shown in Figure 18.2, where fluid appears at some point in the flow at the flow rate m and moves radially in all directions, then at radius r from the source, using the polar coordinate definitions gives

Figure 18.1   Uniform flow streamlines.

$$\phi = \frac{m}{2\pi}\ln(r) \quad \psi = \frac{m\theta}{2\pi} = \frac{m}{2\pi}\tan^{-1}\left(\frac{z}{x}\right)$$

$$q_r = \frac{m}{2\pi r} \quad q_\theta = 0 \quad u = \frac{mx}{2\pi r^2} \quad w = \frac{mz}{2\pi r^2}$$

(18.6)

The sink is the exact opposite of the source, with flow disappearing into some point at rate m; all of the expressions in Equation (18.6) are then preceded by a minus sign. Note that sources and sinks are singular points with an infinite flow velocity where they occur.

18.1.3 Source–Sink Pair

Consider a source and sink of equal strength m placed at distances d and $-d$ respectively from the origin on the x axis, as shown in Figure 18.3. The velocity potential and stream function are found by superposition to be

$$\phi = \frac{m}{2\pi}[\ln(r_1) - \ln(r_2)] \quad \psi = \frac{m}{2\pi}(\theta_1 - \theta_2)$$

$$u = \frac{m}{2\pi}\frac{x+x_0}{(x+x_0)^2 + z^2} - \frac{m}{2\pi}\frac{x-x_0}{(x-x_0)^2 + z^2} \quad w = \frac{m}{2\pi}\frac{z}{(x+x_0)^2 + z^2} - \frac{m}{2\pi}\frac{z}{(x-x_0)^2 + z^2}$$

(18.7)

The streamlines for the combined flows are circles passing through the source and sink as shown in Figure 18.4. When $(\theta_1 - \theta_2) = \pi/2$ the streamline is a semicircle with centre O.

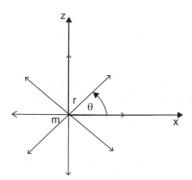

Figure 18.2   Source streamlines.

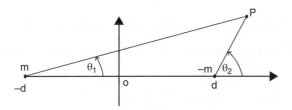

Figure 18.3   Combined source and sink.

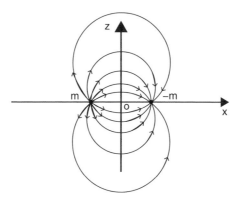

Figure 18.4 Streamlines for a source–sink pair.

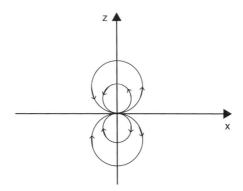

Figure 18.5 Doublet streamlines.

18.1.4 Doublet

If the source and sink are allowed to approach each other then $d \rightarrow 0$ and the resulting flow component is known as a doublet. By letting $md = \mu/2$, where μ is a constant known as the doublet strength such that m increases as d increases, then it can be shown that the velocity potential and stream function become

$$\phi = \frac{-\mu \cos\theta}{2\pi} \frac{}{r} \qquad \psi = \frac{\mu}{2\pi} \frac{z}{x^2 + z^2} = \frac{\mu}{2\pi} \frac{\sin\theta}{r} \tag{18.8}$$

The resulting streamlines are circles that are tangential to each other with centres on the z axis, as shown in Figure 18.5.

18.1.5 Source–Sink Pair in a Uniform Flow (Rankine Oval)

If a uniform flow of V in the x direction is combined with a source–sink pair, then the velocities are found as

$$\phi = Vx + \frac{m}{2\pi}\left[\ln(r_1) - \ln(r_2)\right] \qquad \psi = Vz + \frac{m}{2\pi}(\theta_1 - \theta_2)$$

$$u = V + \frac{m}{2\pi} \frac{x + x_0}{(x + x_0)^2 + z^2} - \frac{m}{2\pi} \frac{x - x_0}{(x - x_0)^2 + z^2} \qquad w = \frac{m}{2\pi} \frac{z}{(x + x_0)^2 + z^2} - \frac{m}{2\pi} \frac{z}{(x - x_0)^2 + z^2} \tag{18.9}$$

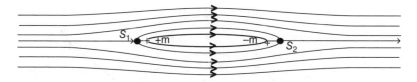

Figure 18.6 Source–sink pair in a uniform flow – streamlines (Rankine Oval).

The so-called stagnation points, where the velocity is zero, are found by determining where both velocities u and w are zero. The streamline passing through the stagnation points is known as the dividing streamline and is defined by $\psi = 0$. The resulting shape of the dividing streamline is the so-called Rankine oval shown in Figure 18.6; here there are two stagnation points S_1 and S_2 on the x axis. The flow inside the oval is of no interest and consequently the oval can be replaced by a solid body without changing the flow outside the oval. This is an example of how potential flow analysis can be used to determine the flow around a body.

18.1.6   Doublet in a Uniform Flow

The velocity potential and stream function of a doublet in a uniform flow can be found from superposition of the relevant results above. Taking the doublet strength as $-\mu$, then in polar coordinates

$$\phi = \cos\theta\left(Vr + \frac{\mu}{2\pi r}\right) \quad \psi = \sin\theta\left(Vr - \frac{\mu}{2\pi r}\right) \tag{18.10}$$

where θ is defined as above and the dividing streamline ($\psi = 0$) is found to be a circle of radius $R = \sqrt{\mu/(2\pi V)}$ with the centre at (0,0), as shown in Figure 18.7. Thus, to achieve a dividing streamline of a given radius R, the required doublet strength is defined as

$$\mu = 2\pi V R^2 \tag{18.11}$$

Once again, the flow inside the dividing streamline may be ignored and the streamline may be replaced by a cylindrical body. For the flow around a cylinder of radius R,

$$\phi = V\cos\theta\left(r + \frac{R^2}{r}\right) \quad \psi = V\sin\theta\left(r - \frac{R^2}{r^2}\right)$$

$$q_r = V\cos\theta\left(1 - \frac{R^2}{r^2}\right) \quad q_\theta = -V\sin\theta\left(1 + \frac{R^2}{r^2}\right) \tag{18.12}$$

and on the cylinder surface $q_r = 0$ and $q_\theta = -2V\sin\theta$.
   The pressure acting around the cylinder can be calculated using Bernoulli's equation (see Chapter 4) so that

$$p_\infty + \frac{\rho V^2}{2} = p + \frac{\rho q_\theta^2}{2} \quad \text{giving} \quad p_\infty - p = \frac{\rho V^2}{2}\left(1 - 4\sin^2\theta\right) \tag{18.13}$$

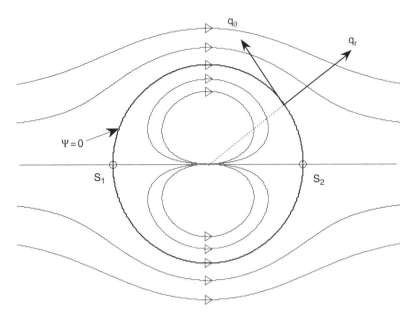

Figure 18.7 Doublet in a uniform flow – streamlines.

on the cylinder surface and a pressure coefficient distribution around the cylinder of

$$C_p = \frac{2(p-p_\infty)}{\rho V^2} = 1 - 4 \sin^2\theta \qquad (18.14)$$

The lift and drag per unit span on the cylinder can be calculated by integrating the forces acting in the vertical and horizontal directions due to the pressure acting on the element $R\,d\theta$ on the cylinder surface. In this case both the lift and drag are zero, due to the symmetry in both the vertical and horizontal planes. In practice the flow around a cylinder will separate from the cylinder surface, which is not predicted using the above analysis, demonstrating one of the limitations of inviscid theory.

18.2 Inclusion of Vorticity

So far any particles in the flow have been assumed to maintain their orientation and so are irrotational. A more accurate representation of any fluid motion can be provided by including rotational flow; this can be defined in terms of the vorticity, which is a measure of the angular velocity or spin within the fluid.

18.2.1 Vortices

Consider the 'swirling' flow shown in Figure 18.8 with the vortex positioned at (x_0, z_0). The radial velocity q_r at each point is zero and the tangential velocity q_θ is constant at each radius. Defining the strength, or circulation, of the vortex Γ acting in a clockwise direction as

$$\Gamma = 2\pi r q_\theta \qquad (18.15)$$

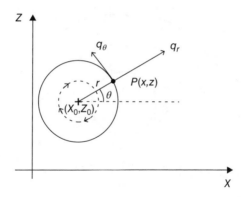

Figure 18.8   Velocities around a vortex.

then the circulation has units of m²/s in a two-dimensional flow. The consequence of this equation is that the tangential velocity around a vortex reduces in proportion to $1/r$. The velocity potential and stream functions at point P (x, z) for a vortex at point (x_0, z_0) can be shown to be

$$\phi = -\frac{\Gamma}{2\pi} \tan^{-1}\left(\frac{z-z_0}{x-x_0}\right) \quad \psi = -\frac{\Gamma}{2\pi} \ln r \qquad (18.16)$$

where r is the distance between the vortex and point P. The velocities at point P can then be found as

$$u = \frac{\Gamma}{2\pi} \frac{(z-z_0)}{(x-x_0)^2 + (z-z_0)^2} \quad w = -\frac{\Gamma}{2\pi} \frac{x-x_0}{(x-x_0)^2 + (z-z_0)^2}, \quad q_\theta = -\frac{\Gamma}{2\pi r} \quad q_r = 0 \qquad (18.17)$$

18.2.2   Flow past a Cylinder with a Vortex at the Centre

Consider the previous case of a doublet of strength $\mu = 2\pi R^2$ in a uniform flow, but this time adding a vortex of strength $-\Gamma$ at the origin. The velocity potential and stream function are found by superposition to be

$$\phi = V \cos\theta \left(r + \frac{R^2}{r}\right) - \frac{\Gamma}{2\pi}\theta \quad \psi = V \sin\theta \left(r - \frac{R^2}{r}\right) + \frac{\Gamma}{2\pi} \ln r \qquad (18.18)$$

giving radial and tangential velocities of

$$q_r = V \cos\theta \left(1 - \frac{R^2}{r^2}\right) \quad q_\theta = -V \sin\theta \left(1 + \frac{R^2}{r^2}\right) - \frac{\Gamma}{2\pi r} \qquad (18.19)$$

On the cylinder surface $(r = R)$ the velocities are

$$q_r = 0 \quad q_\theta = -2V \sin\theta - \frac{\Gamma}{2\pi R} \qquad (18.20)$$

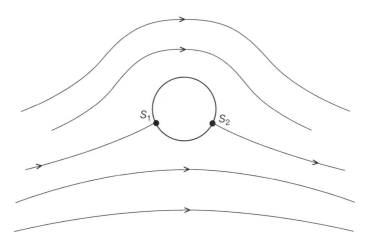

Figure 18.9   Flow around a cylinder with added vortex.

The stagnation points can be found when $q_\theta = 0$ and thus occur where

$$\sin \theta = -\frac{\Gamma}{4\pi R V} \qquad (18.21)$$

Therefore, provided $\Gamma \le 4\pi R V$ a solution exists and the stagnation points S_1 and S_2 lie on the dividing streamline which is a circle, as shown in Figure 18.9.

The lift per unit span (positive upwards) is found by integrating around the entire circumference the elemental force acting in the z direction due to the pressure on element $R\,d\theta$ such that

$$L = \int_0^{2\pi} -pR\sin\theta\,d\theta \qquad (18.22)$$

Remembering from Bernoulli's equation that

$$p_\infty + \frac{\rho V^2}{2} = p + \frac{\rho q_\theta^2}{2} \quad \rightarrow \quad p = p_\infty + \frac{\rho V^2}{2} - \frac{\rho}{2}\left(2V\sin\theta + \frac{\Gamma}{2\pi R}\right)^2 \qquad (18.23)$$

then the lift per unit span (units N/m) can be shown to be

$$L = \rho V \Gamma \qquad (18.24)$$

This is a very important result which shows that the lift is proportional to the strength of the vortex. Since the flow considered forms a closed loop around the cylinder, Γ is often referred to as the circulation. Note, however, that even with the inclusion of the vortex, the drag in this two-dimensional example is still zero; this is because viscosity is neglected.

18.3   Numerical Steady Aerodynamic Modelling of Thin Two-dimensional Aerofoils

It is now possible to start modelling flows around two-dimensional aerofoils. Assuming a two-dimensional incompressible and inviscid steady flow around a thin aerofoil at a small angle of attack, then a number of different components can be employed to model the flow, namely uniform flow, source, sink, doublet, vortex etc. The zero normal flow boundary condition on the solid boundary of the aerofoil is used to determine the strengths of these components. The problem must be modelled in two parts: firstly, a symmetric aerofoil of the same thickness at zero incidence and, secondly, a thin plate of the camber line shape at the prescribed incidence. A further condition that must be met is the so-called Kutta condition, where the flow must leave the sharp trailing edge of the aerofoil smoothly, implying that the velocity there must be finite. As an example, taking the vortex function described earlier as the fundamental building block, the flow can be modelled as a distribution of vortices along the centre line in a uniform flow; the vortices will have differing strengths.

18.3.1 Aerofoil Flow Modelled using a Single Element

To illustrate the process, consider the thin aerofoil shown in Figure 18.10, without camber but at incidence α and air speed V; a single vortex component is placed at the 1/4 chord to help model the flow; it is superimposed upon a uniform flow. It can be shown that the zero normal flow boundary condition needs to be enforced at the 3/4 chord control point in order to model the flow correctly. This constraint also ensures that the Kutta condition is met. The aerofoil is thus modelled by considering it as a single element.

Since there is zero normal flow at the control point, the contribution to the downwash from the vortex must be equal and opposite to that of the freestream component; hence, from Equation (18.17) it can be shown that

$$\frac{\Gamma_1\ ^c/_2}{2\pi\left(^c/_2\right)^2} = \frac{\Gamma_1}{\pi c} = V\sin\alpha \approx V\alpha \quad \text{so} \quad \Gamma_1 = V\alpha c\pi \tag{18.25}$$

and thus the unknown vortex strength may be found. The resulting lift and pitching moment per unit span are then found as

$$L = \rho V\Gamma = \rho V \times V\alpha c\pi = \frac{1}{2}\rho V^2 2\pi c\alpha \quad \text{and} \quad M = \frac{c}{4}L = \frac{c}{4}\rho V \times V\alpha c\pi = \frac{1}{8}\rho V^2 2\pi c^2\alpha \tag{18.26}$$

which are the same results as would be found using strip theory.

If more than a single vortex is used, then the aerofoil can divided into several chordwise elements, each with a vortex and control point. By setting up a series of flow equations from the vortex velocity components that meet the boundary conditions at each control point, the vortex strengths can be determined. Once the vortex distribution has been found, the overall lift and pitching moment can be obtained (the drag is zero for the two-dimensional inviscid case).

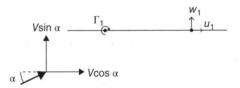

Figure 18.10 2D aerofoil modelled using a single element with a vortex component on ¼ chord and control point on ¾ chord.

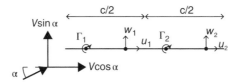

Figure 18.11 2D aerofoil modelled using two elements.

18.3.2 *Aerofoil Flow Modelled using Two Elements*

Consider now the 2D aerofoil in a uniform flow at incidence α, modelled with two equal size elements, each with its own vortex and control point, as shown in Figure 18.11.

The downwash at each of the control points, w_1 and w_2, has a contribution from both of the vortices, and these can be expressed in terms of the unknown vortex strengths Γ_1, Γ_2 as

$$\mathbf{w} = \left\{ \begin{matrix} w_1 \\ w_2 \end{matrix} \right\} = \begin{bmatrix} \dfrac{\Gamma_1}{2\pi c/4} - \dfrac{\Gamma_2}{2\pi c/4} \\ \dfrac{\Gamma_1}{2\pi 3c/4} + \dfrac{\Gamma_2}{2\pi c/4} \end{bmatrix} = \begin{bmatrix} \dfrac{2}{\pi c} & \dfrac{-2}{\pi c} \\ \dfrac{2}{3\pi c} & \dfrac{2}{\pi c} \end{bmatrix} \left\{ \begin{matrix} \Gamma_1 \\ \Gamma_2 \end{matrix} \right\} = \begin{bmatrix} \psi_{11} & \psi_{12} \\ \psi_{21} & \psi_{22} \end{bmatrix} \left\{ \begin{matrix} \Gamma_1 \\ \Gamma_2 \end{matrix} \right\} = \mathbf{\Psi}\mathbf{\Gamma} \quad (18.27)$$

Each of the ψ_{ij} terms are known as the *influence coefficients*, and define the normal flow (downwash) at the ith control point resulting from the jth unit strength vortex. Once the free stream flow is included there are zero normal flow conditions at both control points, namely

$$\frac{\Gamma_1}{2\pi c/4} - \frac{\Gamma_2}{2\pi c/4} = V\alpha \quad \text{and} \quad \frac{\Gamma_1}{2\pi 3c/4} + \frac{\Gamma_2}{2\pi c/4} = V\alpha \qquad (18.28)$$

These equations can be combined into a matrix form to set up simultaneous equations to find the unknown vortex strengths such that

$$\begin{bmatrix} -\dfrac{2}{c\pi} & \dfrac{2}{c\pi} \\ -\dfrac{2}{3c\pi} & -\dfrac{2}{c\pi} \end{bmatrix} \left\{ \begin{matrix} \Gamma_1 \\ \Gamma_2 \end{matrix} \right\} = -V \left\{ \begin{matrix} \alpha \\ \alpha \end{matrix} \right\} = -V \left\{ \begin{matrix} 1 \\ 1 \end{matrix} \right\} \alpha \quad \Rightarrow \quad \left\{ \begin{matrix} \Gamma_1 \\ \Gamma_2 \end{matrix} \right\} = V\pi c \left\{ \begin{matrix} 3/4 \\ 3/4 \end{matrix} \right\} \alpha \quad (18.29)$$

If camber were to be present, then the angle at each of the control points would need to be included in the flow boundary conditions; the incidence of each element would differ due to the camber.

In a more general form this becomes

$$V\alpha + \mathbf{\Psi}\mathbf{\Gamma} = 0 \qquad \Rightarrow \qquad \mathbf{\Gamma} = -V\mathbf{\Psi}^{-1}\alpha \qquad (18.30)$$

where α is a vector of the angles of incidence of each element. Having found the vortex strengths, the lift on each element L_i is defined as $\rho V\Gamma_i$; thus the total lift per unit span is the sum of the lift on the two elements

$$L = \sum_{i=1}^{2} L_i = \rho V \sum_{i=1}^{2} \Gamma_i = \rho V^2 \pi c \alpha \qquad (18.31)$$

and the pitching moment about the mid-chord is found as

$$M = \sum_{i=1}^{2} M_i = \rho V^2 \alpha \pi \left(\frac{3c}{4} \frac{3c}{8} - \frac{c}{4} \frac{c}{8} \right) = \frac{1}{4} \rho V^2 \alpha \pi c^2 \qquad (18.32)$$

Both the lift and pitching moment expressions are exactly the same as those obtained by using strip theory for a unit span aerofoil of chord c and a lift-curve slope of 2π.

It is useful to re-write the above expression in terms of the lift acting on each element in terms of the angle of incidence of each element, such that

$$\left\{ \begin{matrix} L_1 \\ L_2 \end{matrix} \right\} = \rho V \left\{ \begin{matrix} \Gamma_1 \\ \Gamma_2 \end{matrix} \right\} = \frac{\rho V^2}{2} \begin{bmatrix} \frac{3c\pi}{4} & \frac{3c\pi}{4} \\ -\frac{c\pi}{4} & \frac{3c\pi}{4} \end{bmatrix} \left\{ \begin{matrix} \alpha \\ \alpha \end{matrix} \right\} = \frac{\rho V^2}{2} \begin{bmatrix} AIC_{11} & AIC_{12} \\ AIC_{21} & AIC_{22} \end{bmatrix} \left\{ \begin{matrix} \alpha \\ \alpha \end{matrix} \right\} \qquad (18.33)$$

In this important equation, the AIC_{ij} terms are known as the *aerodynamic influence coefficients* (AICs). They relate the lift on each element to its angle of incidence and also the dynamic pressure. Care must be taken not to confuse the definitions for the *influence coefficients* and the *aerodynamic influence coefficients* as they are quite different, yet closely related, quantities.

The general form of Equation (18.33) is given by

$$L = \rho V \Gamma = -\rho V \Psi^{-1} \alpha = \frac{\rho V^2}{2} AIC\alpha \qquad (18.34)$$

and comparison with Equation (18.30) shows that $AIC = -2\Psi^{-1}$.

Although only discrete vortices have been used in the above example, it is possible to use any combination of flow components, such as sources, sinks or doublets. Note also that they do not have to be discrete, and can take the form of continuous distributions (e.g. linear or quadratic) along the length of each element.

18.4   Steady Aerodynamic Modelling of Three-Dimensional Wings using a Panel Method

The approach developed above for two-dimensional flows can be extended to the three-dimensional case with the wing in a uniform flow being divided up into chordwise and spanwise panels (or elements), each with an aerodynamic component such as a vortex. The influence coefficients are calculated for whatever distribution of flow components is chosen, and are then used together with the surface boundary conditions at the control points to obtain the strengths of each component. It is then possible to calculate the lift, pitching moment and, for the three-dimensional case, the induced drag.

18.4.1   *Vortex Filaments and the Biot–Savart Law*

Assume that it is possible to have a continuous line, or filament, of vortices of constant strength Γ per unit length that rotate about that line in three-dimensional space. It can be shown that the vortex filament cannot end in a fluid, but must either form a closed loop or extend to

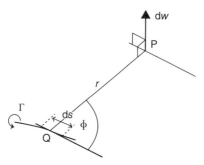

Figure 18.12   Induced velocity at point P due to the vortex filament.

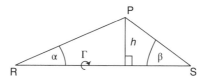

Figure 18.13   Downwash at point P due to the vortex filament RS of strength Γ.

the boundaries of the fluid. A real-life example of this is the trailing tip vortices that form from both wing tips of an aircraft. These vortices extend for many miles behind a commercial aircraft, eventually breaking down due to friction losses, and have a big influence upon the spacing between aircraft allowed by Air Traffic Control.

The Biot–Savart law defines the induced velocity at some point P due to an element of vortex filament. Consider the vortex filament of strength Γ and length ds passing through point Q in Figure 18.12; the filament is at angle ϕ to PQ. Then the induced velocity dw at point P in the direction shown is

$$dw = \frac{\Gamma \sin \phi \, ds}{4 \pi r^2} \tag{18.35}$$

This law can be extended to consider the straight vortex line RS in Figure 18.13 of strength Γ and perpendicular distance h from point P. It can be shown by integrating along the vortex line that the induced velocity at point P is

$$w = \frac{\Gamma}{4 \pi h}(\cos \alpha + \cos \beta) \quad \Rightarrow \quad w = \frac{\Gamma}{2 \pi h} \quad \text{as} \quad \alpha \to 0, \beta \to 0 \tag{18.36}$$

where $(\alpha \to 0, \beta \to 0)$ is the condition for an infinite vortex line.

18.4.2 *Finite Span Wing – Modelled with a Single Horseshoe Vortex*

It is common to model finite wings with vortex components that tend to infinity and therefore to include the effect of the wake. Consider the wing of span $2s$ and chord c at a small angle of incidence α as shown in Figure 18.14. The wing is modelled as a *horseshoe vortex* that consists of a bound vortex of length $2s$ positioned along the 1/4 chord and two trailing vortices that tend

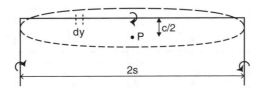

Figure 18.14 Wing modelled as a single horseshoe vortex.

to infinity. The closing section of the vortex is assumed to have no effect upon the downwash at the wing.

The total downwash w at the control point P, positioned a distance of $3/4c$ from the leading edge at the mid-span, has three contributions, namely those due to the bound vortex and the two trailing vortices. Making use of the Biot–Savart Law gives

$$w = \frac{\Gamma}{4\pi s}\left(\frac{c/2}{\sqrt{s^2+(c/2)^2}}+1\right) + \frac{\Gamma}{4\pi s}\left(\frac{c/2}{\sqrt{s^2+(c/2)^2}}+1\right) + \frac{\Gamma}{4\pi c/2}\frac{2s}{\sqrt{s^2+(c/2)^2}}$$

(18.37)

$$= \frac{\Gamma}{2\pi s}\left(\frac{c}{2\sqrt{s^2+(c/2)^2}}+1\right) + \frac{\Gamma}{\pi c}\frac{s}{\sqrt{s^2+(c/2)^2}} = \Psi\Gamma$$

where Ψ (units 1/m) is the influence coefficient (in this case a scalar value for a single vortex). At the control point, the resulting downwash and the velocity contribution from the airflow (assuming small angles) must equal zero (cf. Equation (18.28)), so giving

$$V\alpha + \Psi\Gamma = 0$$

(18.38)

from which the vortex strength Γ (units m^2/s) can be found for a given air speed and incidence.

Considering the forces acting upon a wing, the lift and drag are found by integrating the force on the element dy due to the vorticity across the span such that

$$L = \int_{-s}^{s} \rho V\Gamma \, dy \quad \text{and} \quad D = \int_{-s}^{s} \rho V\Gamma\alpha \, dy$$

(18.39)

where Γdy is the strength of the vortex filament element. Note that this is the induced drag and not the total drag, which also has a friction (viscous) drag component that increases with Mach number (see Chapter 4); for a three-dimensional wing, as the span tends to infinity the induced drag tends to zero.

For this case, a single horseshoe vortex has been used, which implies that there is a constant lift distribution across the entire span. In order to obtain a result in which the lift falls off towards the tip, vorticity needs to be shed all along the wing.

18.4.3 Finite Span Wing – Modelled with a Vortex Lattice

In practice, the variation of lift along the span can be modelled by using a number of horseshoe vortices placed in a 'lattice' arrangement on a series of panels in both the spanwise and

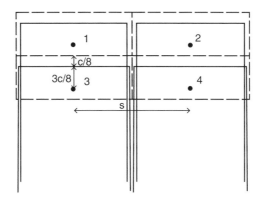

Figure 18.15 Four panel wing with horseshoe vortices.

chordwise directions; the vortices are placed side by side or behind each other along the 1/4 chord of each panel. Since the total normal velocity on the control point for each panel, due to all the vortices and the overall flow, must be zero, it is possible to determine the influence coefficients and hence the strength of the vortices for a particular incidence. The following example shows the process when using a lattice of horseshoe vortices.

Consider the wing of span $2s$ and chord c with four equal sized aerodynamic panels each containing a horseshoe vortex, as shown in Figure 18.15. On each panel, the bound vortex runs along the 1/4 chord and there is a control point at the 3/4 chord on the panel mid-span. Each horseshoe vortex affects all the control points, and thus the downwash at each control point has a contribution from each vortex, which can be written using superposition as

$$w = \begin{bmatrix} w_1 \\ w_2 \\ w_3 \\ w_4 \end{bmatrix} = \begin{bmatrix} \psi_{11} & \psi_{12} & \psi_{13} & \psi_{14} \\ \psi_{21} & \psi_{22} & \psi_{23} & \psi_{24} \\ \psi_{31} & \psi_{32} & \psi_{33} & \psi_{34} \\ \psi_{41} & \psi_{42} & \psi_{43} & \psi_{44} \end{bmatrix} \begin{bmatrix} \Gamma_1 \\ \Gamma_2 \\ \Gamma_3 \\ \Gamma_4 \end{bmatrix} = \Psi\Gamma \qquad (18.40)$$

where the elements of matrix Ψ are the influence coefficients, so ψ_{ij} is the downwash at the ith control point due to the jth horseshoe vortex of unit strength. These influence coefficients can be determined by application of Equation (18.36) for all of the vortices and panels.

As before, the zero normal flow boundary condition at the control points is satisfied via the relation

$$\Psi\Gamma + V\alpha = 0 \qquad (18.41)$$

where α is a vector of the angles of incidence of each panel. Equation (18.40) can be solved to find the unknown vortex strengths, but it can also be rearranged to determine the vector of lifts on each panel such that

$$L = \rho V S\Gamma = \frac{\rho V^2}{2} \text{AIC}_R \alpha \qquad (18.42)$$

where AIC_R is the fully-populated 4×4 matrix of aerodynamic influence coefficients with $\text{AIC}_R = -2S\Psi^{-1}$ for the 3D case and $S = \text{diag}[s\,s\,s\,s]$ is the 4×4 diagonal matrix of the spans of each panel, equal in this example. For this steady case the AIC matrix will be real; however, the subscript R is used since in the unsteady case there will be both real and imaginary terms. The equation is related to the two-dimensional version for an aerofoil section in Equation (18.34).

If the same example were solved using *strip theory*, then two strips of width s and chord c are considered (i.e. one strip for each wing) and the lift curve slope is assumed to be 2π. The lift vector for the two strips is given by

$$L = \frac{\rho V^2}{2} \begin{bmatrix} 2\pi cs & 0 \\ 0 & 2\pi cs \end{bmatrix} \alpha \tag{18.43}$$

and now the AIC matrix is diagonal as the flow over one strip does not affect the other strip.

Panel method aerodynamic methods are formulated for the entire wing span. If only the semi-span is being considered, such as for a fixed root wing, then a further transformation is required (Katz and Plotkin, 2001) to allow for only one half of the aircraft being considered. This ensures that there is a non-zero lift at the wing root.

As in the two-dimensional case, it is also possible to use one of a wide range of different aerodynamic components that are available on each element, such as sources, sinks and doublets. It is also feasible to use ringed components (e.g. a vortex ring) rather than the horseshoe modelling approach, and these may contain a distribution of aerodynamic components across the entire area. In general, the same approach as that above is used to determine the strengths of the aerodynamic components and then to calculate the lift and drag.

18.5   Unsteady Aerodynamic Modelling of Wings undergoing Harmonic Motion

So far in this chapter only steady aerodynamics has been considered. However, for dynamic aeroelastic systems, the motion of the lifting surface needs to be taken into account. It was shown in Chapter 9 how the Wagner and Theodorsen functions could be used to model the unsteady effects of the motion of a two-dimensional aerofoil for either a general motion (using convolution) or harmonic motion, and also how the use of the Küssner and Sears functions allowed representation of gust-dependent unsteady aerodynamics. In this section, the panel method approach is extended in concept to model the aerodynamic forces resulting from unsteady harmonic motion.

18.5.1   Harmonic Motion of a Two-dimensional Aerofoil

Consider the two-dimensional aerofoil modelled using two vortices as before, as shown in Figure 18.16, but now the angle of incidence θ and the heave z vary in a harmonic manner. Once a steady state condition is reached, the resulting lift and pitching moment vary sinusoidally, but with a different amplitude and phase from the quasi-steady predictions (depending

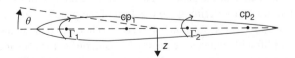

Figure 18.16   Oscillatory motion of a two-dimensional aerofoil.

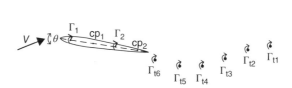

Figure 18.17 Progression of vortices in a time-marching solution for harmonic motion of the aerofoil.

upon the reduced frequency); the lift on each element will in general have a different amplitude and phase. Note also that the wake behind the aerofoil also moves in a sinusoidal manner.

The changing aerodynamic forces due to a predefined structural motion of the aerofoil can be modelled in a 'time-marching' manner by defining a series of vortex rings in the wake (or other aerodynamic components) and, at each time step, changing the strengths of the vortices on the aerofoil, and also the position of the wake vortices in relation to the motion of the structure. Figure 18.17 shows an example of this sort of process six time instants into the computation (after Katz and Plotkin, 2001). For a given instant in time, the strengths of the aerofoil vortices are determined from the boundary conditions at the control points, then the wake vortices (which cannot change their strength) are moved spatially (which will include roll-up effects) to the next time point and the strength of a new wake vortex close to the wing is determined. The aerofoil itself is then moved in time, and the process repeated.

Whereas such an approach could be applied for any general motion of the aerofoil, it does not lend itself to an efficient aeroelastic analysis and will not be considered further. However, a fully coupled methodology based upon harmonic motion does lead to an effective means of determining the aeroelastic response of an aircraft without calculating the development of the wake and is widely used in industry (Albano and Rodden, 1969; Blair, 1994; Rodden, 2011).

Consider once again the oscillating two vortex aerofoil model shown in Figure 18.16. The panel method approach can be extended to include the unsteady effects by allowing the influence coefficients and Aerodynamic Influence Coefficients to become complex and functions of the reduced frequency. Consequently, as the incidence and heave position of the aerofoil vary harmonically with time, the vortex strengths also vary harmonically with time. The zero normal flow condition at each control point still applies at each instant, but now there is a component due to the structural motion that must be included.

The calculation of the complex Aerodynamic Influence Coefficient matrix for vortex or other aerodynamic components (e.g. doublets) is beyond the scope of this book. However, the final form of the relationship between the lift on the elements and the motion of the control points is given by the expression

$$\tilde{L}(k) = \frac{\rho V^2}{2} \mathbf{AIC}(k)_{\text{Unsteady}} h \tag{18.44}$$

where $\mathbf{AIC}(k)_{\text{Unsteady}}$ is the complex AIC matrix (a function of reduced frequency k), $\tilde{L}$ is the vector of lift forces on each panel (with the symbol ~ indicating that it is a complex amplitude; see Chapters 1 and 2) and h is the vector of displacements at the control points (related to the angles of incidence and heave of the panels). Note that there are a number of different ways of writing the AICs and the vector defining the position of the panels; e.g. sometimes the results for panels arranged in a streamwise fashion behind each other are amalgamated into an effective strip. This variation of lift with reduced frequency is analogous to the lift variation with reduced frequency of an oscillating two-dimensional aerofoil described by the Theodorsen function in Chapter 9.

The expression in Equation (18.44) may be rewritten by separating out the real and imaginary parts of the AIC matrix and introducing the reduced frequency such that the complex lift amplitude vector is

$$\tilde{L} = \frac{\rho V^2}{2} \text{AIC}_R h + i\omega \frac{\rho V b}{2 k} \text{AIC}_I h \qquad (18.45)$$

This equation may also be written in a time domain form with structural velocity present such that

$$L = \frac{\rho V^2}{2} \text{AIC}_R h + \frac{\rho V}{2} \frac{b}{k} \text{AIC}_I \dot{h} \qquad (18.46)$$

and this is similar to the expression involving aerodynamic stiffness and damping for a particular reduced frequency as shown in Chapter 9. The matrices would normally be transformed into modal space for aeroelastic and loads calculations.

18.5.2   Harmonic Motion of a Three-Dimensional Wing

The extension to the three-dimensional case, as shown in Figure 18.18 for an oscillating three-dimensional wing modelled with a single horseshoe vortex, follows the same approach as for the steady case. The vortex filaments on the wing oscillate in strength, whereas the trailing vortices vary in both strength and position.

The most common three-dimensional unsteady panel approach is the Doublet Lattice (DL) method (Albano and Rodden, 1969; Blair, 1994; Rodden, 2011). In the DL method, the aerodynamic forces resulting from the unsteady motion are modelled using so-called acceleration potential doublets along the 1/4 chord of each panel. These doublets lead to the calculation of complex AICs that relate the lift acting along the doublet line to the displacement of each panel.

The steady forces corresponding to the underlying steady flow are calculated using the steady Vortex Lattice method as described above, and Figure 18.19 shows a typical panel set-up with both doublets and vortices. The derivation of the Doublet Lattice method is well beyond the scope of this book, but the interested reader is directed to Blair (1994) and Rodden (2011) for a comprehensive explanation.

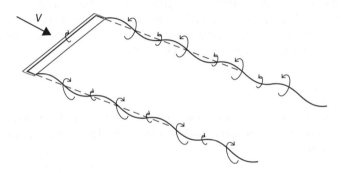

Figure 18.18   Horseshoe vortex on a three-dimensional oscillating wing.

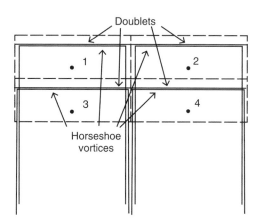

Figure 18.19   Typical doublet and horseshoe vortex set-up for the Doublet Lattice method.

18.6   Aerodynamic Influence Coefficients in Modal Space

Up to this point, the use of a panel aerodynamics method has been illustrated using a simple two-dimensional aerofoil and three-dimensional wing model based on physical displacements of the structure. In practice, the aerodynamic forces are calculated for the whole aircraft and it is more efficient and common to consider aeroelastic and loads analysis in terms of the rigid body and flexible modes of the structure.

In this section, the application of AICs to the aeroelastics and loads equations based in modal space is introduced for steady aerodynamics. Comments are also made about the form of the results for the unsteady case. The aerodynamic forces arising from heave, pitch and flexible mode motions for the free–free aircraft are obtained and related to the aerodynamic derivatives employed for manoeuvres and gust encounters earlier in Chapters 12 to 15; the approach is also relevant to static aeroelasticity and flutter for the whole aircraft.

18.6.1   Heave Displacement

Consider firstly the effect of all the panels undergoing heave displacement z (positive downwards), effectively a displacement in the heave rigid body mode. There will be no resulting change of incidence due to this change of position (in the steady case) and therefore no resulting forces on the NP panels. The net force (downwards) and pitching moment (nose up positive) are therefore zero and so, in essence, are the aerodynamic derivatives, namely $Z_z = M_z = 0$, where, for example, $Z_z = \partial Z / \partial z$ is the downwards force per heave displacement derivative. Likewise, there will be no net force in the elastic mode equations and so the corresponding elastic derivatives for heave displacement are $\mathbf{Q}_z = \partial \mathbf{Q} / \partial z = \mathbf{0}$.

18.6.2   Pitch Angle

Now consider the effect of the wing, and therefore all the panels, undergoing a pitch θ (nose up), effectively a rigid body pitch mode displacement. The resulting change of incidence vector will be given by

$$\boldsymbol{\alpha} = \{1 \quad 1 \quad \dots \quad 1\}^{\mathrm{T}} \theta = \boldsymbol{e}\theta \tag{18.47}$$

where e is a unit column vector. Therefore, from Equation (18.42), using the form where AICs are related to the panel incidence, the vector of lift forces on the panels will be given by

$$L = \frac{\rho V^2}{2} \mathbf{AIC}_R e \theta \tag{18.48}$$

The total heave force (positive downwards) on the wing will be given by the summation of the individual panel forces, so

$$Z = -\sum_{k=1}^{NP} L_k = -\{1 \;\; 1 \;\; \cdots \;\; 1\} L = -e^T L = -\frac{\rho V^2}{2} e^T \mathbf{AIC}_R e \theta \tag{18.49}$$

Thus the *heave force per pitch angle derivative* $Z_\theta = \partial Z / \partial \theta$ will be given in terms of the AIC matrix by

$$Z_\theta = -\frac{\rho V^2}{2} e^T \mathbf{AIC}_R e \tag{18.50}$$

Following a similar approach, the total pitching moment on the wing due to a nose up pitch angle is given by

$$M = \sum_{k=1}^{NP} L_k x_k = \{x_1 \;\; x_2 \;\; \cdots \;\; x_{NP}\}^T L = x^T L = \frac{\rho V^2}{2} x^T \mathbf{AIC}_R e \theta \tag{18.51}$$

where x_k is the distance of the kth panel reference grid point forward of the axis, about which the pitching moment is being evaluated. The *pitching moment per pitch angle derivative* is then given by

$$M_\theta = \frac{\rho V^2}{2} x^T \mathbf{AIC}_R e \tag{18.52}$$

For a flexible aircraft, the lift forces per panel due to the pitch angle introduce a modal force in the elastic mode. If there are M free–free elastic modes of vibration to be considered together with the rigid body heave and pitch modes, then the modal force vector Q due to the pitch angle is given by pre-multiplying the lift force vector by the transpose of the modal matrix defined at the panel grid points (i.e. Φ is an $NP \times M$ modal matrix), so

$$Q = -\Phi^T L = -\frac{\rho V^2}{2} \Phi^T \mathbf{AIC}_R e \theta \tag{18.53}$$

Therefore the modal force per pitch angle derivative vector is given by

$$Q_\theta = -\frac{\rho V^2}{2} \Phi^T \mathbf{AIC}_R e \tag{18.54}$$

18.6.3   Flexible Mode Motion

The final case to consider is what net lift force, pitching moment and elastic modal force are generated when an elastic mode deformation q occurs. The corresponding derivatives will then be obtained. Firstly, the change in incidence for each panel due to a modal deformation will need to be found using the slope in the mode shape. Consider the jth mode and the lth panel. The corresponding change in incidence at that panel will be given by the modal slope multiplied by the modal coordinate, so

$$\alpha_{lj} = -\left.\frac{\partial \phi}{\partial x}\right|_{lj} q_j \tag{18.55}$$

where the negative sign occurs because the mode shape is defined as positive downwards. The vector of all the changes of incidence due to modal displacements in all the modes will then be given by

$$\boldsymbol{\alpha} = -\boldsymbol{\Phi}_x \boldsymbol{q} \tag{18.56}$$

where $\boldsymbol{\Phi}_x$ is the NP $\times$ M matrix of all the modal slopes at the panels. Using Equations (18.42) and (18.56), it can be seen that the vector of lift forces and net heave force due to the modal displacements are given by

$$\boldsymbol{L} = -\frac{\rho V^2}{2}\mathbf{AIC}_R\boldsymbol{\Phi}_x\boldsymbol{q} \quad \Rightarrow \quad Z = -\boldsymbol{e}^\mathrm{T}\boldsymbol{L} = \frac{\rho V^2}{2}\boldsymbol{e}^\mathrm{T}\mathbf{AIC}_R\boldsymbol{\Phi}_x\boldsymbol{q} \tag{18.57}$$

Therefore the row vectors of heave force and pitching moment per elastic mode derivatives are given by

$$Z_q = \frac{\rho V^2}{2}\boldsymbol{e}^\mathrm{T}\mathbf{AIC}_R\boldsymbol{\Phi}_x \quad \text{and} \quad M_q = -\frac{\rho V^2}{2}\boldsymbol{x}^\mathrm{T}\mathbf{AIC}_R\boldsymbol{\Phi}_x \tag{18.58}$$

Finally, using Equation (18.57) and recognizing that the modal force vector is obtained by pre-multiplying the lift forces by the modal matrix, it may be seen that

$$\boldsymbol{Q} = -\boldsymbol{\Phi}^\mathrm{T}\boldsymbol{L} = \frac{\rho V^2}{2}\boldsymbol{\Phi}^\mathrm{T}\mathbf{AIC}_R\boldsymbol{\Phi}_x\boldsymbol{q} \tag{18.59}$$

Thus the matrix of elastic modal force per modal displacement derivatives is given by

$$\mathbf{Q}_q = \frac{\rho V^2}{2}\boldsymbol{\Phi}^\mathrm{T}\mathbf{AIC}_R\boldsymbol{\Phi}_x \tag{18.60}$$

18.6.4 Summary of Steady Aerodynamic Terms

The full set of steady aerodynamics may be expressed as a vector of aerodynamic forces, moments and elastic forces and is given in terms of the AICs or the derivatives by

$$\left\{\begin{array}{c} Z \\ M \\ \hline Q \end{array}\right\}_{\text{Steady}} = \frac{\rho V^2}{2}\left[\begin{array}{cc|c} 0 & -\boldsymbol{e}^\mathrm{T}\mathbf{AIC}_R\boldsymbol{e} & \boldsymbol{e}^\mathrm{T}\mathbf{AIC}_R\boldsymbol{\Phi}_x \\ 0 & \boldsymbol{x}^\mathrm{T}\mathbf{AIC}_R\boldsymbol{e} & -\boldsymbol{x}^\mathrm{T}\mathbf{AIC}_R\boldsymbol{\Phi}_x \\ \hline 0 & \boldsymbol{\Phi}^\mathrm{T}\mathbf{AIC}_R\boldsymbol{e} & \boldsymbol{\Phi}^\mathrm{T}\mathbf{AIC}_R\boldsymbol{\Phi}_x \end{array}\right]\left\{\begin{array}{c} z \\ \theta \\ \hline q \end{array}\right\} = \left[\begin{array}{cc|c} 0 & Z_\theta & Z_q \\ 0 & M_\theta & M_q \\ \hline 0 & Q_\theta & Q_q \end{array}\right]\left\{\begin{array}{c} z \\ \theta \\ \hline q \end{array}\right\} \tag{18.61}$$

where the $\mathbf{AIC}_R$ terms are at zero reduced frequency for this steady case.

The similarity between this derivative matrix and the representation employed in the equilibrium manoeuvre model used earlier in Chapter 12 may be seen clearly, except that only a single mode was employed there and the aerodynamic terms were derived from integrations over strips as opposed to summations over panels. If the wing alone were considered, as in the earlier Chapter 10 on flutter, then only the elastic generalized coordinates q would be included.

In practice, the AICs are not usually used to determine the aircraft rigid body aerodynamic derivatives, as these are obtained more accurately from other calculations with wind tunnel test adjustments. However, the AIC results may be used to assist in correcting the rigid aircraft derivatives for elastic effects and adjusting the lift and moment distributions over the span.

18.6.5   Unsteady Aerodynamics

In the analysis shown above, only the steady aerodynamics case was examined, with the AIC matrix being real. Once unsteady aerodynamics behaviour is considered, panel methods yield a complex AIC matrix that is reduced frequency dependent. The real part defines the in-phase aerodynamic component and would be in the same form as Equation (18.61), whereas the imaginary part, $\mathbf{AIC_I}$, represents the quadrature component. It should be noted that during dynamic motion, there would be an effective incidence equal to the heave velocity component of any panel divided by the air speed; this means that there would no longer be a slope term $\mathbf{\Phi}_x$ in the matrix expressions for the quadrature aerodynamic force but instead the modal matrix $\mathbf{\Phi}$ would be used. Also the quadrature forces would be proportional to V not V^2.

The quadrature terms now take the form

$$
\left\{ \begin{matrix} \mathbf{Z} \\ \mathbf{M} \\ \overline{\mathbf{Q}} \end{matrix} \right\}_{\substack{Quad \\ Aero}} = i\omega \frac{\rho V}{2} \frac{b}{k} \left[\begin{array}{cc|c} -e^{\mathrm{T}}\mathbf{AIC}_{\mathrm{I}}e & e^{\mathrm{T}}\mathbf{AIC}_{\mathrm{I}}x & -e^{\mathrm{T}}\mathbf{AIC}_{\mathrm{I}}\mathbf{\Phi} \\ x^{\mathrm{T}}\mathbf{AIC}_{\mathrm{I}}e & -x^{\mathrm{T}}\mathbf{AIC}_{\mathrm{I}}x & x^{\mathrm{T}}\mathbf{AIC}_{\mathrm{I}}\mathbf{\Phi} \\ -\mathbf{\Phi}^{\mathrm{T}}\mathbf{AIC}_{\mathrm{I}}e & \mathbf{\Phi}^{\mathrm{T}}\mathbf{AIC}_{\mathrm{I}}x & -\mathbf{\Phi}^{\mathrm{T}}\mathbf{AIC}_{\mathrm{I}}\mathbf{\Phi} \end{array} \right] \left\{ \begin{matrix} z \\ \theta \\ q \end{matrix} \right\} = i\omega \left[\begin{array}{c|c|c} Z_{\dot z} & Z_{\dot \theta} & \mathbf{Z}_{\dot q} \\ M_{\dot z} & M_{\dot \theta} & \mathbf{M}_{\dot q} \\ \mathbf{Q}_{\dot z} & \mathbf{Q}_{\dot \theta} & \mathbf{Q}_{\dot q} \end{array} \right] \left\{ \begin{matrix} z \\ \theta \\ q \end{matrix} \right\} \quad (18.62)
$$

where $\mathbf{AIC_I}$ is the quadrature/imaginary part of the AIC matrix. Note that there would be rate-dependent derivatives, all of which are nominally nonzero. The total aerodynamic terms for a particular reduced frequency are the sum of the expressions in Equations (18.61) and (18.62). Similar terms are seen in the gust response model in Chapter 14.

18.6.6   Gust-dependent Terms

All the above analysis has been derived for response-dependent aerodynamic terms. However, for flight through gusts and turbulence, the aerodynamic forces due to gust velocity need to be evaluated. The panel methods allow these terms to be obtained, with gust penetration lags present in the frequency domain analysis to allow for the spatial separation of panels on different parts of the wing, and also on the tailplane (Hoblit, 1988).

18.7 Examples

1. For a two-dimensional aerofoil, develop programs to determine the lift and pitching moment about the mid-chord for distributions of vortices, sources and sinks, and doublets. Investigate the effect of increasing the number of aerodynamic elements included along the chord.
2. Determine the resultant downwash at the 3/4 chord control point of a vortex ring with each side of the ring having the same strength of vortex.
3. Develop programs to determine the steady lift distribution for a three-dimensional wing with either horseshoe vortices or vortex rings. Investigate the effect of increasing the number of aerodynamic panels in both the spanwise and chordwise directions.

19

Coupling of Structural and Aerodynamic Computational Models

The majority of the book so far has shown how aeroelastic and loads models can be developed using continuous approximation models of the structure and aerodynamics. Such an approach is fine for a simplistic aircraft representation; however, real-life structures are non-uniform and consequently are impossible to model accurately using approaches such as the Rayleigh–Ritz method. Instead, industry makes use of discrete approximation methods, such as Finite Elements (see Chapters 17 and 21), to produce detailed models of the aircraft structure. Similarly, numerical 3D panel methods, such as the Vortex or Doublet Lattice methods (see Chapters 18 and 21), are often used to represent the aerodynamic forces acting on the aircraft. Although more sophisticated Computational Fluid Dynamics (CFD) methods have been developed and used, for example, in performing accurate drag calculations and analysis in the transonic flight regime, the vast majority of the aeroelastic and loads analyses carried out in industry for commercial aircraft are performed using 3D panel methods (sometimes the panel methods are used to correct the rigid aircraft aerodynamics for flexible effects). Such an approach enables the structure and the aerodynamic models to be combined in a very efficient manner so that the static/dynamic aeroelastic and loads behaviour can be determined.

This chapter shows how potential flow aerodynamics can be combined with a structural model to produce static and dynamic aeroelastic (or loads) models. A 2D rigid aerofoil in pitch and heave/pitch is modelled using two elements with vortex components (see Chapter 18). Also, a simple 3D built-in wing structure is modelled using only two beam finite elements and four aerodynamic panels in order to illustrate the process of generating a coupled model for both static and dynamic aeroelastic analyses. Finally, the development of a state-space aeroelastic model using a Rational Function Approximation to the reduced frequency dependent aerodynamics is described.

Introduction to Aircraft Aeroelasticity and Loads, Second Edition. Jan R. Wright and Jonathan E. Cooper.
© 2015 John Wiley & Sons, Ltd. Published 2015 by John Wiley & Sons, Ltd.

19.1   Mathematical Modelling – Static Aeroelastic Case

Before considering simple examples, a general form of the static analysis will be presented for the coupling of 3D panel and Finite Element representations. The static deflection behaviour of a Finite Element model relates the displacement vector r to the force vector R via the overall stiffness matrix $\mathbf{K}$ (see Chapter 17), such that

$$R = \mathbf{K}r \tag{19.1}$$

From Chapter 18, in the general case where varying sized panels are used, the aerodynamic panel methods lead to a vector of lift forces, distributed along the 1/4 chord of each panel, namely

$$L = \rho V \mathbf{S} \Gamma \tag{19.2}$$

where $\mathbf{S}$ is the diagonal matrix whose elements are the spans for each panel. The elements of vector Γ are the strengths of the vortices acting on each panel, determined from the zero normal flow boundary conditions

$$\Psi \Gamma = -V(\theta + \theta_0) \tag{19.3}$$

where Ψ is the matrix of influence coefficients, V is the free-stream air speed, θ is the vector of angles of incidence for each panel due to elastic twist (assuming small angles) and θ_0 is the vector of initial angles of incidence of each panel. The initial angle of incidence allows the wing twist to be determined (see Chapter 7).

The aerodynamic terms depend solely upon the incidence of each panel relative to the free-stream, whereas deflection and initial (wind off) vectors r and r_0 for the structure contain translation and rotation terms in all degrees of freedom (DoF). In order to make the two models compatible, Equation (19.3) can be rewritten as

$$\Psi \Gamma = -V \mathbf{T}_1 (r + r_0) \tag{19.4}$$

where the transformation matrix $\mathbf{T}_1$ relates the angles of incidence of each panel to the structural displacements and rotations and so allows a full set of structural DoFs to be included in the equations.

The aerodynamic forces on the panels cause an equivalent set of applied forces and moments to act at the nodes of the FE model and are a function of the vortex strengths so

$$R = \rho V \mathbf{T}_2 \mathbf{S} \Gamma \tag{19.5}$$

where $\mathbf{T}_2$ is another transformation matrix that maps forces between the aerodynamic and the structural models. In practice, it is usual for the fluid and structure meshes to be different in terms of density, orientation and node position; consequently the mappings of the panel deflections to the structural displacements and of the fluid forces to the structural grid are not straightforward and must be achieved via some form of interpolation.

Combining equations (19.1), (19.4) and (19.5) gives

$$R = \mathbf{K}r = \rho V \mathbf{T}_2 \mathbf{S} \Gamma = -\rho V^2 \mathbf{T}_2 \mathbf{S} \Psi^{-1} \mathbf{T}_1 (r + r_0) = \frac{\rho V^2}{2} \mathbf{AIC}(r + r_0) \tag{19.6}$$

where the Aerodynamic Influence Coefficient matrix **AIC** is defined in Chapter 18 and relates the forces to the deflections such that

$$\mathbf{AIC} = -2\mathbf{T}_2\mathbf{S}\boldsymbol{\Psi}^{-1}\mathbf{T}_1 \tag{19.7}$$

Thus, Equation (19.6) becomes

$$\left(\rho V^2\mathbf{T}_2\mathbf{S}\boldsymbol{\Psi}^{-1}\mathbf{T}_1 + \mathbf{K}\right)r + \rho V^2\mathbf{T}_2\mathbf{S}\boldsymbol{\Psi}^{-1}\mathbf{T}_1 r_0 = 0$$

or $\tag{19.8}$

$$\left(\frac{\rho V^2}{2}\mathbf{AIC} + \mathbf{K}\right)r + \frac{\rho V^2}{2}\mathbf{AIC}r_0 = 0$$

which is in the form of the classic static aeroelastic equations

$$\rho V^2\mathbf{C}(r + r_0) + \mathbf{E}r = 0 \tag{19.9}$$

Solution of these equations leads to the aeroelastic deflections r from which the vortex strengths can be calculated using Equation (19.4). Then the lift and induced drag can be calculated. The divergence speed can also be determined from Equation (19.9), as in Chapter 10, by solving the determinant $|\rho V^2\mathbf{C} + \mathbf{E}| = 0$ or via the eigenvalues of $-\mathbf{C}^{-1}\mathbf{E}/\rho$.

19.2   2D Coupled Static Aeroelastic Model – Pitch

Consider the 2D symmetric aerofoil of chord c, modelled aerodynamically using the two elements with vortex components of unknown strengths Γ_1 and Γ_2 shown in Figure 19.1; the aerofoil rotates on a torsional spring of stiffness k_θ per unit span at the mid-chord. The total incidence is the sum of the initial angle θ_0 and the elastic twist θ.

The vortex strengths are calculated from the zero normal flow boundary condition at the control points using the influence coefficient matrix but now the unknown elastic twist θ must be included so (for small angles)

$$\begin{bmatrix} -\dfrac{2}{c\pi} & \dfrac{2}{c\pi} \\[2mm] -\dfrac{2}{3c\pi} & -\dfrac{2}{c\pi} \end{bmatrix} \begin{Bmatrix} \Gamma_1 \\ \Gamma_2 \end{Bmatrix} = -V \begin{Bmatrix} \theta_0 + \theta \\ \theta_0 + \theta \end{Bmatrix} \Rightarrow \begin{Bmatrix} \Gamma_1 \\ \Gamma_2 \end{Bmatrix} = \begin{Bmatrix} \dfrac{3}{4} \\[2mm] \dfrac{1}{4} \end{Bmatrix} \pi c V(\theta_0 + \theta) \tag{19.10}$$

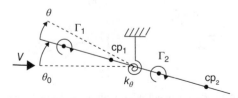

Figure 19.1   Two element wing with a torsional spring.

The aerodynamic moment is balanced by the structural restoring moment so

$$M = \rho V \Gamma_1 \frac{3c}{8} - \rho V \Gamma_2 \frac{c}{8} = \left\{ \frac{3c\rho V}{8} - \frac{c\rho V}{8} \right\} \left\{ \begin{matrix} \Gamma_1 \\ \Gamma_2 \end{matrix} \right\} = k_\theta \theta \tag{19.11}$$

Combining equations (19.10) and (19.11) leads to

$$\left\{ \frac{3c\rho V}{8} - \frac{c\rho V}{8} \right\} \left\{ \begin{matrix} \frac{3}{4} \\ \frac{1}{4} \end{matrix} \right\} \pi c V (\theta_0 + \theta) = k_\theta \theta \Rightarrow \left(k_\theta - \frac{\rho V^2 \pi c^2}{4} \right) \theta = \frac{\rho V^2 \pi c^2}{4} \theta_0 \tag{19.12}$$

and thus the elastic angle of twist can be expressed in terms of the initial incidence as

$$\theta = \frac{\rho V^2 \pi c^2}{(4 k_\theta - \rho V^2 \pi c^2)} \theta_0 \tag{19.13}$$

The vortex strengths may then be determined from Equation (19.10), hence the lift and pitching moment. The divergence speed occurs when the twist becomes infinite and so is given by

$$V_{\text{div}} = \frac{2}{c} \sqrt{\frac{k_\theta}{\rho \pi}} \tag{19.14}$$

which is exactly the same result as found using strip theory.

19.3 2D Coupled Static Aeroelastic Model – Heave/Pitch

Consider the same model as above, but a spring of stiffness k_z per unit span in the heave direction (z positive downwards) is also included, as seen in Figure 19.2. The total lift from both vortex components is

$$L = -(\rho V \Gamma_1 + \rho V \Gamma_2) = -\rho V \{1 \quad 1\} \left\{ \begin{matrix} \Gamma_1 \\ \Gamma_2 \end{matrix} \right\} \tag{19.15}$$

The pitching moment about the mid-chord is given as before in Equation (19.11) and the vortex strengths are found from the zero normal flow boundary condition in Equation (19.10).

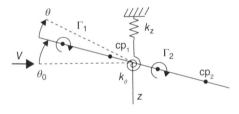

Figure 19.2 Two-element wing with heave and torsional springs.

The lift and pitching moment are related to the structural deflections via the stiffness matrix so

$$\left\{ \begin{matrix} L \\ M \end{matrix} \right\} = \begin{bmatrix} k_z & 0 \\ 0 & k_\theta \end{bmatrix} \left\{ \begin{matrix} z \\ \theta \end{matrix} \right\} \tag{19.16}$$

Combining equations (19.10), (19.11) and (19.15) into (19.16) gives the equivalent of Equation (19.9), namely

$$\begin{bmatrix} -\rho V & -\rho V \\ \dfrac{3\rho Vc}{8} & -\dfrac{\rho Vc}{8} \end{bmatrix} \left\{ \begin{matrix} \Gamma_1 \\ \Gamma_2 \end{matrix} \right\} = \begin{bmatrix} -\rho V & -\rho V \\ \dfrac{3\rho Vc}{8} & -\dfrac{\rho Vc}{8} \end{bmatrix} \left\{ \begin{matrix} \dfrac{3\pi c V}{4} \\ \dfrac{\pi c V}{4} \end{matrix} \right\}(\theta_0 + \theta) = \begin{bmatrix} k_z & 0 \\ 0 & k_\theta \end{bmatrix} \left\{ \begin{matrix} z \\ \theta \end{matrix} \right\} \tag{19.17}$$

which after some rearranging leads to

$$\begin{bmatrix} k_z & \pi \rho c V^2 \\ 0 & \left(k_\theta - \dfrac{\pi \rho c^2 V^2}{4} \right) \end{bmatrix} \left\{ \begin{matrix} z \\ \theta \end{matrix} \right\} = \rho V^2 \left\{ \begin{matrix} -\pi c \\ \dfrac{\pi c^2}{4} \end{matrix} \right\} \theta_0 \tag{19.18}$$

This equation is in the classical static aeroelastic form. The elastic heave and torsional deflections are then found as

$$\left\{ \begin{matrix} z \\ \theta \end{matrix} \right\} = \left\{ \begin{matrix} -\dfrac{4 k_\theta p}{k_z (4 k_\theta - cp)} \\ \dfrac{cp}{(4 k_\theta - cp)} \end{matrix} \right\} \theta_0 \tag{19.19}$$

where $p = \pi \rho c V^2$. The same results for lift, pitching moment and divergence speed are found as for strip theory.

19.4   3D Coupled Static Aeroelastic Model

In this example, the coupling of panel method aerodynamics with an FE model is considered. The cantilevered wing model in Figure 19.3 has chord c and semi-span $s = 2 L$. It consists of two beam elements of length L along the mid-chord to allow for bending and torsion, and four aerodynamic panels whose AICs can be defined from a distribution of vortex components placed on the panels (see Chapter 18). This model is rather idealized, but is employed to illustrate some ideas involved in the solution process. Note that in practice many more finite elements and panels, both chordwise and spanwise, have to be used in order to achieve accurate results.

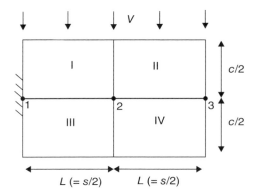

Figure 19.3   Wing represented by two beam elements with four aerodynamic panels.

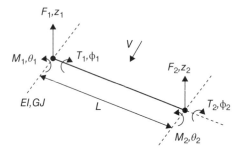

Figure 19.4   Two-node beam element.

19.4.1   Structural Model

Consider the general 2-node beam bending/torsion element (see Chapter 17) shown in Figure 19.4. At each node, a vertical deflection z, bending rotation θ and twist ϕ are shown, with corresponding normal force F, moment M and torque T. The element stiffness matrix, for coordinates in the order $(z_1, \theta_1, \phi_1, z_2, \theta_2, \phi_2)$, is

$$
k = \begin{bmatrix}
12\dfrac{EI}{L^3} & 6\dfrac{EI}{L^2} & 0 & -12\dfrac{EI}{L^3} & 6\dfrac{EI}{L^2} & 0 \\[2ex]
6\dfrac{EI}{L^2} & 4\dfrac{EI}{L} & 0 & -6\dfrac{EI}{L^2} & 2\dfrac{EI}{L} & 0 \\[2ex]
0 & 0 & \dfrac{GJ}{L} & 0 & 0 & -\dfrac{GJ}{L} \\[2ex]
-12\dfrac{EI}{L^3} & -6\dfrac{EI}{L^2} & 0 & 12\dfrac{EI}{L^3} & -6\dfrac{EI}{L^2} & 0 \\[2ex]
6\dfrac{EI}{L^2} & 2\dfrac{EI}{L} & 0 & -6\dfrac{EI}{L^2} & 4\dfrac{EI}{L} & 0 \\[2ex]
0 & 0 & -\dfrac{GJ}{L} & 0 & 0 & \dfrac{GJ}{L}
\end{bmatrix}
\tag{19.20}
$$

Remembering to partition out the terms relating to the nodal degrees of freedom at the wing root (see Chapter 17), the overall system of structural equations for the two elements combined may then be shown to be

$$
R = \begin{Bmatrix} F_2 \\ M_2 \\ T_2 \\ F_3 \\ M_3 \\ T_3 \end{Bmatrix} = \begin{bmatrix} 24\dfrac{EI}{L^3} & 0 & 0 & -12\dfrac{EI}{L^3} & 6\dfrac{EI}{L^2} & 0 \\[2mm] 0 & 8\dfrac{EI}{L} & 0 & -6\dfrac{EI}{L^2} & 2\dfrac{EI}{L} & 0 \\[2mm] 0 & 0 & 2\dfrac{GJ}{L} & 0 & 0 & -\dfrac{GJ}{L} \\[2mm] -12\dfrac{EI}{L^3} & -6\dfrac{EI}{L^2} & 0 & 12\dfrac{EI}{L^3} & -6\dfrac{EI}{L^2} & 0 \\[2mm] 6\dfrac{EI}{L^2} & 2\dfrac{EI}{L} & 0 & -6\dfrac{EI}{L^2} & 4\dfrac{EI}{L} & 0 \\[2mm] 0 & 0 & -\dfrac{GJ}{L} & 0 & 0 & \dfrac{GJ}{L} \end{bmatrix} \begin{Bmatrix} z_2 \\ \theta_2 \\ \phi_2 \\ z_3 \\ \theta_3 \\ \phi_3 \end{Bmatrix} = K\,r \qquad (19.21)
$$

and the notation needs to be converted later from L for the element to s for the wing, using $L = s/2$.

19.4.2 Aerodynamic Model

The vortex strengths $\Gamma_{I, II, II, IV}$ may be found from the rearrangement of Equation (19.3) for this example, namely

$$
\Psi \begin{Bmatrix} \Gamma_I \\ \Gamma_{II} \\ \Gamma_{III} \\ \Gamma_{IV} \end{Bmatrix} = -V \begin{Bmatrix} \alpha_I + \alpha_{0I} \\ \alpha_{II} + \alpha_{0II} \\ \alpha_{III} + \alpha_{0III} \\ \alpha_{IV} + \alpha_{0IV} \end{Bmatrix} \qquad (19.22)
$$

where Ψ is the (4×4) influence coefficient matrix determined using the approach described in Chapter 18. The α and α_0 terms are the angles of incidence of each panel due to elastic and initial deflections respectively.

The total angle of incidence depends upon the twist of the beam ϕ which is defined at each node. Therefore, for convenience, take the angle of incidence of each panel to be the average twist between the nodes at each end of the beam that they are attached to. Since there is no chordwise bending of the wing, the angles are equal for the front and rear panels at each spanwise location. Also, the FE model twist is defined as positive nose down so there will be a negative sign present in the relationship between twist and incidence. Thus

$$
\alpha_I + \alpha_{0I} = \alpha_{III} + \alpha_{0III} = -\frac{\phi_1 + \phi_{10} + \phi_2 + \phi_{20}}{2} \qquad \alpha_{II} + \alpha_{0II} = \alpha_{IV} + \alpha_{0IV} = -\frac{\phi_2 + \phi_{20} + \phi_3 + \phi_{30}}{2}
$$

$$
(19.23)
$$

Since the wing has a fixed root boundary condition and the FE equations have been partitioned, then the twist at node 1 must be removed from Equation (19.23). Obviously the elastic twist at the root is zero, but it will be assumed for simplicity that the initial incidence at the root is also zero. Thus, Equation (19.22) becomes

$$
\mathbf{\Gamma} = \begin{Bmatrix} \Gamma_{\mathrm{I}} \\ \Gamma_{\mathrm{II}} \\ \Gamma_{\mathrm{III}} \\ \Gamma_{\mathrm{IV}} \end{Bmatrix} = -V\mathbf{\Psi}^{-1} \begin{bmatrix} 0 & 0 & -\dfrac{1}{2} & 0 & 0 & 0 \\[2mm] 0 & 0 & -\dfrac{1}{2} & 0 & 0 & -\dfrac{1}{2} \\[2mm] 0 & 0 & -\dfrac{1}{2} & 0 & 0 & 0 \\[2mm] 0 & 0 & -\dfrac{1}{2} & 0 & 0 & -\dfrac{1}{2} \end{bmatrix} \left(\begin{Bmatrix} z_2 \\ \theta_2 \\ \phi_2 \\ z_3 \\ \theta_3 \\ \phi_3 \end{Bmatrix} + \begin{Bmatrix} z_{20} \\ \theta_{20} \\ \phi_{20} \\ z_{30} \\ \theta_{30} \\ \phi_{30} \end{Bmatrix} \right) = -V\mathbf{\Psi}^{-1}\mathbf{T}_1(r + r_0)
$$

$$(19.24)$$

where $\mathbf{T}_1$ is the transformation matrix. Again, in practice, the transformation between the FE displacements and the panel deflections will be much more complicated and is likely to require an interpolation process.

19.4.3 Transformation of Aerodynamic Forces to Structural Model

From Equation (19.2), since each panel has the same span $s/2$, the vector of lift forces on each panel is

$$
\mathbf{L} = \begin{Bmatrix} L_{\mathrm{I}} \\ L_{\mathrm{II}} \\ L_{\mathrm{III}} \\ L_{\mathrm{IV}} \end{Bmatrix} = \rho V \begin{bmatrix} \dfrac{s}{2} & 0 & 0 & 0 \\[2mm] 0 & \dfrac{s}{2} & 0 & 0 \\[2mm] 0 & 0 & \dfrac{s}{2} & 0 \\[2mm] 0 & 0 & 0 & \dfrac{s}{2} \end{bmatrix} \begin{Bmatrix} \Gamma_{\mathrm{I}} \\ \Gamma_{\mathrm{II}} \\ \Gamma_{\mathrm{III}} \\ \Gamma_{\mathrm{IV}} \end{Bmatrix} = \rho V \mathbf{S} \mathbf{\Gamma} \qquad (19.25)
$$

The equivalent forces and moments acting at the element nodes, as shown in Figure 19.5, may be determined based on the idea of kinematically equivalent nodal forces (see Chapter 4). This is a transformation of distributed aerodynamic forces (acting along the ¼ chord of the panels) onto the structural model. It can be shown that the equivalent of Equation (19.5) is

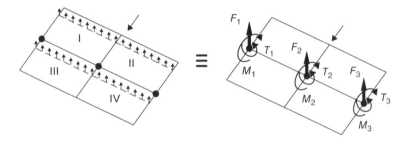

Figure 19.5 Equivalent forces and moments at the beam nodes.

$$
R = \begin{Bmatrix} F_2 \\ M_2 \\ T_2 \\ F_3 \\ M_3 \\ T_3 \end{Bmatrix} = \rho V
\begin{bmatrix}
\dfrac{1}{2} & \dfrac{1}{2} & \dfrac{1}{2} & \dfrac{1}{2} \\[2mm]
-\dfrac{s}{24} & \dfrac{s}{24} & -\dfrac{s}{24} & \dfrac{s}{24} \\[2mm]
-\dfrac{3c}{16} & -\dfrac{3c}{16} & \dfrac{c}{16} & \dfrac{c}{16} \\[2mm]
0 & \dfrac{1}{2} & 0 & \dfrac{1}{2} \\[2mm]
0 & -\dfrac{s}{24} & 0 & -\dfrac{s}{24} \\[2mm]
0 & -\dfrac{3c}{16} & 0 & \dfrac{c}{16}
\end{bmatrix}
\begin{bmatrix}
\dfrac{s}{2} & 0 & 0 & 0 \\[2mm]
0 & \dfrac{s}{2} & 0 & 0 \\[2mm]
0 & 0 & \dfrac{s}{2} & 0 \\[2mm]
0 & 0 & 0 & \dfrac{s}{2}
\end{bmatrix}
\begin{Bmatrix} \Gamma_{\mathrm{I}} \\ \Gamma_{\mathrm{II}} \\ \Gamma_{\mathrm{III}} \\ \Gamma_{\mathrm{IV}} \end{Bmatrix}
= \rho V \mathbf{T}_2 \mathbf{S} \boldsymbol{\Gamma}
\qquad (19.26)
$$

where $\mathbf{T}_2$ is the relevant transformation matrix.

19.4.4 Assembly of Aeroelastic Model

All the equations may now be combined to generate a coupled aeroelastic system in the form of Equation (19.6), namely

$$
\begin{bmatrix}
192\dfrac{EI}{s^3} & 0 & 0 & -96\dfrac{EI}{s^3} & 24\dfrac{EI}{s^2} & 0 \\[3mm]
0 & 16\dfrac{EI}{s} & 0 & -24\dfrac{EI}{s^2} & 4\dfrac{EI}{s} & 0 \\[3mm]
0 & 0 & 4\dfrac{GJ}{s} & 0 & 0 & -2\dfrac{GJ}{s} \\[3mm]
-96\dfrac{EI}{s^3} & -24\dfrac{EI}{s^2} & 0 & 96\dfrac{EI}{s^3} & -24\dfrac{EI}{s^2} & 0 \\[3mm]
24\dfrac{EI}{s^2} & 4\dfrac{EI}{s} & 0 & -24\dfrac{EI}{s^2} & 8\dfrac{EI}{s} & 0 \\[3mm]
0 & 0 & -2\dfrac{GJ}{s} & 0 & 0 & 2\dfrac{GJ}{s}
\end{bmatrix}
\begin{Bmatrix} w_2 \\ \theta_2 \\ \phi_2 \\ w_3 \\ \theta_3 \\ \phi_3 \end{Bmatrix} =
$$

$$
-\rho V^2
\begin{bmatrix}
\dfrac{1}{2} & \dfrac{1}{2} & \dfrac{1}{2} & \dfrac{1}{2} \\[2mm]
-\dfrac{s}{24} & \dfrac{s}{24} & -\dfrac{s}{24} & \dfrac{s}{24} \\[2mm]
-\dfrac{3c}{16} & -\dfrac{3c}{16} & \dfrac{c}{16} & \dfrac{c}{16} \\[2mm]
0 & \dfrac{1}{2} & 0 & \dfrac{1}{2} \\[2mm]
0 & -\dfrac{s}{24} & 0 & -\dfrac{s}{24} \\[2mm]
0 & -\dfrac{3c}{16} & 0 & \dfrac{c}{16}
\end{bmatrix}
\mathbf{S}\boldsymbol{\Psi}^{-1}
\begin{bmatrix}
0 & 0 & -\dfrac{1}{2} & 0 & 0 & 0 \\[2mm]
0 & 0 & -\dfrac{1}{2} & 0 & 0 & -\dfrac{1}{2} \\[2mm]
0 & 0 & -\dfrac{1}{2} & 0 & 0 & 0 \\[2mm]
0 & 0 & -\dfrac{1}{2} & 0 & 0 & -\dfrac{1}{2}
\end{bmatrix}
\left(\begin{Bmatrix} w_2 \\ \theta_2 \\ \phi_2 \\ w_3 \\ \theta_3 \\ \phi_3 \end{Bmatrix}
+ \begin{Bmatrix} w_{20} \\ \theta_{20} \\ \phi_{20} \\ w_{30} \\ \theta_{30} \\ \phi_{30} \end{Bmatrix} \right)
$$

$$
(19.27)
$$

A similar approach is taken when more elements and aerodynamic panels are used and also when the control points of the aerodynamic model and nodes of the structural model do not coincide. However, the interpolation process between the grids for displacements and aerodynamic forces will be far more involved.

19.5   Mathematical Modelling – Dynamic Aeroelastic Response

When the dynamic case is considered, the FE model relates the displacement and acceleration vectors to the force vector R via the mass and stiffness matrices (ignoring structural damping) such that

$$R = \mathbf{M}\ddot{r} + \mathbf{K}r \tag{19.28}$$

For some defined harmonic oscillation at frequency ω (and by implication reduced frequency k), the vectors of forces acting at the 1/4 chord of each panel is defined in terms of the influence coefficients (usually derived from the Doublet Lattice method for the dynamic case) such that

$$\mathbf{\Psi}\mathbf{\Gamma} = -V\mathbf{T}_1 r \tag{19.29}$$

where the matrix $\mathbf{\Psi}$ and the vector of circulation $\mathbf{\Gamma}$ are now complex as mentioned in Chapter 18. In this case the flow boundary condition is expressed in terms of the structural displacements and a transformation matrix $\mathbf{T}_1$ as before. The initial deformed shape r_0 is not included here since only the flutter solution is sought.

The lift forces on the aerodynamic panels are related to equivalent forces and moments acting at the FE model nodes as before via

$$R = \rho V \mathbf{T}_2 \mathbf{S}\mathbf{\Gamma} = -\rho V^2 \mathbf{T}_2 \mathbf{S}\mathbf{\Psi}^{-1}\mathbf{T}_1 r \tag{19.30}$$

Combining equations (19.28) gives

$$\mathbf{M}\ddot{r} + \mathbf{K}r = R = \frac{\rho V^2}{2}\mathbf{AIC}\, r \tag{19.31}$$

where the AIC matrix is now complex and is a function of reduced frequency $k = \omega b/V$. Equation (19.31) can be written in terms of the real and imaginary parts of the AIC matrix (see Chapter 18) as

$$\mathbf{M}\ddot{r} + \mathbf{K}r = \frac{\rho V}{2}\frac{b}{k}\mathbf{AIC}_\mathrm{I}\dot{r} + \frac{\rho V^2}{2}\mathbf{AIC}_\mathrm{R}r \tag{19.32}$$

This equation can be transformed into modal coordinates q if required (see Chapter 2) via the modal matrix $\mathbf{\Phi}$, then using $r = \mathbf{\Phi}q$ and pre-multiplying by $\mathbf{\Phi}^\mathrm{T}$ yields

$$\mathbf{M}_\mathrm{q}\ddot{q} + \mathbf{K}_\mathrm{q}q = \frac{\rho V}{2}\frac{b}{k}\mathbf{\Phi}^\mathrm{T}\mathbf{AIC}_\mathrm{I}\mathbf{\Phi}\dot{q} + \frac{\rho V^2}{2}\mathbf{\Phi}^\mathrm{T}\mathbf{AIC}_\mathrm{R}\mathbf{\Phi}q \tag{19.33}$$

Note that the equations do not apply in this form for quasi-steady motion (zero reduced frequency), so the apparent singularity as $k \to 0$ does not occur. Inspection of Equation (19.33) shows that it is in the form of the classic aeroelastic equation with reduced frequency dependent aerodynamic terms such that

$$\mathbf{A}\ddot{q} + \rho V\mathbf{B}\dot{q} + \left(\rho V^2\mathbf{C} + \mathbf{E}\right)q = 0 \tag{19.34}$$

which is sometimes written in the form

$$\mathbf{A}\ddot{q} + \mathbf{E}q = \frac{\rho V^2}{2}\mathbf{Q}q \qquad (19.35)$$

where the complex matrix $\mathbf{Q}$ contains the reduced frequency dependent terms, and frequency matching methods discussed in Chapter 10, such as the 'p–k' method, must be used.

19.6   2D Coupled Dynamic Aeroelastic Model – Bending/Torsion

Consider the same aeroelastic 2D aerofoil model as in Section 19.2 but now with numerical values used and taking dynamic effects into account. Then the equations of motion become

$$\begin{bmatrix} M & 0 \\ 0 & I_G \end{bmatrix}\begin{Bmatrix} \ddot{z} \\ \ddot{\theta} \end{Bmatrix} + \begin{bmatrix} k_z & 0 \\ 0 & k_\theta \end{bmatrix}\begin{Bmatrix} z \\ \theta \end{Bmatrix} = \frac{\rho V}{2}\frac{b}{k}\mathbf{AIC_I}\begin{Bmatrix} \dot{z} \\ \dot{\theta} \end{Bmatrix} + \frac{\rho V^2}{2}\mathbf{AIC_R}\begin{Bmatrix} z \\ \theta \end{Bmatrix} \qquad (19.36)$$

where M and I_G are the mass and pitch moment of inertia of the aerofoil per unit span. For this example, use numerical values and take the modal mass and stiffness matrices as $\mathbf{M_q} = \text{diag} [1\ 1]$ and $\mathbf{K_q} = \text{diag} [852\ 33070]$. The mode shapes are pure heave and pitch respectively, with undamped natural frequencies of 4.65 and 28.95 Hz.

The four AIC values are complex and were calculated in modal format using the Doublet Lattice method from a well known commercial code. The AICs are dependent upon the reduced frequency and are shown in Figure 19.6. The points on each plot indicate the values at which the AICs are explicitly calculated; other intermediate values required when performing a 'p–k' flutter solution may be found by interpolation.

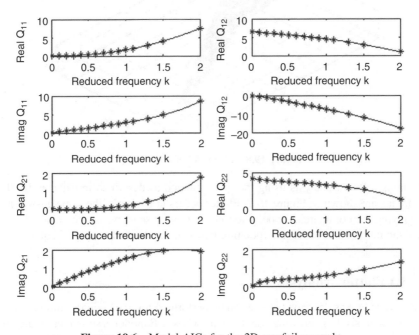

Figure 19.6   Modal AICs for the 2D aerofoil example

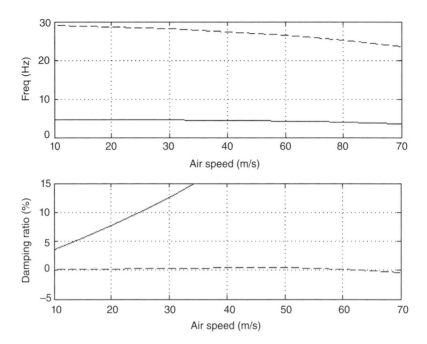

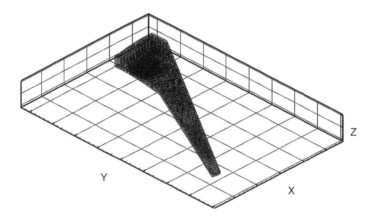

Figure 19.7 Frequency and damping ratio trends for the 2D aerofoil example.

Figure 19.8 FE model of a 3D wing.

Performing a flutter analysis using the 'p–k' method as described in Chapter 10, then $V\omega$ and Vg trends are shown in Figure 19.7. Flutter occurs at 64.1 m/s, the same answer as given when applying the commercial code. The trends are very sensitive to the values of the AICs and care should be taken when interpolating between the defined reduced frequency values.

19.7 3D Flutter Analysis

The same procedure as described above can be performed for a 3D wing or full aircraft. As an example, an analysis was performed on the wing whose FE model is shown in Figure 19.8 and aerodynamic panels shown in Figure 19.9 as part of a study (Taylor *et al.*, 2006) to compare

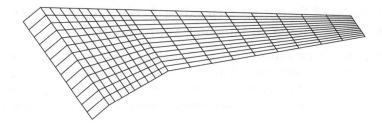

Figure 19.9 Aerodynamic panels for the 3D wing.

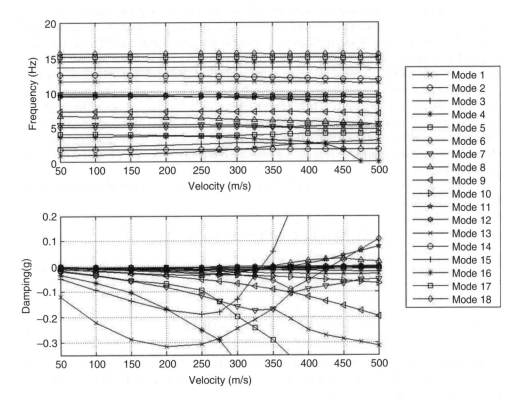

Figure 19.10 Frequency and damping ratio trends for 3D wing.

the flutter boundary predictions between potential flow aerodynamics (Doublet Lattice) and Euler and Navier-Stokes CFD methods.

The resulting frequency and damping ratio trends from a 'p–k' analysis, using AIC matrices for unsteady aerodynamics obtained using the Doublet Lattice method, are shown in Figure 19.10. This is much more complicated to interpret than the 2DoF flutter results as there are many more modes; however, it can be seen that flutter occurs at 332 m/s. Note that the damping results are plotted using the convention that positive damping is unstable (see 'k' method in Chapter 10).

In practice, this analysis would be applied to the whole aircraft model including rigid body modes, thus enabling the response to flight/ground manoeuvres, gust/turbulence encounters and control inputs as well as static aeroelastic deflections to be obtained.

19.8   Inclusion of Frequency Dependent Aerodynamics for State–Space Modelling – Rational Function Approximation

It has been shown that the effects of reduced frequency dependent aerodynamics need to be taken into account when performing a *frequency* domain aeroelastic analysis (Chapter 10) or modelling the response to continuous turbulence (Chapter 14); this is possible using the Theodorsen and Sears functions for 2D strip theory or using the relevant AICs from a 3D panel method.

In the same way, if *time* domain models are required for calculations in which non-linear effects are to be examined, then the reduced frequency dependency of the aerodynamics needs to be accounted for; one possibility is to use a convolution employing the Wagner and Aerodynamic Influence Coefficient functions, but these are based on a 2D strip theory approach. If the results from a 3D panel approach are to be used in the time domain, a different method is required. For state-space time domain models (see Chapters 6 and 11), this is achieved through the use of a so-called Rational Function Approximation (RFA) for aerodynamics, which approximates the AICs obtained from either Doublet Lattice modelling or experimental measurements at a range of reduced frequencies for a given flight condition. The approach described here (Eversman and Tewari, 1991) is a variation of the classical approach (Roger, 1977) that doesn't require the computational effort of the well known minimum state method (Karpel, 1982). The way in which the RFA is implemented in the state-space model is described in this section.

Consider the classical aeroelastic model but with the aerodynamic terms placed on the right hand side such that

$$\mathbf{A}\ddot{q} + \mathbf{D}\dot{q} + \mathbf{E}q = -\rho V \mathbf{B}\dot{q} - \rho V^2 \mathbf{C}q = Q_{\text{Aero}}(t) \tag{19.37}$$

where Q_{Aero} is the aerodynamic generalized force vector. Alternatively, in the Laplace domain (see Chapter 6)

$$\left(\mathbf{A}s^2 + \mathbf{D}s + \mathbf{E}\right)q(s) = \frac{\rho V^2}{2}\mathbf{Q}(s)q(s) \tag{19.38}$$

where $\mathbf{Q}(s)$ is the Rational Function Approximation to the AIC matrix expressed in modal space. The matrix $\mathbf{Q}(s)$ is expanded using a Rational Function Approximation in terms of the Laplace variable $s = i\omega$, so that

$$\mathbf{Q}(s) = \mathbf{A}_0 + \mathbf{A}_1\frac{sb}{V} + \mathbf{A}_2\left(\frac{sb}{V}\right)^2 + \frac{V}{b}\sum_{n=1}^{N_L}\frac{\mathbf{A}_{n+2}}{\left(s + \frac{V}{b}p_n\right)} \tag{19.39}$$

or

$$\mathbf{Q}(ik) = \mathbf{A}_0 + \mathbf{A}_1 ik + \mathbf{A}_2(ik)^2 + \sum_{n=1}^{N_L}\frac{\mathbf{A}_{n+2}}{(ik + p_n)}$$

noting that $sb/V = ik$. Here p_n are the N_L poles (or lag parameters) used for the approximation of the unsteady aerodynamic matrix $\mathbf{Q}(s)$ and $\mathbf{A}_i$, $i = 0,1, \ldots N_L + 2$ are unknown matrices to be found.

Now, if $\mathbf{AIC}(k_m)$, $m = 1, 2 \ldots N_k$ is a given set of AICs obtained using a panel method or experimental measurements at N_k reduced frequencies, then the square of the error between the RFA (shown here with two aerodynamic lag terms) and $\mathbf{AIC}(k_m)$ for the rsth element is written as

$$
\begin{aligned}
\varepsilon_{rs} &= \sum_{m=1}^{N_k} \left(\mathbf{Q}_{rs}(ik_m) - \mathbf{AIC}_{rs}(ik_m) \right)^2 \\
&= \sum_{m=1}^{N_k} \left(\mathbf{A}_0 + \mathbf{A}_1(ik_m) + \mathbf{A}_2(ik_m)^2 + \frac{\mathbf{A}_3}{(ik_m + p_1)} + \frac{\mathbf{A}_4}{(ik_m + p_2)} - \mathbf{AIC}(k_m) \right)^2_{rs}
\end{aligned}
\tag{19.40}
$$

where it should be noted that the subscript 'rs' indicates that the rsth element of each matrix is considered so Equation (19.40) is actually written in scalar terms. The least squares minimization between the given data and the Rational Function Approximation model for each element is defined as

$$
\left(\frac{\partial \varepsilon}{\partial \mathbf{A}_n} \right)_{rs} = 0 \quad \text{for} \quad n = 0, 1, 2, \ldots, N_L + 2
\tag{19.41}
$$

which leads to the equations

$$
\begin{bmatrix}
1 & ik_1 & -k_1^2 & \dfrac{1}{ik_1 + p_1} & \dfrac{1}{ik_1 + p_2} \\
1 & ik_2 & -k_2^2 & \dfrac{1}{ik_2 + p_1} & \dfrac{1}{ik_2 + p_2} \\
\vdots & \vdots & \vdots & \vdots & \vdots \\
1 & ik_{N_k} & -k_{N_k}^2 & \dfrac{1}{ik_{N_k} + p_1} & \dfrac{1}{ik_{N_k} + p_2}
\end{bmatrix}
\begin{Bmatrix}
A_0 \\ A_1 \\ A_2 \\ A_3 \\ A_4
\end{Bmatrix}_{rs}
=
\begin{Bmatrix}
\mathrm{AIC}(ik_1) \\ \mathrm{AIC}(ik_2) \\ \vdots \\ \vdots \\ \mathrm{AIC}(ik_{N_k})
\end{Bmatrix}_{rs}
\tag{19.42}
$$

and these can be solved using least squares to find the unknown A_{rs} values. Note that the curve fit is performed term by term in the AIC matrix using the same lag values throughout; however a single step matrix approach could be used. The pole values must be positive but their choice is not straightforward and a range of different values should be tried. Better results are obtained if a good distribution of k values is taken over the range of interest but it is usual to have more points at low values of k as these are more important for flutter.

Figure 19.11 shows a typical curve-fit using the above Rational Function Approximation approach with four aerodynamic lag terms on a sample element (1, 2) of the Aerodynamic Influence Coefficient matrix (the 'truth'). It can be seen that a reasonable fit can be achieved.

Having estimated the unknown matrix parameters, Equation (19.38) can be written in the time domain for two aerodynamic lags and no excitation vector so that

$$
\mathbf{A}\ddot{q} + \mathbf{D}\dot{q} + \mathbf{E}q = \hat{\mathbf{A}}_0 q + \frac{b}{V}\hat{\mathbf{A}}_1 \dot{q} + \left(\frac{b}{V}\right)^2 \hat{\mathbf{A}}_2 \ddot{q} + \hat{\mathbf{A}}_3 q_{a_1} + \hat{\mathbf{A}}_4 q_{a_2}
\tag{19.43}
$$

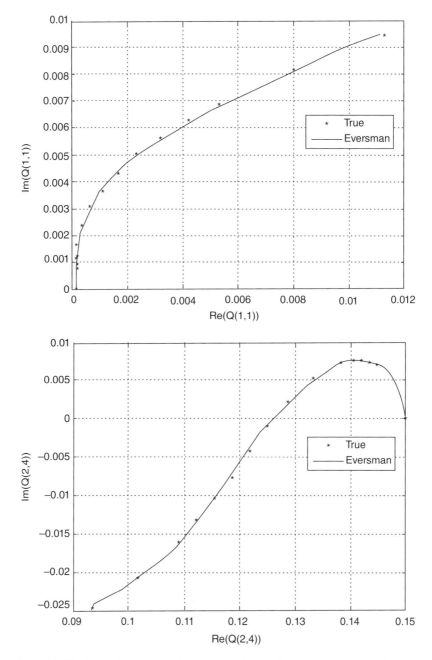

Figure 19.11 Curve fit of AIC data elements using Rational Function Approximation.

where $\hat{\mathbf{A}}_n = \frac{\rho V^2}{2}\mathbf{A}_n$.

Rewriting this expression yields

$$\left(\mathbf{A} - \left(\frac{b}{V}\right)^2 \hat{\mathbf{A}}_2\right)\ddot{\boldsymbol{q}} + \left(\mathbf{D} - \frac{b}{V}\hat{\mathbf{A}}_1\right)\dot{\boldsymbol{q}} + \left(\mathbf{E} - \hat{\mathbf{A}}_0\right)\boldsymbol{q} = \hat{\mathbf{A}}_3 \boldsymbol{q}_{a_1} + \hat{\mathbf{A}}_4 \boldsymbol{q}_{a_2} \qquad (19.44)$$

or

$$\tilde{\mathbf{A}}\ddot{q} + \tilde{\mathbf{D}}\dot{q} + \tilde{\mathbf{E}}q = \hat{\mathbf{A}}_3 q_{\mathrm{a}_1} + \hat{\mathbf{A}}_4 q_{\mathrm{a}_2}$$

where the so-called augmented states, arising from the convolution applied to the inverse Laplace transform of the $\mathbf{Q}(s)$ matrix, are defined as

$$q_{\mathrm{a}_n} = \int_0^t q e^{-\frac{V}{b}p_n(t-\tau)} d\tau \quad 1 \le n \le N_L \quad \dot{q}_{\mathrm{a}_n} = q - \frac{V}{b} p_n q_{\mathrm{a}_n} \tag{19.45}$$

but neither term needs to be calculated explicitly. Equation (19.43) can be combined with Equation (19.45) to write the system in the classical state-space formulation of

$$\dot{x} = \mathbf{A}_\mathrm{s} x \quad x = \begin{bmatrix} \dot{q}^\mathrm{T} & q^\mathrm{T} & q_{\mathrm{a}_1}^\mathrm{T} & q_{\mathrm{a}_2}^\mathrm{T} \end{bmatrix}^\mathrm{T} \tag{19.46}$$

The system matrix $\mathbf{A}_\mathrm{S}$ is

$$\mathbf{A}_\mathrm{s} = \begin{bmatrix} -\tilde{\mathbf{A}}^{-1}\tilde{\mathbf{D}} & -\tilde{\mathbf{A}}^{-1}\tilde{\mathbf{E}} & \tilde{\mathbf{A}}^{-1}\hat{\mathbf{A}}_3 & \tilde{\mathbf{A}}^{-1}\hat{\mathbf{A}}_4 \\ \mathbf{I} & 0 & 0 & 0 \\ 0 & \mathbf{I} & -\frac{V}{b}p_1\mathbf{I} & 0 \\ 0 & \mathbf{I} & 0 & -\frac{V}{b}p_2\mathbf{I} \end{bmatrix} \tag{19.47}$$

Thus the state space equations of motion can be solved as before (see Chapter 11) but clearly there are additional states present. The model approximates the reduced frequency dependency of the aerodynamics in the time domain by effectively replacing the use of convolution with the Wagner and Küssner functions for 2D strip theory.

Part III

Introduction to Industrial Practice

20

Aircraft Design and Certification

The aim of this final part of the book is to outline briefly some of the processes employed in the commercial aerospace industry for aeroelastics and loads analysis with reference made to the certification specifications. However, it is recognized that practice across companies, and even different aircraft projects, will differ so the treatment described here is not unique. Where relevant, the processes will be related to the earlier chapters of the book where a simplified approach was taken.

20.1 Aeroelastics and Loads in the Aircraft Design Process

Aeroelastic and loads considerations play a part across much of the design and development of an aircraft. The aeroelastic and loads behaviour of the aircraft have an impact upon the concept and detailed structural design, aerodynamic characteristics, weight, jig shape, FCS design, handling qualities, control surface design, propulsion system, performance (effect of flight shape on drag), landing gear design, structural tests, etc.

It can be convenient to think of the design and development cycle as comprising a number of phases:

a. A *concept phase* in which the objective is to determine the best viable aircraft concept to meet the design aims. Estimates of loads play an important role here in the process of estimating the structural weight of each concept under consideration and so influencing the design trade-offs between alternative options. Aeroelastics is also important in order to eliminate concepts that are likely to be aeroelastically unacceptable, or to assist in positioning engines, for example. At this stage of aircraft development the engineering team is relatively small and their need is for a rapid turnaround of analyses. Hence the availability of detailed aircraft data is sparse and, instead, there must be a heavy reliance on theoretical analysis calibrated, if at all, mainly by empirical data or data scaled from previous designs in the same family of aircraft. Likewise, the focus of attention will be on those few loading cases that have been identified or are judged to be the most likely design drivers. The analysis methods too are likely to be as simple as they can be, while still capable of delivering worthwhile results, and this will often allow them to be more automated than will be the case later

Introduction to Aircraft Aeroelasticity and Loads, Second Edition. Jan R. Wright and Jonathan E. Cooper.
© 2015 John Wiley & Sons, Ltd. Published 2015 by John Wiley & Sons, Ltd.

in the design process when practical constraints arising from the detailed design often have to be taken into account. For aircraft that employ an FCS, it is well known that the FCS will have an important effect on loads, but at this early stage of development the FCS concept design will have barely started and there would certainly not be sufficient detail to take rational account of it. Hence it will be necessary to make allowance for the FCS in some other way.

b. A *pre-design phase* would follow in which the top level architecture of the chosen concept is fleshed out. The start of pre-design usually marks a significant ramp-up of engineering resource on the project (although still an order of magnitude below the expenditure once industrialization gets under way). Reasonable loads estimates are important here in order to determine billet and forging sizes of long-lead items and to contribute to the choice of alternative structural architectures (landing gear attachment scheme, position of the primary structural elements within the external contour, two- or three-spar wing design, etc.). In many ways, the pre-design phase is one of transition from quick-turnaround methods used for concept selection towards the more rigorous and detailed mainstream methods that are necessary to underpin the loads to be used for detailed design. Pre-design is also just about the last opportunity for aeroelastic drivers to contribute to the fundamental structural concept, so much attention will be paid to ensuring that sufficient analysis is performed to give good confidence that flutter margins will be achievable. Attention will also be paid to ensuring that sufficient control surface static effectiveness is available throughout the design envelope, so as to achieve the required handing qualities and controllability, and to provide any load alleviation that may be part of the design concept.

c. A *detailed design phase* follows, involving a big ramp-up of resourcing costs. This will only follow a formal industrial go-ahead with full financing in place. Once this decision is made, a completed detailed design is required as soon as possible, so a good quality set of loads on which to base the design is needed almost immediately – hence there is a high reliance on the preparatory mathematical modelling undertaken in pre-design. Ideally these loads should not be allowed to increase, although in practice there is often considerable pressure to seek decreases in order to facilitate weight saving. It is evident that as the detailed design progresses, the 'best knowledge' of mass properties, aerodynamics, stiffness and systems will change – all potentially influencing calculated loads and aeroelastics characteristics. The focus of attention will therefore naturally turn towards refining the mathematical models to track the effects of the evolving design and towards developing solutions to any problems that are identified. Design is naturally an iterative process, and the level of detail and confidence in the models employed for analysis will improve cycle by cycle as more information becomes available. The processes by which these design cycles are managed differ from manufacturer to manufacturer and play an important role in the eventual success of the project. (Note that it is quite possible for an iterative process to become divergent).

d. Finally a *validation and certification phase* follows in which the aircraft (or components) have been built and used in ground and flight testing in order to validate whether the characteristics built into the loads and aeroelastics models are correct. Any adjustments resulting from this and from any late design changes (to the FCS, for example) must be taken into account in producing a set of loads on which the structural certification can be based. There may have to be some late design changes, even at this stage, driven by the evolution of loads or by the need to demonstrate sufficient aeroelastic margins, but the aim is to confine these changes to items that can be changed relatively quickly and without major industrial repercussions, e.g. control law software or fine tuning of control surface mass balance (where used).

There are a number of different customers (inside the company) of loads, each with their own specific requirements. The two that deserve most attention are design and certification. The customers for *design* (and structural justification) need the loads in order to determine extreme stress levels in their structure or to estimate fatigue damage or damage tolerance. For both purposes, they will wish to apply loads cases to their detailed structural models. This usually comprises a post-process involving the decomposition of correlated loads cases, supplied from the result of loads calculations, into some form of nodal loading that can be applied to specific structural elements. It is therefore important to this analysis that the applied loads are well in balance with constraints. Hence the supplied loads cases must themselves be in balance, e.g. for a specific time instant in a dynamic gust or manoeuvre response. This implies that the loads group must be prepared to deliver as many load cases as are identified as potentially critical on some part of the structure.

For *certification* justification (of the loads themselves), the focus of attention can be more limited, e.g. to the envelope values of the loading quantities of primary interest (such as shear force, bending moment and torque variations along each of the main components).

20.2 Aircraft Certification Process

20.2.1 Certification Authorities

Commercial aircraft whose weight is above 5700 kg (12 500 lb) are certified under the 'large aircraft' certification requirements, denoted by the number 25 in the regulation description (e.g. CS-25 and FAR-25). The certification of aircraft is a complex process and depends upon where in the world the aircraft is manufactured, and where it is to be purchased and operated. The US certification is looked after by the FAA (Federal Aviation Administration) and European certification by EASA (European Aviation Safety Agency); until 2003, the European certification of aircraft was overseen by the JAA (Joint Airworthiness Authorities) and prior to that many countries certified aircraft through their own agencies, e.g. CAA (Civil Airworthiness Authority) in the UK.

Any new aircraft model requires type certification by the airworthiness agency corresponding to the area of the world in which the manufacturer is based; this is primary certification. If the aircraft is to be exported to other parts of the world, then secondary certification by the relevant agencies is also required; such an exercise will be eased by any bilateral agreements existing between the agencies.

20.2.2 Certification Requirements

To cover the type certification of large commercial aircraft, the FAA and EASA issue FAR-25 and CS-25 documents respectively. The FAR-25 contains basic FARs (Federal Aviation Regulations, e.g. FAR 25.491 Taxi, Take-off and Landing Roll) and any additional material is contained within ACs (Advisory Circulars, prefixed AC 20 or 25, e.g. AC 25.491). The CS-25 Book 1 contains CS (Certification Specifications) which are standard technical interpretations of the essential airworthiness requirements (e.g. CS 25.491 Taxi, Take-off and Landing Roll); the book is divided into a number of subparts together with appendices. The key subparts for loads and aeroelasticity issues are Subpart B (Flight), C (Structure) and D (Design and Construction). Aeroelastics requirements are fairly well concentrated in CS 25.629, which is within Subpart D. Loads requirements are mainly in Subpart C as the prescription of the loading cases that have to be accounted for in design is a primary prerequisite for assuring structural integrity over the operating environment of the aircraft. Most of the structural airworthiness requirements are related to static (as opposed to fatigue) loads.

CS-25 Book 2 contains what are known as AMCs (Additional Means of Compliance, e.g. AMC 25.491) which are non-exclusive means of demonstrating compliance with airworthiness codes or implementing rules. The AMC (previously termed ACJ when published by JAA) is related to the AC in FAR-25.

Over recent years, there has been considerable effort to rationalize the US and European codes and for most issues they are now the same, though there have been different rates of incorporating changes. An extremely valuable contribution in the field relevant to this book was made by the Loads and Dynamics Harmonisation Working Group (LDHWG), a mix of technical specialists from the US and Europe. The group considered revisions to the codes to assist harmonization and incorporate improvements; a particular area where new work was needed was for large commercial aircraft with more than two main landing gears, where the redundant gear layout meant that some of the ground loads requirements were inappropriate and more rational approaches were required. Many of these changes are already in the codes and others are awaiting incorporation.

When revisions are proposed to the requirements, an NPA (Notice of Proposed Amendment) is issued and after consultation it becomes part of the code (e.g. NPA 11/2004 in Europe led to the issue of Appendix K on Interaction of Systems and Structures). For Europe, the current rule-making plans are on the EASA website.

There is a degree of flexibility in meeting the requirements and some room for the manufacturer proposing, and discussing with the certification authorities, an alternative approach to a particular issue. The manufacturer can produce a CRI (Certification Review Item) and Special Conditions may be agreed to meet a particular requirement prior to the codes being altered. The need for Special Conditions may be driven by the manufacturer, due to unusual design features in their aircraft, prompting the need for analysis justification beyond the basic requirement, or by the Authority, who often wish to apply a pending NPA as a Special Condition ahead of its adoption into the general requirements.

The certification process depends upon whether an aircraft has 'new structure' (e.g. changes in the design philosophy regarding structures and loads have been made, or the manufacturer has not built an aircraft of this type before), 'similar new structure' (utilizes similar design concepts to an existing tested aircraft) or 'derivative/similar structure' (uses structural design concepts almost identical to those on which analytical methods have been validated). Obviously, the more similarity a particular aircraft type has with an existing design or design approach, the simpler the design and certification process. The process of analysis and test depends on what previous test evidence is available. A very useful introduction to the compliance with different loading conditions is given in CS/AMC 25.307 Proof of Structure; e.g. the appropriateness of standard methods and formulae, and the use of the Finite Element method for complex structures are discussed together with the need for testing.

20.2.3 Design Envelope

It is important for a manufacturer to put the specific loads requirements into the context of the design envelope. CS 25.321(c) requires that enough points on or within the boundary of the design envelope are investigated to ensure that the most extreme loads for each part of the aircraft structure are identified. In this context, the design envelope encompasses the

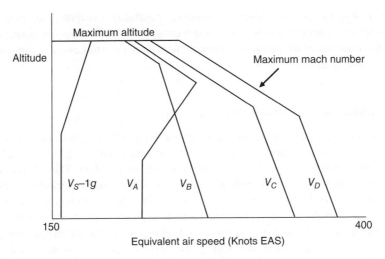

Figure 20.1 Flight envelope – design speed versus altitude. Reproduced by permission of Airbus (redrawn with modifications).

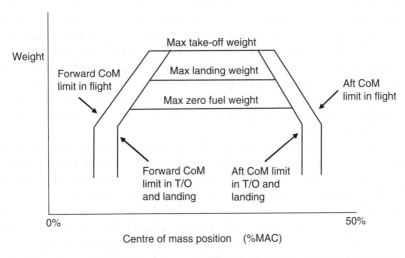

Figure 20.2 Weight versus centre of mass envelope. Reproduced by permission of Airbus (redrawn with modifications).

respective ranges of permitted mass/centre of mass envelopes (MTOW, maximum take off weight; MZFW, maximum zero fuel weight; MLW, maximum landing weight; OWE, operating weight empty; etc.), ground design speeds, flight design speed envelopes versus altitude and aircraft (aerodynamic) configuration, flight control law or autopilot mode, etc. Sample envelopes for air speed against altitude and centre of mass/weight against mean aerodynamic chord (MAC) are shown in Figure 20.1 and Figure 20.2. The forward centre of mass limit is often limited by the ability to apply sufficient control and trim, whereas the aft limit depends upon the stability of the aircraft, sensitivity of controls and danger of the aircraft tipping over.

For loads to be analyzed for a large number of different flight and load condition combinations, this would result in an enormous number of calculations – a prospect that could not be considered in bygone days of limited computing capability. Nevertheless, even with the computational power now available, it is still appropriate to take an intelligent approach to case selection, focusing detail where initial calculations or previous studies show the potential for cases to become significant.

20.2.4 Bookcase and Rational Load Cases

The load cases for certification are sometimes classified as either (a) bookcase or (b) rational. The *bookcase* types of load case are often somewhat artificial in that applied loads are assumed and reacted by inertia loads, leading to a static equilibrium problem. They therefore provide simple rules for design but not necessarily a realistic flight case, especially in early stages of the FCS design. The load cases may not necessarily be reproduced on the real aircraft; they tend to have been developed when computation methods were in their relative infancy and the loads assumed have been found over the years to lead to aircraft that are judged strong enough. A number of clear examples are shown in the ground loads cases (e.g. turning or braked roll; see Chapters 15 and 24).

On the other hand, *rational* types of load case are those where an attempt is made to model the loads and dynamics of the aircraft as realistically as possible. They have developed as computational methods and the understanding of aircraft behaviour have matured, and as non-linear effects have become more important. In some cases, rational methods have become essential, e.g. where redundant landing gear layouts are introduced by employing more than two main landing gear units. In other cases, use of rational modelling may have allowed reduced loads to be justified (e.g. braking). The rational methods may be classed as 'fully rational' where the full dynamic condition is sought (e.g. landing) or 'quasi-rational' where a trimmed/equilibrium condition is sought from the fully rational (often nonlinear) dynamic model in order to determine a rational balanced condition (e.g. a steady braking case where the static landing gear characteristics are represented).

The bookcase is most useful early on in the design cycle (pre-development and sizing) when the aircraft is at a less mature stage of development and realistic calculations are impractical and probably inaccurate. On the other hand, the rational approach can be used later in the design cycle for loads certification, sensitivity studies, study of failure cases, etc.

Attention to detail is necessary in reading and interpreting the requirements defining load cases for certification. In some cases, loads cases are defined in such a way that they can only be applied as a 'bookcase' or as a 'rational' analysis, whereas in other cases, the requirement for rational analysis is limited to certain circumstances only. In yet other cases, the manufacturer is given the choice of applying a conservative 'bookcase' or, alternatively, of investing additional time and resources in a more detailed rational analysis in the expectation (not always realized) that it will produce less conservative loads and so allow the aircraft to be more competitive.

20.2.5 Limit and Ultimate Loads

The certification requirements include a number of general requirements that are helpful for the engineer to be aware of. Strength requirements (CS 25.301, 25.303 and 25.305) are specified in terms of:

a. Limit loads are the maximum loads to be expected in service and which the structure needs to support without 'detrimental permanent deformation'.

b. Ultimate loads (limit loads multiplied by a factor of safety, normally 1.5 unless otherwise stated) are loads that the structure must be able to support without failure/rupture (for at least 3 seconds).

Limit loads might realistically happen once per lifetime ($\sim 10^5$ hours) for a single aircraft, but once the probabilities of the aircraft actually being at the critical flight state/condition when the event occurs are accounted for, then the probability of experiencing the limit load in practice should be much lower than once per lifetime. The loads specified in CS-25 are almost entirely limit loads. The requirements also specify that compliance with the strength and deformation conditions needs to be demonstrated for each critical loading condition (CS 25.307); how this may be shown by analysis and test is discussed in CS/AMC 25.307 Proof of Structure.

A relatively recent addition to CS-25 is a requirement on the interaction of systems and structures (CS 25.302/Appendix K) where the structural performance following the possible failure of systems that may influence structural loads must be considered. This requirement is unusual in specifying factors of safety for consideration of loads at the time of occurrence of the failure and factors of safety for consideration of loads for continuation of the flight, which reduce below 1.5 as a function of reducing the probability of occurrence or exposure.

20.2.6   Fatigue and Damage Tolerance

Although often less visible than static design loads, the requirement to provide loads information for fatigue and damage tolerance assessment is nevertheless an important aspect of aircraft design. However, it was decided not to cover this aspect of loads in detail earlier in the book, since it falls into the category of loads post-processing, whereas the main focus here has been on response and loads determination.

The structural analysis is aimed at an evaluation of an appropriately factored 'life' for a structural element relative to the design life goal of the aircraft or against the maintenance inspection interval schedule. In both cases, a 'stress spectrum' must be built for each structural element, which can be derived directly from a 'loads spectrum' provided by the loads group representing the anticipated usage (or 'mission') of the aircraft. A typical mission (e.g. short- medium- or long-haul) is broken down into multiple segments (e.g. taxi, braked roll, take-off, climb, cruise, descent, gear down, approach, landing, braking, reverse thrust) with each segment having a statistical model defining the occurrence of some global driving parameter such as gust velocity, and with each segment also having a set of 'unit loads' (or 'fatigue loads') per unit driving parameter. A load spectrum defining load vs probability of occurrence may then be formed for each segment and summed to yield a total probability of occurrence for all segments. Alternatively the loads group may supply just the 'statistical model' and the 'unit loads' for each segment, and the 'stress spectrum' may then be formed directly by the structural analysis group without recourse to a loads spectrum. Determination of the statistical model is usually based on operational statistics from aircraft in service. The 'mission' definition of an aircraft will be directly linked to the top level aircraft requirements and design characteristics (weights, maximum payload, payload fractions, fuel capacities, speed/altitude profiles, etc.).

The 'fatigue loads' are normally generated using the same mathematical model formulations as used in generating static design loads for stylized conditions. For example, within the airborne part of the mission, they may be provided for the datum 1 g condition in each segment, plus the incremental loads due to unit root-mean-square (RMS) gust velocity applied in either some form of discrete gust profile, or alternatively from a continuous turbulence analysis that will directly facilitate the application of the 'mission analysis' method (see Chapter 23).

In the provision of 'fatigue loads', it is often necessary to identify in some way the frequency of loads occurrences relative to the global parameter on which the statistical model is based. If a deterministic model is used for 'fatigue loads', then the contribution to fatigue damage from the second, third and subsequent peak loads in the event time history may need to be accounted for. If a stochastic model is used (such as continuous turbulence), then the characteristic frequency of the load needs to be identified in order to apply the mission analysis exceedance formula. With this final point in mind, a further useful requirement to be aware of is CS/AMC 25.571 *Damage-Tolerance and Fatigue Evaluation of Structures*, which includes considerable discussion of how damage and fatigue issues should be handled.

21

Aeroelasticity and Loads Models

In this chapter, the building blocks that make up the aeroelastic and manoeuvre/gust load models will be introduced. Where possible, the models will be related to those used earlier in Part II of the book. Various comments on the requirements for structural and aerodynamic models are made for gusts and flutter in CS/AMC 25.341 and 25.629. Note that what is presented here is one way of developing suitable mathematical models for aeroelastics and loads, but other approaches could be used.

21.1   Structural Model

21.1.1   Introduction

The basic mathematical model of the aircraft must be able to represent its vibration behaviour over the frequency range of interest, typically 0–40 Hz for large commercial aircraft and 0–60 Hz for small commercial aircraft. Thus, the model will need to yield natural frequencies, modal masses and normal mode shapes in these frequency ranges. The model should adequately represent the aircraft complexity, including control surface and engine behaviour, and generate sufficiently accurate mode shapes.

In the following sections the mass properties and idealization will firstly be defined and then the different possible stiffness models and resulting structural models considered.

21.1.2   Mass Properties

A considerable degree of detail is known for the mass distribution on the aircraft, but only part of the mass is structural (typically 30%). A significant proportion of the overall mass is linked with non-structural elements such as fuel, payload, equipment, systems etc. Thus, it is not appropriate to try to link the non-structural mass to any stiffness model generated in a detailed manner; a different approach is therefore required for an aircraft structure. In this section, the mass idealization will be considered and, in the following sections, the way in which the mass and stiffness representations are connected to generate structural models will be discussed.

What is commonly done when dynamic modelling a commercial aircraft with high aspect ratio wings, is that the wing, for example, is divided into a number of streamwise sections

Introduction to Aircraft Aeroelasticity and Loads, Second Edition. Jan R. Wright and Jonathan E. Cooper.
© 2015 John Wiley & Sons, Ltd. Published 2015 by John Wiley & Sons, Ltd.

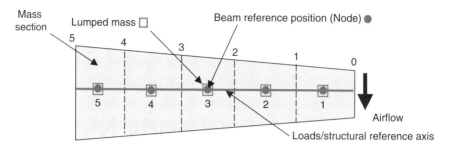

Figure 21.1 Mass idealization.

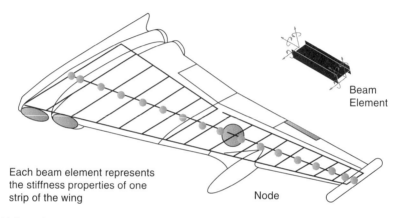

Figure 21.2 Wing model using a 'beam-like' representation. Reproduced by permission of Airbus.

(or strips) as shown in Figure 21.1. A reference axis is defined along the wing and the total mass, centre of mass, mass-moments and moments of inertia for each section are associated with the corresponding mass reference position on the reference axis. The displacements and rotations at the reference points are the unknowns in the final structural model.

21.1.3 Structural Model 1 – 'Stick' Representation

A classical approach to developing a mathematical model for aircraft with high aspect ratio wings is to treat the aircraft as an assembly of sticks (or beams), placed along the reference axes for each major aircraft component (e.g. wing, front fuselage, rear fuselage, tailplane, fin). Each reference axis is taken to be the elastic axis if straight; the elastic axis is the locus of shear centres (see Chapter 3). The beams are capable of bending, shear, torsional and axial deformations. In such an approach, each beam is divided into several sections as shown in Figure 21.2 for the wing example. The variation of flexural rigidity EI and torsional rigidity GJ along each beam are traditionally estimated from the member section properties by classical structural analysis methods. Such a stiffness model is essentially 'stick-like' or 'beam-like'.

An early approach to determining the stiffness model for each component was to use the method based on Flexibility Influence Coefficients to assemble the flexibility matrix (Rao, 1995). Alternatively, the stiffness model was assembled using the Finite Element method to obtain the stiffness matrix (the inverse of the flexibility matrix). This type of model was introduced in Chapter 17.

In order to obtain a structural model for each component, the idealized mass distribution explained earlier is then connected to the stiffness model developed for the stick representation, taking account of the mass, centre of mass, mass-moments and moments of inertia for each section by using a rigid link.

21.1.4   Structural Models – 'Box-Like' Representation

The problem with employing beam elements directly for an aircraft structure is that for such a complex thin-walled structure, the calculated stiffness distribution is rather inaccurate. Such an approach may possibly be suitable for an aircraft in the early design phase, where the detailed structure has not yet been defined, and where scaled stiffness and mass properties from previous aircraft might be employed; however, it would not be appropriate at the later stages of design and certification where structural detail is available and important. Therefore, in recent years, the Finite Element approach has been used to set up the structural stiffness characteristics for the aircraft by representing its detailed 'box-like' construction.

The semi-monocoque 'box-like' aircraft structure (Niu, 1988; Megson, 1999) is composed of discrete stiffeners (e.g. spar booms, stringers) and thin-walled panels (e.g. cover skins, webs for spar/rib). The structure is typically modelled using a combination of beam and shell finite elements. The beam elements are employed to model discrete stiffeners (e.g. spar boom, stringers) in axial, bending, torsional and shear deformations. The shell elements represent both in-plane membrane and plate bending/twisting effects, and are therefore used to model cover skins and rib/spar webs. By using such elements, the 'overall' bending and twisting of the wing, fuselage etc. is represented, as well as 'local' bending and twisting of the cover skins, stiffeners and spar/rib webs. If, however, a more economical model employing only membrane and bar elements is employed, only overall bending and twisting will have been catered for.

Figure 21.3 shows an example of a Finite Element model for wing and pylons. It should be pointed out that the model employed for dynamics purposes may not always be as detailed in structural representation as the model used for stress analysis, though the level of sophistication employed in industry is continually growing. Additional structure to support landing gear and control surfaces needs to be represented as accurately as possible since local stiffness values are important for aeroelastics and loads calculations.

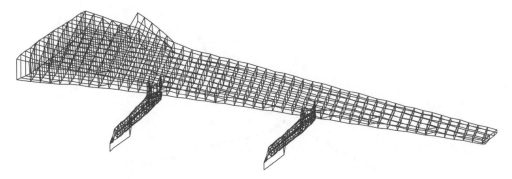

Figure 21.3   Finite Element wing model using a 'box-like' representation. Reproduced by permission of Airbus.

In the next two subsections, the connection of a relatively small number of lumped masses to the 'box-like' Finite Element model having a large number of nodes will be considered, so yielding two other possible structural models. In each case a reference axis is defined along each member but this is no longer the elastic axis but rather an axis to use for subsequent loads values.

21.1.4.1   Structural Model 2 – Concentrated Mass attached to a Condensed FE Model

One approach to linking a moderate number of mass points (order 10^2) to a much larger number of Finite Element nodes (order 10^5) is to condense the stiffness model down to a 'beam-like' representation; this condensation process reduces the stiffness model to correspond with a limited number of structural reference (or nodal) points lying on the reference axes (or elsewhere as necessary).

The condensation is typically carried out using a method such as Guyan reduction (see Appendix D) where the Finite Element stiffness model of original order N is reduced to a significantly smaller set of master degrees of freedom N_m corresponding to the chosen reference axis (and any other chosen points). The master responses, once they have been calculated from the reduced model, may be used to obtain the responses at the slave degrees of freedom N_s ($= N - N_m$) that were condensed out. An example of a condensed beam-like model is shown in Figure 21.4 where the lumped masses are shown and where additional condensation points are employed to represent the engine pylon; similar arrangements could be used for the landing gear support points.

The representation of the control surface behaviour depends upon the case being considered. For dynamic loads calculations, the control surface modes may be ignored and control rotation simply treated as imposing forces and moments on the reference axes. However, for aeroelastic calculations such as flutter, the control modes need to be represented and so any reduced model would be extended to include suitable condensation points in the region of the control surface; mass stations would still be linked rigidly to chosen grid points.

The outcome of this process is that the condensed beam model degrees of freedom and mass property data are consistent in size. Then, for each lumped mass and section, a concentrated mass element (see Section 17.6) may be employed and connected to the condensed stiffness model using a rigid body element with an infinite constraint (see Section 17.8.1) i.e. each lumped mass is rigidly linked to the condensed model. This has the effect of yielding a structural model of moderate size, but one which was formed from a 'box-like' Finite Element model.

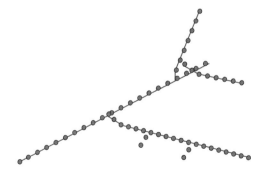

Figure 21.4   Finite Element model condensed onto structural reference axes. Reproduced by permission of Airbus.

21.1.4.2 Structural Model 3 – Concentrated Mass attached to a 'Box-Like' FE Model

A more recent approach to connecting a moderate number of lumped masses to a large 'box-like' Finite Element model is to take advantage of the rigid body element employing an interpolation constraint (see Section 17.8.2). In this case the full Finite Element stiffness model is not condensed.

The mass associated with each wing or fuselage bay is attached to a node positioned on the relevant loads reference axis. This mass node is then connected to selected reference structural nodes for the bay in question and the adjacent bays. The deflection of these reference nodes should be representative for the bay; an example for an appropriate reference wing node is the junction of a spar, rib and upper wing cover. The connection between the mass node and the reference nodes is achieved using a special finite element, which does not stiffen the FE model and assigns the average reference node deflection to the mass node.

The use of this interpolation element has the effect of 'constraining' each lumped mass to follow the motion of the surrounding local structure in a least squares approximate manner, whilst retaining the full structural representation of the Finite Element model.

21.1.5 Modal Model

Once a Finite Element structural model is available, the equations of motion governing the modal behaviour of the free–free structure may be determined. The free–free normal modes of vibration may then be calculated (see Chapters 2 and 3), leading to natural frequencies, mode shapes and modal masses for each mode. Both whole aircraft rigid body and flexible modes are generated.

The diagonal modal mass and modal stiffness matrices may then be obtained using a transformation employing the modal matrix, exactly as originally introduced in Chapter 2. The number of desired flexible modes N_d ($<<N$) to be employed in the aeroelastic or loads analyses must be selected to cover the frequency range of interest with sufficient accuracy; the transformation process effectively discards the excess modes in the uncoupled modal equations. Thus, the large set of N degrees of freedom in the box-like representation will have been reduced to a small set of N_d uncoupled single degree of freedom modal equations; typically N_d is of the order of $N^{1/3}$ and perhaps only 30–100 modes are retained.

In Figures 21.5 and 21.6, an example of symmetric wing bending and wing torsion mode shapes is shown; the sections on the aircraft (i.e. fuselage, wing, fin, tailplane) are defined by the mass boundaries while the engine is shown as a generic shape scaled for length and diameter. In the case of the condensed structural model 2, then once mode shapes are known for the condensed (master) degrees of freedom, the values of the even greater number of slave degrees of freedom may be determined and the mode shapes plotted for the entire structural grid if desired. For model 3, the mode shapes will be known for the full Finite Element mesh.

Note that separate analyses are often performed for symmetric and antisymmetric behaviour of the aircraft and for different fuel and payload states.

21.1.6 Damping Model

It is not possible to model the distributed damping for an aircraft structure in any realistic way. What is normally done is to include an assumed level of modal damping per flexible mode

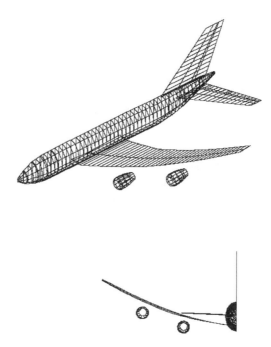

Figure 21.5   Sample symmetric wing bending mode shape. Reproduced by permission of Airbus.

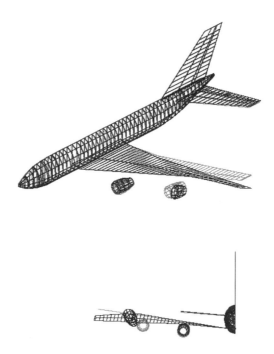

Figure 21.6   Sample symmetric wing torsion/outboard engine lateral mode shape. Reproduced by permission of Airbus.

(typically 1 % critical viscous damping or 0.02 structural damping) based on experience, and then to update this assumption, particularly if lower or higher values are obtained using results from the ground vibration test (see Chapter 25). If an actuator or flutter damper model is incorporated, then a more accurate local damping representation would be available. Sometimes, for flutter calculations, structural damping is ignored but then a small level of negative damping above V_D would be acceptable as the inherent structural damping would eliminate this flutter in practice.

If a non-linear model for damping, where damping increased with response level, could be justified by test, then it would be possible to include such a model in loads calculations and so, potentially, to reduce loads values and allow a more efficient structure to be designed.

21.1.7 Rigid Aircraft Model

A model for the rigid aircraft is required for loads and aeroelastics calculations, as well as for flight mechanics studies. A linear small angle rigid aircraft model is referenced to inertial axes and is effectively composed of the rigid body heave, fore-and-aft and lateral displacements and pitch, roll and yaw rotations. This is essentially the same as a modal model represented by rigid body modal coordinates and associated rigid body mode shapes (see Appendix A for the longitudinal case).

The alternative rigid aircraft model is the large angle non-linear flight mechanics model referenced to a body fixed axes system where rigid body velocities are defined relative to these fixed axes, as introduced in Chapter 13.

In either model, the key mass data required are the mass, centre of mass position, moments of inertia and product moments of inertia for the whole aircraft.

21.2 Aerodynamic Model

The aerodynamic models used in loads and aeroelastics calculations are vital in allowing the aeroelastic mechanisms to be explored and sufficiently accurate deformations and load distributions to be estimated. Different manufacturers tend to adopt different practices in detail, though there will be similar core features.

21.2.1 Aerodynamic Model for Flight Mechanics

The aerodynamic models for flight mechanics calculations must include force and moment derivatives for the rigid aircraft at a range of flight conditions (e.g. high incidence). Such information is obtained from a combination of design formulae, data sheets, previous experience on similar configurations, CFD (Computational Fluid Dynamics) models and adjustments from wind tunnel measurements (especially where non-linear or transonic effects are important); the process will depend upon the complexity of the configuration being considered.

In order to allow for static aeroelastic effects on the derivatives and on the distribution of loads, the results from the three-dimensional panel method analysis at zero reduced frequency (or frequency parameter) may be used to determine suitable corrections such that the rigid aircraft model would be in effect quasi-flexible, i.e. will have derivatives and distribution corrected for flexibility effects. The rigid aircraft aerodynamics, corrected for flexible effects, need to be linked to the panel method grid. Such a model could be employed for dynamic manoeuvres (see Chapter 23) if flexible mode coordinates were not to be included.

21.2.2   Aerodynamic Model for Aeroelastics and Gusts

An unsteady aerodynamic model that represents the forces acting on a flexible structure is required for conditions where aeroelastic effects are present (static aeroelastics, flutter, gusts, etc.). Both response-dependent and gust-dependent aerodynamic models are required (see Chapter 9). The modelling sophistication will depend upon the complexity of the aircraft configuration, the dynamic motions expected, the extent to which the flight envelope includes the transonic regime and finally the stage in the design process. Flutter is the phenomenon that requires the most accurate and careful aerodynamic modelling.

Either two-dimensional unsteady strip theory or three-dimensional unsteady panel methods are required by the regulations CS/AMC 25.341 and 25.629. However, the panel methods are much more accurate and therefore more widely used. Basic compressibility effects are included as necessary, but these two approaches do not represent transonic behaviour accurately. Since transonic effects are particularly crucial in flutter prediction for commercial jet aircraft, the correction of the calculated two- or three-dimensional aerodynamic results using output from steady CFD and wind tunnel studies, where transonic effects can be included, is important.

The strip theory approach is a relatively crude method, though modifications can be introduced to account approximately for tip effects; however, the method is sometimes used, particularly in early design calculations and for less challenging dynamic conditions (e.g. low speed aircraft). The basic idea of strip theory was introduced earlier in the book (see Chapter 4) and has been adopted throughout the aeroelastics and loads studies for simplicity of approach. The unsteady strip theory results will be a function of reduced frequency, and displacements/rotations at the structural and aerodynamic reference axes must be related in order to assemble a coupled aerodynamic/structural model in modal space. When performing gust response calculations in the time domain, the indicial lift effects (via Wagner and Küssner functions) should be used for two-dimensional strip analysis, together with penetration lags to account for the position of the strips on swept wings and for tailplane effects. This idea was introduced using the examples in Chapter 14.

For more complex configurations (i.e. most commercial aircraft), aerodynamic forces are usually calculated using a three-dimensional unsteady panel approach, such as the Doublet Lattice method (DL method), though there is an increased direct use of CFD methods to cater for transonic effects (see later). The three-dimensional panel method (see Chapter 18) makes allowance for handling the interference between multiple lifting surfaces, all of which are represented by panels (e.g. wing, winglet, tailplane, fin, nacelle and fuselage). A sample Doublet Lattice method panel grid for the whole aircraft is shown in Figure 21.7.

The fact that the structural and aerodynamic data are represented on different grids means that a three-dimensional interpolation/spline approach is required to link the mode shapes with the aerodynamic forces on each panel when assembling a coupled aerodynamic/structural model in modal space (see Chapter 19). The use of a panel method on a simple wing structure was introduced in Chapters 18 and 19. Alternatively, and more simply for convenience, the Doublet Lattice results (in the form of the Aerodynamic Influence Coefficient (AIC) matrix) may be effectively condensed down on to strips on the loads reference axis, thus effectively generating lift and moment coefficients per strip. The approach could be extended when control modes are incorporated.

The aerodynamic model must cater for unsteady effects (i.e. attenuation and phase lag of the aerodynamic forces with respect to the structural motions) so the resulting Aerodynamic Influence Coefficient (AIC) matrix derived from the Doublet Lattice method for oscillatory motion is complex and is a function of the reduced frequency (see Chapter 18); the AIC matrix must therefore be evaluated at a range of reduced frequencies and Mach numbers. Gust dependent

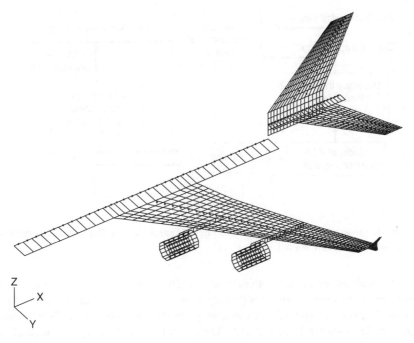

Figure 21.7 Aerodynamic panel grid for the Doublet Lattice method. Reproduced by permission of Airbus.

effects are based on a harmonic gust input, and the AIC matrix from the panel method will allow for gust penetration effects in the frequency domain. When performing gust response calculations in the time domain using a three-dimensional panel method, the frequency domain results for a number of reduced frequencies need to be converted into a time domain representation using the Rational Function Approximation discussed in Chapter 19. Unsteady CFD studies can be used additionally to obtain frequency-dependent adjustments to the results.

Aerodynamic matrices for a range of reduced frequencies are required for flutter and gust/turbulence calculations, whereas the zero reduced frequency (quasi-steady) aerodynamics are used for static aeroelastics and manoeuvres.

The aerodynamic model needs to include terms for both rigid body and flexible modes, and thus at zero reduced frequency, the linear rigid aircraft derivatives may be determined (as shown in Chapter 18) and the AIC results corrected using the flight mechanics aerodynamics model. Control surface aerodynamic terms may require greater correction and the sensitivity of results to these values should be studied.

Finally, there is a growing interest in performing aeroelastic calculations using coupled CFD/structure (FE) models. Such a model may be developed for static aeroelastic and flutter calculations (see Chapter 22).

21.3 Flight Control System

A non-linear Flight Control System (FCS) model will be developed for handling the aircraft (see Chapter 13); it will be coupled with the aircraft dynamic model for a modern commercial aircraft. The FCS commands movements of the aircraft control surfaces depending upon the feedback of information about the aircraft states (e.g. accelerations, rates and air data sensors)

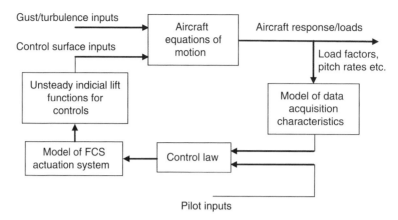

Figure 21.8   Representation of aircraft plus the Flight Control System. Reproduced by permission of Airbus (redrawn with modifications).

and these demands are combined with pilot inputs. The FCS is likely to influence the aircraft loads and aeroelastics behaviour considerably as it will alter the dynamic behaviour of the aircraft (see Chapters 11 and 13). A block diagram showing how the FCS integrates with the aircraft model is shown in Figure 21.8. The control law will incorporate multiple control loops as required.

The FCS model non-linear features may be retained when calculations are performed in the time domain, but it will need to be linearized if a frequency domain solution is to be carried out. Apart from the main function of the FCS of providing 'envelope protected demand control piloting', it may also sometimes be used to perform manoeuvre and/or gust load alleviation functions. A Gust Load Alleviation (GLA) system uses measurements of the vertical acceleration of the aircraft to deploy ailerons and spoilers so as to cancel out a proportion of the outer wing gust force generated. A Manoeuvre Load Alleviation (MLA) system uses a similar approach, but is slower acting in order to bring the effective centre of pressure inboard during a manoeuvre. Both have the effect of reducing the wing root bending moment and shear force.

21.4   Other Model Issues

In addition to the stiffness, mass, aerodynamic and control terms, the following models should be available:

a. a non-linear large angle landing gear model with oleo, tyre, braking, steering, runway profile, flexible modes, etc. (see Chapters 15 and 24 for more detail),
b. an engine model with thrust, momentum drag and gyroscopic effects and
c. a pilot model to allow for the rate of control application.

21.5   Loads Transformations

In Chapter 16, it was shown how a transformation could link the response accelerations, velocities and displacements with the control and gust input (if any) to generate the desired internal loads via an auxiliary equation. Part of the required model for loads requires these matrices to be formed for all the quantities that are of interest to the designer.

22

Static Aeroelasticity and Flutter

In this chapter, an outline of typical industrial practice for static aeroelasticity and flutter calculations will be given and related to the earlier chapters in Part II. The certification requirements covering the issues in this chapter are seen primarily in CS/AMC 25.629 under Subpart D, Design and Construction. However, the requirements as written are dominated by considerations of flutter.

22.1 Static Aeroelasticity

There is little specific mention of static aeroelastic phenomena in the certification specifications. The aeroelastic stability requirement (CS 25.629) simply states that 'divergence, control reversal and any undue loss of stability and control as a result of structural deformation' should be evaluated. In CS/AMC 25.629, it is stated that non-oscillatory aeroelastic instabilities (divergence and control reversal) should be analyzed to show compliance with CS 25.629 and that loss of control effectiveness should be investigated, but little detail is specified.

However, static aeroelasticity appears indirectly in a number of situations via the effect of structural deformation on the rigid aircraft aerodynamic model, and therefore upon gust and manoeuvre loads, etc. It is an important issue that must be considered.

22.1.1 Aircraft Model for Static Aeroelasticity

The aircraft model is traditionally very much the same as that employed for flutter and gust considerations, for example a whole aircraft full or condensed Finite Element model, but because the calculations for static aeroelastic effects are steady, no mass or damping matrices are needed. There is no requirement for the control surface modal behaviour to be represented, except where it can contribute significantly to the loss of effectiveness through its distortion (including control circuit stiffness). Such a model would be based on rigid body/flexible modal coordinates, with the output converted into physical space where required; however, it is possible to perform calculations directly in physical coordinates by expressing the model differently.

Introduction to Aircraft Aeroelasticity and Loads, Second Edition. Jan R. Wright and Jonathan E. Cooper.
© 2015 John Wiley & Sons, Ltd. Published 2015 by John Wiley & Sons, Ltd.

The aerodynamic model is usually formulated using a three-dimensional panel method such as the Doublet Lattice method, but a modified two-dimensional strip theory is permissible. Only the zero reduced frequency terms are required for this steady case, with the AIC results scaled to match wind tunnel corrected rigid body aerodynamic data (and steady CFD results if available) to correct for transonic effects. Control surface aerodynamic terms need to be included. The right-hand side load vectors are required for aerodynamic terms of zero incidence and control rotation, together with a $1g$ inertial load. In Chapter 12, the equations for a combined rigid body/flexible aircraft illustrated this model, with such load vectors being included.

22.1.2 Control Effectiveness and Reversal

The certification specifications require that the aircraft be adequately controllable over the flight envelope so that there is adequate effectiveness at the boundary of the flight envelope. For control effectiveness/reversal calculations, there are two possible approaches (illustrated in Chapter 8 for a simple wing/control model).

In the first approach, the structural model is fully constrained at the wing root (using the wing as an example) and then an aileron control surface rotation is applied. For the *flexible* aircraft, the incremental deformation of the wing component due to control rotation is obtained from the solution of the equations of motion using the right-hand side control vector; an internal load (such as the wing bending moment) at the constraint point would be determined using, for example, the force summation method (described in Chapter 16). However, for the *rigid* aircraft, the internal loads would be derived directly from the distribution of aerodynamic forces due to control rotation. The ratio between the flexible and rigid aircraft internal load results then indicates the control effectiveness under this manoeuvre initiation; the variation of this ratio with air speed will yield the reversal condition where there is zero rolling moment.

In the second approach, the aircraft model is unconstrained and allowed to roll at a constant rate with linear downwash effects due to roll motion included. Here the ratio of the resulting angular rates for the flexible and rigid aircraft will indicate the control effectiveness.

The two approaches will yield the same reversal speed (corresponding to zero effectiveness) but will have different behaviour at lower air speeds. Note that the control effectiveness could be explored further using the dynamic manoeuvre model with FCS (see Chapter 13). Note that the elevator (or stabilizer) effectiveness could be explored by considering the fuselage pitching moment and pitch rate per elevator.

22.1.3 'Jig Shape' – Flexible Deformation and Effect on Loads Distribution

An important issue to address for an aircraft is the relationship between the so-called 'jig shape' (i.e. the unstrained aircraft shape when supported in the jigs during manufacture, with no inertial/aerodynamic forces acting) and the symmetric 'flight shape' that is desired at an optimal (i.e. reference) cruise flight condition; the flight shape has an important influence on the drag acting. The difference between the two shapes is due to elastic deformation under aerodynamic and inertia loads.

Assuming the jig shape were to be known and taken as a reference, the trim solution of the static aeroelastic equations with a right-hand side vector (a combination of zero incidence and inertial terms) yields the flexible deformation and hence the flight shape, as well as the elevator (or stabilizer) angle for trim. The distribution of the aerodynamic forces may then be compared to those for the rigid aircraft and the effects of flexibility seen. Of course, the jig shape is actually unknown initially, though the desired final flight shape and the aerodynamics associated with the design point will be known; an inverse calculation is performed to determine the jig shape and stiffness distribution required to achieve the desired flight shape for minimum weight, subject to a range of constraints. The difference between jig and flight shapes is calculated by applying the aerodynamic and inertia loading to the aircraft stiffness representation. Having established the jig shape needed for the optimal flight shape, the calculation of other suboptimal flight points at other flight conditions can be undertaken.

Errors in the predicted jig shape can arise from a number of sources, e.g. errors in the stiffness and mass representation in the structural model and errors in aerodynamic load prediction. Indeed, however refined the aerodynamic theory is (and this includes CFD), then any errors in the structural model can spoil the results because of an imperfect prediction of flexibility effects. An inaccurate jig shape will lead to errors in the flight shape during flight test; this can mean a greater flight drag than was anticipated when the flight shape was proposed and therefore a suboptimal cruise performance. It is therefore essential to model the structure as accurately as possible.

The influence of flexible effects in altering the lift distribution was shown in Chapter 7 for a simplified wing. Then the trim calculation for a whole aircraft with rigid body/flexible modes was presented in Chapter 12; this process permitted determination of the trim state, airframe deformation and aerodynamic load distribution.

22.1.4   Correction of Rigid Body Aerodynamics for Flexible Effects

In the previous section, the use of the rigid body/flexible aircraft model to calculate the flexible deformation in steady flight was described and the resulting change of load distribution indicated. In some cases, it could be desirable to use the rigid body equations without the flexible modes (so reducing the model size significantly), so it would then be helpful to correct the rigid aircraft derivatives for flexible effects. By taking the rigid body/flexible equations and removing the flexible degrees of freedom using an approach akin to that of Guyan reduction (see Appendix D), a set of rigid body equations with steady derivatives corrected for flexible effects would be available.

Having corrected the derivatives themselves, the shape of the rigid aircraft aerodynamic load distribution can be modified to correspond to that found for the flexible aircraft. A quasi-flexible 'rigid' aircraft model would then be available which represented the total forces/moments as well as the distributions, but with flexible effects included. Such a set of corrected derivatives and distributions may also be incorporated in, for example, the flight mechanics model as well as the static aeroelastic model.

22.1.5   Divergence

The divergence condition must be sought using a combined rigid body/flexible model as the phenomenon may occur for either a flexible or rigid body mode (such as the short period mode) or combination of the two. As flight speed increases, a pair of oscillatory roots reduces to zero frequency and becomes two stable real roots on the root locus diagram (see Chapter 6). With a further increase in speed, one root moves towards $-\infty$ and the other root moves towards zero; on the flutter diagram, both roots would have 100% damping and zero frequency. Divergence occurs when one real root crosses the imaginary axis (i.e. becomes positive); in the flutter 'world' this would correspond to −100% damping and divergence might be seen on the flutter plot as a curve moving from 100 to −100% in a single speed step. Divergence for a simple wing example is considered in Chapters 7 and 10. The concept is mentioned for a whole aircraft with rigid body/flexible modes in Chapter 13.

22.1.6   CFD Calculations

It was pointed out in Chapter 21 that steady CFD calculations allow correction of aerodynamic results for transonic effects. CFD is becoming more widespread in its use. For example, a coupled FE/CFD model could be employed to obtain an improved accuracy in determining the flight shape and drag for suboptimal conditions.

22.2   Flutter

The flutter requirements are covered in CS/AMC 25.629 and should be consulted for detailed explanations.

22.2.1   Aircraft Model for Flutter

For linear flutter calculations, a modal model composed of whole aircraft rigid body and free–free flexible modes is normally used, either based on a beam or condensed FE model with a lumped mass distribution, as described in Chapter 21. In particular, the model should represent control surface rotation (including bending and torsional flexible modes) as well as accurate wing/pylon/engine attachment characteristics, since flutter often involves these components. Control surface actuators may be modelled at several levels, e.g. a spring, spring/damper or full linearized hydraulic model. Also, areas where stiffness may be reduced locally should be modelled adequately (e.g. cut-outs/doors) as stiffness is a key parameter in flutter.

An unsteady aerodynamic model is required with terms evaluated at a range of reduced frequencies. A three-dimensional panel method is usually employed, although a modified two-dimensional strip theory could be permitted, depending upon the aircraft configuration, though the latter is much more accurate, especially for intersecting lifting surfaces (e.g. wing/winglet). In such a case, the effect of steady aerodynamic forces and flexible deformations should be included to allow for in-plane coupling effects. Also, the control surface aerodynamic representation is very important for flutter calculations, and careful adjustments to hinge moment terms are carried out based on wind tunnel test results. Once again, steady CFD and wind tunnel results are used to adjust the aerodynamic results to make allowance for transonic effects. In addition, unsteady oscillatory CFD calculations enable corrections to be made to the frequency-dependent aerodynamic behaviour.

The influence of damping is represented, in essence, either by a value per mode assumed at 0.02 structural (or 1% critical viscous) damping or by values obtained from test (see Chapter 25). Thus far, the flutter model is similar to that employed for gusts. The FCS model used for flutter considerations is generally linearized, though the model will still be frequency (and therefore reduced frequency) dependent; the FCS introduces additional equations into the aeroelastic modal model by involving feedback between structural response and control demand.

22.2.2   *Flutter Boundary – Normal and Failure Conditions*

The aircraft must be designed to be free from aeroelastic instability for all normal configurations and conditions (i.e. fuel, ice, FCS, thrust settings) within the aeroelastic flight envelope (CS/AMC 25.629); the flight envelope considered for flutter is the normal V_D/M_D versus altitude envelope, but enlarged by 15% in EAS at both constant Mach number and constant altitude. This enlarged envelope provides a margin of safety for what is an extremely difficult phenomenon to predict accurately. There should be an adequate margin of stability at the V_D/M_D boundary and no rapid reduction in stability near to it. The freedom from flutter must normally be demonstrated by both calculation and flight test (see Chapter 25). There is no quantitative prescription of what defines an acceptable minimum damping in flight for any mode.

Freedom from flutter must also be demonstrated by calculation for a number of failure cases, described fully in CS/AMC 25.629, namely critical fuel loading conditions, failure of any passive flutter control system, inadvertent ice accumulation, failure of a key structural element supporting a large mass element, various engine failure conditions and potential damage scenarios. Because of the limited probability of occurrence of such failures, the fail-safe clearance envelope within which freedom from flutter must be demonstrated is less extreme than for normal conditions.

Detailed guidance is given in regard to various types of control surface/actuator failure in CS/AMC 26.629. The aircraft must be free from flutter in the presence of freeplay in the control system (represented in a linear model by a stiffness reduction). The FCS and structure should not interact to produce an aeroelastic (or aeroservoelastic) instability. Also, special conditions are specified in CS 25.302, Appendix K, for consideration of flutter in the face of a systems failure likely to impact upon the aircraft dynamic behaviour.

Guidance is given in regard to the incorporation of balance weights and passive flutter dampers. Also, there is sometimes a requirement for investigation of whirl flutter involving gyroscopic effects from engines.

22.2.3   *Flutter Calculations*

The stability of the aeroelastic equations in modal space is investigated using a suitable approach that allows for the frequency dependency of the aerodynamics (and also the linearized FCS), such as the '*p–k*' method; the flutter condition is 'matched' (see Chapter 10) to the aerodynamic behaviour at each air speed depending upon the solution approach adopted. The calculations are performed for a range of weight, fuel, ice state, centre of mass, engine location, control system characteristics, failure cases, etc. The sensitivity of the flutter results to variation in key parameters, such as the control surface aerodynamics, is examined.

Plots of frequency and damping per mode against air speed or Mach number are examined to study the margins of stability around the edge of the flight envelope as defined above. A sample set of results for a multi-mode system was shown in Chapters 10 and 19. Such solution issues were considered in Chapter 10 for simplified two or three degree of freedom models; however, to keep the model complexity to a minimum, no rigid body modes were included there. The combination of rigid body and flexible modes was considered in Chapter 10 and Chapters 12 to 15.

22.2.4   *Aeroservoelastic Calculations*

In CS/AMC 25.629 it is stated that if control/structure coupling is a potential problem, then aeroelastic stability analyses should include the control system, actuator characteristics, etc.

The aeroservoelastic behaviour of the aircraft will need to be investigated using the airframe equations extended to include the linearized Flight Control System equations; control law stability margins should be determined, together with a study of robustness and potential failure cases. Time domain analyses allow the study of nonlinear FCS behaviour, plus failure case scenarios (CS 25.302, Appendix K), but then a time domain unsteady aerodynamic representation (e.g. the Rational Function Approximation approach described in Chapter 19) will be required. Chapter 11 provides a brief introduction to aeroservoelastic analysis.

22.2.5 Non-linear Aeroelastic Behaviour

There is an increasing interest in predicting the aeroelastic behaviour when non-linear effects are present; e.g. structural non-linearities include freeplay in the control surfaces, root stiffening of engine pylons and geometric stiffening for large deflections, aerodynamic non-linearities involve oscillatory motion of the shock waves in transonic flow and control system nonlinearities include rate limits on the motion of the control surfaces. Chapter 10 provides a brief description. The prediction of non-linear aeroelastic behaviour, particularly in the transonic flight regime, is an area of current research interest using CFD/FE coupled models with time-marching solutions, though more efficient reduced-order modelling methods are also being explored.

23

Flight Manoeuvre and Gust/Turbulence Loads

In this chapter, the different ways in which flight manoeuvres and gusts are handled in industry when seeking type certification are discussed. The wide range of different cases that have to be considered is examined. From these bookcase and rational design calculations, the internal loads can be determined which then are used to estimate the stresses.

In Chapter 24, the corresponding issue of ground manoeuvre loads will be considered. Also in Chapter 24, the calculation of stresses from internal loads (via loads acting upon components) and loads sorting will be considered briefly. Once again, it should be pointed out that the treatment in this book stops at the generation of shear forces, bending moments, torques, etc., and does not cover estimation of stresses in any detail.

23.1   Evaluation of Internal Loads

In all the following manoeuvre/gust cases, the internal loads and other interesting quantities may be extracted from the response and force time histories using the 'mode displacement', 'mode acceleration' or 'force summation' approaches (CS/AMC 25.341). The 'force summation' method was illustrated in Chapter 16, showing how suitable transformation matrices allowed summation of inertial and aerodynamic effects via an auxiliary equation when generating internal loads.

It should be noted that in this book, 'internal' loads (as far as the whole aircraft is concerned) are taken to be bending moments, axial forces, shear forces and torques ('MAST' loads, which are 'internal' because the structure has to be 'cut' to expose them; see Chapter 5). The loads and aeroelastics department (or equivalent) generates such internal loads from aerodynamic and inertia forces, calculated using the aircraft dynamic response. The issue of terminology, where these internal loads are sometimes later referred to as external loads, will be addressed later in Chapter 24.

23.2   Equilibrium/Balanced Flight Manoeuvres

In this book, the distinction has been made between equilibrium and dynamic manoeuvres, but in some cases there is a somewhat 'fuzzy' boundary in that some load cases may be solved using an equilibrium balanced analysis or, alternatively, the same requirements may be met

---

Introduction to Aircraft Aeroelasticity and Loads, Second Edition. Jan R. Wright and Jonathan E. Cooper.
© 2015 John Wiley & Sons, Ltd. Published 2015 by John Wiley & Sons, Ltd.

using a rational calculation (see Chapter 20). However, the distinction is considered helpful and will be made as clear as possible, given the differences in practice likely to be found within the industry.

Symmetric equilibrium manoeuvres involve steady motion of the aircraft with a steady pitch rate (zero pitch acceleration). In Chapter 12, it was shown that a pull-out from a dive, a push-down and a steady banked turn, all fall into this category. However, in the certification process, such cases are not considered as separate manoeuvres; instead, the balanced equilibrium flight case involving the application of d'Alembert's principle and the introduction of a load factor is considered in conjunction with the flight manoeuvre envelope. The equilibrium manoeuvre case is in effect a bookcase, since it does not require a full response simulation; however, it is not completely artificial since such a balanced flight case is realizable in practice unless aerodynamic or FCS limitations prevent large load factors being achieved. This case is particularly useful early in the design process when the Flight Control System (FCS) design is not available or mature.

It is also possible to analyse *asymmetric* equilibrium manoeuvres. In particular, a steady roll rate or a steady sideslip may be viewed as balanced equilibrium steady-state cases where the moment due to control application is balanced by aerodynamic restoring moments. If aerodynamic and inertia couplings between the yaw/sideslip/roll motions are permitted, then the steady roll rate and steady sideslip are realizable conditions (and could be determined by a trim solution of the full equations, applying all control surfaces); however, if the approach neglects coupling terms (e.g. pure roll is assumed) then the condition is somewhat artificial. In addition, for cases where a control rotation is suddenly applied, the moment due to the control may be balanced by an inertia couple. This is not a steady-state condition, but is an artificially generated balanced condition for manoeuvre initiation which leads to conservative angular acceleration and loads estimates. In essence, all these cases are bookcases; however, when using a full dynamic model, some cases are also rational since they can actually be achieved (see later).

Note that the symmetric equilibrium manoeuvre case was considered extensively in Chapter 12 in order to illustrate the effect of introducing a flexible mode. The asymmetric rolling and yawing cases were considered in Chapter 13 for a rigid aircraft with reduced models (Lomax, 1996) to illustrate the concepts of solving for the bookcase manoeuvres.

23.2.1 Aircraft Model for Equilibrium Manoeuvres

The rigid body characteristics for the whole aircraft including the control influences are represented for flight manoeuvres, possibly with symmetric and asymmetric equations considered separately. In CS 25.301 the statement is made that: 'if deflections under load would significantly change the distribution of internal or external loads, this redistribution must be taken into account'. This comment means that the static aeroelastic effects must be accounted for in the model used; i.e. the effect on the aerodynamics of distortion under load must be included; the model is in essence a whole aircraft static aeroelastic model. The model should include thrust and drag representations for the symmetric pitching case and allow for these forces to be out-of-line. Also, non-linear aerodynamic and FCS effects that would influence the trimmed condition are included, which means that a non-linear solution of the balanced case may be required. The approach shown in Chapter 12 is similar to that employed in industry for the symmetric manoeuvre. The asymmetric manoeuvre involves a simplified model with certain lateral coupling terms neglected, as illustrated briefly in Chapter 13.

An alternative approach might be to employ a trimmed solution of the large angle non-linear flight mechanics model introduced in Chapter 13, with all coupling terms present and steady aeroelastic effects included.

23.2.2   *Equilibrium Flight Manoeuvres – Pitching*

The certification specifications for symmetric balanced pitching manoeuvres (CS 25.321 and 25.331(b)) require that the aircraft limit load strength is met at each combination of air speed and load factor, both on and within the flight manoeuvre envelope (as defined in CS 25.333, based on design speeds defined in CS 25.335 and load factors in CS 25.337), for critical altitudes, weights, centres of mass and thrusts. CS 25.337 allows for the possibility of considering reduced load factors where physical limitations (such as aircraft stall) would prevent the prescribed load factors being achieved for specific aircraft conditions. Sufficient points on the flight manoeuvre envelope (see Chapter 12) must be investigated to ensure that the maximum load on each part of the aircraft is found.

When the manoeuvres involve the aircraft flying in a region where the aerodynamic behaviour is non-linear, the solution must take this into account. Also, the impact of the FCS on loads must also be accounted for, including potential failure cases. The FCS and aerodynamic limitations (such as stall) may prevent certain balanced manoeuvre conditions on the boundary of the flight envelope from being achieved, and these may be used to limit the load developed within the 'target conditions' set by the airworthiness requirements, provided the manufacturer can justify that the constraints applied really exist.

23.2.3   *Equilibrium Flight Manoeuvres – Rolling*

Here, the rolling manoeuvre conditions described in CS 25.349(a) and Lomax (1996) will be considered using a balanced analysis. The unsymmetrical gust case is covered later. Rolling manoeuvres are assumed to be superimposed upon steady symmetric flight cases defined by load factors of zero and two-thirds of the maximum load factor for design; the latter case, for example, could correspond to a manoeuvre in which the direction of turn of the aircraft is reversed while remaining at a constant load factor.

Note that cross-coupling effects between roll and yaw degrees of freedom may be neglected for this requirement, though they would certainly not be ignored for more fully rational calculations. Torsional flexible deformation should be allowed for. The rolling conditions specified in the certification requirements are therefore nominally of the bookcase type, but the option for a rational calculation is available. The roll manoeuvre contributes to wing and empennage design considerations.

The first condition to be examined is that for a steady roll rate. This is in essence a steady-state solution of the lateral equations with zero roll acceleration and with the rolling moment due to aileron balanced by the aerodynamic moment from the roll damping effect. The other case considered is that of maximum roll acceleration; in the absence of a more rational calculation the rate of roll is assumed to be zero, with the aileron rolling moment balanced by a roll inertia couple. This roll initiation condition is conservative. The treatment corresponds to the statement in the requirement that unbalanced aerodynamic moments must be reacted by inertia effects in a rational or conservative manner.

The aileron inputs are defined such that at V_A, the aileron is deflected suddenly to its maximum value, whereas at V_C and V_D the aileron application should yield respectively a roll rate equal to, or one third of, the roll rate obtained for V_A.

These two cases (i.e. steady roll rate and maximum roll acceleration) may therefore be solved as bookcases via the balanced conditions described above, using, for example, simplified models such as the one in Lomax (1996) where a roll-only equation of motion is shown, with approximate aeroelastic effects included. The basic principles are described in Chapter 13.

Alternatively, the possibility exists to perform a more rational simulation using a mathematical model with five or six rigid degrees of freedom, representing each of the previously mentioned conditions of roll initiation, maximum achieved roll rate and reversal. Indeed, such an approach may be the most practical for an aircraft with an electronic FCS in which the control surface deflections are not directly proportional to the pilot demand through the cockpit control device. A problem to be solved for such a rational simulation is how to represent the degree of piloting control imposed on the non-roll degrees of freedom.

23.2.4   Equilibrium Flight Manoeuvres – Yawing

Here, the yaw manoeuvre conditions described in CS 25.351 and Lomax (1996) will also be considered using a balanced analysis. The manoeuvres involve abrupt application of the rudder followed by a sideslip response and are considered for air speeds between V_{MC} (i.e. the minimum control speed with the critical engine inoperative) and V_D. The aircraft is initially in steady flight, with wings level and at zero angle of yaw, when the rudder is applied. In essence a 'flat' manoeuvre is performed where the wings remain level to maximize sideslip by suitable adjustment of the aileron angle; the roll velocity and acceleration are therefore assumed to be zero. This is a somewhat artificial manoeuvre and is a bookcase condition. In calculating the tail loads, the yawing velocity may be assumed to be zero.

The tail loads are obtained for the following scenarios, focusing in turn upon each of the following four aspects that the aircraft is considered to experience: (a) a sudden application of the rudder to its maximum value (or to a value limited by rudder pedal force) and this rudder deflection is maintained, so generating (b) an overswing (or overshoot) sideslip angle, then (c) a steady equilibrium sideslip condition is reached with the maximum rudder deflection maintained and finally (d) the rudder is returned suddenly to neutral (i.e. zero deflection).

Clearly, this sequence could be simulated rationally (see later), but as a bookcase it would actually be solved via four separate calculations. For the sudden rudder applications in (a) and (d) above, the applied yawing moment has to be balanced by a yaw inertia couple (rather as for the maximum roll acceleration case above) and this would be conservative. Thus, any unbalanced aerodynamic moments would have been reacted by inertia effects in a rational or conservative manner. The steady sideslip condition (c) would be met via the balance of the aerodynamic yawing moments from control and sideslip angles. Finally, the overswing condition (b) is obtained by assuming a dynamic overswing factor applied to the steady sideslip result (see Chapter 1).

This case may therefore be solved as four bookcases via the balanced conditions described above, using, for example, simplified models such as the one in Lomax (1996) where two and three degree of freedom equations of motion are shown, with approximate aeroelastic effects included. The basic principles for steps (a) and (c) are described in Chapter 13.

It should be noted that yawing manoeuvre loads are sensitive to multiple excitation cycles due to low natural damping; this point emerged in the American Airlines A300 accident in

New York, where the pilot applied a sequence of rudder reversals which had the effect of building the low damping Dutch roll mode response considerably above the levels that could be a achieved by the existing single half-cycle manoeuvre of CS 25.351. Some consideration is currently being given to promoting greater robustness in the requirements to protect the aircraft against such inappropriate pilot inputs; there could be a widening of scope of the existing CS 25.351 requirement, or possibly even a new additional requirement.

23.2.5   Other Load Cases

A further bookcase to cover unsymmetrical loads on the empennage (CS 25.427) involves applying 100% of the maximum loading from the symmetric manoeuvre on one side of the tailplane and 80% on the other side. A similar requirement exists for the gust case (see later).

23.3   Dynamic Flight Manoeuvres

In the above section, the cases of balanced equilibrium manoeuvres involving pitch, roll and yaw were considered as bookcases. In this section, the treatment of the aircraft when undergoing some form of dynamic manoeuvre will be considered. The airworthiness requirements are somewhat inconsistent on whether a steady bookcase and/or a dynamic rational simulation is necessary – both are explicitly required in the pitch axis, but not in the roll and yaw axes. The various dynamic manoeuvres must be considered at a range of altitudes, air speeds, weights, centre of mass positions, thrusts, etc. Sufficient points on, and within, the boundaries of the design envelope must be investigated to ensure that the maximum load on each part of the aircraft is obtained. Modern commercial aircraft require many thousands of load cases to be considered, given the combination of multiple loading actions, multiple speed/altitude combinations and multiple mass/centre of mass cases. Efficient loads processing then becomes very important.

As seen later for ground loads, the flight manoeuvre load cases are a mixture of bookcase and rational types, with the former described earlier under the equilibrium manoeuvres case. The rational manoeuvres will normally be solved in the time domain, giving the potential to undertake the solution with non-linear aerodynamics and/or with a non-linear Flight Control System. Some rational manoeuvres are effectively replacing the bookcases with a more realistic simulation involving the FCS, whereas other rational cases allow investigation of failure cases, FCS design, parametric studies, flight test scenarios and any other manoeuvres of interest to the designers in ensuring a safe aircraft. While very little is said about the rolling and yawing manoeuvres in CS-25, manufacturers have built up experience in design calculations and much is captured in Special Conditions documentation (see Chapter 20).

It is worth noting that the dynamic manoeuvres described by the airworthiness requirements are not necessarily the most severe manoeuvres that can be envisaged, but they have been shown to be sufficiently severe to ensure that the aircraft is designed to be strong enough to withstand the sort of extreme events that may occur in operational service, however improbable. Manufacturers may well choose to consider manoeuvres more severe than those covered by the requirements in order to provide additional protection; any such manoeuvres should bear in mind the training given to pilots on how the aircraft should be flown. If, during the flight test, the aircraft is to be intentionally subject to manoeuvres that may potentially be more severe, then the manufacturer/operator must take additional precautions.

23.3.1 Aircraft Model for Dynamic Manoeuvres

There is a number of ways in which the aircraft may be modelled for calculation of dynamic manoeuvres. The first requirement of the aircraft model is that it should represent the rigid body behaviour of the aircraft following deployment of control surface deflection, adjustments to thrust, engine failure, etc. Also, any non-linear Flight Control System must be included since it is likely to influence significantly the response dynamics in the manoeuvre by effectively filtering the control inputs; failure cases will also be examined. A pilot model would be included.

Some manufacturers use a linear aircraft model where motion is referred to an inertial axes system (rather as used for the gust encounter in Chapter 14), whereas others may employ the non-linear large angle flight mechanics model with motion referred to a body fixed axes system, as introduced in Chapter 13; the latter model is more accurate for manoeuvres where changes in altitude and incidence may be significant.

The aircraft model should include the rigid body aerodynamics, validated by appropriate means, such as the wind tunnel test or increasingly by Computational Fluid Dynamics (CFD) calculation, and with non-linear effects included if deemed important. If the flexible mode natural frequencies are significantly higher than the rigid aircraft dynamic stability modes (e.g. short period, Dutch roll), the rigid aircraft model will often suffice for loads calculations, though correcting the rigid body aerodynamics for static aeroelastic effects is necessary if the influence of such terms is important. Both the coefficients and distributions should be corrected for flexible effects to ensure a sufficiently accurate distribution of loads (particularly spanwise). Thus the aim is for the aerodynamic models used for handling qualities/flight mechanics and loads studies to be as consistent as possible through the design process.

Where the flexible modes are closer in frequency to the rigid body dynamic stability modes, and the dynamic flexible response is thought to influence the loads significantly, the flexible modes may be combined with the rigid body modes, leading to additional degrees of freedom, i.e. rigid body motions are combined with modal coordinates as described earlier in Chapter 13. Such a model could be used for the landing calculations where flexible mode effects need to be included (see later) or for flight manoeuvres where flexible dynamic effects are expected. The use of an electronic FCS can mean that the frequency range excited by the control surfaces in flight is reduced and therefore that the flexible mode effects are less important. However, if the flexible modes are part of the model, then both rigid and flexible aerodynamics may be included, most likely for quasi-steady aerodynamics (zero frequency parameter). Creating a coherent set of aerodynamic forces that match the unsteady forces and moments from panel methods with wind tunnel derived terms is not straightforward, especially for the rigid body/flexible aerodynamic coupling terms.

The pitch and roll cases considered in Chapter 13 are rational and indicate the kind of approach used, though without any FCS present. A flexible mode is included together with rigid body motions.

23.3.2 Dynamic Manoeuvres – Pitching

The certification specifications cover abrupt symmetric pitching manoeuvres (CS 25.331(c); Lomax, 1996), which fall into the mandatory dynamic rational category. There are two types of manoeuvre, with the movement of the pitch control often termed 'unchecked' and 'checked'.

The abrupt *unchecked* 'avoidance' manoeuvre involves the aircraft in steady level flight at V_A, when the cockpit pitch control (and hence also the elevator in a conventionally controlled aircraft) is suddenly pulled straight back so as to yield an extreme nose up pitching acceleration; typically, the elevator motion may be idealized as a ramp input up to the maximum value feasible. The aim is to provide a design case for the vertical load at the tail. The manoeuvre need not proceed beyond the point where the positive limit load factor (on the manoeuvre envelope) or the peak tail load is reached. The aircraft response must be considered so this is nominally a rational case (although peak tail load usually arises from the initial elevator displacement and is reached quite quickly).

The other category is the *checked* manoeuvre with the aircraft in steady level flight between V_A and V_D. In essence, the cockpit pitch control device is moved sinusoidally through three-quarters of a cycle at the natural frequency of the short period motion, though the effect of varying frequency may be explored. The nature of the manoeuvre is such that high pitch accelerations are generated; a pitch motion is initiated and then a recovery begun, in such a way that causes the aircraft to respond significantly in its short period motion. It should be noted that the specifications do not require the pilot to move the elevator back and forth over more than three-quarters of a cycle, nor to consider a short period of greater than 4 s. For aircraft with an electronic FCS, the prescribed input will not necessarily translate into a severe elevator command. There may also be some ambiguity about what constitutes the 'short period motion' due to the interaction of aircraft and FCS; it may be more appropriate in these circumstances to undertake a frequency search.

For a nose up checked pitching manoeuvre, the amplitude of the cockpit control motion may be scaled down so that the aircraft just reaches the positive limit load factor. Also, for a nose down checked pitch manoeuvre, the input may be scaled down such that the aircraft does not go below a normal acceleration of $0g$ at the centre of mass. If the defined elevator motion for three-quarters of a cycle at the maximum feasible value does not cause the aircraft to reach the positive limit load factor, then the elevator may be allowed to remain at its maximum value after a quarter of a cycle for a period of up to 5 s (i.e. a 'stretched' sine), until the limit positive load factor is reached, before completing the remaining half cycle of motion. In these checked manoeuvre cases, conditions are given for when the simulation should stop. It is assumed that the final return to the trimmed condition is carried out smoothly.

In Chapter 13, an example of such a flexible heave/pitch model that could be used for a simple rational calculation was considered.

23.3.3   Dynamic Manoeuvres – Rolling

The rolling manoeuvres described in CS 25.349(a) and Lomax (1996) were considered earlier as equilibrium balanced manoeuvres, with steady roll rate and maximum roll acceleration cases. The rolling conditions specified were therefore nominally of the bookcase type, but the option for a rational calculation was said to be available.

A fully rational manoeuvre (Lomax, 1996) would use a model where FCS and roll/yaw/ sideslip coupling effects are represented. Then, for example, the dynamic response to a ramp up/constant/ramp down aileron deflection profile could be determined. The steady roll rate condition would be reached and the resulting roll accelerations would then be lower than for the simple roll initiation bookcase. Thus the rational calculation would yield less conservative results.

Other cases may also be examined (see later). In Chapter 13, an example of such a roll model that could be used for a simple rational calculation was considered.

23.3.4   Dynamic Manoeuvres – Yawing

Here, the yaw manoeuvre conditions described in CS 25.351 and Lomax (1996) involve application of the rudder followed by sideslip overswing, steady sideslip and return to neutral. These were treated as four separate bookcase manoeuvres, but the option for a rational calculation is available.

One possibility for a rational calculation (Lomax, 1996) is to determine, using a simplified two or three DoF model (with aeroelastic effects included), the dynamic response to an abrupt rudder input, leading to a dynamic sideslip response including the overswing, followed by the rudder return to neutral. No examples of a dynamic yaw/sideslip model were shown in Part II of this book.

In practice, a fully rational manoeuvre would use a model where FCS and roll/yaw/sideslip coupling effects are represented; such an advanced model might be employed in, for example, simulating the abrupt rudder application bookcases in a rational manner to gain less conservative results. An issue here is to decide how to represent appropriate pilot action to maintain appropriate control in the roll and pitch axes. Other cases may also be examined (see later).

23.3.5   Engine Failure Cases

The unsymmetrical load conditions arising from failure of what is deemed to be the critical engine are considered in CS 25.367. Failures due to fuel flow interruption (a limit case) and mechanical failure of the engine (an ultimate case) must be investigated. Firstly, following an engine failure, there are two relevant steady-state conditions, namely maximum sideslip reached with zero rudder and rudder angle required for zero sideslip. Secondly, a corrective application should be considered, with the rudder applied when maximum yaw velocity is reached (but not earlier than after 2 s). The simulation should include thrust decay and drag build-up effects (more severe for the ultimate failure case) as well as the FCS. Thus, in essence, bookcase or rational calculations may be carried out; in Lomax (1996) simplified linear equations are presented.

23.3.6   Other Load Cases

A fully rational calculation may be carried out to study other proposed failure cases or flight test scenarios, to examine more realistic control actions, to assist in FCS design and to carry out parametric studies (e.g. sensitivity), as well as to perform certification calculations. One issue of growing importance is any failure case involving the FCS and causing control surface runaway or continuous oscillation. Such scenarios would be investigated by rational calculation (CS 25.302, Appendix K), considering firstly loads at the time of occurrence and secondly modified loads that could occur during continuation of the flight.

Other flight cases, such as those specified in CS-25 (Subpart B) or flight cases of concern, may be investigated by rational calculation as well as by flight test. The models may be validated by comparison with flight test data (including loads; see Chapter 25).

23.4 Gusts and Turbulence

A considerable effort over many years has been put into deciding how to define the certification requirements for loads in gusts and turbulence, with the methodologies becoming ever more rational and complex (Hoblit, 1988; Lomax, 1996). An extensive experimental programme involving measurements on a number of in-service aircraft, namely CAADRP (Civil Aircraft Airworthiness Data Recoding Program), was carried out to help decide upon design gust and turbulence levels using a statistical basis, e.g. the amplitude of the discrete gust is targeted to the 1 in 70 000 flying hours probability level and the variation in amplitude with wavelength adjusted to maintain an equal probability of occurrence. In this section, the main gust and turbulence cases required for certification will be considered (CS/AMC 25.341).

Real gusts and turbulence in the atmosphere tend to occur in patches and to be isotropic in all three dimensions. Airworthiness requirements in general do not require this level of detail in modelling. Rather, for a conventional aircraft at least, the main lifting surfaces respond predominantly to inputs in one or other directions: wings and horizontal tailplane respond to the symmetric vertical component; fins and rear fuselage to the lateral component; and high drag devices, such as flaps, to the fore-and-aft component. Consequently, airworthiness requirements simplify the required analyses to a one-dimensional form only, subject to certain safeguards introduced by 'round-the-clock' gusts, etc. Furthermore, the range of input turbulence manifestations is covered only by the need to consider the two extremes of a single, isolated, 'discrete gust' and of a continuous Gaussian turbulence field. Patches of real turbulence are highly unlikely to be either 'discrete' or 'continuous', but by covering these extremes it is possible to assure robustness of the aircraft structure to gusts and turbulence.

There have been attempts over the years to postulate a gust and turbulence regime that could replace the separate 'discrete' and 'continuous' requirements under one umbrella. One such example, the statistical discrete gust (Jones, 1989), allows a range of equiprobable gust patterns, containing the potential to represent complex turbulence shapes for those parts of the aircraft susceptible to more continuous excitation, while allowing for more discrete inputs to those parts of the aircraft more susceptible to these. The so-called matched filter theory (Scott et al., 1993) provides the worst case combination of gust velocities when an aircraft is considered to be linear; however, a search algorithm is required when non-linearities are present. All such approaches, while being academically attractive, have so far had practical implementation difficulties, which have caused them to fail to gain widespread support.

(As an aside it may be of interest to note that limit gust and turbulence have been defined so that they represent the level of severity that a single aircraft might expect to encounter once per lifetime. This implies that the limit load should be encountered less frequently than once per lifetime, allowing for the probability that the aircraft is unlikely to be at its most critical mass/centre of mass/flight condition at the time of the encounter. See also Chapter 20.)

23.4.1 Aircraft Model for Gusts and Turbulence

The aircraft model will include rigid body and flexible modes referenced to the inertial axes system, with the response generally solved in modal coordinates. Structural damping would normally be included (see the model for flutter in Chapter 22). The response dependent and gust-dependent aerodynamics will be unsteady and available at a range of reduced frequencies; typically modified two-dimensional strip theory or a three-dimensional panel method such as the DL (Doublet Lattice) method would be employed (incompressible/compressible, depending upon the complexity of the aerodynamic configuration and flight case). The three-dimensional panel approach (see Chapter 18) will account for the penetration effects of the swept wing and of the tailplane, whereas the two-dimensional strip theory would require the penetration effect to be included explicitly. The aerodynamic model was discussed in Chapter 21.

For the response to continuous turbulence, the aerodynamic matrices evaluated at different reduced frequencies need to be interpolated to allow frequency domain spectral calculations to be carried out (see Chapter 14). On the other hand, when calculating the response to a discrete gust in the time domain, either two-dimensional indicial lift functions (i.e. Wagner and Küssner functions) may be used with two-dimensional strip theory, or else the frequency domain aerodynamic forces may be transformed into a time domain representation using a 'Rational Function Approximation' (see Chapter 19).

The aircraft model should include models for all relevant modes of any Flight Control System if any coupling with the structural response is possible. It should also include thrust and gyroscopic effects for the engines.

23.4.2 Discrete Gust Loads

In the certification requirements for discrete gusts, the aircraft is considered to be in level $1g$ flight and subject to symmetric vertical and lateral gusts of the '1-cosine' type introduced earlier in Chapter 14. The incremental loads due to the gust would subsequently be combined with the steady $1\,g$ flight load solved using a model allowing for static aeroelastic effects (e.g. equilibrium manoeuvre model). Gusts in both positive and negative senses would be considered.

Dynamic response calculations are performed either by direct simulation in the time domain (most suitable if non-linear effects need to be considered), by convolution or by employing the Fourier Transform and Impulse Response Function. These approaches were introduced in Chapters 1 and 2. The approaches are rational and similar in principle to the calculations introduced in Chapter 14. Limit loads must be determined for critical altitudes, weights, centres of mass, air speeds, thrusts, etc. A sufficient number of gust gradient distances H (equal to half the gust wavelength) in the range 9 m (30 ft) to 107 m (350 ft) need to be investigated such that critical responses are found for each load quantity. When a stability augmentation system is employed, any significant system non-linearity should be taken into account.

The design gust velocity U_{ds} (i.e. the maximum value of the '1-cosine' gust) is expressed in terms of a reference velocity U_{ref}, a flight profile alleviation factor F_g and the gradient distance H (in m) via the formula

$$U_{ds} = U_{ref} F_g \left(\frac{H}{107} \right)^{1/6} \tag{23.1}$$

Thus the longer gusts have a larger gust velocity. For air speeds between V_B and V_C, the reference gust velocity reduces linearly from 17.07 m/s (or 56 ft/s) EAS at sea level to 13.41 m/s (44 ft/s) EAS at 4572 m (15 000 ft), and again to 6.36 m/s (20.86 ft/s) EAS at 18 288 m (60 000 ft); these values are halved at V_D, with values at intermediate air speeds given by linear interpolation. The flight profile alleviation factor F_g increases linearly from the sea level value (a function of weight and maximum operating altitude information) to 1.0 at the operating altitude. The intent of this flight profile alleviation factor is to place reduced weighting on those altitudes within the flight envelope where the aircraft is less likely to fly at the design gust speeds.

If a balanced load distribution for the aircraft is required at any instant of time during the response to a discrete gust, then time-correlated results must be used, i.e. all responses and internal loads (shear forces, etc.) are extracted at the same time in the simulation (see Chapter 16).

Finally, note that a specific requirement is included for wing-mounted engines, using a discrete gust at different angles normal to the flight path, and also a pair of vertical and lateral gusts in the most severe combination or as a 'round-the-clock' gust.

23.4.3   Continuous Turbulence Loads

The aircraft dynamic response to vertical and lateral continuous turbulence needs to be taken into account for certification. The approach is a rational one and is similar to the methodology introduced in Chapters 14 and 16. A frequency domain power spectral analysis is carried out using the von Karman turbulence Power Spectral Density (PSD) together with a transfer function relating the output response (or load) quantity to the amplitude of a harmonically oscillating gust field. The PSD and root-mean-square value of the response and loads may then be obtained.

The method currently employed in CS-25 for continuous turbulence is the so-called 'design envelope analysis', similar to the way in which discrete gusts are handled; limit loads must be determined for critical altitudes, weights, centres of mass, air speeds, thrusts, etc. An alternative approach, previously popular in the USA is 'mission analysis'; here, particular mission profiles are set up and analysed in segments using a frequency of exceedance model (Hoblit, 1988). A load spectrum may then be built up by summation of the contributions from all the mission segments and a limit load identified by a nominated frequency of exceedance on that spectrum. The use of one or both of these criteria has been a source of debate between Europe and the USA for many years. In fact a 'harmonization agreement' has been reached via the LDHWG (see Chapter 20) to implement what is now the CS-25 text, hence eliminating the mission analysis option. The issue is discussed in detail in Hoblit (1988), but in this book only the design envelope analysis practice will be considered further.

As the name suggests, design envelope analysis is similar to other load requirements in needing multiple calculations of loads to identify critical mass/centre of mass/flight conditions at which a defined turbulence field is most critical, whereas mission analysis investigates the response to a 'typical' usage of the aircraft. In fact, as mission analysis is both intimately linked to atmospheric turbulence statistics and is capable of representing 'typical' usage of an aircraft, mission analysis remains well suited to producing gust and turbulence load spectra for fatigue and damage tolerance analysis. Its use is expected to continue in this area, at the choice of the manufacturer.

The limit load P_{Limit} for any load of interest (e.g. the wing root bending moment) is determined using the expression

$$P_{\text{Limit}} = P_{1g} \pm U_\sigma \bar{A} \tag{23.2}$$

where P_{1g} is the corresponding steady 1 g load for the relevant condition, $\bar{A}$ is the ratio of the root-mean-square incremental load to the root-mean-square (RMS) turbulence velocity and U_σ is the limit turbulence intensity; both positive and negative incremental loads need to be considered. Thus the term $U_\sigma \bar{A}$ is the incremental load due to turbulence in the limit condition. These issues are discussed in Chapters 14 and 16.

The limit turbulence intensity U_σ is a function of air speed and altitude, rather like the reference gust velocity, and may be expressed as the product of a reference turbulence intensity $U_{\sigma ref}$ and the flight profile alleviation factor F_g, namely

$$U_\sigma = U_{\sigma ref} F_g \tag{23.3}$$

For air speeds between V_B and V_C, the reference turbulence intensity reduces linearly from 27.43 m/s (or 90 ft/s) EAS at sea level to 24.08 m/s (79 ft/s) EAS at 7315 m (24 000 ft), and then remains constant up to 18 288 m (60 000 ft). These values are halved at V_D, with values at intermediate air speeds given by linear interpolation. In essence, the large values of turbulence intensity chosen allow for a probability of exceedance corresponding to the limit condition.

It is worth noting that identifying the 'reference turbulence intensity' for any air speed and altitude does not imply that real continuous turbulence would have an RMS of this magnitude. Rather, U_σ can be considered to comprise the product of two scalars, one to represent the RMS of the patch of turbulence most likely to generate a 'limit' gust encounter and another to account for the probability factor to define an appropriate 'extreme event' point on the tail of the assumed Gaussian probability function relative to the RMS (i.e. peak to RMS ratio).

If balanced correlated load distributions are required from a continuous turbulence analysis, then they may be obtained using equiprobable solution results using cross-correlation coefficients (CS-25; Hoblit, 1988). If critical stress values depend upon more than one internal load, then an extension of the correlation coefficient approach is required. The whole statistical processing of results from continuous turbulence is complex and more advanced texts should be consulted if further information is required (Hoblit, 1988).

23.4.4 Handling Aircraft with Non-linearities

With the advent of widespread use of non-linear automatic flight control and/or load alleviation systems containing thresholds, authority limits, rate limits, digital logic and non-linear control surface actuators, there has been a pressing need for manufacturers to be able to address such non-linearities in the calculation of discrete gust and turbulence loads. In fact, for the calculation of discrete gust loads in the time domain there is no big issue; all that is required is to be able to represent the non-linear system within the mathematical model used for the simulation.

However, to address continuous turbulence is more difficult and requires an understanding of how the atmosphere model presented in CS 25.341(b) has been simplified. It is particularly important to recognize that it is not the RMS of a load response that is most physically significant for limit loads, rather it is the more extreme peaks or exceedances within the response (after all, if the RMS were the limit load, then the structure would have failed long before an RMS could be established). For a linear Gaussian model (as assumed in the basic airworthiness requirement), this distinction is not important because there will be a constant linear relationship between the RMS and the 'tail' of the Gaussian distribution of the response. However, for a non-linear aircraft the 'tail' might be stretched or compressed relative to a Gaussian distribution. A large number of different approaches have been studied in the treatment of non-linearities for determination of continuous turbulence loads, mostly falling into one of two categories:

a. Linearization methods, e.g. the 'equivalent gain' method, which, for a limit class of 'symmetric' non-linearities of the non-linear gain, seek to identify a linear model with appropriate gain to minimize some error function of the response. This linear model may then be used to produce loads as for a conventional linear formulation.

b. Timeplane stochastic methods utilize a non-linear mathematical model of the aircraft to generate a time response to a 'stochastic' gust history, which has been generated to be compatible with the continuous turbulence atmosphere definition. This approach requires a number of issues to be resolved, namely generation of the stochastic gust history itself (for example: is the turbulence Gaussian?, what is its RMS value?, how much data must be considered for it to be statistically significant? etc.); the process to identify limit loads in the response (by a level exceedance criterion or by peak counting); and how to generate correlated loads that may be used for structural analysis. CS/AMC 25.341 describes one such solution that has been found acceptable in the past.

23.4.5   Other Gust Cases

The approaches described above are the main gust and turbulence cases that have to be considered and are rational analyses. However, there are a number of other special areas where gusts need to be considered for certification, e.g. loads on high lift devices and other aerodynamic control surfaces, rolling conditions and unsymmetrical loads, fuel loads etc. Some examples will be considered briefly here.

A requirement under the 'Rolling Conditions' heading (CS 25.349) is for the aircraft to be subjected to unsymmetrical vertical gusts, where limit air loads are obtained using the maximum load on the wing from the discrete gust analysis, with 80% and 100% of this load applied on the two sides of the aircraft respectively. The aircraft is then balanced using inertia forces and so in effect this is a bookcase condition. A very similar bookcase under the 'unsymmetrical loads' heading (CS 25.427) applies to the fin and tailplane. These are of course bookcases designed to give the wing and empennage attachments to the fuselage sufficient robustness to deal with asymmetric loading.

24

Ground Manoeuvre Loads

In this chapter, the treatment of ground loads will be outlined in a somewhat similar manner to that considered for flight manoeuvre and gust loads in Chapter 23. In the certification requirements (CS-25; Lomax, 1996), the load cases are essentially subdivided into the two categories of *landing* and *ground handling*, the latter covering taxi, take-off and landing roll, braked roll, turning, jacking, towing, etc. The calculations lead to airframe internal loads, together with landing gear ground, attachment and component loads; the methodologies are similar to those for flight loads except that discrete landing gear loads are also present. As in the case of flight loads, an alternative subdivision would be into 'bookcases' (mainly ground handling) and 'rational' cases requiring flexible or quasi-flexible dynamic modelling (dynamic landing, dynamic taxi, dynamic braked roll, etc.). The calculation of stresses from internal loads (via loads acting upon components) and loads sorting will also be considered.

24.1 Aircraft/Landing Gear Models for Ground Manoeuvres

As explained earlier in Chapter 15, the landing gears provide a vital function for landing and ground handling of the aircraft. Landing gears are highly non-linear for the shock absorber, tyre and gear mechanism. A non-linear gear model is required for all rationally based requirements, whereas bookcases may (in most cases) use a simpler model as their requirement is usually only to represent a reasonable static closure geometry under the prevailing loading.

Where rational calculations are required, particularly specified in CS-25 for landing and taxi, take-off and landing roll, a non-linear landing gear model is developed. This model will include the constituent mechanisms of the gear, an oleopneumatic shock absorber with non-linear stiffness and damping, a dynamic model of the tyre and unsprung mass, and models for braking and steering. A Finite Element model of the gear should be available to provide modal properties for different shock absorber closures, particularly to cover the spin-up and spring-back behaviour on landing, though a simplified beam model may suffice (see Chapter 15). The ability to interface the model with a runway profile is needed, both for variations in elevation along the runway and also to cater for a convex profile across the runway; the latter is only relevant for aircraft with more than two main gears, such as the Airbus A380 and Boeing 747. Such landing gear models are developed by both the airframe and gear manufacturers. Nowadays, more use is being made of advanced COTS ('commercial-off-the-shelf') mechanisms software.

Introduction to Aircraft Aeroelasticity and Loads, Second Edition. Jan R. Wright and Jonathan E. Cooper.
© 2015 John Wiley & Sons, Ltd. Published 2015 by John Wiley & Sons, Ltd.

The aircraft model depends upon the load case. The landing case employs a flexible aircraft model (often a non-linear large angle rigid body, i.e. flight mechanics model, based on body fixed axes; see Chapter 13) with quasi-steady aerodynamics for both rigid and flexible deformations (clean wing and high lift) and a fully non-linear main gear model. On the other hand, the ground handling cases employ a linear airframe model (based on inertial axes) with rigid body and flexible modes, quasi-steady aerodynamics (rigid body and flexible contributions) and a non-linear gear model.

24.2   Landing Gear/Airframe Interface

The gear to airframe attachment arrangement is important in terms of how the loads are transferred. For conventional main gears, the attachment to the airframe via two pintles (providing restraint in $x/y/z$ and y/z respectively) and pinned side brace (providing a tension/compression restraint force) will usually be considered as statically determinate. For nose gears the attachment may be statically indeterminate (or redundant) as the two pintles will provide restraint in $x/y/z$ and x/z, but the drag brace will be attached at two points on either side of the aircraft plane of symmetry; even then, the arrangement will be statically determinate for symmetric loads. Other less conventional gears may have a redundant attachment and so the attachment loads will also depend upon the local airframe flexibility. Any weight benefits gained by structural redundancy will need to be offset against additional analysis costs.

Where rational calculations are carried out, a coupled flexible aircraft/landing gear model will be employed, so the attachment arrangement may be represented whether it is statically determinate or redundant. However, where bookcase calculations are required, any gear model required will depend upon whether the gear/airframe interface is statically determinate or redundant. At best for a statically determinate attachment, no dynamic or flexible model of the gear is required, with specified ground loads simply being transferred to the airframe attachment points via geometric data and simple equilibrium considerations. The shock absorber closure needs to be specified because of its effect on the overall leg length. For a redundant gear/airframe attachment, a gear stiffness model with a simple shock absorber representation coupled to the airframe flexible model is needed in order to allow the attachment loads due to applied ground reactions to be calculated.

24.3   Ground Manoeuvres – Landing

The first category for ground load certification cases is that of *landing* (CS-25; Lomax, 1996). The fundamental calculation is a fully rational dynamic case (CS 25.473, 25.479 and 25.481) employing a flexible aircraft model (often non-linear, large angle) and non-linear landing gear. The aircraft impacts the ground at a prescribed vertical velocity e.g. 10 ft/s (or 3 m/s) for maximum landing weight (MLW) and 6 ft/s at maximum take-off weight (MTOW); wings are level and a range of air speeds must be considered for level and tail down landings. (A trimmed attitude corresponding to the $1g$ condition for a range of air speeds would be a more rational initial condition and then a variety of attitudes could be explored; as it is, a variety of wind speeds would need to be applied in order to match the required initial speed/attitude combinations.)

The dynamic response and loads are calculated using a non-linear time-stepping solution. Important considerations include the contributions of shock absorber stiffness and damping to the vertical load and their correlation with the 'spring back' of the leg due to wheel 'spin-up' drag and the dynamic 'spring-forward' of the leg fore-and-aft bending modes on completion of wheel spin-up. The basic idea, albeit highly simplified in terms of the landing gear model used, was explained in Chapter 15. Note that drop tests (CS/AMC 25.723) on full-scale main landing gears, with and without opposite wheel spin imposed, are carried out to validate the dynamic model.

Apart from the rational landing calculations providing some design loads for the airframe and gear, the critical vertical leg reaction (i.e. the worst case from all the rational calculations) provides a reference load for bookcases to cover lateral drift and one gear landing (CS 25.483), as well as drag and side load cases (CS 25.485). For aircraft with more than two main gears, and to cater for the case where the gears impact the ground at different times in a non-level landing, a rational rolled landing at a reduced descent velocity is carried out.

Additional certification requirements exist to cover emergency landing, and crash simulations are often carried out.

24.4   Ground Manoeuvres – Ground Handling

The second and more extensive category for ground load certification is ground handling (CS-25; Lomax, 1996).

24.4.1   Taxi, Take-off and Roll Case

At present, the only fully rational dynamic ground handling case required for aircraft with two main gears is that for *taxi, take-off and roll* (CS/AMC 25.491). In this case, which has been treated historically in many different ways, the current recommended process is to carry out dynamic response calculations using a linear flexible airframe model with quasi-steady aerodynamics (rigid body and flexible contributions) and a non-linear gear model; again the dynamic response and loads will be calculated using a non-linear time-stepping solution. The basic 'random' case employs the San Francisco 28R runway (as surveyed in the 1960s, prior to more recent resurfacing), being seen as a traditionally 'rough' runway that was known to lead to high loads and pilot complaints; runs are carried out at a range of steady ground speeds and with zero and maximum thrust (or reverse thrust/braking), depending upon whether take-off or landing is being considered. The basic idea of runway response calculations, albeit significantly simplified, was explained in Chapter 15.

A further rational discrete load case may be carried out by taxiing the aircraft over specified double '1-cosine' bumps, though an alternative bookcase approach is optional. Key considerations here are aircraft ground speed/bump spacing conditions, which give rise to synchronized vertical excitation through the landing gear of the aircraft in heave or pitch or in specific flexible aircraft modes. Note that a bookcase is also specified in order to allow investigation of combined vertical, side and drag loads on the main gear.

24.4.2   Braked Roll, Turning and Other Ground Handling Cases

The remaining load cases for ground handling of aircraft with two main gears tend to fall into the bookcase category, though it should be emphasized that the story is far from simple and only an outline will be given here. No aerodynamic model is usually required (or permitted, noting that aerodynamics normally act to relieve the loading on the landing gears) and the specified ground reaction loads (vertical/side/drag) are in most cases reacted by inertia forces (and sometimes moments) distributed over the airframe according to the mass distribution (see Chapter 16).

For aircraft with more than two main gears, the approach has had to be altered in many cases, either by adapting the bookcase prescribed ground reaction loads, or by carrying out so-called quasi-rational cases where the equilibrium/trim condition is obtained from the model of the quasi-static flexible aircraft (i.e. flexible aircraft and gear stiffness effects represented but no dynamics), with a simple non-linear gear model including leg flexibility.

The *braked roll* bookcase (CS 25.493) allows calculation of the ground reactions at the gears when a braking force is applied at the braked wheels; both steady and dynamic cases are catered for by specifying different load magnitudes. The steady braking case was shown in Chapter 15. However, a fully rational dynamic or quasi-rational steady approach (based on an aircraft with a shock absorber/tyre model and a brake torque/time input, confirmed using information from brake test data where appropriate) may be employed to justify lower braking loads and to cater for the aircraft with more than two main gears. A reversed braking bookcase (CS 25.507) is also specified.

For *turning*, a 0.5g bookcase is specified (CS 25.495), with side loads (equal to 0.5 static gear reaction) applied at all wheels and reacted by vertical and side inertia loads distributed over the airframe; this case was shown in Chapter 15. For aircraft with more than two main gears, a rational approach to estimating the distribution of tyre loads in the turn may be required in order to protect against the possibility that any one gear may be subject to a disproportionate loading due to its stiffness characteristics. While it is convenient for the authorities to define and retain the proven conservatism of the turning bookcase, a problem for the manufacturer is that 'real' tyres used as part of a rational model may not be able to react the loads implied by the 0.5g turn.

Additional bookcases cover *nose wheel yaw/steering* (CS 25.499) by considering the nose gear side load when braking is applied on one side at the main gear, and also separately by considering steering torque. *Pivoting* (where brakes on one side are locked while differential thrust attempts to turn the aircraft, so loading the main gear in torsion) is an additional bookcase related to turning (CS 25.503); this case may need to be revised for redundant gear arrangements. Other bookcases cover *towing* (CS 25.509) and *jacking* and *tie-down* (CS 25.519). Again, in some cases a quasi-rational treatment is required to handle the aircraft with more than two main gears.

As if the picture is not complicated enough, the cases where one or two tyres are deflated (CS 25.511) need to be considered for many of the ground handling cases by applying specified reduction factors. Local redundancy calculations at the wheels may be necessary for the distribution of loads across tyres and bogie units; this is not a straightforward issue. Additional requirements are included to cover the retracting mechanism, wheels, tyres and brakes/braking systems (CS 25.729, 25.731, 25.733, 25.735).

Shimmy studies using a linearised and/or non-linear model may be carried out where appropriate; calculations are supplemented by tests.

24.5   Loads Processing

It should perhaps be emphasized at this stage that the quantities of interest to the designer from loads calculations are more than loads alone but include response accelerations and rates (especially where accelerometer and rate gyros for the FCS are positioned), strains, control forces, values within the FCS loop, aerodynamic pressures, flight mechanics parameters, oleo pressures, etc.

24.5.1 Loads Sorting

The generation of internal loads is a major step in the analysis of the flexible aircraft under static or dynamic loading. As indicated in earlier chapters, a range of calculations will need to be employed so as to cover the many combinations of flight envelope points and mass distributions that could lead to dimensioning loads (i.e. load cases responsible for sizing a particular piece of structure somewhere on the aircraft; see Chapters 16 and 20). Once the appropriate moment, axial force, shear force and torque ('MAST') distributions have been determined, often as a function of time, then they need to be processed further ('sorted') in order to obtain critical load cases. It is important to use correlated load sets, i.e. loads extracted at the same instant of time to ensure that the loads are balanced in any subsequent analysis; note that correlated loads for turbulence data need special treatment (Hoblit, 1988).

The loads sorting process aims to extract those correlated load sets that identify the most positive, or most negative, internal ('MAST') load value somewhere on the aircraft; this process involves huge amounts of data. The idea of using one-, two- and multi-dimensional load envelopes to determine critical cases is introduced in Chapter 16. The resulting loads sets are then stored in a database and issued to the stress office or internal/external 'customers' for design or certification. These internal loads are then converted to component loads (see the next section) to allow subsequent analyses that provide data to allow individual structural elements to be stressed.

24.5.2 Obtaining Stresses from Internal Loads

The next stage in the loads determination process is now outlined briefly. There are essentially two stages of calculation undertaken for the determination of loads/stresses in an aircraft structure (alluded to in Chapter 5):

a. In the first stage of calculation, externally applied loads are considered to act on the whole aircraft structure via distributed aerodynamic and inertia effects (including, in the case of ground manoeuvres, discrete landing gear loads acting as reactions). As a result, internal loads (or so-called 'stress resultants' – moment, axial, shear and torque) are generated at chosen cross-sections of each slender member such as a wing or fuselage. Worst case load conditions are selected by sorting (see above). This stage has been the subject of this section so far.

b. In the second stage of loads calculation, the classical approach (see Chapter 5) is to determine stresses directly from these internal loads using, for example, a formula based methodology, but this assumes that load paths are well defined. However, for more complex semi-monocoque (i.e. stiffened skin) aircraft structures where the load paths are ill defined, this classical approach is inappropriate as there will be no simple formulae linking internal 'MAST' loads to stresses in a structural element such as a wing rib or spar on a swept wing. Instead, the internal 'MAST' loads from the first stage of the analysis will need to be decomposed (or converted) into an equivalent set of 'new' balanced external loads acting, for example, at nodal points on the reference axes of the condensed Finite Element model (see Chapters 20 and 21). Then, because the FE model will represent the internal load paths, the FE analysis will yield 'new' internal loads (i.e. different to overall 'MAST' loads) and stresses acting on the internal structural elements. Further, more detailed FE models or data sheet/formula approaches on subcomponents may then be employed. Once maximum stress values for different load cases are known, the static strength and fatigue/damage tolerance considerations may be examined for individual structural elements.

Thus, when such subsequent analyses are performed, what have so far been called *internal* loads ('MAST') may sometimes be referred to as *external* loads since they will actually determine the distributed external loads acting at the nodes in the FE model for the component considered. Clearly, this can give rise to some confusion of terminology. Since the main focus of the book is on the first stage (namely obtaining moment, axial, shear and torque ('MAST') loads), these have been referred to, in the classical way, as internal loads. The context of a particular external load usage will indicate which type of load is being considered.

Note that the process outlined above, where the internal 'MAST' loads are decomposed on to the reference axis of a condensed beam FE model for further analysis, is only one of a number of possible approaches to what is a complex problem. By making use of knowledge of the chordwise mass and aerodynamic pressure distributions (and making assumptions about the fuel mass distribution), the nodal loadings may be further refined by distributing them over the uncondensed 'box-like' FE model (see Chapter 20). Clearly, the more accurate the analysis, the more precise can be the calculations of stress for design.

25

Testing Relevant to Aeroelasticity and Loads

25.1 Introduction

The clearance process relating to aeroelastic and loads issues during the design and certification phase of a new aircraft is a combination of numerical modelling backed up by testing. An outline of the main tests used to validate various elements of the aircraft mathematical model for aeroelastic issues is shown in Figure 25.1. Note how there is an opportunity to update the numerical models of each element based upon the test results. A similar series of tests are undertaken to certify aircraft for ground and flight loads. In this chapter, only a very brief outline of the methods will be given.

It must be emphasized that no one test is capable of providing the information for full validation of the mathematical models used for certification. That must inevitably come by building up a range of test results on different aspects (e.g. structural stiffness, mass, mass distribution, centre of mass, wind tunnel tests, systems tests, etc.). The ground tests can be quite accurate and are backed up by checks that show, when assembled together, that the structural dynamic and the flight response properties are both reasonable. However, tests performed inflight to demonstrate aeroelastic stability and validate flight loads are subject to a number of uncertainties. The test set up for the Flight Flutter Test is much less ideal (e.g. noisy environment, inadequate excitation) than that for the Ground Vibration Test, whereas flight loads have to be determined from strain gauge readings and these calibrations can be problematic.

25.2 Wind Tunnel Tests

There are two types of wind tunnel test that are particularly relevant to aeroelasticity and dynamic loads, namely determination of rigid aircraft aerodynamic derivatives and flutter model testing. A wind tunnel model is shown in Figure 25.2.

Aircraft configurations are often complex and accurate prediction of rigid aircraft aerodynamic derivatives is difficult, especially when non-linearity is important (e.g. near to the stall condition) or when the aircraft is in the transonic regime. Therefore model tests for the rigid scale model of a whole aircraft are performed to measure pressure distributions, net forces and moments and so estimate the aerodynamic derivatives at different flight conditions. Values obtained are used to

Introduction to Aircraft Aeroelasticity and Loads, Second Edition. Jan R. Wright and Jonathan E. Cooper.
© 2015 John Wiley & Sons, Ltd. Published 2015 by John Wiley & Sons, Ltd.

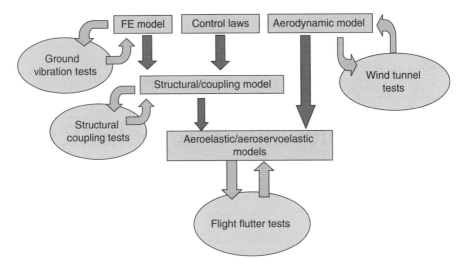

Figure 25.1   Analysis and test certification process for aeroelasticity.

Figure 25.2   Wind tunnel test. Reproduced by permission of ONERA.

validate and possibly update results calculated using design formulae, data sheets, CFD, etc., and also to scale calculated unsteady aerodynamic results at zero reduced frequency.

A wind tunnel flutter test program (AMC 25.629) may sometimes be undertaken using a dynamically scaled model. However, such testing is not sufficiently reliable to use in clearing the aircraft for flutter; instead it helps in the main certification calculations by validating unsteady aerodynamic methodologies, performing parametric studies, studying new configurations, investigating interference and compressibility effects, etc.

25.3   Ground Vibration Test

The Ground Vibration Test or GVT, sometimes called a modal test (Ewins, 1995), is performed on the prototype aircraft to obtain estimates of the whole aircraft normal modes (natural frequencies, damping ratios, mode shapes and modal masses). These modal data may then be used to confirm, or adjust, the calculated normal modes used in the critical flutter

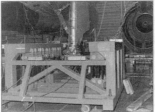

Figure 25.3 Ground vibration test – aircraft under test, main gear bungee support and flap excitation. Reproduced by permission of Airbus/ONERA/DLR.

calculations, as well as providing substantiated estimates of damping. Different fuel and hydraulics configurations are studied.

The aircraft is usually supported on soft springs (e.g. air bags, elastic bungees or deflated tyres) so that it behaves as near to the free–free condition as possible; alternatively, where this is not possible (e.g. where the aircraft has very low frequency modes), a stiff support may be employed. Whatever the support arrangement, it may be included in the dynamic model when comparing measurements with calculations.

The aircraft is instrumented with, typically, many hundreds of accelerometers to allow adequate mode shape definition, and a number of electrodynamic exciters (typically up to eight) are used to excite vibration of the aircraft at a sufficiently high level. Normally, multiple exciters are applied simultaneously to distribute energy adequately over the structure and to allow modes close in frequency to be identified; exciter positions may need to be varied to excite modes successfully. The excitation signals used to drive the exciters are usually sinusoidal or random, depending upon the test methodology being employed. An example set up for a GVT is shown in Figure 25.3.

There are two main approaches to testing. In one approach, using a *phase separation* method, a broadband multipoint uncorrelated random excitation or swept sine correlated excitation is applied to the structure using several exciters. Then a matrix of Frequency Response Functions (FRFs) is estimated and a frequency or time domain identification (i.e. a curve fit) is employed to identify the modal properties.

In an alternative approach, using a *phase resonance* method, sinusoidal excitation forces are applied at each natural frequency (estimated from an initial broadband test) in turn, and the amplitude and phase of the forces are determined from the FRFs or else adjusted iteratively until a normal mode is being excited (indicated by the forces and responses being monophase, with a 90° phase shift between excitation and response such that the structure behaves in essence like a single degree of freedom system). The mode shape is then measured and damping/modal mass estimated usually by varying the excitation frequency around resonance. The latter approach is more suitable for studying important non-linear effects such as control freeplay, pylon stiffening, etc. However, the identification of multiple-mode non-linear systems is a challenging area of research.

25.4 Structural Coupling Test

In AMC 25.629, it is stated that 'the automatic Flight Control System should not interact with the airframe to produce an aeroelastic instability' and that 'when analyses indicate possible adverse coupling, tests should be performed'. These tests include structural coupling tests, which are in effect an extension of the GVT in which additional measurements are performed to include the control system characteristics; e.g. the open loop transfer functions between the response sensors for the control system (e.g. rate gyro/accelerometer) and actuator drive signal (via the flight control computer) would be measured. This would allow the aeroservoelastic model (see Chapter 11) to be checked on the ground at zero air speed where control inertia effects are dominant. Other tests could be carried out to determine the dynamic characteristics of the actuation and systems components.

It is also possible to undertake open and closed loop transfer function measurements in flight to check the flight analytical model; here the aerodynamic effects of the control surface become equally, if not more, important than inertial effects.

25.5   Flight Simulator Test

A significant amount of testing is undertaken prior to the first flight to assist in FCS design via examination of the aircraft handling qualities; later on, simulators are used for in-service pilot training. The simulator is controlled to match the characteristics of a rigid aircraft dynamic (flight mechanics) model. However, because of the impact of static aeroelastic effects upon the rigid body aerodynamic derivatives, especially for large flexible aircraft, it is important to incorporate static aeroelastic corrections into the aerodynamic model used for the flight simulator. It is not simple to keep the aerodynamic models for flight mechanics and loads/ aeroelastics in step as the aircraft develops.

It is also worth mentioning the so-called 'Iron Bird' tests, which are a form of simulation not so much directed towards handling qualities, but as validation of the systems concepts and the real system performance. Typically 'Iron Bird' tests include hardware for hydraulic/electrical systems, the 'real' flight control computers and simulation of the natural aircraft to close the loop. These are of most value in loads and aeroelastics issues by providing system transfer functions and performance constraints for modelling flight control systems.

25.6   Structural Tests

In Chapters 23 and 24, some of the loads requirements in CS-25 were outlined. Most of the structural tests (CS/AMC 25.307) are aimed at demonstrating the limit and ultimate load requirements, and so are strength related; tests are not specifically related to dynamic loads except that certain critical cases may be dynamic in origin and that excessive deformation should not occur. Therefore, only brief mention is made here of structural testing.

The amount of testing required will depend upon the classification of the aircraft; e.g. the most extensive test program would be set in place for a new structure, with full scale subcomponent (e.g. spar), component (e.g. wing) and whole aircraft tests carried out to limit and ultimate conditions whereas, for a derivative aircraft, considerably less testing would be required. Also, significant testing at detail, subcomponent, component and whole aircraft level would be carried out in relation to fatigue and damage tolerance conditions, as described in CS/AMC 25.571. Examples of structural strength testing for two different aircraft under near-ultimate wing loading conditions are shown in Figures 25.4(a) and (b), the former showing the loading arrangement and the latter showing the considerable wing deformation that is possible prior to failure.

(a)   (b)

Figure 25.4 Static strength testing of the wings under near-ultimate load conditions; (a) view of the wing loading arrangement and (b) rear view showing wing deformation prior to failure.Reproduced by permission of DGA/CEAT.

25.7 Flight Flutter Test

The uncertainty in the aeroelastic model used for flutter calculations, and especially the unsteady aerodynamics, means that calculated flutter speeds will almost certainly be inaccurate to some extent, especially in the transonic region. It is therefore a requirement of the certification process (AMC 25.629) to validate the flutter behaviour and demonstrate freedom from aeroelastic instability over the flight envelope in a Flight Flutter Test (FFT).

On the basis of calculations, a nominal flight envelope is cleared to permit a first flight to take place. Thereafter, the FFT program precedes every other flight test at each flight envelope point because of the safety critical nature of flutter. The basic FFT philosophy seeks to gradually extend the flight envelope by assessing the flutter stability of the aircraft at progressively increasing speed and Mach number.

It is normal to assess the flutter stability by identifying the frequency and damping of the complex/damped modes (see Chapters 2 and 10) of the aircraft at each test point. The allowable flight envelope is expanded from an initially agreed boundary by examining the results along lines of increasing EAS at constant altitude and lines of constant Mach number, as indicated in Figure 25.5.

The procedure at each test point is: (a) to excite vibration of the aircraft over the frequency range of interest and to measure its response, (b) to curve-fit the excitation and response signals in the time or frequency domain and so to identify the model parameters and (c) to determine whether it is safe to proceed to the next test point.

A variety of excitation devices can be used, namely (i) control surface movement via stick/pedal input or explosive charges, (ii) control surface movement via a signal from the Flight Control System, (iii) movement of an aerodynamic vane fitted to the aircraft flying surface or engine/store or (iv) inertia exciter mounted in the fuselage. The aircraft response is measured using typically around 20–100 accelerometers, far fewer than for a GVT.

The most common excitation signals are pulse (via stick/pedal or explosive charge) and chirp (a fast frequency sweep applied as a signal to the control surface, vane or inertia exciter). Where excitation devices are available on both sides of the aircraft, then excitation may be applied in or out of phase in order to exploit symmetry/antisymmetry; doing this will simplify the analyses and provide results with more confidence. Occasionally a random excitation signal

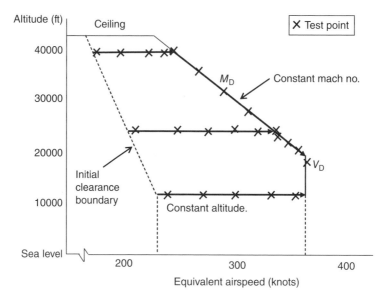

Figure 25.5   Typical flight flutter test clearance envelope.

is employed. Sometimes the response of the aircraft to natural turbulence alone is used (but this is not recommended because the excitation is not 'white' and is not guaranteed to excite adequately all the modes of interest). Where possible, it is preferable that the excitation signal is recorded in order to improve the identification accuracy, but analysis methods are available if this is not the case. Each excitation sequence will probably only last a maximum of 60 s because of the difficulty of holding the aircraft on condition, especially near to the limit of the flight envelope. The test may be repeated and some form of averaging employed to improve the data quality.

Once the test is completed, the results are processed on the ground, either during or after the flight. FRFs are computed, or else the raw time data are used directly. Time or frequency domain identification algorithms (i.e. curve-fits) are then employed in order to identify the frequency and damping values for each mode in the data. Again, this process is very similar to modal testing but the levels of noise on the data are far more severe since turbulence (an unmeasured excitation) is exciting the aircraft during the test. The test time is also limited.

The damping values may be compared to results from previous test points and the damping trends for each mode extrapolated to allow progression to the next test point (i.e. by defining a permitted increment in speed or Mach number). Other techniques such as the flutter margin (Zimmermann and Weisenberger, 1964) method can be used to estimate the flutter speed based upon the measured test data. The test process is complete when the extrapolated damping values are still positive at a margin (typically 15%) above the design dive speed for each Mach number.

These results are compared to the predictions from the aeroelastic flutter model and some basic attempts may be made to reconcile any differences. Any flutter problem (e.g. the anticipated flutter speed from test is too low) will require urgent design action (e.g. additional mass balance) and could prove to be extremely costly. Note that in AMC 25.629, the evaluation of any phenomena not amenable to analysis (e.g. buffet, buzz, etc.) should be investigated during the flight test program.

25.8 Flight Loads Validation

In CS 25.301(b) it is stated that '*methods used to determine load intensities and distribution must be validated by flight load measurement unless the methods used for determining those loading conditions are shown to be reliable*'. The definition of such a flight loads test program is considered in AMC 25.301 and depends upon a comparison of design features with previous aircraft (i.e. new features/configurations will require assessment), the manufacturer's experience in load validation and proven accuracy of analysis methods, etc.

The aim of the flight loads programme is to demonstrate that the loads calculation and prediction process produces reliable loads for the flight certification cases. In some cases where analytical methods are inadequate, such as for buffet, then flight loads are used for design purposes. Clearly, gust loads cannot be validated as the gust conditions may not be arranged, and also failure cases etc., are not subject to flight test for loads validation purposes. The nature of the flight loads programme is explained in AMC 25.301.

The predicted loads need to be validated by flight manoeuvres, both equilibrium and dynamic, and ground manoeuvres. Typically, load cases up to 85% limit load will be explored for separate symmetric and asymmetric manoeuvres. The critical Interesting Quantities (IQs) for which validation is required must be selected prior to flight test and suitable instrumentation installed. Flight loads are not measured directly, but rather determined through correlation of predictions with measured strains, accelerations, pressures and flight mechanics parameters.

In particular, internal loads (namely bending moment, shear force, torque) will be required for comparison with the model predictions at a number of spanwise stations on the major components (e.g. wings, horizontal tailplane, fin). Since these internal loads quantities cannot be measured directly in flight, they must be inferred from an array of strain gauge measurements.

Such a calibration relationship between internal loads and strain measurements may be obtained from load tests carried out on the ground before first flight. Externally applied static 'point' loads of increasing magnitude are applied at different spanwise and chordwise locations on each component and the corresponding internal loads determined knowing the component geometry. An influence coefficient matrix linking these internal loads to the corresponding measured strains may then be determined via a least squares type process; there are a number of different approaches to selection of the best strain gauges to be used in the loads model, but typically around ten might be employed.

Once suitable calibrations have been obtained, then internal loads may be estimated under different manoeuvres and compared to results of an onboard loads monitoring model based on a dynamic manoeuvre model. Any significant differences in results would be reconciled by adjusting this model.

Appendices

A

Aircraft Rigid Body Modes

A.1 Rigid Body Translation Modes

The aircraft heave motion may often be represented by a free–free rigid body heave mode (subscript h) instead of using the heave displacement of the centre of mass as a physical coordinate. The mode shape is normalized to have a unit downwards displacement at all points on the aircraft as shown in Figure A.1(a). Thus the mode shape is $\kappa_h(x, y) = 1$ and the corresponding generalized coordinate is q_h. The heave mode modal mass $m_h = m$ (i.e. aircraft mass) is found by equating the kinetic energy expressed in terms of the physical and modal masses/coordinates, so

$$T = \frac{1}{2}m_h\dot{q}_h^2 = \frac{1}{2}\int_{A/c}\dot{z}^2 dm = \frac{1}{2}\int_{A/c}[\kappa_h(x,y)\dot{q}_h]^2 dm = \frac{1}{2}m\dot{q}_h^2. \tag{A.1}$$

The other two translational rigid body modes are similar, namely the sideslip and fore-and-aft modes.

A.2 Rigid Body Rotation Modes

The pitch, roll and yaw motions may also be represented by rigid body modes of rotation rather than via the pitch, roll and yaw angles. The rigid body pitch mode shape involves a nose up

(a) (b)

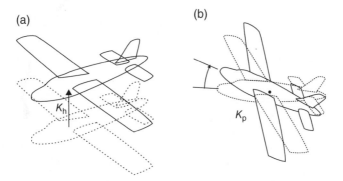

Figure A.1 Rigid body: (a) heave and (b) pitch modes.

Introduction to Aircraft Aeroelasticity and Loads, Second Edition. Jan R. Wright and Jonathan E. Cooper.
© 2015 John Wiley & Sons, Ltd. Published 2015 by John Wiley & Sons, Ltd.

rotation about the aircraft centre of mass, as shown in Figure A.1(b), and if the mode shape is normalized as $\kappa_p(x, y) = -x$ (i.e. 1 radian rotation) then the pitch generalized coordinate q_p will be equal to the pitch angle. The pitch mode modal mass $m_p = I_y$ (i.e. the aircraft pitch moment of inertia) may also be found by equating kinetic energies

$$T = \frac{1}{2} m_p \dot{q}_p^2 = \frac{1}{2} \int_{A/c} \dot{z}^2 \, dm = \frac{1}{2} \int_{A/c} \left[\kappa_p(x, y) \dot{q}_p \right]^2 dm = \frac{1}{2} I_y \dot{q}_p^2 \tag{A.2}$$

The other two rotational rigid body modes are similar, namely the roll and yaw modes. Product moments of inertia are considered to be zero for this case where the axes chosen are principal axes.

B

Table of Longitudinal Aerodynamic Derivatives

In the main text, two different axes types have been considered for manoeuvres and gust encounters, namely inertial axes fixed in space and wind/body axes fixed in the aircraft. Inertial axes are appropriate for flutter, equilibrium manoeuvres, most ground manoeuvres and gust/turbulence encounters, where small excursions in angle and displacement about a datum position are considered. The use of derivatives was primarily to yield a convenient and compact representation of the equations. However, wind/body axes are used for dynamic manoeuvres in flight (and possibly landing) and the corresponding aerodynamic stability derivatives are obtained by considering small perturbations.

When longitudinal derivatives are obtained for these two different axes types, differences are present in some derivatives, particularly where perturbations in velocity are considered, but they are relatively small. Flexible derivatives are the same for the two cases when using quasi-steady strip theory aerodynamics. Table B.1 presents the derivatives used in the calculation of symmetric manoeuvre loads; the wind axes results were converted into the present notation from Cook (1997).

Table B.1 Derivatives used in the calculation of symmetric loads

	Inertial axes (Chapters 12 and 14)		Wind axes (fixed to aircraft) (Chapter 13)
Z_0	$-\dfrac{1}{2}\rho V^2[S_W a_W - S_T a_T k_\varepsilon]\alpha_0$		
Z_α	$-\dfrac{1}{2}\rho V^2[S_W a_W + S_T a_T(1-k_\varepsilon)]$		
$Z_{\dot{z}}$	$-\dfrac{1}{2}\rho V[S_W a_W + S_T a_T(1-k_\varepsilon)]$	Z_w	$-\dfrac{1}{2}\rho V_0[S_W a_W + S_T a_T(1-k_\varepsilon) + S_W C_D]$
Z_q	$-\dfrac{1}{2}\rho V S_T a_T l_T$	Z_q	$-\dfrac{1}{2}\rho V_0 S_T a_T l_T$ (tailplane only)
Z_η	$-\dfrac{1}{2}\rho V^2 S_T a_E$	Z_η	$-\dfrac{1}{2}\rho V_0^2 S_T a_E$

(continued overleaf)

Introduction to Aircraft Aeroelasticity and Loads, Second Edition. Jan R. Wright and Jonathan E. Cooper.
© 2015 John Wiley & Sons, Ltd. Published 2015 by John Wiley & Sons, Ltd.

Table B.1 (*continued*)

	Inertial axes (Chapters 12 and 14)		Wind axes (fixed to aircraft) (Chapter 13)
Z_e	$\frac{1}{2}\rho V^2[-S_W a_W J_1 - S_T a_T \gamma_{eT}]$	Z_e	$\frac{1}{2}\rho V_0^2[-S_W a_W J_1 - S_T a_T \gamma_{eT}]$
$Z_{\dot{e}}$	$-\frac{1}{2}\rho V S_T a_T \kappa_{eT}$	$Z_{\dot{e}}$	$-\frac{1}{2}\rho V_0 S_T a_T \kappa_{eT}$
Z_{gW}	$-\frac{1}{2}\rho V S_W a_W$		
Z_{gT}	$-\frac{1}{2}\rho V S_T a_T (1-k_\varepsilon)$		
M_0	$\frac{1}{2}\rho V^2 S_W c C_{M_{0W}} - \frac{1}{2}\rho V^2 [S_W a_W l_W + S_T a_T k_\varepsilon l_T]\alpha_0$		
M_α	$\frac{1}{2}\rho V^2 [S_W a_W l_W - S_T a_T (1-k_\varepsilon) l_T]$		
$M_{\dot{z}}$	$\frac{1}{2}\rho V (S_W a_W l_W - S_T a_T (1-k_\varepsilon) l_T)$	M_W	$\frac{1}{2}\rho V_0 (S_W a_W l_W - S_T a_T (1-k_\varepsilon) l_T)$
M_q	$-\frac{1}{2}\rho V S_T a_T l_T^2$	M_q	$-\frac{1}{2}\rho V_0 S_T a_T l_T^2$ (tailplane only)
M_η	$-\frac{1}{2}\rho V^2 S_T a_E l_T$	M_η	$-\frac{1}{2}\rho V_0^2 S_T a_E l_T$
M_e	$\frac{1}{2}\rho V^2 [S_W a_W l_W J_1 - S_T a_T l_T \gamma_{eT}]$	M_e	$\frac{1}{2}\rho V_0^2 [S_W a_W l_W J_1 - S_T a_T l_T \gamma_{eT}]$
$M_{\dot{e}}$	$-\frac{1}{2}\rho V S_T a_T l_T \kappa_{eT}$	$M_{\dot{e}}$	$-\frac{1}{2}\rho V_0 S_T a_T l_T \kappa_{eT}$
M_{gW}	$\frac{1}{2}\rho V S_W a_W l_W$		
M_{gT}	$-\frac{1}{2}\rho V S_T a_T l_T (1-k_\varepsilon)$		
Q_0	$\frac{1}{2}\rho V^2 [S_W a_W J_2 - S_T a_T k_\varepsilon \kappa_{eT}]\alpha_0$		
Q_α	$\frac{1}{2}\rho V^2 [-S_W a_W J_2 - S_T a_T (1-k_\varepsilon)\kappa_{eT}]$		
$Q_{\dot{z}}$	$\frac{1}{2}\rho V [-S_W a_W J_2 - S_T a_T (1-k_\varepsilon)\kappa_{eT}]$	Q_W	$\frac{1}{2}\rho V_0 [-S_W a_W J_2 - S_T a_T (1-k_\varepsilon)\kappa_{eT}]$
Q_q	$-\frac{1}{2}\rho V S_T a_T l_T \kappa_{eT}$	Q_q	$-\frac{1}{2}\rho V_0 S_T a_T l_T \kappa_{eT}$
Q_η	$-\frac{1}{2}\rho V^2 S_T a_E \kappa_{eT}$	Q_η	$-\frac{1}{2}\rho V_0^2 S_T a_E \kappa_{eT}$
Q_e	$\frac{1}{2}\rho V^2 [-S_W a_W J_3 - S_T a_T \gamma_{eT}\kappa_{eT}]$	Q_e	$\frac{1}{2}\rho V_0^2 [-S_W a_W J_3 - S_T a_T \gamma_{eT}\kappa_{eT}]$
$Q_{\dot{e}}$	$-\frac{1}{2}\rho V S_T a_T \kappa_{eT}^2$	$Q_{\dot{e}}$	$-\frac{1}{2}\rho V_0 S_T a_T \kappa_{eT}^2$
Q_{gW}	$-\frac{1}{2}\rho V S_W a_W J_2$		
Q_{gT}	$-\frac{1}{2}\rho V S_T a_T (1-k_\varepsilon)\kappa_{eT}$		

When inertial axes were used in the main text, the symbol V was used for true air speed, with no air speed perturbations considered. However for wind axes, the steady equilibrium value was taken as V_0, with V taken as the perturbed total velocity (in the limit $V \cong V_0$).

Note also that

$$J_1 = \frac{1}{s}\int_{y=0}^{s} \gamma_e \, dy \quad J_2 = \frac{1}{s}\int_{y=0}^{s} (\kappa_e - l_A \gamma_e) \, dy \quad J_3 = \frac{1}{s}\int_{y=0}^{s} (\kappa_e - l_A \gamma_e)\gamma_e \, dy$$

where $\gamma_e = \gamma_e(y)$, $\kappa_e = \kappa_e(y)$ are functions describing the wing flexible mode shape (see Appendix C and Chapter 13).

C

Aircraft Symmetric Flexible Modes

In the treatment of manoeuvres and gust encounters in this book, the whole free–free aircraft is considered, initially as a rigid body, and then with a single flexible mode added to illustrate the important effect of flexibility. The aim of the flexible mode representation is to keep the mathematics as basic as possible while illustrating the impact that flexibility might have. In particular, the flexible mode shapes described in this appendix are used in Chapters 12 to 15 for manoeuvres (flight and ground) and gust/turbulence encounters; only one flexible mode is used in each case, but its shape is derived from a 'master' mode template and may be altered. The mode is first introduced in Chapter 12 but some of the content is repeated here for convenience, with modal parameters being derived. In this appendix, the constituent flexible mode shapes and modal parameters will be introduced in order to avoid significant digressions in the main chapters of the book. The idea is to define the flexible mode shapes geometrically, with some unknown parameters that are determined by applying conditions of orthogonality to the rigid body modes. Then it is possible to determine the modal masses for the flexible modes, with each modal stiffness found simply by choosing the desired natural frequency. Therefore no modal analysis needs to be performed explicitly.

C.1 Aircraft Mass Model

The free–free aircraft is shown in Figure C.1. It has unswept and untapered flexible wings with the mass distributed uniformly along the span and with the mass axis aft of the elastic axis. The tailplane is rigid and the engines are mounted on the rear fuselage (but not shown) to leave the wings 'clean'. The fuselage is flexible and for convenience its mass is discretized at three locations: front, centre and rear. The aircraft is somewhat contrived in order to keep it as simple as possible. The essence of the cases considered in this book should not be significantly affected by making the model more realistic, except for the addition of wing sweep, commented on in Chapter 12.

The wings have semi-span s and chord c, with the wing elastic axis a distance l_A aft of the wing aerodynamic centre axis (which is itself at a quarter chord) and a distance l_E ahead of the wing mass axis. The key fuselage dimensions from the aircraft centre of mass to the tail, wing

---

Introduction to Aircraft Aeroelasticity and Loads, Second Edition. Jan R. Wright and Jonathan E. Cooper.
© 2015 John Wiley & Sons, Ltd. Published 2015 by John Wiley & Sons, Ltd.

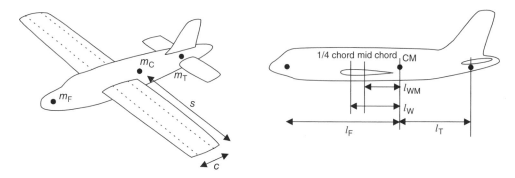

Figure C.1   Free–free flexible aircraft model.

aerodynamic centre, wing mass axis and front fuselage position are l_T, l_W, l_{WM}, l_F respectively. The physical width dimension of the fuselage will be neglected in any spanwise integration over the wings.

The wing mass per unit span is μ_W so the total wing mass is $m_W = 2\mu_W s$. The front, centre and rear fuselage discrete masses, m_F, m_C, m_T, are located at the assumed front fuselage position (e.g. pilot or nose gear), the aircraft centre of mass and the tail respectively; the tail centre of mass is assumed to be coincident with the tailplane aerodynamic centre for convenience. The total mass of the aircraft is $m = m_F + m_W + m_C + m_T$.

By setting the first moment of mass about the centre of mass to zero for the whole aircraft, it may be seen by reference to Figure C.1 that the masses and their positions must be related by the expression

$$m_F l_F + m_W l_{WM} - m_T l_T = 0 \tag{C.1}$$

The pitch moment of inertia of the wing per unit span is χ_W so the total value for the wings about their mass axis is $I_W = 2\chi_W s$. The total pitch moment of inertia of the aircraft about its centre of mass is given by

$$I_y = \left(I_W + m_W l_{WM}^2\right) + m_F l_F^2 + m_T l_T^2 \tag{C.2}$$

where the wing term employs the parallel axis theorem and the centre fuselage mass term does not appear.

C.2   Symmetric Free–Free Flexible Mode

C.2.1   Description of the Flexible Mode Shape

A free–free flexible (or elastic) mode shape (subscript e), where the wing can deform in bending and twist and the fuselage can deform in bending, is shown in Figure C.2. The flexible mode will be defined by a modal coordinate q_e. The wing bending and twist deformations are defined relative to the wing elastic axis by the functions $\kappa_e(y)$ (downwards positive) and $\gamma_e(y)$ (nose up positive) respectively, where the values at the wing root are $\kappa_{e0} = \kappa_e(0)$ and $\gamma_{e0} = \gamma_e(0)$.

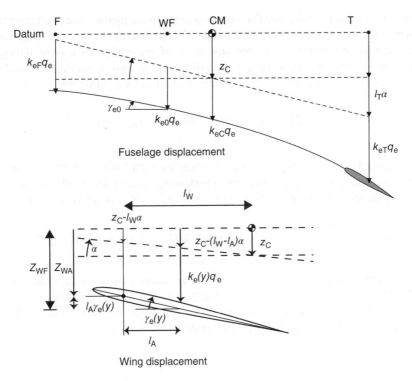

Figure C.2 Aircraft symmetric free–free flexible mode.

The fuselage modal deformation (downwards) is defined by displacement values κ_{eF}, κ_{e0}, κ_{eW}, κ_{eC}, κ_{eT} at the front fuselage, wing elastic axis, wing mass axis, aircraft centre of mass (coincident with centre fuselage) and tail positions respectively. The centre section of fuselage that includes the wing cross-section and the aircraft centre of mass is assumed to behave rigidly and so will heave κ_{e0} and pitch nose up with angle γ_{e0}, relative to the wing elastic axis position. Thus the displacements at the wing and aircraft centres of mass are given by

$$\kappa_{eW} = \kappa_{e0} + l_E \gamma_{e0} \qquad \kappa_{eC} = \kappa_{e0} + (l_{WM} + l_E)\gamma_{e0} \tag{C.3}$$

The nose up pitch of the tailplane γ_{eT} can be estimated by assuming that the rear and front fuselage deformations in the mode shape vary quadratically; based on this assumption and knowing the displacement and slope of the centre section, it may be shown that the tailplane pitch is given by

$$\gamma_{eT} = 2\left\{\frac{\kappa_{eT} - \kappa_{eC}}{l_T}\right\} - \gamma_{e0} \tag{C.4}$$

C.2.2 Conditions for Orthogonality with Rigid Body Modes

A free–free symmetric mode is, by definition, orthogonal (see Chapters 2 and 3) to the rigid body heave and pitch modes (described in Appendix A). One way of imposing this orthogonality condition is to ensure that there is no net vertical inertia force or pitching moment in the flexible mode deformation. As an example, the physical inertia force associated with the tail mass modal acceleration is an upwards force of $m_T(\kappa_{eT}\ddot{q}_e)$. When the orthogonality conditions are imposed and the common generalized acceleration terms $\ddot{q}_e$ are cancelled, then it may be shown that the condition for zero net inertia force (upwards positive) is

$$m_F\kappa_{eF} + 2\int_0^s \mu_W[\kappa_E(y) + l_E\gamma_E(y)]dy + m_C\kappa_{eC} + m_T\kappa_{eT} = 0 \quad (C.5)$$

where the second term is an integral of the wing bending contribution per strip dy. Note that the motion at the wing mass axis is a function of both wing bending and twist. The condition for zero net inertia moment about the aircraft centre of mass (nose down positive) is

$$-m_F\kappa_{eF}l_F - 2\int_0^s \mu_W[\kappa_E(y) + l_E\gamma_E(y)]dy\,l_{WM} + 2\int_0^s \chi_W\gamma_E(y)dy + m_T\kappa_{eT}l_T = 0 \quad (C.6)$$

where the third term is a torsion term which is an integral of the wing twist contribution per strip. The tailplane pitch angle does not appear in this equation because no moment of inertia for the tail was included. These expressions allow the free–free modal parameters to be determined for different types of flexible mode.

C.2.3 Wing Deformation Shapes

The wing bending and twisting shapes relative to the wing elastic axis now need to be defined. To simplify the analysis, simple assumed polynomial expressions will be employed (see Chapter 3). Thus, the wing *bending* deformation can be written as a quadratic function

$$\kappa_E(y) = \kappa_{e0}\left[1 + A\left(\frac{y}{s}\right)^2\right] \quad (C.7)$$

where A is an unknown constant defining the amount of bending and the spanwise position of any nodal line running across the wing. The wing *torsional* deformation may be written as a linear function, so

$$\gamma_E(y) = \gamma_{e0}\left[1 + B\left(\frac{y}{s}\right)\right] \quad (C.8)$$

where B is an unknown constant defining the amount of twist present at the tip relative to the root. The displacement at the wing tip (trailing edge) will be used later to normalize the mode shapes and is given by

$$\kappa_{Tip\_TE} = \kappa_{e0}(1 + A) + \gamma_{e0}(1 + B)l_{TE} \quad (C.9)$$

where $l_{TE} = 3c/4 - l_A$ is the distance from the elastic axis to the wing trailing edge.

Now the orthogonality expressions defined earlier may be used to help define the unknown values in the mode shapes. When these polynomial functions are substituted into the orthogonality expressions (C.5) and (C.6), assuming that the aircraft deforms to some degree in both the wing and fuselage, then the result is

$$m_F \kappa_{eF} + m_W \left(1 + \frac{A}{3}\right) \kappa_{e0} + m_W l_E \left(1 + \frac{B}{2}\right) \gamma_{e0} + m_C \kappa_{eC} + m_T \kappa_{eT} = 0$$

$$-m_F \kappa_{eF} l_F - m_W \left(1 + \frac{A}{3}\right) \kappa_{e0} l_{WM} + (I_W - m_W l_E l_{WM}) \left(1 + \frac{B}{2}\right) \gamma_{e0} + m_T \kappa_{eT} l_T = 0$$

(C.10)

In order to satisfy these equations, several unknowns need to be chosen and this does not appear to be straightforward or even possible. However, if special cases are selected, then the choice of values is not too difficult. The three cases to be considered are where fuselage bending, wing bending and wing twist each become the dominant component in the mode shape. This approach will then allow each effect to be examined separately.

C.2.4 Mode with Fuselage Bending Dominant

Firstly, consider the case where the wing is treated as completely rigid, so $A = B = 0$, but the fuselage is flexible. As the mode does not involve wing pitch, it will be assumed that $\gamma_{e0} = 0$ and therefore, from Equation (C.3), $\kappa_{eC} = \kappa_{e0}$ and the rigid centre section will simply heave an amount κ_{e0}. When these values are substituted into the orthogonality expressions (C.10), and the result simplified, then it may be shown that

$$m_F \kappa_{eF} + m_T \kappa_{eT} = -(m_W + m_C) \kappa_{e0} \qquad -m_F l_F \kappa_{eF} + m_T l_T \kappa_{eT} = m_W l_{WM} \kappa_{e0} \qquad \text{(C.11)}$$

Solving these equations leads to the front fuselage and tailplane modal displacements κ_{eF}, κ_{eT} expressed in terms of κ_{e0}. The tailplane pitch is then determined using Equation (C.4). The mode shape can be defined by using a normalization process where, for example, κ_{e0} or the wing tip trailing edge displacement κ_{Tip_TE} are unity.

C.2.5 Mode with Wing Bending Dominant

Now, consider the dominant wing bending case, where the wing is rigid in torsion and $B = 0$. The fuselage is rigid but allowed to pitch by γ_{e0} about the wing elastic axis in order to satisfy orthogonality; however, because the fuselage is rigid, the tailplane pitch γ_{eT} is equal to γ_{e0} and the modal displacements at the front fuselage, centre of mass and tail positions must be a geometric function of the heave and pitch of the fuselage centre section thus

$$\kappa_{eF} = \kappa_{e0} - (l_F - l_{WM} - l_E)\gamma_{e0} \qquad \kappa_{eC} = \kappa_{e0} + (l_{WM} + l_E)\gamma_{e0} \qquad \kappa_{eT} = \kappa_{e0} + (l_T + l_{WM} + l_E)\gamma_{e0} \quad \text{(C.12)}$$

If these terms are substituted into the orthogonality equations, the result may be simplified to

$$\left(m + m_W \frac{A}{3}\right) \kappa_{e0} + m(l_E + l_{WM})\gamma_{e0} = 0 \qquad -\left(m_W \frac{A}{3}\right) l_{WM} \kappa_{e0} + I_y \gamma_{e0} = 0 \qquad \text{(C.13)}$$

where the earlier expressions for moment of inertia and first moment of mass have been employed. Both Equations (C.13) yield a value for the ratio γ_{e0}/κ_{e0} as a function of A, and by equating these ratio expressions for a consistent result, it may be shown that A must be given by

$$\frac{m_W}{m}\left[1 + \frac{l_{WM}(l_E + l_{WM})}{l_y^2}\right]A = -3 \tag{C.14}$$

where $I_y = m l_y^2$ and l_y is the aircraft pitch radius of gyration. The ratio γ_{e0}/κ_{e0} may then be determined from the value of A. Values of the mode shape for the front, centre and rear fuselage positions can then be determined in terms of κ_{e0} by substituting this ratio into Equations (C.12). The mode shape is then defined as before, based on a normalization process where, for example, κ_{e0} or the wing tip displacement κ_{Tip_TE} are set to unity.

C.2.6 Mode with Wing Twist Dominant

Finally, consider the case where the wing is flexible in twist but rigid in bending and so $A = 0$. Also, the fuselage is rigid in bending but allowed to heave and pitch so the geometric relationships in Equation (3.12) still apply. The tailplane pitch γ_{eT} again equals γ_{e0}. Substituting these terms into the orthogonality Equations (3.10) gives

$$\kappa_{e0} + \left(l_E + l_{WM} + \frac{m_W}{m}\frac{B}{2}l_E\right)\gamma_{e0} = 0 \quad \left(I_y + \bar{I}_W\frac{B}{2}\right)\gamma_{e0} = 0 \tag{C.15}$$

where $\bar{I}_W = I_W - m_W l_E l_{WM}$. Since the fuselage pitch γ_{e0} is nonzero, then $B = -2I_y/\bar{I}_W$ and the ratio γ_{e0}/κ_{e0} may be determined. Values of the mode shape for the front, centre and rear fuselage positions may then be determined in terms of κ_{e0} by substituting this ratio into Equations (C.12) and the mode shape normalized.

C.2.7 Modal Mass Values for the Flexible Aircraft

The modal mass m_e for the whole aircraft flexible mode may be defined by writing the kinetic energy associated with motion of the physical masses in the flexible mode deformation and equating it to the value expressed in terms of the modal mass. The general kinetic energy term includes integral expressions for the bending and twisting of the wing, as well as energy terms for each discrete mass, so

$$T = \frac{1}{2}m_e\dot{q}_e^2 = \frac{1}{2}m_F(\kappa_{eF}\dot{q}_e)^2 + \frac{1}{2}2\int_0^s \mu_W\{[\kappa_e(y) + l_E\gamma_e(y)]\dot{q}_e\}^2 dy$$
$$+ \frac{1}{2}2\int_0^s \chi_W[\gamma_e(y)\dot{q}_e]^2 dy + \frac{1}{2}m_C[\kappa_{eC}\dot{q}_e]^2 + \frac{1}{2}m_T[\kappa_{eT}\dot{q}_e]^2 \tag{C.16}$$

After carrying out the integrations and cancelling the common factor $\frac{1}{2}\dot{q}_e^2$, a general expression for the modal mass may be shown to be

$$m_e = m_F\kappa_{eF}^2 + m_W\left(1 + \frac{2A}{3} + \frac{A^2}{5}\right)\kappa_{e0}^2 + \left(I_W + m_W l_e^2\right)\left(1 + B + \frac{B^2}{3}\right)\gamma_{e0}^2$$
$$+ 2m_W l_E\left[1 + \frac{A}{3} + \frac{B}{2} + \frac{AB}{4}\right]\kappa_{e0}\gamma_{e0} + m_C\kappa_{eC}^2 + m_T\kappa_{eT}^2 \tag{C.17}$$

The modal mass may then be estimated for the relevant mode shape, bearing in mind that the numerical value is not unique, but will depend upon the mode shape normalization assumed (see Chapter 2).

Knowing the relevant modal mass m_e and assuming a natural frequency ω_e (rad/s), the modal stiffness may be determined using $k_e = \omega_e^2 m_e$, so permitting the flexible mode frequency to be adjusted easily without needing to alter values of bending or torsional stiffness and then having to perform a modal calculation.

C.2.8   Example Data

The following parameter values will be assumed for calculating example values for the mode shape and modal mass in the symmetric manoeuvres and gust encounters: $c = 2.0$ m, $l_{WM} = 0.1$ m, $l_E = 0.25$ m, $l_A = 0.25$ m, $l_F = 6.8$ m, $l_T = 7$ m, $m = 10\,000$ kg, $m_F = 0.15\,m$, $m_W = 0.3\,m$, $m_C = 0.4\,m$, $m_T = 0.15\,m$, $I_y = 144\,000$ kg m^2 and $I_W = 1330$ kg m^2. Using different parameters here would lead to different relative fuselage and wing motion in the mode shapes that follow. In each case the modes are arbitrarily normalized with respect to a unit displacement at the wing tip (trailing edge) κ_{Tip_TE}, obtained via Equation (3.9).

C.2.8.1   Fuselage Bending Dominant

For the chosen parameter values it may be shown that

$$\{\kappa_{eF},\ \kappa_{e0},\ \kappa_{eC},\ \kappa_{eT}\} = \{-2.382, 1.000, 1.000, -2.285\}$$
$$\{\gamma_{e0},\ \gamma_{eT}\} = \{0, -0.939\} \quad \{\kappa_{Tip\_LE},\ \kappa_{Tip\_TE}\} = \{1.000, 1.000\}$$

(C.18)

The resulting mode shape may be seen in Figure C.3 so the fuselage bends in the 'sagging' sense but there is no wing twist or bending. The modal mass is $m_e = 23\,340$ kg.

C.2.8.2   Wing Bending Dominant

For the chosen parameter values then it may be shown that $A = -9.98$. The mode shape values are given by

$$\{\kappa_{eF},\ \kappa_{e0},\ \kappa_{eC},\ \kappa_{eT}\} = \{-0.116, -0.111, -0.111, -0.106\}$$
$$\{\gamma_{e0},\ \gamma_{eT}\} = \{0.00077, 0.00077\} \quad \{\kappa_{Tip\_LE},\ \kappa_{Tip\_TE}\} = \{0.999, 1.000\}$$

(C.19)

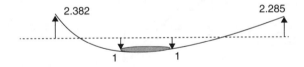

Figure C.3   Aircraft symmetric flexible mode – fuselage bending dominant.

The resulting mode shape is shown in Figure C.4 so the wings bend downwards and the fuselage heaves upwards with minimal nose up pitch. The modal mass is $m_e = 616\,\text{kg}$.

C.2.8.3 Wing Torsion Dominant

For the chosen parameter values then $B = -\,229$, being a large value since the unswept wing pitch moment of inertia is a relatively small proportion of the value for the whole aircraft. The mode shape values are given by

$$\{\kappa_{eF}, \kappa_{e0}, \kappa_{eC}, \kappa_{eT}\} = \{0.0222, -0.0004, -0.0017, -0.0262\}$$

$$\{\gamma_{e0}, \gamma_{eT}\} = \{-0.0035, -0.0035\} \quad \{\kappa_{\text{Tip\_LE}}, \kappa_{\text{Tip\_TE}}\} = \{-0.600, 1.000\} \tag{C.20}$$

The resulting mode shape may be seen in Figure C.5 so the wings twist nose up with a small fuselage nose down pitch and minimal upwards heave at the centre of mass. The modal mass is $m_e = 325\,\text{kg}$.

C.2.9 'J' Integrals

In the expressions for aerodynamic derivatives associated with the flexible aircraft shown in Chapters 12 to 15 and Appendix B, three integrals involving the wing bending and twisting shapes were quoted. These integrals need to be evaluated and modal quantities substituted when calculating the flexible aircraft aerodynamic derivatives. Using the assumed shapes defined in Equations (C.7) and (C.8) above, then

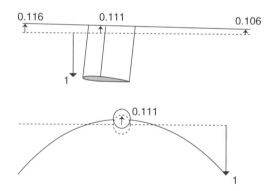

Figure C.4   Aircraft symmetric flexible mode – wing bending dominant.

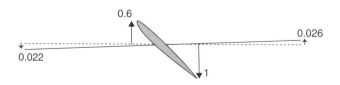

Figure C.5   Aircraft symmetric flexible mode – wing twist dominant.

$$J_1 = \frac{1}{s}\int_{y=0}^{s} \gamma_e \, dy = \left(1 + \frac{B}{2}\right)\gamma_{e0} \tag{C.21}$$

$$J_2 = \frac{1}{s}\int_{y=0}^{s} (\kappa_e - l_A \gamma_e) \, dy = \left(1 + \frac{A}{3}\right)\kappa_{e0} - l_A\left(1 + \frac{B}{2}\right)\gamma_{e0} \tag{C.22}$$

$$J_3 = \frac{1}{s}\int_{y=0}^{s} (\kappa_e - l_A \gamma_e)\gamma_e \, dy = \left(1 + \frac{A}{3} + \frac{B}{2} + \frac{AB}{4}\right)\kappa_{e0}\gamma_{e0} - l_A\left(1 + B + \frac{B^2}{3}\right)\gamma_{e0}^2 \tag{C.23}$$

C.2.10   Other Mode Shapes

If a mode shape had been sought using the above approach with the aim of obtaining combinations of fuselage bending, wing bending and wing twisting motions, it would have been found that insufficient equations were available to obtain a solution. However, a feasible approach is to combine two or more of the above three mode shapes with suitable weightings to create a new mode; if two or more modes are individually orthogonal to the rigid body modes, then a summation of these modes will also be orthogonal to them. The modal mass of the combined mode would be obtained by adding the weighted mode shapes within each term in Equation (C.16). It should be noted that the three flexible modes defined above will not be orthogonal to each other, only to the rigid body modes.

D

Model Condensation

The idea of model condensation is that the size of a static or dynamic model in either physical or modal space may be reduced in size and a faster solution obtained with a limited penalty in accuracy. In this appendix, the basic ideas will be illustrated primarily for a physical model (see Chapter 21 for application), but the concept extends to a modal model and this will be commented upon.

D.1 Introduction

For complex structures such as aircraft, the FE model can be extremely large; a static solution may be readily carried out, but a dynamic solution, such as determining eigenvalues or time response, is often preceded by some form of physical 'condensation', or model order reduction.

Firstly, consider the physical equations of motion for an FE model of order N:

$$\mathbf{M}\ddot{r} + \mathbf{K}r = R \tag{D.1}$$

where $\mathbf{M}$, $\mathbf{K}$ are the mass and the stiffness matrices ($N \times N$), r is the displacement vector and R is the vector of applied forces. Now, consider dividing the displacements into 'master' (N_m) and 'slave' (N_s) degrees of freedom r_m, r_s, where normally $N_m \ll N$. The master DoFs are those that are to be retained in the analysis, while the slave DoFs are to be removed by physical condensation for the initial analysis; however, estimates of the slave DoFs can be recovered later. There are two types of condensation, static and dynamic.

For modal condensation, then clearly in the above equations the matrices are modal mass and stiffness matrices and the vectors are vectors of modal coordinates (see later).

D.2 Static Condensation

Consider the static part of Equation (D.1) and write it in partitioned form, separating master and slave DoFs, so

$$\begin{bmatrix} \mathbf{K}_{mm} & \mathbf{K}_{ms} \\ \mathbf{K}_{sm} & \mathbf{K}_{ss} \end{bmatrix} \begin{Bmatrix} r_m \\ r_s \end{Bmatrix} = \begin{Bmatrix} R_m \\ R_s \end{Bmatrix} \tag{D.2}$$

Introduction to Aircraft Aeroelasticity and Loads, Second Edition. Jan R. Wright and Jonathan E. Cooper.
© 2015 John Wiley & Sons, Ltd. Published 2015 by John Wiley & Sons, Ltd.

Taking the second (or slave) equation, then the slave DoF may be written in terms of the master DoF as

$$r_s = \mathbf{K}_{ss}^{-1} \{ R_s - \mathbf{K}_{sm} r_m \} \tag{D.3}$$

Substituting the slave DoF expression in the first (or master) equation in (D.2) yields an expression involving the master DoF such that

$$\left[\mathbf{K}_{mm} - \mathbf{K}_{ms} \mathbf{K}_{ss}^{-1} \mathbf{K}_{sm} \right] r_m = \left\{ R_m - \mathbf{K}_{ms} \mathbf{K}_{ss}^{-1} R_s \right\} \tag{D.4}$$

or

$$\mathbf{K}_c r_m = R_c \tag{D.5}$$

where the subscript c indicates a 'condensed' matrix or vector. It may be seen that the N equations have been condensed into N_m equations and therefore that the master DoF responses may be determined from a reduced (or condensed) model. The slave DoF results may then be determined from the transformation in Equation (D.3), noting that no approximations are involved. A static solution of an FE model does not normally require condensation, but the approach may be used in some circumstances for reducing an aeroelastic model.

D.3   Dynamic Condensation – Guyan Reduction

Dynamic condensation is normally carried out using Guyan Reduction (Cook *et al.*, 1989). The main assumption in the approach is that for the lower frequency modes, the inertia and damping forces for the slave DoF are less important than the elastic forces associated with the master DoF, i.e. the slave DoF are assumed to behave quasi-statically in response to the master DoFs. The partitioned equations for the damped dynamic case are

$$\begin{bmatrix} \mathbf{M}_{mm} & \mathbf{M}_{ms} \\ \mathbf{M}_{sm} & \mathbf{M}_{ss} \end{bmatrix} \begin{Bmatrix} \ddot{r}_m \\ \ddot{r}_s \end{Bmatrix} + \begin{bmatrix} \mathbf{D}_{mm} & \mathbf{D}_{ms} \\ \mathbf{D}_{sm} & \mathbf{D}_{ss} \end{bmatrix} \begin{Bmatrix} \dot{r}_m \\ \dot{r}_s \end{Bmatrix} + \begin{bmatrix} \mathbf{K}_{mm} & \mathbf{K}_{ms} \\ \mathbf{K}_{sm} & \mathbf{K}_{ss} \end{bmatrix} \begin{Bmatrix} r_m \\ r_s \end{Bmatrix} = \begin{Bmatrix} R_m \\ R_s \end{Bmatrix} \tag{D.6}$$

In the static condensation, a relationship between the slave and master DoFs was determined from the slave equation; however, in the dynamic case, the external forces are time varying and so cannot be employed in the transformation as they were for the static case. Therefore, in this case, a similar idea is used but the mass, damping and external forcing terms are temporarily neglected in determining the transformation such that there is no net elastic force at the slave DoF and

$$r_s = -\mathbf{K}_{ss}^{-1} \mathbf{K}_{sm} r_m = \mathbf{T}_s r_m \tag{D.7}$$

Then, adding the identity $r_m = r_m$ and combining these two expressions in matrix form yields the transformation

$$\begin{Bmatrix} r_m \\ r_s \end{Bmatrix} = \begin{bmatrix} \mathbf{I} \\ -\mathbf{K}_{ss}^{-1} \mathbf{K}_{sm} \end{bmatrix} r_m = \begin{bmatrix} \mathbf{I} \\ \mathbf{T}_s \end{bmatrix} r_m = \mathbf{T} r_m \tag{D.8}$$

where $\mathbf{I}$ is the identity matrix and $\mathbf{T}$ is an $(N \times N_m)$ transformation matrix.

This transformation expression may be substituted into Equation (D.6) and the resulting equation pre-multiplied by the transpose of $\mathbf{T}$, so leading to the condensed equations of motion

$$\mathbf{M}_c\ddot{r}_m + \mathbf{D}_c\dot{r}_m + \mathbf{K}_c r_m = R_c \tag{D.9}$$

where

$$\mathbf{M}_c = \mathbf{T}^T\mathbf{M}\mathbf{T} \quad \mathbf{D}_c = \mathbf{T}^T\mathbf{D}\mathbf{T} \quad \mathbf{K}_c = \mathbf{T}^T\mathbf{K}\mathbf{T} \quad R_c = \mathbf{T}^T R \tag{D.10}$$

These are quite complex expressions when written out fully, with the condensed mass matrix involving the stiffness matrix.

The eigenvalue and response calculations may now be carried out for the master DoFs using the relevant parts of Equation (D.9). Estimates of the slave DoF results may then be determined from the master DoF results using Equation (D.7).

Some errors will be present in both the master and slave DoF solutions because of the assumptions made in the transformation. Therefore care must be taken that sufficient master DoFs are included to adequately model the structure over the frequency range of interest and so minimise errors. The analyst has the choice of how many master DoFs to use and where they are located; alternatively, the analysis program can select the master DoFs automatically from the desired frequency range and from the mass and stiffness matrices (Cook *et al.*, 1989).

D.4   Static Condensation for Aeroelastic Models

In Chapter 21, it is explained how a 'beam-like' model may be developed for the aircraft dynamic behaviour. In this case, the static Finite Element model is 'box-like' (i.e. a reasonably representative structural model) and is condensed statically down to a model where the master structural DoFs lie along suitable reference axes for the wing/fuselage, etc. Thus, in essence the static condensed model becomes a 'beam-like' structure. Because much of the aircraft mass is non-structural, then masses are effectively lumped at the master DoFs on the structural reference axis with centre of mass offsets allowed for; mass properties are therefore condensed only in the sense that the mass is discretized at suitable structural locations. Thus the condensed dynamic equations of motion become

$$\mathbf{M}_{mm}\ddot{r}_m + \left[\mathbf{K}_{mm} - \mathbf{K}_{ms}\mathbf{K}_{ss}^{-1}\mathbf{K}_{sm}\right]r_m = R_m \tag{D.11}$$

where $\mathbf{M}_{mm}$ is a mass matrix (not diagonal because of the mass offset; see Chapter 17) representing the effect of the offset masses on the structural DoF. It is clear that any applied forces at the slave positions are neglected. Aerodynamic forces may then be added later based on this reduced model.

D.5   Modal Condensation

So far in this appendix, the focus has been upon condensation based on physical DoFs. For modal condensation, then clearly in the above equations the matrices and vectors would be based in modal space.

The concept of static condensation could be employed to reduce a rigid body/flexible static aeroelastic or equilibrium manoeuvre model expressed in modal coordinates to a rigid body model, so effectively yielding the rigid body aerodynamic derivatives corrected for flexible effects. Also, the dynamic condensation approach may be applied to a dynamic aeroelastic model to reduce the model order for dynamic response calculations; in this case the slave DoFs are flexible modes to be condensed out.

D.6   Modal Reduction

When a physical Finite Element model has been transformed into modal space by using the modal matrix as a transformation matrix, this process can achieve a considerable reduction in the model order by simple truncation of the model, that is, by discarding some of the modal coordinates (see Chapter 2). Sufficient modes need to be retained in order to represent the structural behaviour over the frequency range of interest. This truncation is normally carried out prior to calculating the unsteady aerodynamic forces in modal coordinates.

E

Aerodynamic Derivatives in Body Fixed Axes

In Chapter 13, the idea of aerodynamic stability derivatives was introduced for a body fixed (or wind) axes system; these allow the effect on aerodynamic forces and moments, for a small perturbation about the equilibrium condition, to be defined. In this appendix, an example of calculating three such derivatives will be shown. Other references should be consulted for more detail and additional derivative examples (Cook, 1997; ESDU Data Sheets).

E.1 Longitudinal Derivative Z_w

In this section, the longitudinal derivative Z_w will be obtained for use in Chapter 13, along with many other derivatives that are not derived here.

E.1.1 Perturbed state

The aircraft is in steady level flight at velocity V_0 (equilibrium value) and experiences a small perturbation as shown in Figure E.1; wind axes are used and the lift, drag, pitching moment and thrust are all perturbed quantities. In the *perturbed* condition, the total velocity is V with components U, W along the oxz axes. Then

$$V^2 = U^2 + W^2 \tag{E.1}$$

and

$$U = U_e + u = V \cos \theta \quad W = W_e + w = V \sin \theta \tag{E.2}$$

For wind axes, the pitch attitude perturbation θ is equal to the incidence perturbation α and so for small angles

$$\theta = \alpha = \frac{W}{U} \tag{E.3}$$

Introduction to Aircraft Aeroelasticity and Loads, Second Edition. Jan R. Wright and Jonathan E. Cooper.
© 2015 John Wiley & Sons, Ltd. Published 2015 by John Wiley & Sons, Ltd.

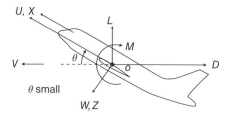

Figure E.1   Perturbed wind axes in heave/pitch.

The lift and drag forces produced in the perturbation may be resolved (transformed) into the disturbed aircraft axes, so yielding the perturbed axial/normal forces in the wind axes direction and also the moment, so

$$X = L\sin\theta - D\cos\theta + \tau \quad Z = -L\cos\theta - D\sin\theta \tag{E.4}$$

where τ is the thrust. The moment M requires no transformation. These quantities are then used to find the required derivatives.

E.1.2   Derivative for Normal Force due to Normal Velocity Perturbation

Consider, as an example, the normal force derivative due to normal velocity W. Using Equation (E.4) this is

$$Z_w = \frac{\partial Z}{\partial W} = -\frac{\partial}{\partial W}\left[\frac{1}{2}\rho V^2 S_W (C_L\cos\theta + C_D\sin\theta)\right] \tag{E.5}$$

where C_L, C_D are the whole aircraft lift and drag coefficients referenced to the wing area. The calculation of this derivative requires certain partial derivatives to be defined, namely from Equations (E.1) to (E.3) then

$$\frac{\partial V}{\partial U} = \frac{U}{V} = \cos\theta \approx 1 \quad \frac{\partial V}{\partial W} = \frac{W}{V} = \sin\theta \approx 0 \quad \frac{\partial \alpha}{\partial W} = \frac{\partial \theta}{\partial W} = \frac{1}{U} = \frac{1}{V\cos\theta} \approx \frac{1}{V} \tag{E.6}$$

and also

$$\frac{\partial C_L}{\partial W} = \frac{\partial C_L}{\partial \alpha}\frac{\partial \alpha}{\partial W} = \frac{1}{V}\frac{\partial C_L}{\partial \alpha} \quad \frac{\partial(\sin\theta)}{\partial W} = \frac{1}{U} = \frac{1}{V\cos\theta} = \frac{1}{V} \tag{E.7}$$

Thus, carrying out the differentiation in Equation (E.5), treating angles as small and substituting the above results yields the normal force due to the heave velocity derivative

$$Z_w = -\frac{1}{2}\rho V_0 S_W\left(\frac{\partial C_L}{\partial \alpha} + C_D\right) \tag{E.8}$$

where in the limit the total perturbation velocity tends to the equilibrium value and so $V \cong V_0$ has been used. Note that the value tabulated in Appendix B, and used in Chapter 13, is written in terms of separate wing and tailplane lift curve slopes.

Other derivatives may be found in similar ways once changes in lift and moment have been expressed in terms of the velocity perturbations.

E.2   Lateral Derivatives L_p, L_ξ

Lateral derivatives involve similar principles to longitudinal derivatives but also require integration along the wing and fin using strip theory (Cook, 1997). These derivatives are not accurate and better estimates may be found in data sheets (ESDU). In this section, the lateral derivatives L_p, L_ξ will be obtained for use in Chapter 13.

E.2.1   Rolling Moment Derivative due to the Roll Rate

This important damping derivative arises largely from the wing, with smaller contributions from the tailplane, fin and fuselage. Consider the aircraft in steady level flight at velocity V_0 and trim incidence α_e. When the aircraft experiences a perturbation in the roll rate p, then there is an effective change of incidence on each wing strip dy, as shown in Figure E.2 for a strip on the starboard wing ($y \geq 0$). The effective increase in incidence for the elemental strip at position y is given by

$$\alpha' = \frac{py}{V_0} \quad py \ll V_0 \tag{E.9}$$

The strip lift and drag in the perturbed state are aligned normal to and along the perturbed velocity vector, so

$$dL'_W = \frac{1}{2}\rho V_0^2 c\, dy\, a_W\left(\alpha_e + \frac{py}{V_0}\right) \quad dD'_W = \frac{1}{2}\rho V_0^2 c\, dy\, C_D \tag{E.10}$$

Referring to Figure E.2, the normal force in the wind axes direction, corresponding to the perturbed state, will be

$$dZ = -dL'_W \cos\alpha' - dD'_W \sin\alpha' \approx -dL'_W - dD'_W\, \alpha' \tag{E.11}$$

The elemental contribution to the rolling moment will be

$$dL = y\, dZ = \left(-dL'_W - dD'_W\alpha'\right)y = -\frac{1}{2}\rho V_0^2\left[a_W\alpha_e + (a_W + C_D)\frac{py}{V_0}\right]cy\, dy \tag{E.12}$$

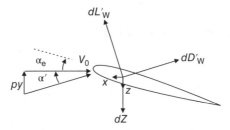

Figure E.2   Wing strip incidence in rolling flight – perturbed state.

which can be integrated over the wing to determine the rolling moment. When the equivalent expression is obtained for a strip on the port side and the two rolling moment contributions added, the terms involving the trim incidence α_e cancel out as they cause no net roll effect. The total rolling moment may then be written as

$$L = -2 \int_0^s \frac{1}{2} \rho V_0^2 c(a_W + C_D) \frac{py}{V_0} y \, dy \qquad (E.13)$$

Evaluating the integral and recognizing that $L = L_p p$ yields the aerodynamic rolling moment due to the roll rate derivative

$$L_p = -\frac{1}{2} \rho V_0 \left[\frac{S_W (a_W + C_D) s^2}{3} \right] \qquad (E.14)$$

The value for L_p agrees with that defined in Cook (1997) for wind axes, when neglecting fin/tailplane effects, but the wing roll damping included here is the most significant term.

E.2.2 Rolling Moment Derivative due to Aileron

In this case, the rolling moment due to deploying the full span aileron through a perturbation in ξ needs to be determined. The lift force perturbation developed on a strip on the starboard wing due to control rotation is

$$dL_W = -\frac{1}{2} \rho V_0^2 c \, dy \, a_C \xi \qquad (E.15)$$

where ξ is the aileron angle (positive trailing edge down) and a_C is the sectional lift coefficient per control angle. The normal force perturbation is then given by $dZ = -dL_W$ since the wind axes are not perturbed. The rolling moment from each wing is the same and so, by integration, the total rolling moment is given by

$$L = 2 \int_0^s \frac{1}{2} \rho V_0^2 c a_C \xi y \, dy \qquad (E.16)$$

After evaluating the integral, the rolling moment due to aileron derivative is given by

$$L_\xi = \frac{1}{2} \rho V_0^2 \left(\frac{S_W a_C s}{2} \right) \qquad (E.17)$$

A more accurate derivative may be found in ESDU.

References

Advisory Group for Aerospace Research and Development (AGARD) *Manual of Aeroelasticity 1956–1970*, NATO (North Atlantic Treaty Organization), Brussels.

Albano, E. and Rodden, W.P. (1969) A doublet-lattice method for calculating lift distributions on oscillating surfaces in subsonic flows. *AIAA Journal*, **7** (2), 279–285.

Anderson, A. (2001) *Fundamentals of Aerodynamics*, 3rd edn, McGraw-Hill, New York.

Babister, A.W. (1980) *Aircraft Dynamic Stability and Response*, Pergamon Press, Oxford.

Bairstow, L. and Fage, A. (1916) *Oscillations of the Tailplane and Body of an Aircraft in Flight*. ARC R&M (Aeronautical Research Council Reports and Memoranda) 276, part 2AERADE Reports Archive, Cranfield.

Benham, P.P., Crawford, R.J. and Armstrong, C.G. (1996) *Mechanics of Engineering Materials*, 2nd edn, Longman, London.

Bishop, R.E.D. and Johnson, D.C. (1979) *The Mechanics of Vibration*, Cambridge University Press, Cambridge.

Bisplinghoff, R.L., Ashley, H. and Halfman, R.L. (1996) *Aeroelasticity*, Dover (Addison Wesley, 1955), New York.

Blair, M. (1994) *A Compilation of the Mathematics Leading to the Doublet Lattice Method*. WL-TR-95-3022.

Blevins, R.D. (2001) *Formulas for Natural Frequency and Mode Shape*, Kreiger, Malabar, FL, USA.

Broadbent, E.G. (1954) The elementary theory of aeroelasticity. *Aircraft Engineering and Aerospace Technology*, **26** (3), 70–79.

Bryan, G.H. (1911) *Stability in Aviation*, Macmillan, London.

Chen, P.C. (2000) A damping perturbation method for flutter solution: the g-method. *AIAA Journal*, **38** (9), 1519–1524.

Collar, A.R. (1978) The first fifty years of aeroelasticity. *Aerospace*, **5** (2), 12–20.

Collar, A.R. and Simpson, A. (1987) *Matrices and Engineering Dynamics*, Ellis Horwood, Harlow.

Cook, M.V. (1997) *Flight Dynamics Principles*, Arnold, London.

Cook, R.D., Malkus, D.S. and Plesha, M.E. (1989) *Concepts and Applications of Finite Element Analysis*, John Wiley & Sons, Ltd., Chichester, UK.

Cooper, J.E., Lind R. and Wright, J.R. (2010) *Aeroelastic Testing and Certification* Encyclopedia of Aerospace Engineering. John Wiley & Sons, Ltd, Chichester, UK.

Craig, R.R. and Kurdila, A.J. (2006) *Fundamentals of Structural Dynamics*, John Wiley & Sons, Ltd, Chichester, UK.

CS-25: Certification Specifications for EASA (European Aviation Safety Agency); http://www.easa.europa.eu

---

Introduction to Aircraft Aeroelasticity and Loads, Second Edition. Jan R. Wright and Jonathan E. Cooper.
© 2015 John Wiley & Sons, Ltd. Published 2015 by John Wiley & Sons, Ltd.

Currey, N.S. (1988) *Aircraft Landing Gear Design: Principles and Practices*, AIAA Education Series, Reston, VA, USA.

Davies, G.A.O. (1982) *Virtual Work in Structural Analysis*, John Wiley & Sons, Ltd, Chichester, UK.

Den Hartog, J.P. (1984) *Mechanical Vibrations*, Dover, New York.

Donaldson, B.K. (1993) *Analysis of Aircraft Structures: An Introduction*, McGraw-Hill, New York.

Dorf, R.C. and Bishop, R.H. (2004) *Modern Control Systems*, 10th edn, Prentice-Hall, New Jersey.

Dowell, E.H., Edwards, J.W. and Strganac, T.W. (2003) Nonlinear Aeroelasticity. *Journal of Aircraft*, **40** (5), 857–874.

Dowell, E.H., Clark, R., Cox, D., Curtiss, H.C., Edwards, J.W., Hall, K.C., Peters, D.A., Scanlon, R., Simiu, E., Sisto, F. and Strganac, T.W. (2004) *A Modern Course in Aeroelasticity – Solid Mechanics and Its Applications*, 4th revised and enlarged edn, Springer, Dordrecht.

Duncan, W.J, Thom, A.S. and Young, A.D. (1962) *The Mechanics of Fluids*, Arnold, London.

Edwards, J.W. and Weiseman, C.D. (2003) *Flutter and Divergence Analysis Using the Generalized Aeroelastic Analysis Method*. In: 44th AIAA Conference on *Structures, Structural Dynamics and Materials*, 2003, AIAA paper 2003–1489.

ESDU (Engineering Sciences Data Unit) Data Sheets, London.

Etkin, B. and Reid, L.D. (1996) *Dynamics of Flight*, John Wiley & Sons, Ltd, Chichester.

Eversman, W. and Tewari, A. (1991) Consistent rational function approximation for unsteady aerodynamics. *Journal of Aircraft*, **28**, 545–552.

Ewins, D.J. (1995) *Modal Testing: Theory and Practice*, Research Studies Press, Baldock, UK.

FAR-25: Certification Specifications for FAA (Federal Aviation Administration); http://www.faa.gov

Flomenhoft, H.I. (1994) Brief history of gust models for aircraft design. *Journal of Aircraft*, **31** (5), 1225–1227.

Flomenhoft, H.I. (1997) *The Revolution in Structural Dynamics*, Dynaflo Press, Palm Beach Gardens, FL, USA.

Forsching, H.W. (1974) *Grundlagen der Aeroelastik*, Springer Verlag, Berlin.

Frazer, R.A. and Duncan, W.J. (1928) *The Flutter of Aeroplane Wings*. ARC R&M (Aeronautical Research Council, Reports and Memoranda) 1155, AERADE Reports Archive, Cranfield, UK.

Frazer, R.A., Duncan, W.J. and Collar, A.R. (1938) *Elementary Matrices*, Cambridge University Press, Cambridge.

Friedmann, P.P. (1999) Renaissance of aeroelasticity and its future. *Journal of Aircraft*, **36** (1), 105–121.

Fuller, J.R. (1995) Evolution of airplane gust loads design requirements. *Journal of Aircraft*, **32** (2), 235–246.

Fung, Y. (1969) *An Introduction to the Theory of Aeroelasticity*, Dover (original 1955), New York.

Garrick, I.E. and Reed, W.H. (1981) Historical development of aircraft flutter. *Journal of Aircraft*, **18** (11), 897–912.

Golub, G.H. and van Loan, C.F. (1989) *Matrix Computations*, 2nd edn, John Hopkins Press, Baltimore.

Graupe, D. (1972) *Identification of Systems*, Van Nostrand-Reinhold, New York.

Hancock, G.J. (1995) *An Introduction to the Flight Dynamics of Rigid Airplanes*, Ellis Horwood, Harlow, Essex, UK.

Hancock, G.J., Simpson, A. and Wright, J.R. (1985) On the teaching of classical flutter. *Aeronautical Journal*, **89**, 285–305.

Hassig, H.J. (1971) An approximate true damping solution of the flutter equation by determinant iteration. *Journal of Aircraft*, **8**, 885–889.

Hoblit, F.M. (1988) *Gust Loads on Aircraft: Concepts and Applications*, AIAA Education Series, Reston, VA, USA.

Hodges, D.H. and Pierce, G.A. (2011) *Introduction to Structural Dynamics and Aeroelasticity*, 2nd edn. Cambridge University Press, Cambridge.

Houghton, E.L. and Brock, A.E. (1960) *Aerodynamics for Engineering Students*, Edward Arnold, London.

Houghton, E.L. and Carpenter, P.W. (2001) *Aerodynamics for Engineering Students*, 5th edn, Butterworth Heinemann, Oxford.

Howe, D. (2004) *Aircraft Loading and Structural Layout*, John Wiley & Sons, Ltd., Chichester, UK.

Inman, D.J. (2006) *Vibration with Control*, John Wiley & Sons, Ltd., Chichester, UK.

Jones, J.G. (1989) Statistical discrete gust method for predicting aircraft loads and dynamic response. *Journal of Aircraft*, **26** (4), 382–392.

Karpel, M. (1982) Design for active flutter suppression and gust alleviation using state space aeroelastic modelling. *Journal of Aircraft*, **19** (3), 221–227.

Katz, J. and Plotkin, A. (2001) *Low Speed Aerodynamics*, 2nd edn, Cambridge University Press, Cambridge.

Kuo, B.C. (1995) *Digital Control Systems*, Oxford University Press, Oxford.

Lanchester, F.W. (1916) *Torsional Vibration of the Tail of an Airplane*. ARC R&M (Aeronautical Research Council, Reports and Memoranda) 276, part 1, AERADE Reports Archive, Cranfield, UK.

Librescu, L. (2005) Advances in the linear/nonlinear control of aeroelastic structural systems. *Acta Mechanica*, **178**, 147–186.

Livne, E. (2003) Future of airplane aeroelasticity. *Journal of Aircraft*, **40** (6), 1066–1092.

Lomax, T.L. (1996) *Structural Loads Analysis for Commercial Transport Aircraft: Theory and Practice*. AIAA Education Series, Reston, VA, USA.

Megson, T.H.G. (1999) *Aircraft Structures for Engineering Students*, 3rd edn, Arnold, London.

Meirovitch, L. (1986) *Elements of Vibration Analysis*, McGraw-Hill, New York.

Meriam, J.L. (1980) *Engineering Mechanics, Volume 2: Dynamics*, John Wiley & Sons, Ltd., Chichester, UK.

Milne, R.D. (1964) Dynamics of the deformable aeroplane. (Aeronautical Research Council, Reports and Memoranda) ARC R & M 3345, AERADE Reports Archive, Cranfield, UK.

Hellen, T.K. and Becker, A.A., (2013) *Finite Element Analysis for Engineers – a Primer*, NAFEMS, East Kilbride, UK.

Newland, D.E. (1987) *An Introduction to Random Vibrations and Spectral Analysis*, Longman, Oxford.

Newland, D.E. (1989) *Mechanical Vibration Analysis and Computation*, Longman, Oxford.

Niblett, L.T. (1998) A guide to classical flutter. *Aeronautical Journal*, **92**, 339–354.

Niu, M.C.Y. (1988) *Airframe Structural Design*, Conmilit Press, Hong Kong.

Pacejka, H.B. (2005) *Tyre and Vehicle Dynamics*, Butterworth-Heinemann, Oxford.

Pratt, R.W. (ed.) (2000) Flight Control Systems: Practical Issues in Design and Implementation*, IEE (Institution of Electrical Engineers) Control Engineering Series*, Stevenage, UK.

Rao, S.S. (1995) *Mechanical Vibrations*, Addison Wesley, Boston, MA, USA.

Raven, F.H. (1994) *Automatic Control Engineering*, 5th edn, McGraw-Hill, New York.

Rodden, W.P. (2011) *Theoretical and Computational Aeroelasticity*. Crest Publishing, Burbank, CA, USA.

Roger, K.L. (1977) Airplane math modelling and active aeroelastic control design. Advisory Group for Aerospace Research and Development AGARD-CP-228, pp. 4.1–4.11, NATO (North Atlantic Treaty Organization), Brussels.

Russell, J.B. (2003) *Performance and Stability of Aircraft*, Butterworth-Heinemann, Oxford.

Scanlan, R.H. and Rosenbaum, R. (1960) *Introduction to the Study of Aircraft Vibration and Flutter*, Macmillan, London.

Schmidt, L.V. (1998) *Introduction to Aircraft Flight Dynamics*, AIAA Education Series, Reston, VA, USA.

Scott, R.C., Pototzky, A.S. and Perry, B. (1993) Computation of maximised gust loads for nonlinear aircraft using matched filter based schemes. *Journal of Aircraft*, **30** (5), 763–768.

Stengel, R.F. (2004) *Flight Dynamics*, Princeton University Press, Princeton, NJ, USA.

Stroud, K.A. and Booth, D.J. (2007) *Engineering Mathematics*, 6th edn, Industrial Press, South Norwalk, CT, USA.

Sun, C.T. (2006) *Mechanics of Aircraft Structures*, John Wiley & Sons, Ltd., Chichester, UK.

Taylor, N.V., Vio, G.A., Rampurawala, A.M., Allen, C.B., Badcock, K.J., Cooper, J.E., Gaitonde, A.L., Henshaw, M.J., Jones, D.P. and Woodgate, M.A. (2006) Aeroelastic simulation through linear and non-linear analysis, A summary of flutter prediction in the PUMA DARP. *Aeronautical Journal*, **110** (1107), 333–343.

Theodorsen, T. (1935) *General Theory of Aerodynamic Instability and the Mechanism of Flutter*. National Advisory Committee for Aeronautics NACA Report 496, AERADE Reports Archive, Cranfield, UK.

Thomson, W.T. (1997) *Theory of Vibration with Applications*, 5th edn, Chapman and Hall, London.

Tse, F.S., Morse, I.E. and Hinkle, R.T. (1978) *Mechanical Vibrations: Theory and Applications*, Allyn and Bacon, Boston, MA, USA.

Waszak, M.R. and Schmidt, D.K. (1988) Flight dynamics of aeroelastic vehicles. *Journal of Aircraft*, **25** (6), 563–571.

Weisshaar, T.A. (2010) *Static and Dynamic Aeroelasticity*. Encyclopedia of Aerospace Engineering, John Wiley & Sons, Ltd, Chichester, UK.

Wells, D.A. (1967) *Theory and Problems of Lagrangian Dynamics*, Schaum's Outline Series, McGraw-Hill, New York.

Whittle, P. (1996) *Optimal Control, Basics and Beyond*, John Wiley & Sons Ltd, Chichester, UK.

Wright, J.R., Wong, J., Cooper, J.E. *et al.* (2003) On the use of control surface excitation in flutter testing, *Proc. Instn Mech. Engrs, Part G: J. Aerospace Engineering*, **217** (96), 317–332.

Yates, E.C. (1966) Modified strip analysis method for predicting wing flutter at subsonic to hypersonic speeds. *Journal of Aircraft*, **3** (1), 25–29.

Young, W.C. (1989) *Roark's Formulas for Stress and Strain*, 6th edn, McGraw-Hill, New York.

Zimmermann, H. (1991) Aeroservoelasticity. *Computer Methods in Applied Mechanics and Engineering*, **90** (13), 719–735.

Zimmermann, N.H. and Weissenberger, J.T. (1964) Prediction of flutter onset speed based on flight testing at subcritical speeds. *Journal of Aircraft*, **1** (4), 190–202.

Index

Introduction to Aircraft Aeroelasticity and Loads, Second Edition. Jan R. Wright and Jonathan E. Cooper.
© 2015 John Wiley & Sons, Ltd. Published 2015 by John Wiley & Sons, Ltd.